Bio-Assays for Oxidative Stress Status
(BOSS)

Bio-Assays for Oxidative Stress Status (BOSS)

Edited by

William A. Pryor

Thomas and David Boyd Professor
Director
Biodynamics Institute
Louisiana State University

ELSEVIER

AMSTERDAM – LONDON – NEW YORK – OXFORD – PARIS – SHANNON – TOKYO

ELSEVIER SCIENCE B.V.
Sara Burgerhartstraat 25
P.O. Box 211, 1000 AE Amsterdam, The Netherlands

First edition 2001

Library of Congress Cataloging in Publication Data
A catalog record from the Library of Congress has been applied for.

ISBN: 0-444-50957-7

Transferred to digital print on demand, 2006
Printed and bound by CPI Antony Rowe, Eastbourne

INTRODUCTION TO
OXIDATIVE STRESS STATUS (OSS)

WILLIAM A. PRYOR

Biodynamics Insitute, Louisiana State University, Baton Rouge, LA, USA

The series of chapters included in this volume were originally published as articles in a FORUM, a symposium-in-print that is a feature of the journal *Free Radical Biology & Medicine,* a journal that Dr. Kelvin Davies and I and four outstanding scientists, edit. The FORUM was entitled OSS – OXIDATIVE STRESS STATUS and was the longest and largest FORUM we had yet published – it encompassed 7 issues of *FRBM* and included 31 articles by some of leading authorities in the field. [In 2001, the name FORUM was changed to CONTINUING REVIEWS and, because of the over whelming success of *FRBM*, some modest page-length restrictions were placed on these serial reviews.] Generally, a leading authority is chosen to act as organizer, editor, and facilitator of these serial reviews; for the OSS review, I served in this role.

Many of us in the field of oxidative biology have attempted to design – indeed, have dreamed of the day when it will be routine to use – a method for measuring OSS. This OSS method would give the instantaneous level of oxidative stress in an organism. Why might that be useful? Here are a number of observations relevant to OSS:

- Diseases such as cancer, cataract, and heart disease are known to involve oxidation of a number of types of biomolecules, and therefore detecting the levels of this oxidation could provide an early warning signal.
- Trials of drugs or antioxidant vitamins often take years to develop sufficient case data to allow statistically robust conclusions. If OSS measurements tracked with the decease being studied, it might be possible to use OSS as an early end point.
- Since the oxidized biological molecular products from oxidative stress are generally more stable than the oxidant itself - i.e., oxidized lipids, proteins, and nucleic acids are more stable than the free radicals that effected their oxidation, OSS measurements often turn out to involve determining the level of a Biomarker of Oxidative Stress Status (a BOSS). Thus, one obtains evidence of the activities of an elusive animal by its footprints and usually does not expect to see the creature itself.

- In an article that introduced the first group of OSS FORUM articles, I explained the historical reference for the acronym "OSS" and cited the career of "Wild Bill" Donovan, who founded the Office of Strategic Services (OSS) at the time of World War II. [See *FRBM* 27, 1135-11365(1999); that introduction is also included in this volume].
- We know that all classes of biomolecules are oxidized, but we do not know the relative average rates (or even if there is an "average rate of oxidation of different classes of biopolymer molecules").
- Which oxidation products are best related to a given disease state? For example, are oxidized DNA bases such as 8-oxodeoxyguanosin most closely associated with cancer? Is oxidized LDL most associated with heart disease?
- How much do measures of OSS vary in a group of humans?
- Can a non-invasive, reliable, repeatable measure of OSS be identified?
- Does OSS decrease as a result of life-change factors? From taking vitamin E or other antioxidant vitamins or substances?
- Does OSS increase with age? With disease? With stress? If OSS does increase with age, does that "prove" Denham Harman's theory that free radical oxidation "causes" aging?

We, the editors and publishers of *FRBM* hope that this collection of articles will reveal to the general reader the status of studies of OSS, currently used examples of BOSS, and answers to at least some of the questions posed above.

September 2001

CONTENTS

OXIDATIVE STRESS STATUS:
OSS, BOSS, AND "WILD BILL" DONOVAN

WILLIAM A. PRYOR

Biodynamics Institute, Louisiana State University, Baton Rouge, LA, USA

The Forum that makes its first appearance in this issue, *Methods to Measure Oxidative Stress Status*, will be the most extensive yet published by *FRBM*. In this issue we present the first seven articles on this subject; we will publish other groups of papers for this Forum in the coming months, with a total number of contributions near 30.

Almost a decade ago [1,2], Susan Godber and I took upon ourselves the task of searching the literature to discover if many methods for measuring the oxidative stress status (OSS) of individuals had been published, and, if so, if there was a consensus on which methods were the most useful and reliable for measuring what we called OSS.[1] Sad to say, the answers appeared to be "a huge bunch" and "no," respectively. By 1993, the situation had not changed much [4], although the efforts of the Vanderbilt group with isoprostanes, "non-natural" isomers resulting from the nonenzymatic, free radical bicyclization of arachidonate [5], looked very promising [6–12].

This year, the National Institutes of Health (NIH) again visited the question of OSS methods and their reliability. A group of rats were given carbon tetrachloride, a toxin known to act via free radical reaction mechanisms, and the tissue was distributed to a number of groups with the plan that they all would try OSS measurements, meet at the hitching post near the OK Corral at the end of the day, and compare results. Some of those groups, in fact, will report their findings as part of this Forum. The NIH, proving again the Darwinian notion that evolution brings benefits, named the methods

"BOSS," for Biomarkers of Oxidative Stress Status. Of course, as readers of *FRBM* are only too aware, radicals generally have only a fleeting existence in vivo, and "footprints" of these elusive species must be discovered—hence, BOSS. Well-recognized examples of BOSS are oxidized lipids (such as aldehydes and isoprostanes), oxidized amino acid residues (such as carbonyl-labeled), or oxidized DNA bases (such as 8-oxo-dG). The faithful readers of this journal need not be told the names of the many extremely talented groups working in this area and the various methodologies that have been tested and are under development.

The concept that diseases raise the OSS of an individual, or at least of some tissue in that individual, is both obvious and attractive. Many a free radical biologist has dreamed of the day when a simple blood or urine test would divulge the OSS of an individual and the prediction that some organ or tissue might need closer examination. If a marker, a BOSS, could be developed that would predict that an individual has a greater chance of developing disease, or already has a disease, that clearly would be both a confirmation of the concept that oxidative processes can lead to pathologic results and a very useful medical advance. There is some hope, for example, that lower levels of antioxidant nutrients in low-density lipoprotein (LDL) and/or the greater oxidizability of an individuals's LDL might be predictive of cardiovascular disease [13,14].

Originally, our intent in this Forum was to stress BOSS methods that are easily applied techniques that could be used, at least with further development, noninvasively, and thus, could ultimately make a contribution to clinical medicine. However, as authors suggested contributions, and as this Forum developed, the scope of the Forum was expanded. Thus, you will find that all of the contributions to this Forum do not present methods that are now, or perhaps ever could lead to, practical, noninvasive methods for measuring OSS. Indeed, if they did,

---

Address correspondence to: Dr. William A. Pryor, Biodynamics Institute, Louisiana State University, Baton Rouge, LA 70803

[1]Methods of oxidative stress measures attempt to "spy" out the oxidative status of a subject. By using the acronym "OSS", we remember and honor America's first "big-time" spy, William ("Wild Bill") Donovan [3], the founder of what was called the Office of Strategic Services during World War II (and which later became the CIA—but that is another story).

Reprinted from: *Free Radical Biology & Medicine, Vol. 27, Nos. 11/12, pp. 1135-1136, 1999*

there probably would be no need for the Forum! Instead, we hope that by providing a collection of methods, discussed by their developers and protagonists, that readers will both learn the state of the field and, also, get ideas as to how one or more methods could be made more useful, and therefore, a greater contribution to human health.

REFERENCES

[1] Pryor, W. A.; Godber, S. S. Noninvasive measures of oxidative stress status in humans. *Free Radic. Biol. Med.* **10:**177–184; 1991.

[2] Pryor, W. A.; Godber, S. S. Oxidative stress status: an introduction. *Free Radic. Biol. Med.* **10:**173; 1991.

[3] Brown, A. C. *The last hero. Wild Bill Donovan.* New York: Times Books; 1982.

[4] Pryor, W. A. Measurement of oxidative stress status in humans. *Cancer Epidemiol. Biomarkers Prev.* **2:**289–292; 1993.

[5] Pryor, W. A.; Stanley, J. P. A suggested mechanism for the production of malonaldehyde during the autoxidation of polyunsaturated fatty acids. Nonenzymatic production of prostaglandin endoperoxides during autoxidation. *J. Org. Chem.* **40:**3615–3617; 1975.

[6] Morrow, J. D.; Awad, J. A.; Boss, H. J.; Blair, I. A.; Roberts, L. J., II. Non-cyclooxygenase-derived prostanoids (F_2-isoprostanes) are formed *in situ* on phospholipids. *Proc. Natl. Acad. Sci. USA* **89:**10721–10725; 1992.

[7] Morrow, J. D.; Harris, T. M.; Roberts, L. J., II. Noncyclooxygenase oxidative formation of a series of novel prostaglandins: analytical ramifications for measurement of eicosanoids. *Anal. Biochem.* **184:**1–10; 1990.

[8] Morrow, J. D.; Roberts, L. J. Mass spectrometry of prostanoids: F_2-isoprostanes produced by non-cyclooxygenase free radical catalyzed mechanism. *Methods Enzymol.* **233:**163–174; 1994.

[9] Awad, J. A.; Morrow, J. D.; Hill, K. E.; Roberts, L. J.; Burk, R. F. Detection and localization of lipid peroxidation in selenium-and vitamin E deficient rats using F_2-isoprostanes. *J. Nutr.* **124:**810–816; 1994.

[10] Lynch, S. M.; Morrow, J. D.; Roberts, L. J., II; Frei, B. Formation of non-cyclooxygenase-derived prostanoids (F_2-isoprostanes) in plasma and low density lipoprotein exposed to oxidative stress in vitro. *J. Clin. Invest.* **93:**998–1004; 1994.

[11] Morrow, J. D.; Roberts, L. J., II. The isoprostanes. Current knowledge and directions for future research. *Biochem. Pharmacol.* **51:**1–9; 1996.

[12] Mueller, M. J. Radically novel prostaglandins in animals and plants: the isoprostanes. *Chem. Biol.* **5:**R323–R333; 1998.

[13] Holvoet, P. Role of oxidatively modified low density lipoproteins and anti-oxidants in atherothrombosis. *Expert Opin. Invest. Drugs* **8:**527–544; 1999.

[14] Pryor, W. A. Vitamin E and heart disease: biochemistry to human trials. *Free Radic. Biol. Med.* in press; 2000.

Bioassays for Oxidative Stress Status (BOSS). Edited by W.A. Pryor

NOVEL HPLC ANALYSIS OF TOCOPHEROLS, TOCOTRIENOLS, AND CHOLESTEROL IN TISSUE

EUGENIOS KATSANIDIS and PAUL B. ADDIS

Department of Food Science and Nutrition, University of Minnesota, St. Paul, MN, USA

(Received 10 September 1999; Accepted 13 September 1999)

Abstract—Tocopherols and tocotrienols are being increasingly recognized to have an important role in the prevention of atherosclerosis. It has been reported that they protect low-density lipoprotein (LDL) and tissues from oxidative stress and that tocotrienols can reduce plasma cholesterol levels. Two isocratic high-performance liquid chromatography (HPLC) methods for simultaneous analysis of tocopherols, tocotrienols, and cholesterol in muscle tissue were developed. Method A involves basic saponification of the sample, but causes losses of the γ- and δ-homologs of vitamin E. Method B does not involve saponification, thereby protecting the more sensitive homologs. Both permit rapid analysis of multiple samples and neither requires specialized equipment. These methods may provide techniques useful in simultaneous assessment of oxidative stress status (OSS) and cholesterol levels. © 1999 Elsevier Science Inc.

Keywords—Tocopherols, Tocotrienols, Free radical, Cholesterol, HPLC

INTRODUCTION

It is well established that lipid oxidation can have deleterious effects on human health [1–5]. Antioxidants play a major role in controlling lipid oxidation in vivo and in food. The usefulness of supplementation of human diet with antioxidants is still being evaluated [6], but it is generally accepted that increased levels of antioxidants can have beneficial effects on human health by increasing resistance of tissues and lipoproteins to oxidative stress. Therefore, antioxidant quantification provides one method for determining oxidative stress status (OSS).

Vitamin E is one of the most important naturally occurring primary antioxidants and it plays a very important role in protecting LDL from oxidative modification, which has been implicated in the development of atherosclerosis [7].

Vitamin E is a generic term that includes all entities that exhibit the biological activity of α-tocopherol. In nature, eight substances have been found to have vitamin E activity: d-α-, d-β-, d-γ-, and d-δ-tocopherol; and d-α-, d-β-, d-γ-, and d-δ-tocotrienol. Each of these forms of vitamin E has a different biopotency [8]. α-Tocopherol is traditionally reported to have the highest biological activity of all homologs, based on the fetal absorption test or the hemolysis test [9]. However, these are not the most critical tests and their relevance to physiologic importance and health benefits has been questioned [10]. Tocotrienols, on the other hand, exhibit characteristics that would render them more important than originally thought [10–12]. Tocotrienols exhibit a higher recycling efficiency in microsomes and liposomes in the presence of reduced nicotinamide adenine dinucleotide phosphate or ascorbate; they have a more uniform distribution in membrane bilayer; and cause a stronger disordering of membrane lipids, making interaction with lipid radicals

Address correspondence to: Dr. Paul B. Addis, Department of Food Science and Nutrition, University of Minnesota, 1334 Eckles Avenue, St. Paul, MN 55108, USA; Tel: (612) 624-7704; Fax: (612) 625-5272; E-Mail: paddis@che2.che.umn.edu

Eugenios Katsanidis received his B.S. degree in Food Sciences & Nutrition at the Aristotle University of Thessaloniki, Greece, with emphasis in Food Engineering in 1992. He received his M.S. (1995) and Ph.D. (1999) degrees in Food Science at the University of Minnesota, emphasizing antioxidant and pro-oxidant systems in food products. Dr. Katsanidis is currently working at the Pillsbury Company in Minneapolis, MN.

Paul Addis received his B.S. degree from Washington State University in 1962 and his Ph.D. degree in Food Science from Purdue University in 1967. He joined the University of Minnesota in January of 1967 and was granted a leave for research at the Max Plancx Institute at Marienesee. He has studied at the Unviersity of Washington, University of California at Davis and at San Diego. His research has included muscle protein and enzymes, lipids, lipid oxidation, and more recently, the conversion of cellulose into a soluble fiber.

more efficient than tocopherols. Tocopherols and tocotrienols not only inhibit radical-induced lipid oxidation, but they can also quench singlet oxygen ($O_2^1 \triangle g$) both in vivo and in vitro [13].

Tocotrienols have been known to help lower plasma cholesterol levels by regulating HMG-CoA reductase in the liver [14–17] and, thus, decrease the risk for cardiovascular disease in hypercholesterolemic humans [18]. They have also been shown to reduce the concentration of apolipoprotein B, thromboxane B_2, and platelet factor 4 in pigs with inherited hyperlipidemias [19].

Muscle tissue contains 2–3 ppm of α-tocopherol and traces of α-tocotrienol [20] and, thus, there are not many methods for separating and quantifying tocopherols and tocotrienols in muscle tissues. The recent interest in the use of the other vitamin E homologs as antioxidants and the cholesterol reducing effects of tocotrienols create the need for the development of a rapid method for the analysis of these compounds in muscle. Several HPLC methods for the analysis of α-tocopherol in meat have been developed [21,22]. The method developed by Liu et al. [22] involves a 15-min saponification step and it is a very time-efficient one, allowing for the analysis of many samples per day. There are several HPLC methods for the analysis of all eight homologs of vitamin E for cereal products [23–25], but they are not applicable in muscle tissue. This is mostly due to the fact that muscle contains higher levels of protein, which interferes in the process of extraction (foaming and formation of emulsions).

Cholesterol analysis is usually done by enzymatic [26] or chromatographic methods. The most common chromatographic method is gas chromatography [27,28], although there are some HPLC methods [29]. Most of these chromatographic methods include a laborious saponification-extraction procedure, usually overnight, and a chromatographic run of 15–25 min. Because both vitamin E and cholesterol are nonpolar compounds that absorb in the ultraviolet (UV) range, it is possible that they could be analyzed simultaneously.

Our objective was to develop a rapid method for the efficient extraction, separation, and quantification of vitamin E homologs and cholesterol in muscle tissue.

MATERIALS AND METHODS

Tocopherol and cholesterol standards were purchased from Supelco (Bellefonte, PA, USA). Tocotrienol standards were a gift from Eastman (Kingsport, TN, USA) and the Palm Oil Research Institute of Malaysia (Kuala Lumpur, Malaysia). All solvents were HPLC grade.

Two extraction methods were developed. Method A involved saponification whereas method B did not. The chromatographic conditions were the same for both methods.

Method A

One gram of sample was placed in a screw-cap tube and 0.25-g ascorbic acid and 7.3-ml saponification solution were added. The saponification solution was 55% ethanol in distilled water (v/v) with 11% potassium hydroxide (w/v). The tubes were placed in a shaking water bath for 15 min at 80°C. After the saponification, the tubes were cooled in tap water for 1 min, and 4-ml hexane and 2-ml distilled water were added. The added water increases the polarity of the aqueous phase and improves partitioning of vitamin E and cholesterol into the organic (hexane) phase. Tubes were vortexed and the upper layer (hexane) was collected in a screw-cap vial for analysis by normal phase HPLC.

Method B

Two grams of postmortem bovine muscle were placed in a 100-ml plastic tube, 8-ml of absolute ethanol were added, and the mixture was homogenized for 30 s in a PowerGen 700 (Fisher Scientific, Pittsburgh, PA, USA) homogenizer. Ten milliliters of distilled water were added to the tube and sample was homogenized for 15 s. Finally, 8-ml hexane were added and the sample was homogenized for 15 s. The tubes were capped and centrifuged at 1500 rpm for 10 min. The upper layer (hexane) was collected and analyzed by normal phase HPLC.

Chromatography

The HPLC consisted of a Varian 9010 pump and a Varian 9050 UV detector (Varian Associates, Houston, TX, USA). Twenty microliters were injected into a silica column (Zorbax RX-SIL, 5-μm particle size, 4.6-mm ID $\times$ 25 cm). The mobile phase was hexane-isopropanol (99:1). Flow rate was 1.3 ml/min. The wavelength was programmed at 295 nm for the vitamin E homologs and then it was switched to 202 nm for cholesterol. All eight tocopherols and tocotrienols were eluted within 8 min, the wavelength was changed at 9 min, and cholesterol was eluted at 11 min. A typical run lasted approximately 12 min.

RESULTS AND DISCUSSION

Several extraction procedures and solvents were tested. Methanol is a very good extractant of vitamin E [23,25] in grain and cereal samples, but it also extensively extracts and denatures proteins in muscle, causing foaming, and making volume reduction by rotavaporation impossible. Extraction with methanol:chloroform, 2:1 resulted in poor recovery (~60%). The method of Liu et al. [22] produced the best recoveries.

Method A (which is based on the method of Liu et al. [22]) is useful for the analysis of the α-homologs of vitamin E (α-tocopherol and α-tocotrienol). However, other homologs, and especially the δ-homologs, are rapidly degraded during the saponification process. The recoveries for the homologs after saponification were as follows: α-tocopherol, 95%; α-tocotrienol, 95%; β-tocopherol, 95%; γ-tocopherol, 94%; γ-tocotrienol, 85%; δ-tocopherol, 80%; and δ-tocotrienol, 44%. The recovery of cholesterol was 84%.

It is generally known that vitamin E is less stable in alkaline conditions. There was no reference found in the literature regarding the fact that the δ-homologs, and in particular, δ-tocotrienol, are so much more susceptible to degradation during saponification than the α-homologs. This must be due to the fact that there is only one methyl group on the chromane ring, providing less protection to the hydroxyl group compared with the three methyl groups in the α-homologs. Even if cold saponification was used and α-tocopherol was present at high levels, the δ-homologs were destroyed during saponification. Given the fact that most methods are developed based on α-tocopherol, and that there are no commercially available standards for tocotrienols, it is possible that the contents of the γ- and δ-homologs are being underestimated if the sample is saponified.

It is clear that saponification can be used for α-tocopherol and α-tocotrienol analysis, but not for the other homologs. Method B was developed so that the more sensitive γ- and δ-homologs (especially δ-homologs) could be accurately quantified as all homologs of vitamin E are excellent antioxidants.

For method B, the recovery for vitamin E was approximately 96% for all of the homologs, and for cholesterol was 94%. Cholesterol always exhibits slightly lower recovery than vitamin E because it is more polar and the partitioning coefficient of cholesterol (between the organic and aqueous phase) should be slightly lower than that of vitamin E. The high recovery rates make it clear that virtually no degradation of vitamin E or cholesterol took place during the extraction procedure.

In many of the vitamin E methods published, fluorescence detection was used (excitation at 290 nm and emission at 320 nm). Even though it is far more sensitive than UV detection (at 295 nm), fluorescence detectors are very expensive and in some instances fluorescence quenching by other compounds (i.e., iodine, bromine) can affect the results [9].

In conclusion, method B is a new, rapid method for the extraction, separation, and quantification of all vitamin E homologs and cholesterol in muscle tissue. It is not a laborious method, it does not involve lengthy saponification procedures, and it does not require very specialized equipment, allowing for the analysis of many samples per day with relatively low cost. It is also a useful addition to the current methodology because there are no other good alternatives that combine the foregoing attributes. With further modification, it could be used to analyze for vitamin E and cholesterol content in LDL and other lipoprotein particles in blood.

Acknowledgements — Published as paper no. 991180024 of the contribution series of the Minnesota Agricultural Experiment Station on research conducted under Minnesota Agricultural Experiment Station Project 18-23, supported by Hatch Funds.

REFERENCES

[1] Grootvelt, M. C.; Atherton, M; Sheerin, A. N.; Hawkes, J.; Blake, D. R.; Richens, T. E.; Silwood, C. J. L.; Lynch, E.; Claxson, A. W. D. In vivo absorption, metabolism, and urinary excretion of α,β-unsaturated aldehydes in experimental animals. *J. Clin. Invest.* **101:**1–9; 1998.

[2] Claxson, A.; Hawkes, G. E.; Richardson, D. P.; Naughton, D. P.; Haywood, R. M.; Chander, C. L.; Atherton, M; Lynch, E. J.; Grootvelt, M. C. Generation of lipid peroxidation products in culinary oils and fats during episodes of thermal stressing: a high field ^{1}H NMR study. *FEBS Lett.* **355:**81–90; 1994.

[3] Addis, P. B.; Warner, G. J. The potential health aspects of lipid oxidation products in food. In: Auroma, O. I.; Halliwell, B., eds. *Free radicals and food additives.* London: Taylor and Francis, Ltd.; 1991:77–119.

[4] Guardiola, F.; Codony, R.; Addis, P. B.; Rafecas, M.; Boatella, J. Biological effects of oxysterols: current status. *Food Chem. Toxicol.* **34:**193–211; 1996.

[5] Addis, P. B.; Park, P. W.; Guardiola, F.; Codony, R. Analysis and health effects of cholesterol oxides. In: McDonald, R. E.; Min, D. B., eds. *Food lipids and health.* New York: Marcel Dekker, Inc.; 1996:199–239.

[6] Geise, J. Antioxidants: tools for preventing lipid oxidation. *Food Technol.* **50:**73–81; 1996.

[7] Suzukawa, M.; Ayaori, M.; Shige, H.; Hisada, T.; Ishikawa, T; Nakamura, H. Effect of supplementation with vitamin E on LDL oxidizability and prevention of atherosclerosis. *Biofactors* **7:**51–54; 1998.

[8] Gregory, J. F. Vitamins. In: Fennema, O. R., ed. *Food chemistry, 3rd edition.* New York: Marcel Dekker, Inc.; 1996:553–557.

[9] Bourgeois, C., ed. *Determination of vitamin E: tocopherols and tocotrienols.* New York: Elsevier Applied Science; 1992.

[10] Serbinova, E.; Kagan, V.; Han, D.; Packer, L. Free radical recycling and intermembrane mobility in the antioxidant properties of alpha-tocopherol and alpha-tocotrienol. *Free Radic. Biol. Med.* **10:**263–275; 1991.

[11] Suzuki, Y. J.; Tsuchiya, M.; Wassal, S. R.; Choo, Y. M.; Govil, G.; Kagan, V. E.; Packer, L. Structural and dynamic membrane properties of α-tocopherol and α-tocotrienol: implication to the molecular mechanism of their antioxidant potency. *Biochemistry* **32:**10692–10699; 1993.

[12] Serbinova, E.; Packer, L. Antioxidant properties of α-tocopherol and α-tocotrienol. *Meth. Enzymol.* **234:**354–366; 1994.

[13] Kamal-Eldin, A.; Appelqvist, L. A. The chemistry and antioxidant properties of tocopherols and tocotrienols. *Lipids* **31:**671–701; 1996.

[14] Qureshi, A. A.; Pearce, B. C.; Nor, R. M.; Gapor, A.; Peterson, D. M.; Elson, C. E. Dietary α-tocopherol attenuates the impact of α-tocotrienol on hepatic 3-hydroxy-3-methylglutaryl coenzyme A reductase activity in chickens. *J. Nutr.* **126:**389–394; 1996.

[15] Khor, H. T.; Chieng, D. Y.; Ong, K. K. Tocotrienols inhibit liver HMG CoA reductase activity in the guinea pig. *Nutr. Res.* **15:** 537–544; 1995.

[16] Quershi, N.; Qureshi, A. A. Tocotrienols: novel hypocholesterolemic agents with antioxidant properties. In: Packer, L.; Fuchs,

J., eds. *Vitamin E in health and disease.* New York: Marcel Decker, Inc.; 1993:247–267.

[17] Watkins, T.; Lenz, P.; Gapor, A.; Struck, M.; Tomeo, A.; Bierenbaum, M. γ-Tocotrienol as a hypocholesterolemic and antioxidant agent in rats fed atherogenic diets. *Lipids* **28:**1113–1118; 1993.

[18] Qureshi, A. A.; Bradlow, B. A.; Salser, W. A.; Brace, L. D. Novel tocotrienols of rice bran modulate cardiovascular disease parameters of hypercholesterolemic humans. *Nutr. Biochem.* **8:**290–298; 1997.

[19] Qureshi, A. A.; Qureshi, N.; Hasler-Rapacz, J. O.; Weber, F. E.; Chaudhary, V.; Crenshaw, T. D.; Gapor, A.; Ong, A. S.; Chong, Y. H. Dietary tocotrienols reduce concentrations of plasma cholesterol, apolipoprotein B, thromboxane B$_2$, and platelet factor 4 in pigs with inherited hyperlipidemias. *Am. J. Clin. Nutr.* **53:**1042S–1046S; 1991.

[20] Pyrenean, V.; Syvaoja, E. L.; Varo, P.; Salminen, K.; Koivistoinen, P. Tocopherols and tocotrienols in Finnish foods: meat and meat products. *J. Agric. Food Chem.* **33:**1215–1218; 1985.

[21] Arnold, R. N; Scheller, K. K.; Arp, S. C.; Williams, S. N.; Schaefer, D. M. Dietary α-tocopheryl acetate enhances beef quality in Holstein and beef breed steers. *J. Food Sci.* **58:**28–33; 1993.

[22] Liu, Q.; Scheller, K. K.; Schaefer, D. M. Technical note: a simplified procedure for vitamin E determination in beef muscle. *J. Anim. Sci.* **74:**2406–2410; 1996.

[23] Peterson, D. M.; Qureshi, A. A. Genotype and environment effects on tocols of barley and oats. *Cereal Chem.* **70:**157–162; 1993.

[24] Wang, L.; Xue, Q.; Newman, R. K.; Newman, C. W. Enrichment of tocopherols, tocotrienols, and oil in barley fractions by milling and pearling. *Cereal Chem.* **70:**499–501; 1993.

[25] Budin, J. T.; Breene, W. M.; Putnam, D. H. Some compositional properties of camelina (*camelina sativa L. crantz*) seeds and oils. *JAOCS* **72:**309–315; 1995.

[26] Njeru, C. A.; McDowell, L. R.; Shireman, R. M.; Wilkinson, N. S.; Rojas, L. X.; Williams, S. N. Assessment of vitamin E nutritional status in yearling beef heifers. *J. Anim. Sci.* **73:**1440–1448; 1995.

[27] Park, S. W.; Addis, P. B. Cholesterol oxidation products in some muscle foods. *J. Food Sci.* **52:**1500–1503; 1987.

[28] Guardiola, F.; Codony, R.; Rafecas, M.; Boatella, J. Selective gas chromatographic determination of cholesterol in eggs. *JAOCS* **71:**867–871; 1994.

[29] Csallany, A. S.; Kindom, S. E.; Addis, P. B.; Lee, J. H. HPLC method for quantitation of cholesterol and four of its major oxidation products in muscle and liver tissues. *Lipids* **24:**645–651; 1989.

ABBREVIATIONS

HMG-CoA—3-hydroxy-3-methylglutaryl coenzyme A
HPLC—high-performance liquid chromatography
LDL—low-density lipoprotein(s)
OSS—oxidative stress status

Bioassays for Oxidative Stress Status (BOSS). Edited by W.A. Pryor
© 2001 Elsevier Science B.V. All rights reserved.

BASELINE DIENE CONJUGATION IN LDL LIPIDS:
AN INDICATOR OF CIRCULATING OXIDIZED LDL

MARKKU AHOTUPA and TOMMI J. VASANKARI

MCA Research Laboratory and Paavo Nurmi Center, Department of Physiology, University of Turku, Turku, Finland

(Received 3 September 1999; Accepted 4 September 1999)

Abstract—The wide acceptance of the diene conjugation-method in monitoring low-density lipoprotein (LDL) oxidation ex vivo has led to development of an assay, which measures the amount of baseline diene conjugation (BDC) in circulating LDL, and is an indicator of oxidized LDL in vivo. The LDL-BDC assay is based on precipitation of serum LDL with buffered heparin, and spectrophotometric determination of baseline level of conjugated dienes in lipids extracted from LDL. Compared to existing methods for oxidized LDL, LDL-BDC is fast and simple to perform. Chemical studies by HPLC and NMR have verified that LDL-BDC is a specific indicator of circulating mildly oxidized LDL. Validity of the assay is further indicated by strong correlation with the titer of autoantibodies against oxidized LDL. Clinical studies have shown that LDL-BDC is closely related to coronary, carotid, and brachial atherosclerosis. Moreover, several independent studies have demonstrated surprisingly strong associations between LDL-BDC and known atherosclerosis risk factors (obesity, physical inactivity, hypertension, diabetes, and arterial functions). Indeed, these studies seem to indicate that as an indicator of the risk of atherosclerosis LDL-BDC clearly exceeds sensitivity and specificity of the common lipid markers of atherosclerosis. It is concluded that LDL-BDC is a promising candidate in search for methods for the evaluation of in vivo LDL oxidation and the risk of atherosclerosis. © 1999 Elsevier Science Inc.

Keywords—Antioxidants, Atherosclerosis, Free radical, Diene conjugation, In vivo, Low-density lipoprotein, Oxidation

INTRODUCTION

For several years it has been known that development of atherosclerosis is related to the level of the "noxious" low density lipoproteins (LDL). More recent studies

Address correspondence to: Dr. Markku Ahotupa, Ph.D., MCA Research Laboratory, BioCity, Tykistökatu 6 B, FIN-20520 Turku, Finland; Tel: +358 (2) 241-0142; Fax: +358 (2) 241-0143; E-Mail: res.lab@mca.inet.fi.
Dr. Markku Ahotupa received his Ph.D. degree in biochemistry from the University of Turku, Turku, Finland. He worked from 1984 to 1987 at the International Agency for Research on Cancer, Lyon, France, where he studied the role of free radicals and antioxidant functions in carcinogenesis. He is docent in biochemistry (University of Turku) and toxicology (University of Kuopio, Kuopio, Finland) and is leading a research group at the BioCity Research Park in Turku. His research interests focus on biomarkers and pathophysiology of oxidative stress.
Dr. Tommi Vasankari received his Ph.D. degree from the Department of Physiology, University of Turku. Thereafter, he has worked as a senior scientist at Paavo Nurmi Centre, Turku, and the director of the Medical Department in Sports Institute of Finland, Vierumäki. His research interests focus on lipoprotein metabolism (especially oxidation) in atherosclerosis and oxidative stress, and antioxidant functions in physical exercise.

have revealed that ultimate atherogenic agents are in fact the modified, mainly oxidized, forms of LDL [1–7]. Oxidative damage to LDL may range from the slight structural alterations of mildly oxidized LDL to extensive breakdown of lipids and apolipoprotein B of fully oxidized LDL. In vitro studies as well as studies with experimental animals show that the various oxidized forms of LDL contribute to atherogenic processes by multiple mechanisms [1–7].

Despite convincing evidence from the in vitro and animal studies, data concerning the role of oxidized LDL in development of atherosclerosis in humans have remained limited. An obvious reason for the scantiness of human studies has been the lack of methods for direct measurement of oxidized LDL, applicable for the necessary large-scale epidemiologic studies [8]. Thus far, estimation of in vivo LDL oxidation has been based by large on determination of autoantibodies to oxidized LDL [8,9]. This methodology, however, does not seem to be practical for clinical purposes. In particular, reproducible preparation of antigens (oxidized LDL) is in practice not possible, and this has caused much variability to results, most notably between laboratories [10]. Hence, direct comparison of results from different laboratories is not recommended [9]. Thus, there is "need for a rapid and specific measure of LDL oxidation that could become a part of the laboratory repertoire in the diagnosis and management of atherosclerosis" [7]. In the following, a method will be introduced which measures the amount of baseline diene conjugation in circulating LDL, and is an indicator of oxidized LDL in vivo.

ESTIMATION OF IN VIVO LDL OXIDATION

Due to the heterogeneous nature of the chemistry of LDL oxidation, proper determination of oxidized LDL is problematic. Much of the present knowledge on LDL oxidation comes from studies recording the kinetics of LDL oxidation ex vivo [3]. The estimation of in vivo LDL oxidation has been largely based on immunologic methods. F2-isoprostanes, which seem promising as indicators of whole-body oxidative stress, have also been considered as an alternative analytical means (c.f., [11]). Finally, popularity of the diene conjugation method as an able means to monitor LDL oxidation ex vivo has led to development of an in vivo assay, which measures as an index of oxidized LDL the amount of conjugated dienes present in circulating LDL [12,13].

LDL diene conjugation

Determination of diene conjugation has been for decades one of the basic methods for the measurement of lipid peroxidation. Rearrangement of double bonds in polyunsaturated fatty acids, i.e., the formation of conjugated dienes, is an early event of lipid peroxidation taking place soon after initiation of the chain reaction [14]. The oxidation-induced increase of diene conjugation in LDL lipids is well documented [3], and after introduction of the method by Esterbauer et al. [15], measurement of diene conjugation has become the most popular method to monitor oxidation of LDL ex vivo [8]. For this method of ex vivo oxidizability, LDL is first isolated by ultracentrifugation, then dialyzed, and finally subjected to an oxidizing agent (most commonly copper ions). Proceeding of the LDL oxidation ex vivo can be followed by the various markers of lipid peroxidation [3], but due to its convenience and objectivity, the change of the conjugated diene absorbance at 234 nm has in practice become the most popular means for the monitoring of LDL oxidation. The time course of forced LDL oxidation ex vivo can typically be divided into three consecutive phases: (i) lag phase, (ii) propagation phase,

and (iii) decomposition phase. The lag phase (time before onset of lipid peroxidation) has been considered as the most important feature of oxidizability in this assay system [16] although other measures of the assay (e.g., maximum rate of diene formation, total amount of dienes) are also being used. This experimental system has been extensively used to study factors affecting oxidizability of LDL. It has enabled estimation of the contribution of various endogenous antioxidants in protection of LDL from oxidation [3,17,18] and, consequently, the method has become a popular means in testing of novel antioxidant compounds.

Encouraged by usability of the diene conjugation measurement in ex vivo studies, the LDL baseline diene conjugation method (LDL-BDC) was developed to be used as an indicator of LDL oxidation in vivo [12,13]. This method measures the actual amount of conjugated dienes in LDL, but not diene conjugation during or after chemically induced oxidation of LDL ex vivo. Compared with other methods for oxidized LDL in vivo, the LDL-BDC method is fast and easy to perform with instrumentation generally available in clinical laboratories. Since its development and original validation, this method has been successfully utilized in a number of studies on cardiovascular diseases and risk factors (see below).

The LDL-BDC assay is based on (i) precipitation of serum LDL with buffered heparin and (ii) spectrophotometric determination of the baseline level of conjugated dienes (234 nm) in lipids extracted from LDL [12]. Isolation of LDL by heparin does not affect the level of LDL-BDC [13]. Artefactual oxidation of LDL during sample preparation is prevented by addition of ethylenediaminetetraacetate. Chemical studies by HPLC and NMR show that LDL-BDC is almost exclusively ($> 90\%$) due to conjugated dienes in fatty acids, mainly linoleic acid, esterified to cholesterol and triglycerides (Fig. 1). Importantly, HPLC analyses demonstrated that compounds other than LDL constituents are not involved in LDL-BDC. Moreover, NMR spectra clearly indicated a certain degree of peroxidation (diene conjugation) in polyunsaturated fatty acids (Vasankari, T. J.; Ahotupa, M.; Toikka, J.; et al. Oxidized LDL and thickness of carotid intima-media are associated with coronary atherosclerosis in middle-aged men with coronary atherosclerosis: lower levels of oxidized LDL with statin therapy. Submitted for publication.). Together these findings give chemical evidence for specificity and validity of the LDL-BDC assay as an index of LDL lipid peroxidation.

Further evidence on validity of the LDL-BDC assay as an indicator of in vivo LDL oxidation was obtained from comparison with the autoantibody method for oxidized LDL. In a study with healthy middle-aged men, LDL-BDC showed a strong correlation ($r = 0.57$; $p = .001$) with the autoantibody titer [13].

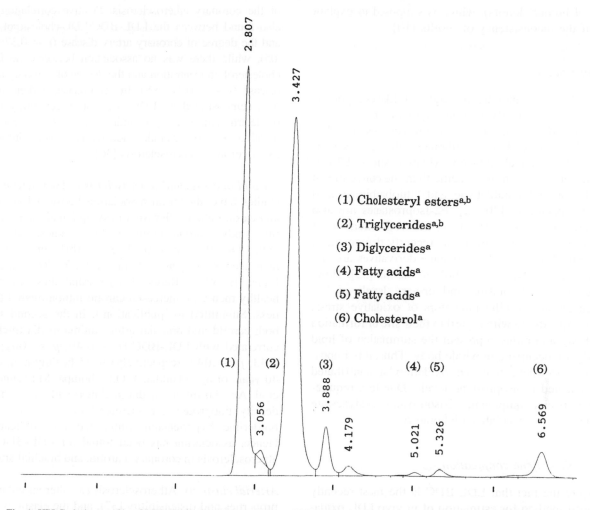

(1) Cholesteryl esters[a,b]

(2) Triglycerides[a,b]

(3) Diglycerides[a]

(4) Fatty acids[a]

(5) Fatty acids[a]

(6) Cholesterol[a]

Fig. 1. HPLC elution profile of LDL lipids absorbing at 234 nm. LDL lipids were first separated and isolated by HPLC (Varian Star 9010 solvent delivery system equipped with the detector Varian 9050, Walnut Creek, CA, USA) and then identified by NMR (Jeol JNM.A500, Tokyo, Japan). The analytical HPLC column was a Chromospher Si, 250 × 4.6 mm, 5 μm, and the preparative one a Lichrospher Si 60, 250 × 10 mm, 5 μm. The HPLC eluent was 4% 1-propanol in hexane (volume/volume) in both cases. The detector was operated at 234 nm, and 0.1 ml (analytical column) or 1 ml (preparative column) aliquots were injected. The flow rate was 1 ml/min or 3 ml/min, respectively. The ^{1}H and ^{13}C NMR spectra were recorded at 27°C on a spectrometer working at 500.16 and 124.78 MHz, respectively. [a]Located by HPLC elution volume correlations with standard samples. [b]Analysis by ^{1}H and ^{13}C NMR spectra of materials separated by preparative HPLC.

LDL OXIDATION MARKERS, ATHEROSCLEROSIS, AND RISK FACTORS

Autoantibodies against oxidized LDL

Upon oxidation LDL becomes immunogenic giving rise to formation of autoantibodies. Demonstration of increased titers of autoantibodies to oxidized LDL is generally regarded as one of the best available indicators of LDL oxidation in vivo (c.f., [8]). Immunologic methods for estimation of oxidized LDL are usually based on use of in vitro oxidized LDL as the antigen. In addition, LDL conjugated with lipid peroxidation products, such as malondialdehyde or 4-hydroxynonenal, have also been used as antigens [3].

Studies with autoantibodies have demonstrated implication of oxidized LDL in human atherosclerosis and related conditions. As reviewed recently [9], some of the studies seem to indicate that autoantibodies against oxidized LDL are predictive for the development of atherosclerosis [19–24], whereas other studies report contradictory results [25–28]. Apparent discrepancies between autoantibody studies from various laboratories have been suggested to result from methodologic differences. Due to the fact that levels of circulating autoantibodies against oxidized LDL are generally low, factors affecting reliability of the assay are especially important. A crucial step is preparation of the antigens (in vitro chemically induced oxidation of LDL isolated from blood of varying groups of human donors), which is supposed to explain much of the inconsistency of results [10].

F2-isoprostanes

The F2-isoprostanes are prostaglandinlike compounds that are formed mainly in nonenzymatic free radical–catalyzed peroxidation of arachidonic acid, esterified to phospholipids in cellular membranes [29–32]. As indicators of free radical–induced oxidative stress, F2-isoprostanes appear far more specific than the conventional methods for malondialdehyde and "thiobarbituric acid reactive substances" [30–32]. F2-Isoprostanes are also produced during cell- or copper-induced oxidation of LDL in vitro [33,34], and it was recently demonstrated that urinary levels of F2-isoprostane derivatives are elevated in patients with hypercholesterolemia [35]. However, even though plasma and urinary levels of F2-isoprostanes are specific indicators of oxidative stress, they lack specificity with regard to their site of formation in the body and rather represent the summation of lipid peroxidation occurring in whole body. Thus, it is impossible to say to what extent LDL oxidation has contributed to the detected F2-isoprostane levels. Due to a requirement for special equipment, F2-isoprostane analyses are possible only for specialized laboratories.

LDL baseline diene conjugation

Despite the fact that LDL-BDC is the most recently developed method for estimation of in vivo LDL oxidation, suitability of the method for clinical studies has enabled fast accumulation of data on implications of LDL-BDC in human atherosclerosis, as well as on relationships between LDL-BDC and the various risk factors (Table 1).

Coronary atherosclerosis. The relation between LDL-BDC and severity of coronary atherosclerosis was investigated in 62 middle-aged men who underwent diagnostic coronary angiography and sonography to measure the carotid intima-media thickness (Vasankari et al., submitted for publication). Most of the subjects had ischemic symptoms before the angiography, while nine men underwent the angiography as a preoperative procedure before valvular reconstruction. Therefore, it was possible to get control subjects with angiographically verified normal coronary arteries. In this study the patients with two- or three-vessel disease who did not use lipid lowering therapy had a 41% higher LDL-BDC:LDL-cholesterol ratio than patients with normal vessels. The regression analysis indicated that the LDL-BDC: LDL-cholesterol ratio and the intima-media thickness were the only factors associated independently with the severity of the coronary atherosclerosis. Positive correlation was also found between the LDL-BDC:LDL-cholesterol ratio and the degree of coronary artery disease ($r = 0.37$; $p < .05$), while there was no association between the LDL-cholesterol concentration and the degree of coronary artery disease ($r = -0.21$; NS). In accordance, earlier studies have demonstrated that LDL oxidation susceptibility, based on determination of copper-induced increase of diene conjugation ex vivo, was independently associated with severity of coronary atherosclerosis [36].

Carotid and brachial atherosclerosis. Two independent studies have shown an association between LDL-BDC and carotid atherosclerosis in asymptomatic men. In the first study, carotid intima-media thickness correlated with LDL-BDC ($r = 0.51$, $p < .001$) in 55 healthy men aged < 45 years (Raitakari, O. T.; Toikka, J. O.; Laine, H.; et al. Reduced myocardial flow reserve in healthy men with increased carotid intima-medial thickness. Submitted for publication.). In the second study, both carotid and brachial artery intima-media thickness correlated with LDL-BDC ($r = 0.49$, $p = .002$; $r = 0.33$, $p = .042$, respectively) in 35 healthy men under 40 years of age (Toikka, J. O.; Ahotupa, M.; Laine, H.; et al. Arterial intima-media thickness and oxidized low-density lipoprotein are increased in young men with borderline hypertension. Submitted for publication.). Hence, association has been found for LDL-BDC and atherosclerosis in coronary, carotid, and brachial arteries.

Arterial elasticity. Atherosclerosis can alter vascular wall properties and distensibility [37], and the elastic properties of arteries can be measured noninvasively in vivo. An in vivo association between LDL-BDC and arterial elasticity was demonstrated in a study where the elastic properties of large arteries in the ascending and descending parts of the thoracic aorta were investigated by using magnetic resonance imaging, and in the extracranial common carotid artery by using high-resolution ultrasound. The compliance of ascending aorta correlated with LDL-BDC ($r = -0.44$, $p = .030$), but not with the traditional lipid risk factors in 25 healthy men < 40 years of age [38]. In univariate analysis the compliance of carotid artery associated with LDL-BDC ($r = -0.49$, $p = .016$) and HDL-cholesterol:total-cholesterol ratio ($r = 0.41$, $p = .040$), and in multivariate regression analyses LDL-BDC was the only independent determinant for compliance of the carotid artery ($p = .016$) [38]. These results strengthen the evidence that oxidized LDL may interact with normal arterial vasodilatory functions [39,40].

In an independent study, endothelium-dependent flow-mediated dilatation of the brachial artery was measured noninvasively using ultrasound in asymptomatic

Table 1. LDL Baseline Diene Conjugation in Clinical Atherosclerosis and Risk Factors

Focus	Population	Study design	Results	Reference/footnote
Coronary atherosclerosis	Male 33–66 years $n = 62$ (cases 52, controls 10) All angiographied	Cross-sectional Severity of CAD evaluated by angiography	Cases with CAD had higher LDL-BDC:LDL ratio ($p = 0.020$)	a
Carotid atherosclerosis	Male Asymptomatic 36 ± 4 years $n = 55$	Cross-sectional Carotid artery intima-media thickness by ultrasound	LDL-BDC correlated with intima media thickness ($r = 0.51$, $p < .001$)	b
Carotid and brachial atherosclerosis	Male Asymptomatic < 40 years $n = 35$	Cross-sectional Carotid & brachial intima-media thickness by ultrasound	LDL-BDC correlated with carotid ($r = 0.49$, $p = .002$) and brachial ($r = 0.33$, $p = .042$) intima-media thickness	c
Arterial elasticity	Male Asymptomatic < 40 years $n = 25$	Cross-sectional Elasticity in aorta by MRI and in carotid artery by ultrasound	LDL-BDC correlated with compliance of aorta ($r = -0.44$, $p = .030$) and carotid artery ($r = -0.49$, $p = .016$)	[38]
Arterial elasticity	Male Asymptomatic < 40 years $n = 20$	Cross-sectional Endothelium-dependent flow-mediated dilation of brachial artery	LDL-BDC correlated with flow-mediated dilatation ($r = -0.56$, $p = 0.013$)	[41]
Arterial elasticity	Male Asymptomatic 36 ± 4 years $n = 55$	Cross-sectional Myocardial flow reserve measured by positron emission tomography	LDL-BDC correlated with myocardial flow reserve ($r = -0.35$, $p = .01$)	b
Heredity and gender	Male and female Healthy 16–57 years $n = 60$	Cross-sectional Heredity studied among 15 families (father, mothers, and male twins)	LDL-BDC correlated between identical twins ($r = 0.81$, 95% CI 0.44–0.95) and between fathers and sons ($r = 0.49$, 95% CI 0.16–0.72) Men had 35% higher LDL-BDC than women	[43]
Gender	Male and female Sedentary 31–58 years $n = 34 + 70$	Intervention 10-month exercise program	Men had 38 and 44% higher LDL-BDC before and after exercise program than women	[44]
Obesity	Female BMI 29–46 kg/m^2 29–46 years $n = 77$	Intervention 12-week weight reduction followed by 9-month weight maintenance	During weight reduction (mean weight loss 13 kg) LDL-BDC decreased 40%	[49]
Hyperglycemia	Male and female 38 ± 12 years $n = 19$	Pilot study among patients with IDDM	HbA1C correlated with LDL-BDC ($r = 0.60$, $p < .05$)	[52]
Hypertension	Male Asymptomatic < 40 years $n = 35$	Cross-sectional Normotensive and borderline hypertensive men	LDL-BDC was 62% higher in borderline hypertensive men compared to controls	c
Physical inactivity	Male Veteran athletes and controls 42–56 years $n = 31$	Cross-sectional Effect of several years of endurance training	Veteran athletes had 37% lower LDL-BDC:LDL ratio than controls ($p = .003$)	[47]
Physical inactivity	Female Gymnasts, runners, and controls 9–15 years $n = 183$	Cross-sectional Effect of sport participation and physical activity	Gymnasts had 15% lower LDL-BDC:LDL ratio than controls ($p = 0.0052$)	[48]
Physical inactivity	Male and female Sedentary 31–58 years $n = 34 + 70$	Intervention 10-month exercise program	LDL-BDC decreased 23% ($p = 0.0010$) in men and 26% ($p < .0001$) in women	[44]

BMI = body mass index; CAD = coronary artery disease; HbA1C = glycosylated hemoglobin; IDDM = insulin dependent diabetes mellitus; MRI = magnetic resonance imaging.

[a] Vasankari, T. J.; Ahotupa, M.; Toikka, J.; et al. Oxidized LDL and thickness of carotid intima-media are associated with coronary atherosclerosis in middle-aged men with coronary atherosclerosis: lower levels of oxidized LDL with statin therapy. Submitted for publication.

[b] Raitakari, O. T.; Toikka, J. O.; Laine, H.; et al. Reduced myocardial flow reserve in healthy men with increased carotid intima-medial thickness. Submitted for publication.

[c] Toikka, J. O.; Ahotupa, M.; Laine, H.; et al. Arterial intima-media thickness and oxidized low-density lipoprotein are increased in young men with borderline hypertension. Submitted for publication.

[d] Vasankari, T.; Fogelholm, M.; Kukkonen-Harjula, K.; et al. Reduced oxidized LDL after weight reduction in obese premenopausal women. Submitted for publication.

men aged < 40 years [41]. In healthy subjects this parameter correlated with LDL-BDC ($r = -0.56$, $p = .013$), total cholesterol ($r = -0.56$, $p = .010$), HDL cholesterol ($r = 0.59$, $p = .006$), and HDL$_2$ cholesterol ($r = 0.62$, $p = .004$). In the same study patients with familial hypercholesterolemia were found to have significantly higher LDL-BDC levels than healthy controls ($p < .001$) [41].

Yet another study demonstrated association between LDL-BDC and myocardial flow reserve, as measured using positron emission tomography (PET). In this study with 55 asymptomatic healthy men < 45 years of age, the myocardial flow reserve correlated negatively with LDL-BDC ($r = -0.35$, $p = .01$) but not with apolipoprotein B, total cholesterol, or HDL cholesterol (Toikka et al., submitted for publication).

In accordance with the above findings, studies with autoantibodies against oxidized LDL have shown that autoantibody titers are inversely related to endothelial dysfunction (measured by forearm blood flow by plethysmography; $p < .002$) [40] and to coronary flow reserve (measured by PET; $p < .02$) [42].

Risk factors of atherosclerosis. In order to investigate contribution of genetic factors to in vivo LDL oxidation, the familial aggregation of LDL-BDC was studied among 15 families (fathers, mothers, and male twins 16–18 years of age). A high association was found to exist between identical twins ($r = 0.81$, 95% confidence intervals 0.44–0.95) and between fathers and sons ($r = 0.49$, confidence intervals 0.16–0.72), but not between mothers and sons ($r = 0.00$, NS) [43]. The fathers had 35% higher concentration of LDL-BDC than the mothers ($p < .05$) [43]. Similar gender difference was seen in another study where 34 sedentary middle-aged men and 70 women were studied before and after a 10-month exercise program [44]. In that study men had 38 to 44% higher LDL-BDC levels than women ($p < .0001$). The gender difference could not be explained by age disparity, because the mean ages were similar. Neither was it explained by a higher concentration of LDL cholesterol, because also the LDL-BDC:LDL-cholesterol ratio was 20 to 25% higher in men. The higher LDL oxidation in men is in line with the generally higher incidence of coronary heart disease (CHD) among men.

Obesity is an independent risk factor for both incidence and mortality of CHD [45,46], but little evidence has existed on possible association between obesity and LDL oxidation. Several studies by the LDL-BDC method have shown that LDL oxidation is related to body mass index. This was demonstrated among veteran athletes and their controls ($r = 0.47$, $p = .008$, $n = 31$) [47], among healthy middle-aged men ($r = 0.65$, $p < .001$, $n = 15$) [43], and in sedentary middle-aged

women ($r = 0.24$, $p < .05$, $n = 70$) [44]. In accordance, the LDL-BDC:LDL-cholesterol ratio correlated positively with the weight ($r = 0.15$, $p = 0.038$), the waist circumference ($r = 0.19$, $p = .0096$), and the hip circumference ($r = 0.18$, $p = .014$) in 183 teenage girls [48]. Importantly, in an intervention study the concentration of LDL-BDC decreased by 40%, and the LDL-BDC:LDL-cholesterol ratio by 33% (both $p < .0001$) during a 12-week weight reduction in 77 obese premenopausal women [49] (Vasankari, T.; Fogelholm, M.; Kukkonen-Harjula, K.; et al. Reduced oxidized LDL after weight reduction in obese premenopausal women. Submitted for publication.). The mean weight loss was 13 kg (from 92 to 79 kg, $p < .0001$), and weight reduction correlated with the decrease of LDL-BDC ($r = 0.24$, $p = .039$).

As arterial elasticity seems to be strongly associated with LDL-BDC [38,41], a study was performed to investigate possible association between hypertension and LDL oxidation. In this study, subjects with borderline hypertension were found to have a significantly higher concentration of LDL-BDC compared with normotensive subjects ($p < .001$). In addition, the 24-h ambulatory systolic blood pressure value correlated positively with LDL-BDC ($r = 0.51$, $p < .0011$) and with the LDL-BDC:LDL-cholesterol ratio ($r = 0.49$, $p = .016$) (Toikka et al., submitted for publication).

A recent study by Bellomo et al. [50] suggested association between hyperglycemia and LDL oxidation, while another recent study [51] failed to confirm this finding. In a cross-sectional pilot study among 19 patients with insulin-dependent diabetes mellitus, it was found that hyperglycemia, as indicated by the level of glycosylated hemoglobin, correlated with LDL-BDC ($r = 0.60$, $p < .05$) and with the LDL-BDC:LDL-cholesterol ratio ($r = 0.56$; $p < .05$) [52].

Physical inactivity is known as an independent risk factor of atherosclerosis, and physical activity is one means of preventing ischemic heart disease [53]. The most favorable type of training is endurance exercise, as shown in former endurance athletes [54]. The cardiovascular benefits of exercise are suggested to be due to decreased concentrations of plasma triglyceride and LDL, and increased concentrations of HDL and apolipoprotein A-I [55]. In order to investigate physical exercise and LDL oxidation, 15 nonsmoking veteran endurance athletes and 16 matched controls were studied [47]. The veteran athletes had 37% lower LDL-BDC:LDL-cholesterol ratio than the controls. LDL-BDC correlated inversely with a leisure time intensive physical activity METs (multiples of resting metabolic rate) ($r = -0.41$, $p = .021$) [47]. Similar results were found in another study, where 183 teenage girls were studied [48]. The physically active group had 15% lower LDL-BDC:

LDL-cholesterol ratio compared with controls ($p =$ .0052), and the difference persisted when the body mass index was included as a covariate (ANCOVA, $p =$ .013). The LDL-BDC:LDL-cholesterol ratio correlated with the physical activity METs ($r = -0.21$, $p =$ .0040). The physically active girls also had a 12% higher HDL-cholesterol:total-cholesterol ratio (AN-COVA, $p = .046$), but no differences were found in other common lipid risk factors between the groups [48].

Further evidence on physical activity and LDL oxidation was obtained in a study where sedentary men and women ($n = 34$ and 70, respectively) participated in a 10-month exercise program [44]. The exercise program increased estimated oxygen uptake by 19% in both men and women (both $p < .0001$). During the exercise program the LDL-BDC value decreased 23% ($p =$ .0010) and 26% ($p < .0001$), and the LDL-BDC:LDL-cholesterol ratio 14% ($p = .016$) and 18% ($p < .0001$) in men and women, respectively. Concurrently, the concentration of HDL cholesterol increased by 15% in men ($p = .0004$) and by 5% in women ($p = .043$), and LDL cholesterol decreased by 10% ($p = .026$) and 11% ($p < .0001$), respectively. Serum total cholesterol and triglyceride concentrations remained unchanged [44]. Acute physical exercise, on the other hand, does not affect LDL-BDC, as demonstrated in sedentary subjects and endurance athletes [56,57]. Thus, in addition to increasing HDL cholesterol and decreasing LDL cholesterol, the endurance-type exercise seems to improve "quality" of circulating LDL (by decreasing the amount of oxidized LDL) and thereby reduce the risk of atherosclerosis.

LDL-BDC AS AN INDICATOR OF MILDLY OXIDIZED LDL

Chemical studies on heparin-precipitated LDL showed no signs of breakdown products of advanced lipid peroxidation (which became apparent after short incubation of LDL with copper in vitro). Moreover, contents of antioxidants (alpha- and gamma-tocopherols, betacarotene, ubiquinol-10) were not altered in samples where LDL-BDC was analyzed (Ahotupa, M. Unpublished results.). Therefore, LDL-BDC is representative of mildly oxidized forms of circulating LDL.

It is generally assumed that native (unoxidized) LDL is taken up by endothelial cells and oxidation of LDL only takes place in subendothelial space. The occurence of heavily oxidized LDL in circulating LDL of healthy human subjects is, indeed, unlikely due to the many antioxidants present in plasma. In addition, heavily oxidized LDL would propably be rapidly removed from circulation by scavenger receptors on hepatic sinusoidal cells. On the other hand, it is known that more than 85%

of LDL will re-enter to circulation after passage through artery wall [58]. It is, therefore, not surprising that subtle degrees of oxidation may occur in circulating LDL. It was recently shown that "minimally oxidized LDL" can circulate in the plasma for a period sufficiently long to enter and accumulate in the arterial intima [59].

As a further indication of the significance of circulating oxidized LDL, it has been recently documented that vascular endothelial cells have a specific receptor (LOX-1) for oxidized LDL [60–63]. This receptor is a membrane protein that is expressed in vascular endothelium and vascular-rich organs, and also in vivo in intact endothelium and atheromatous intima. While oxidized LDL has high affinity for LOX-1, native LDL does not bind to this receptor. Importantly, LOX-1 expression is upregulated by oxidized LDL as shown in human endothelial cells in culture [62].

The existence of a specific high affinity uptake mechanism of endothelial cells for oxidized LDL stresses the potential importance of circulating mildly oxidized LDL. Mildly oxidized LDL is suggested to represent a form of LDL that is primed for more rapid subsequent oxidation in the intima (c.f., [3,4]). In support of this, the amount of preformed peroxides is known to affect the oxidizability of LDL in vitro [64]. This would mean that the level of mildly oxidized LDL in circulation represents a significant risk for the development of atherosclerosis.

PERSPECTIVE: LDL-BDC AS A CLINICAL CHEMICAL ASSAY, AND AS AN INDICATOR OF LDL OXIDATION AND THE RISK OF ATHEROSCLEROSIS

Basic studies on validation of the LDL-BDC assay demonstrated that the method is well suited for clinical purposes, and by the applied procedure artefactual oxidation can be, in practise, eliminated [12,13]. The validity of the assay as an index of LDL oxidation is demonstrated by (i) HPLC studies that show that compounds other than LDL constituents are not involved in LDL-BDC, and (ii) studies by NMR that indicate the presence of diene conjugation in polyunsaturated fatty acids (Fig. 1) (Vasankari et al., submitted for publication). The assay procedure enables efficient analysis of samples and, therefore, the method is applicable even for large-scale epidemiologic studies. Studies on the effects of various risk management programs (exercise program, weight reduction program) showed that LDL-BDC is well suited for monitoring the effectiveness of interventions or treatments [44,49].

The cause and site of oxidation of the minimally oxidized fatty acids, as measured in the LDL-BDC assay by conjugated dienes, is not known. In theory, LDL oxidation can be caused in vivo, e.g., by metal ions, reactive oxygen species, lipo-oxygenases, myeloperoxidase, nitric oxide, thiols, glucose etc. (c.f., [6]). In addition, we cannot rule out the possibility that minimally oxidized lipids present in food could be absorbed and distributed in the body among native lipids. On the other hand, as it seems that circulating mildly oxidized LDL is primed for more rapid subsequent oxidation in arterial intima [3,4,64], not the origin but the amount of the "preoxidized" LDL would be crucial for determining atherogenicity of LDL.

Interference of the LDL-BDC assay by various drugs, as well as influence of drug treatments on LDL-BDC levels, have not been studied systematically. In one study, cholesterol-lowering therapy (statins) was found to decrease significantly the LDL-BDC:LDL-cholesterol ratio among CHD patients (Vasankari et al., submitted for publication), which is in line with the known cardioprotective effect of statin therapy [65]. On the contrary, antioxidant preparations (a combination of 300–400 IU of vitamin E, 500–1000 mg of vitamin C, and 60 mg of ubiquinone or 25 mg of beta-carotene per day), when given for 1 to 4 weeks to healthy young volunteers or endurance athletes, did not affect the LDL-BDC value [12,57].

Until recently, studies on LDL-BDC have come from a single laboratory, and comparisons between laboratories have not been possible. In a recent study, however, atherosclerosis-reversal therapy (intensive exercise, stress management program, and low-fat diet) was found to result in reduction of baseline diene conjugation of LDL [66]. Moreover, in a human dietary intervention study, dietary supplementation of lycopene was found to decrease significantly the amount of LDL-BDC [67]. The above studies are well in accordance with earlier studies on LDL-BDC, and further strengthen the value of LDL-BDC as an indicator of the risk of atherosclerosis.

As can be expected, LDL-BDC is associated with known lipid risk factors such as serum total, LDL-, and HDL-cholesterols as well as triglycerides [41,44,47,48]. Yet, all clinical studies conducted thus far point to the fact that, as an indicator of the risk of atherosclerosis, LDL-BDC clearly exceeds sensitivity and specificity of the common lipid markers of atherosclerosis. In fact, the LDL-BDC has turned out to be surprisingly closely related not only to the disease but also to risk factors (obesity, physical inactivity, hypertension, diabetes, and arterial functions). Altogether, studies by the LDL-BDC method have given strong new evidence on involvement of LDL oxidation in human atherosclerosis.

CONCLUSIONS

The evidence on the key role of oxidized LDL in development of atherosclerosis strongly suggests that a method for oxidized LDL will be, in the future, included in the repertoire of clinical laboratories to be used as an indicator of the risk of atherosclerosis. Due to the heterogenous nature of LDL oxidation, attempts to estimate in vivo LDL oxidation may be based on a variety of methodologies. LDL-BDC is a new clinically applicable method for mildly oxidized circulating LDL, the development of which was stimulated by experience from ex vivo studies. The LDL-BDC assay is fast and simple to perform with equipment generally available in clinical laboratories. Chemical and biological studies have verified that LDL-BDC is a specific measure of oxidized LDL. Clinical applicability of the LDL-BDC-method is indicated by clinical studies that show that LDL-BDC is surprisingly strongly associated with atherosclerosis and risk factors. Indeed, LDL-BDC seems to be far more specific and sensitive as an indicator of the risk of atherosclerosis than the commonly used lipid markers. Taken together, the LDL-BDC assay appears to be a valid indicator of in vivo LDL oxidation and a suitable tool for use in diagnosis, follow-up treatment, and basic research of atherosclerosis.

Acknowledgments — The authors' work has been supported by Juho Vainio Foundation, Finland.

REFERENCES

[1] Steinbrecher, U. P.; Zhang, H.; Lougheed, M. Role of oxidatively modified LDL in atherosclerosis. *Free Radic. Biol. Med.* **9**:155–168; 1990.

[2] Witztum, J.; Steinberg, D. Role of oxidized LDL in atherogenesis. *J. Clin. Invest.* **88**:1785–1792; 1991.

[3] Esterbauer, H.; Gebicki, J.; Puhl, H.; Jurgens, G. The role of lipid peroxidation and antioxidants in oxidative modifications of LDL. *Free Radic. Biol. Med.* **13**:241–290; 1992.

[4] Witztum, J. L. Role of oxidised low density lipoprotein in atherogenesis. *Br. Heart J.* **69**(Suppl.):S12–S18; 1993.

[5] Witztum, J. L. The oxidation hypothesis of atherosclerosis. *Lancet* **344**:793–795; 1994.

[6] Berliner, J. A.; Heinecke, J. W. The role of oxidized lipoproteins in atherogenesis. *Free Radic. Biol. Med.* **20**:707–727; 1996.

[7] Steinberg, D. Low density lipoprotein oxidation and its pathological significance. *J. Biol. Chem.* **272**:20963–20955; 1997.

[8] Jialal, I.; Devaraj, S. Low-density lipoprotein oxidation, antioxidants, and atherosclerosis: a clinical biochemistry perspective. *Clin. Chem.* **42**:498–506; 1996.

[9] Ylä-Herttuala, S. Is oxidized low-density lipoprotein present *in vivo*? *Curr. Opin. Lipidol.* **9**:337–344; 1998.

[10] Devaraj, S.; Jialal, I. Laboratory assessment of lipoprotein oxidation. In: Rifai, N.; Warnick, G. R. A.; Dominiczak, M., eds. *Handbook of lipoprotein testing*. Washington DC: AACC Press; 1997:357–373.

[11] Witztum, J. L. To E or not to E—how do we tell? (editorial). *Circulation* **50**:2785–2787; 1998.

[12] Ahotupa, M.; Ruutu, M.; Mäntylä, E. Simple methods of quantifying oxidation products and antioxidant potential of low density lipoproteins. *Clin. Biochem.* **29**:139–144; 1996.

[13] Ahotupa, M.; Marniemi, J.; Lehtimäki, T.; Talvinen, K.; Raitakari, O.; Vasankari, T.; Viikari, J.; Luoma, J.; Ylä-Herttuala, S. Baseline diene conjugation in LDL lipids as a direct measure of *in vivo* LDL oxidation. *Clin. Biochem.* **31**:257–261; 1998.

[14] Kappus, H. Lipid peroxidation: mechanisms, analysis, enzymology and biological relevance. In: Sies, H., ed. *Oxidative stress*. London: Academic Press; 1985:273–310.

[15] Esterbauer, H.; Striegl, G.; Puhl, H.; Rotheneder, M. Continuous monitoring of in vitro oxidation of human low density lipoprotein. Free Radic. Res. Commun. 6:67–75; 1989.

[16] Kleinveld, H. E.; Ha-Lemmers, H. L. M.; Stalenhoef, A. F. H.; Demacker, P. M. N. Improved measurement of low-density lipoprotein susceptibility to copper-induced oxidation: application of a short procedure for isolating low-density lipoprotein. Clin. Chem. 38:2066–2072; 1992.

[17] Esterbauer, H.; Puhl, H.; Dieber-Rotheneder, M.; Waeg, G.; Rabl, H. Effect of antioxidants on oxidative modification of LDL. Annals. Med. 23:573–581; 1991.

[18] Stocker, R.; Bowry, V. W.; Frei, B. Ubiquinol-10 protects human low density lipoprotein more efficiently against lipid peroxidation than does α-tocopherol. Proc. Natl. Acad. Sci. USA 88:1646–1650; 1991.

[19] Salonen, J. T.; Ylä-Herttuala, S.; Yamamoto, R.; Butlet, S.; Korpela, H.; Salonen, R.; Nyyssönen, K.; Palinski, W.; Witztum, J. L. Autoantibody against oxidised LDL and progression of carotid atherosclerosis. Lancet 339:883–887; 1992.

[20] Maggi, E.; Chiesa, R.; Melissano, G.; Castellano, R.; Aatore, D.; Grossi, A.; Finardi, G.; Bellomo, G. LDL oxidation in patients with severe carotid atherosclerosis: a study of in vitro and in vivo oxidation markers. Arterioscler. Thromb. Vasc. Biol. 14:1892–1899; 1994.

[21] Puurunen, M.; Mänttäri, M.; Manninen, V.; Tenkanen, L.; Alftan, G.; Ehnholm, C.; Vaarala, O.; Aho, K.; Palosuo, T. Antibody against oxidized low-density lipoprotein predicting myocardial infarction. Arch. Intern. Med. 154:2605–2609; 1994.

[22] Ben-Yehuda, O.; Witztum, J. L.; Keaney, J. F., Jr.; Frei, B.; Hankin, B.; Vita, J. A. Autoantibody titer to malondialdehyde modified low-density lipoprotein correlates with extent of coronary artery disease. Circulation 94 (Suppl. 1):1–638; 1996.

[23] Wu, R.; Nityanand, S.; Berglund, L.; Lithell, H.; Holm, G.; Lefvert, A. K. Antibodies against cardiolipin and oxidatively modified LDL in 50-year-old men predict myocardial infarction. Arterioscler. Thromb. Vasc. Biol. 17:3159–3163; 1997.

[24] Lehtimäki, T.; Lehtinen, S.; Solakivi, T.; Nikkilä, M.; Jaakkola, O.; Jokela, H.; Ylä-Herttuala, S.; Luoma, J. S.; Koivula, T.; Nikkari, T. Autoantibodies against oxidized low density lipoprotein in patients with angiographically verified coronary artery disease. Arterioscler. Thromb. Vasc. Biol. 19:23–27; 1999.

[25] Boullier, A.; Hamon, M.; Walters-Laporte, E.; Martin-Nizart, E.; Mackereel, R.; Fruchart, J.-C.; Bertrand, M.; Duriez, P. Detection of autoantibodies against oxidized low density lipoproteins and of IgG-bound low density lipoproteins in patients with coronary artery disease. Clin. Chim. Acta 238:1–10; 1995.

[26] Iribarren, C.; Folsom, A. R.; Jacobs, D. R. Jr.; Gross, M. D.; Belcher, J. D.; Eckfeldt, J. H. Association of serum vitamin levels, LDL, susceptibility to oxidation and autoantibodies against MDA-LDL with carotid atherosclerosis. Arterioscler. Thromb. Vasc. Biol. 17:1171–1177; 1997.

[27] Virella, G.; Virella, I.; Leman, R. B.; Pryor, M. B.; Lopes-Virella, M. F. Anti-oxidized low-density lipoprotein antibodies in patients with coronary heart disease and normal healthy volunteers. Int. J. Clin. Lab. Res. 23:95–101; 1993.

[28] van de Vijver, L. P. L.; Steyger, R.; van Poppel, G.; Boer, J. M. A.; Kruijssen, D. A. C. M.; Seidell, J. C.; Princen, H. M. G. Autoantibodies against MDA-LDL in subjects with severe and minor atherosclerosis and healthy population controls. Atherosclerosis 122:245–253; 1996.

[29] Morrow, J. D.; Hill, K. F.; Burk, R. F.; Nammour, T. M.; Badr, K. F.; Roberts, L. J. A series of prostaglandin F2-like compounds are produced in vivo in humans by a non-cyclooxygenase, free radical-catalyzed mechanism. Proc. Natl. Acad. Sci. USA 87:9383–9387; 1990.

[30] Morrow, J. D.; Roberts, J. L. The isoprostanes: unique bioactive products of lipid peroxidation. Prog. Lipid Res. 36:1–21; 1997.

[31] Patrono, C.; Fitzgerald, G. Isoprostanes. Potential markers of oxidant stress in atherothrombotic disease. Arterioscler. Thromb. Vasc. Biol. 17:2309–2315; 1997.

[32] Roberts, L. J. II; Morrow, J. The generation and actions of isoprostanes. Biochim. Biophys. Acta 1345:121–135; 1997.

[33] Gopaul, N. K.; Nourooz-Zadeh, J.; Mallet, A. I.; Anggard, E. E. Formation of F2-isoprostanes during aortic endothelial cell mediated oxidation of low density lipoprotein. FEBS Lett. 348:297–300; 1994.

[34] Lynch, S. M.; Morrow, J. D.; Roberts, J. D., II; Frei, B. Formation of non-cyclooxygenase-derived prostanoids (F2-isoprostanes) in plasma and low-density lipoprotein exposed to oxidative stress in vitro. J. Clin. Invest. 93:998–1004; 1994.

[35] Reilly, M. P.; Pratico, D.; Delanty, N.; DiMinno, G.; Tremoli, E.; Rader, D.; Kapoor, S.; Rokach, J.; Lawson, J.; FitzGerald, G. A. Increased formation of distinct F2 isoprostanes in hypercholesterolemia. Circulation 98:2822–2828; 1998.

[36] Regnström, J.; Nilsson, J.; Tornvall, P.; Landou, C.; Hamsten, A. Susceptibility of low-density lipoprotein to oxidation and coronary atherosclerosis in man. Lancet 339:1183–1186; 1992.

[37] Salomaa, V.; Riley, W.; Kark, J. D.; Nardo, C.; Folsom, A. R. Non-insulin-dependent diabetes mellitus and fasting glucose and insulin concentrations are associated with arterial stiffness indexes: the ARIC Study: Atherosclerosis Risk in Communities Study. Circulation 91:1432–1443; 1995.

[38] Toikka, J. O.; Niemi, P.; Ahotupa, M.; Niinikoski, H.; Viikari, J. S. A.; Rönnemaa, T.; Hartiala, J. J.; Raitakari, O. T. Large-artery elastic properties in young men: relationships to serum lipoproteins and oxidized low-density lipoproteins. Arterioscler. Thromb. Vasc. Biol. 19:436–441; 1999.

[39] Anderson, T. J.; Meredith, I. T.; Charbonneau, F.; Yeung, A. C.; Frei, B.; Selwyn, A. P.; Ganz, P. Endothelium-dependent coronary vasomotion relates to the susceptibility of LDL to oxidation in humans. Circulation 93:1647–1650; 1996.

[40] Heitzer, T.; Ylä-Herttuala, S.; Luoma, J.; Kurz, S.; Munzel, T.; Just, H.; Olschewski, M.; Drexler, H. Cigarette smoking potentiates endothelial dysfunction of forearm resistance vessels in patients with hypercholesterolemia. Role of oxidized LDL. Circulation 93:1346–1353; 1996.

[41] Toikka, J.; Ahotupa, M.; Viikari, J.; Niinikoski, H.; Taskinen, M.-R.; Irjala, K.; Hartiala, J.; Raitakari, O. Constantly low HDL-cholesterol concentration relates to endothelial dysfunction and increased in vivo LDL oxidation in healthy young men. Atherosclerosis 147:133–138; 1999.

[42] Raitakari, O.; Pitkänen, O.-P.; Lehtimäki, T.; Lahdenperä, S.; Iida, H.; Ylä-Herttuala, S.; Luoma, J.; Mattila, K.; Nikkari, T.; Taskinen, M.-R.; Viikari, J. S. A.; Knuuti, J. In vivo LDL oxidation relates to coronary reactivity in young men. J. Am. Coll. Cardiol. 30:97–102; 1997.

[43] Kujala, U.; Ahotupa, M.; Vasankari, T. J.; Kaprio, J.; Tikkanen, M. Familial aggregation of LDL oxidation. Scand. J. Clin. Lab. Invest. 57:141–146; 1997.

[44] Vasankari, T. J.; Kujala, U. M.; Vasankari, T. M.; Ahotupa, M. Reduced oxidized LDL levels after a 10-month exercise program. Med. Sci. Sports Exerc. 30:1496–1501; 1998.

[45] Jousilahti, P.; Tuomilehto, J.; Vartiainen, E.; Pekkanen, J.; Puska, P. Body weight, cardiovascular risk factors, and coronary mortality. 15-year follow-up of middle-aged men and women in eastern Finland. Circulation 93:1372–1379; 1996.

[46] Hubert, H. B.; Feinleib, M.; McNamara, P. M.; Castelli, W. P. Obesity as an independent risk factor for coronary heart disease: a 26-year follow-up of participants in the Framingham Heart Study. Circulation 67:968–977; 1983.

[47] Kujala, U. M.; Ahotupa, M.; Vasankari, T.; Kaprio, J.; Tikkanen, M. Low LDL oxidation in veteran endurance athletes. Scand. J. Med. Sci. Sports 6:303–308; 1996.

[48] Vasankari, T.; Lehtonen-Veromaa, M.; Möttönen, T.; Ahotupa, M.; Irjala, K.; Heinonen, O.; Leino, A.; Viikari, J. Reduced circulating minimally oxidized LDL in young female athletes. Atherosclerosis in press; 1999.

[49] Vasankari, T.; Fogelholm, M.; Oja, P.; Vuori, I.; Ahotupa, M. Effect of weight reduction on LDL oxidation (abstract). Med. Sci. Sports Exerc. 29:S130; 1997.

[50] Bellomo, G.; Maggi, E.; Poli, M.; Agosta, F. G.; Bollati, P.;

Finardi, G. Autoantibodies against oxidatively modified low-density lipoproteins in NIDDM. *Diabetes* **44:**60–66; 1995.

[51] Uusitupa, M.; Niskanen, L.; Luoma, J.; Mercuri, M.; Rauramaa, R.; Vilja, P.; Ylä-Herttuala, S. Autoantibodies against oxidized LDL do not predict atherosclerotic vascular disease in non-insulin-dependent diabetes mellitus. *Arterioscler. Thromb. Vasc. Biol.* **16:**1236–1242; 1996.

[52] Peltola, V.; Vasankari, T.; Viikari, J.; Ahotupa, M. Lipid peroxidation in diabetic patients: relations to hypertriglyceridemia, obesity and hyperglycemia (abstract). *Diabetologia* **40** (Suppl. 1): A413; 1997.

[53] Powell, K. E.; Thompson, P. D.; Caspersen, C. J.; Kendrick, J. S. Physical activity and the incidence of coronary heart disease. *Ann. Rev. Public Health* **8:**253–287; 1987.

[54] Kujala, U. M.; Kaprio, J.; Taimela, S.; Sarna, S. Prevalence of diabetes, hypertension, and ischemic heart disease in former elite athletes. *Metabolism* **43:**1255–1260; 1994.

[55] Durstine, J. L.; Haskell, W. L. Effects of exercise on plasma lipids and lipoproteins. *Exerc. Sport Sci. Rev.* **22:**477–521; 1994.

[56] Vasankari, T. J.; Kujala, U. M.; Vasankari, T. M.; Vuorimaa, T.; Ahotupa, M. Effects of acute prolonged exercise on serum and LDL oxidation and antioxidant defences. *Free Radic. Biol. Med.* **22:**509–513; 1997.

[57] Vasankari, T. J.; Kujala, U. M.; Vasankari, T. M.; Vuorimaa, T.; Ahotupa, M. Increased serum and LDL antioxidant potential after antioxidant supplementation in endurance athletes. *Am. J. Clin. Nutr.* **65:**1052–1056; 1997.

[58] Schwenke, D. C.; Carew, T. E. Initiation of atherosclerotic lesions in cholesterol-fed rabbits II: selective retention of LDL vs. selective increases in LDL permeability in susceptible sites of arteries. *Arteriosclerosis* **9:**908–918; 1989.

[59] Juul, K.; Nielsen, L. B.; Munkholm, K.; Stender, S.; Nordestgaard, B. G. Oxidation of plasma low-density lipoprotein accelerates its accumulation and degradation in the arterial wall *in vivo*. *Circulation* **94:**1698–1704; 1996.

[60] Sawamura, T.; Kume, N.; Aoyama, T.; Moriwaki, H.; Hoshikawa, H.; Aiba, Y.; Tanaka, T.; Miwa, S.; Katsura, Y.; Kita, T.; Masaki, T. An endothelial receptor for oxidized low-density lipoprotein. *Nature* **386:**73–77; 1997.

[61] Hoshikawa, H.; Sawamura, T.; Kakutani, M.; Aoyama, T.; Nakamura, T.; Masaki, T. High affinity binding of oxidized LDL to mouse lectin-like oxidized LDL receptor (LOX-1). *Biochem. Biophys. Res. Commun.* **245:**841–846; 1998.

[62] Mehta, J. L.; Li, D. Y. Identification and autoregulation of receptor for ox-LDL in cultured human coronary artery endothelial cells. *Biochem. Biophys. Res. Commun.* **248:**511–514; 1998.

[63] Nagase, M.; Hirose, S.; Fujita, T. Unique repetitive sequence and unexpected regulation of expression of rat endothelial receptor for oxidized low-density lipoprotein (LOX-1). *Biochem. J.* **330:**1417–1422; 1998.

[64] Cominacini, L.; Garbin, U.; Pastorino, A. M.; Fratta Pacini, A.; Davoli, A.; De Santis, A.; Campagnola, M.; Faccini, G.; Lo Cascio, V. Role of oxidized low density lipoproteins in the pathogenesis of atherosclerosis. *Eur. J. Lab. Med.* **2:**43–50; 1994.

[65] The Scandinavian Simvastatin Survival Study Group. Randomized trial of cholesterol lowering in 4444 patients with coronary heart disease: the Scandinavian Simvastatin Survival Study (4S). *Lancet* **344:**1383–1389; 1994.

[66] Parks, E. J.; German, J. B.; Davis, P. A.; Frankel, E. N.; Kappagoda, C. T.; Hyson, D. A.; Schneeman, B. O. Reduced oxidative susceptibility of LDL from patients participating in an intensive atherosclerosis treatment program. *Am. J. Clin. Nutr.* **68:**778–785; 1998.

[67] Agarwal, S.; Rao, A. V. Tomato lycopene and low density lipoprotein oxidation: a human dietary intervention study. *Lipids* **33:**981–984; 1998.

ABBREVIATIONS

ANCOVA—analysis of variance with a covariate

BDC—baseline diene conjugation

BMI—body mass index

CAD—coronary artery disease

CHD—coronary heart disease

HDL—high-density lipoproteins

HbA1C—glycosylated hemoglobin

HPLC—high performance liquid chromatography

IDDM—insulin-dependent diabetes mellitus

LDL—low-density lipoproteins

LOX-1—lectinlike oxidized-LDL receptor

MET—multiple of resting metabolic rate

MRI—magnetic resonance imaging

NMR—nuclear magnetic resonance

PET—positron emission tomography

Bioassays for Oxidative Stress Status (BOSS). Edited by W.A. Pryor

STABLE MARKERS OF OXIDANT DAMAGE TO PROTEINS AND THEIR APPLICATION IN THE STUDY OF HUMAN DISEASE

MICHAEL J. DAVIES, SHANLIN FU, HONGJIE WANG, and ROGER T. DEAN

The Heart Research Institute, Camperdown, Sydney, Australia

(Received 7 September 1999; Accepted 8 September 1999)

Abstract—The mechanisms of formation and the nature of the altered amino acid side chains formed on proteins subjected to oxidant attack are reviewed. The use of stable products of protein side chain oxidation as potential markers for assessing oxidative damage in vivo in humans is discussed. The methods developed in the authors laboratories are outlined, and the advantages and disadvantages of these techniques compared with other methodologies for assessing oxidative damage to proteins and other macromolecules. Evidence is presented to show that protein oxidation products are sensitive markers of oxidative damage, that the pattern of products detected may yield information as to the nature of the original oxidative insult, and that the levels of oxidized side-chains can, in certain circumstances, be much higher than those of other markers of oxidation such as lipid hydroperoxides. © 1999 Elsevier Science Inc.

Keywords—Free radical, Oxidants, Protein oxidation, Amino acid oxidation, Hypochlorite, Peroxynitrite, Singlet oxygen

INTRODUCTION

Radical-mediated protein oxidation has been studied since the beginning of this century, though the use of the products of these reactions as specific markers of oxidative damage in vivo has only been developed in the last few years. This has primarily been due to two factors: an absence of specific data on the nature of the products formed, and the lack of sensitive methods for the detection of these materials in complex systems. The use of generic markers of protein oxidation (such as protein carbonyls) as a tool for examining the extent of protein oxidation in vivo has a somewhat

longer history [1]. The use of carbonyls as a general marker of oxidation in vivo has been covered in several reviews [2,3] and will not be covered in detail here.

There is now little doubt that proteins are major targets for radicals and other oxidants when these are formed in both intra- and extracellular environments in vivo. It has been estimated, on the basis of published rate constants and the knowledge of the relative abundance of macromolecules within cells, that proteins can scavenge 50–75% of reactive radicals such as $HO^\bullet$ generated within a cell by γ-radiolysis [4]. Such data, together with the knowledge that some proteins have long half lives and, hence, are likely to accumulate oxidative "hits," suggests that the formation of lesions on proteins may be highly sensitive markers for oxidative damage in mammalian systems [5,6]. The chemistry of reaction of a large number of different radicals with amino acids, peptides, and proteins has been elucidated and recently reviewed [5,7–10]; only the most salient points related to the subject of this article are covered below.

FORMATION OF SPECIFIC PROTEIN OXIDATION PRODUCTS

Reaction of a variety of radicals with proteins and peptides in the presence of O_2 is known to give rise to

Address correspondence to: Dr. Michael J. Davies, The Heart Research Institute, 145 Missenden Road, Camperdown, Sydney, NSW 2050, Australia; Tel: +61 (0) 2 9550-3560; Fax: +61 (0) 2 9550-3302; E-Mail: m.davies@hri.org.au.
Dr. Michael J. Davies leads the Free Radical Group at The Heart Research Institute in Sydney (HRI, see www.hri.org.au); previously, he was a lecturer in chemistry at the University of York, UK. Prof. Roger T. Dean is the Foundation Director and coleader of the Cell Biology group at the HRI; previously, he was Prof. of Cell Biology at Brunel University, UK; he is also an internationally known composer/improviser in music. Michael Davies and Roger Dean have written the only book on protein oxidation induced by free radicals, published by Oxford University Press (*Radical-mediated protein oxidation: from chemistry to medicine*, 1997, ISBN 0-19-850097-1). Drs. Shanlin Fu and Hongjie Wang are senior research fellows at the HRI working on the mechanisms, detection, and biological effects of protein oxidation.

Reprinted from: *Free Radical Biology & Medicine, Vol. 27, Nos. 11/12, pp. 1151-1163, 1999*

alterations to both the backbone and side-chains [5–10]. A number of mechanisms that give rise to cleavage of the protein backbone have been elucidated, and though backbone cleavage can be readily examined with isolated proteins (e.g., using sodium dodecyl sulfate-polyacrylamide gel electrophoresis [SDS-PAGE] or high-performance liquid chromatography [HPLC]), its use as a marker of protein oxidation in vivo is very limited because of the quantity of other proteins present and the potential role of proteases. Thus, backbone fragmentation is rarely used to quantify protein oxidation in complex systems.

Aliphatic side-chains

Reaction of radicals with side-chain residues gives rise to a multitude of products; Table 1 lists some of these materials. Thus, HO• and a range of other radicals can oxidize aliphatic side chains, in the presence of O_2, to hydroperoxides, alcohols and carbonyl compounds, as a result of the formation of an initial carbon-centered species (via hydrogen atom abstraction) and addition of O_2 to give a peroxyl radical [5,8,11,12]. The chemistry of such peroxyl radicals is relatively well understood and gives rise to oxygenated products via standard hydrogen abstraction, fragmentation, and dimerization reactions [5,8,13]. The yields of some of these materials have been studied in some detail [11,14–19], others are less well quantified [20,21]. The hydroperoxides are unstable; these decompose slowly in the absence of light, heat, reducing agents, or metal ions, but are lost rapidly in the presence of any of these agents [4,16,22,23]. These products are, therefore, poor quantitative markers of protein oxidation.

The alcohols that arise either as a result of two-electron reduction of the hydroperoxides [4,16,17,19,23], or directly via standard peroxyl and alkoxyl radical chemistry [5,8,13] are, in the main, stable materials and poorly susceptible to further oxidation or derivatization. A number of these have been used as markers of protein oxidation (Table 1). Not all of these materials are useful species, however, as some are natural products (e.g., 4-hydroxyproline, 4- and 5-hydroxylysines), some undergo further reaction (e.g., the internal cyclization of the two stereoisomers of 5-hydroxyleucine to give methylprolines [19]), and some coelute, under certain conditions, with other products (e.g., the two stereoisomers of 4-hydroxyvaline [17]). Although these products are known to be generated with a range of oxygen-derived radicals (e.g., HO•, alkoxyl, and peroxyl species), much less is known regarding the formation of such materials with other oxidants. Val and Leu alcohols are not generated, for example, by HOCl and long wavelength ultraviolet (UV) light (> 310 nm) [24,25]. Such materials

are also unlikely to be formed to a major extent by peroxynitrite, NO• or $^1O_2$ [5], though this has not been investigated in any detail, and may be complicated in circumstances where secondary (chain) reactions play a significant role. This is not necessarily a disadvantage, as the absence of such products, and the presence of other markers of protein oxidation (see below) may give valuable information as to the nature of the oxidant generating damage.

Carbonyl functions are generated via a number of alkoxyl and peroxyl radical reactions [5,8,13], and such groups have been characterized as oxidation products of a number of aliphatic side chains [2,9,11,15,26]. Some of these materials are, however, prone to further reaction. In particular, aldehydes are prone to further oxidation to carboxylic acids, and both aldehydes and ketones are known to undergo Schiff base reactions with amine functions [5,27]. While the initial stages of the Schiff base reaction are reversible, further rearrangement to Amadori products makes this impossible. Methods have been developed for the assay of total carbonyl yields on proteins (e.g. [2]), and these have been shown to be excellent techniques for examining the overall extent of protein oxidation in in vitro systems. However, the use of these assays in vivo is more problematic due to potential interference from aldehydes/ketones generated from sugars or lipids bound to proteins (e.g., glycoproteins, glycoxidized proteins), and the high levels of damage they imply (see [5]). The presence of such groups on proteins in complex systems, makes the exact quantification of protein damage less certain, though there is little doubt that in appropriate circumstances these methods provide valuable information. Despite the extensive effort that has been placed on the development of assays for total carbonyl yields, there has been little development of specific assays for individual carbonyls.

Heteroatom-containing side chains

Little is known regarding the oxidation products of Ser and Thr, though it has been shown that the latter gives rise to the corresponding carbonyl-containing product [28]; this is as expected on the basis of the known chemistry of the α-hydroxyalkyl radicals formed by hydrogen-atom abstraction [13].

Lysine behaves in a somewhat analogous manner to the aliphatic side chains discussed above, due to the deactivating effect of the side-chain amino group, which limits attack at the C-6 carbon; the low extent of oxidation at this site gives the corresponding aldehyde, adipic semialdehyde [11]. However, this aldehyde also can be generated enzymatically [29]. The alcohols and carbonyl-containing products formed at C-3, C-4, and C-5 have been investigated [14,30], but only the C-3 alcohol is of

Table 1. Oxidative Lesions Detected on Radical- and Oxidant-damaged Proteins and Their Potential Usage as Markers of Oxidative Damage

Substrate and oxidative insult	Product	Potential as marker
Tyr plus HO$^{\bullet}$ or reactive nitrogen species	DOPA	Yes, but susceptible to further oxidation
Tyr plus HOCl	3-Chlorotyrosine 3,5-Dichlorotyrosine	Yes
Tyr plus reactive nitrogen species	3-Nitrotyrosine 3,5-Dinitrotyrosine	Yes
Tyr plus HO$^{\bullet}$, one electron oxidants, or HOCl, followed by radical–radical combination	Di-tyrosine	Yes, but yield dependent on radical flux as a result of radical-radical reactions
Tyr plus $^1O_2$	Tyr endoperoxide	No, unstable
Phe plus HO$^{\bullet}$, one electron oxidants or, reactive nitrogen species	o-, m-Tyrosine	Yes
Phe plus HO$^{\bullet}$ before or after dimerization	Dimers of hydroxylated species	Possible, but complex mixture
Trp plus HO$^{\bullet}$, or one electron oxidants	N-Formylkynurenine 3-Hydroxykynurenine Kynurenine	Can be generated enzymatically, and hence not advised
	2-, 4-, 5-, 6-, or 7-Hydroxytryptophans	Yes
Trp plus $^1O_2$	Trp hydro-/endo-peroxide	No, unstable
His plus HO$^{\bullet}$, or one electron oxidants	2-Oxo-histidine	Possible
His plus $^1O_2$	His hydro-/endo-peroxide	No, unstable
Glu plus HO$^{\bullet}$ in presence of O_2	Glutamic acid hydroperoxides	No, unstable
	4-Hydroxyglutamic acid	Possible
Leu plus HO$^{\bullet}$ in presence of O_2	4- and 5-Hydroperoxyleucines	No, unstable species
	4-Hydroxyleucine	Possible, but co-elutes with other products
	5-Hydroxyleucine	Yes
	α-Ketoisocaproic acid, Isovaleric acid, Isovaleraldehyde, Isovaleraldehyde oxime	Can be generated by other reactions
Val plus HO$^{\bullet}$ in presence of O_2	3- and 4-Hydroperoxyvalines	No, unstable species
	3-Hydroxyvaline	Yes
	4-Hydroxyvaline	Possible, but coelutes with other products
Lys plus HO$^{\bullet}$ in presence of O_2	Lysine hydroperoxides	No, unstable species
	3-Hydroxylysine	Yes
	4- and 5-Hydroxylysines	No, can be generated enzymatically
Pro plus HO$^{\bullet}$ in presence of O_2	Proline hydroperoxides	No, unstable species
	3-Hydroxyproline	Possible
	4-Hydroxyproline	No, can be generated enzymatically
	5-hydroxy-2-aminovaleric acid	Yes, but also from other amino acids
Arg plus HO$^{\bullet}$ in presence of O_2	5-hydroxy-2-aminovaleric acid	Yes, but also from other amino acids
Ile plus HO$^{\bullet}$ in presence of O_2	Isoleucine hydroperoxides	No, unstable species
	Hydroxyisoleucines	Possible, but not fully characterized
Gly: hydrogen atom abstraction from α-carbon followed by reaction with $CO_2^{\bullet-}$ radicals	Aminomalonic acid	Yes, but may arise via successive oxidations of other amino acids (e.g., Ser)
Met plus HO$^{\bullet}$ or one-electron oxidation	Methionine sulfoxide	Yes, but can be enzymatically reduced, and levels may be misleading
Cys plus HO$^{\bullet}$, or other hydrogen atom abstracting species	Cystine, Oxy acids	No, natural product Possible

When HO$^{\bullet}$ attack is indicated, this is not intended to discriminate between alternative HO$^{\bullet}$ generating mechanisms (e.g., metal-ion dependent and independent). Some of the products generated by HO$^{\bullet}$ are also likely to arise from alkoxyl and peroxyl radical reactions. Reactive nitrogen species refers to NO$^{\bullet}$, peroxynitrite, the peroxynitrite-carbon dioxide adduct, and products of interaction of hypochlorite and peroxidase/hydrogen peroxide systems with inorganic nitrite. For further details and selected references see text.

any use as a marker due to the natural abundance of the C-4 and C-5 materials [30]. Little is known about the oxidation products of Glu, Asp, Gln, and Asn, though it is likely that alcohols and carbonyl-containing materials are formed [5,8]. In the case of Glu, a novel fragmentation reaction that results in the loss of the side chain carboxyl group has been reported [31]; this may result in the generation of materials that are unique to the oxidation of this side chain.

Oxidation of Arg has been reported to give 5-hydroxy-2-aminovaleric acid [18,32] and glutamic semialdehyde [33,34] as products; the latter may arise via further oxidation of the former, or directly via peroxyl radical termination reactions. 5-Hydroxy-2-aminovaleric acid is formed via initial hydrogen abstraction at C-6, and subsequent loss of the guanidine group in the presence of O_2, and is a useful marker of protein oxidation [18,32], whereas glutamic semialdehyde can undergo further reaction [34]. Whether these species are also formed by other oxidants is not known.

Most of the oxidation products arising from oxidation of the sulfur-containing amino acids, Cys, Met are likely to be poor markers of oxidative damage in vivo, even though these residues are very readily oxidized and, hence, ought to be sensitive markers of oxidation [13, 35]. The major products arising from these materials are disulfides (either homo- or heterodimers) and oxyacids from Cys (via the formation of an initial thiyl radical; [35]) and the sulfoxide (and to a much lesser extent the sulfone) from Met [36,37]. Assaying disulfide formation is difficult, if not impossible, due to the presence of cystine and enzymes that readily remove and isomerize disulfides. Quantification of the yield of oxyacids during Cys oxidation is also problematic, as the formation of this material depends critically on the concentration of both O_2 and thiol, as reaction of the thiyl radical with both materials are equilibria [35]. The yield of oxyacids, therefore, depends critically on additional factors in addition to the overall radical flux, but it may be possible to use this material as a marker under certain circumstances. Oxidation of Met through to the sulfoxide occurs with both reactive radicals and a number of other oxidants including HOCl, peroxynitrite, and $^1O_2$ [5,37–39]. It can also be generated via molecular (nonradical) reaction with H_2O_2 and alkyl (and lipid) hydroperoxides [37,40,41]. The formation of this species has been used as a tool for examining oxidant stress [42,43], though it is now well established [37,44] that enzymatic processes can repair this lesion (or at least the L stereoisomer [45]). The efficiency of this repair mechanism in mammalian cells, which contain this enzyme [46], has yet to be completely determined. Literature reports suggest that this activity may have a marked effect on the levels of this lesion and, hence, may give misleading data on the extent of oxidation, particularly in chronic low-level oxidant stress, where repair may compete effectively with formation.

Aromatic side-chains

The ease of oxidation of most of the aromatic side chains (Phe, Tyr, Trp, and His) makes these potentially very sensitive markers of oxidation. The chemistry and products of oxidation of Phe, Tyr, and Trp are reasonably well defined; those of His are poorly understood. The major sites of attack in all cases are the aromatic rings and result in ring oxygenation; with Trp this also results in ring cleavage. Thus, Phe gives rise to the 2-, 3-, and 4-hydroxylated amino acids [47]; the former pair (*o*- and *m*-Tyr) are valuable, and seemingly stable, markers [26, 27,48]. The 4-hydroxylated material is Tyr and, hence, this cannot be utilized as a marker. These species can also be generated as a result of direct oxidation of the aromatic ring by powerful one-electron oxidants (i.e., via radical-cation formation and subsequent reaction with water). Thus, the formation of hydroxylated materials is not specific for HO• [5].

Oxidation of Tyr yields mainly dihydroxylated materials, chiefly 3,4-dihydroxyphenylalanine (DOPA) though some of the 2,4-isomer is also generated, and the dimeric material di- (or bi-)tyrosine (di-Tyr) [8,13,16, 48–50]. Di-Tyr is often used as a generic name for both the carbon–carbon dimer and the carbon–oxygen linked material; the former appears to be the major species formed. DOPA and di-Tyr are formed in both the presence and absence of O_2, as they can be generated either via peroxyl radical chemistry or disproportionation reactions [8,13], and, therefore, are not markers of O_2-dependent reactions. Di-Tyr arises via dimerization of phenoxyl radicals, and the yield of this product depends on the radical flux. The absolute levels of this compound, therefore, depend on both the overall extent of insult and the rate of radical formation, and this needs to borne in mind when interpreting data. DOPA, being a catechol, is prone to further oxidation and this can result in the generation of both quinone and cyclized products [49, 51]. This susceptibility to further oxidation, which is particularly marked in the presence of redox active metal ions, can result in further radical formation and damage to other molecules [52]. DOPA quinone is also prone to Michael reactions with nucleophiles such as thiols, and this can result in the formation of adduct species such as 5-cysteinylDOPA [53,54]. All of these further reactions may give rise to misleading data with regard to the true levels of DOPA formed.

Di-Tyr can also be generated by other oxidants; thus peroxidase (such as myeloperoxidase)/H_2O_2 systems can give di-Tyr [50]. The myeloperoxidase/H_2O_2 system in the presence of Cl^- also gives other products such as 3-chloroTyr and 3,5-dichloroTyr [55–57], which appear to be specific markers of chlorinating systems, though the exact nature of the attacking species (chloramines, HOCl, and Cl_2 have all been proposed) is still the subject of debate [55,57,58]. Reaction of HOCl (and MPO systems) with free Tyr also generates a further product, *p*-hydroxyphenylacetaldehyde, which arises from initial reaction at the α-amino function [57,59]. This material is not generated from Tyr residues in proteins [59]. It has been recently shown that 3-bromoTyr is a major product of protein oxidation by eosinophil peroxidase in the presence of H_2O_2 and Br^- [60]. The observation that both 3-chloroTyr and 3-bromoTyr can undergo further reactions to give 3,5-dichloroTyr and 3,5-dibromoTyr respectively [57,60], albeit at relatively low levels, suggests that care must be taken in using the levels of 3-chloroTyr/3-bromoTyr alone as a quantitative markers of HOCl/MPO activity or eosinophil peroxidase activities, respectively.

Reaction of peroxynitrite with Tyr residues gives 3-nitroTyr and di-Tyr, as well as hydroxylated ring products [61–63]. 3-NitroTyr appears to be a specific marker of reactive nitrogen species, though it should be noted that this species can undergo further reaction with excess oxidant to give 3,5-dinitroTyr [64], so care should be taken in the interpretation of measurements of 3-nitroTyr alone. 3-NitroTyr is not specific for peroxynitrite, as the peroxynitrite-bicarbonate adduct can also generate this species, as can other reactive species such as $NO^•$ (via reaction with phenoxyl radicals), $NO_2^•$ and NO_2^+, whose formation may not require peroxynitrite (e.g., via NO_2^- plus peroxidases or NO_2Cl) [63,65,66]. Recent studies have shown that the detection of 3-nitroTyr by GC/MS may be facilitated via reduction to 3-aminoTyr [67].

Oxidation of Trp can be very rapid, due to the low oxidation potential of this compound, and results in the formation of a wide range of materials including ring hydroxylated compounds (at the 2-, 4-, 5-, 6-, or 7-positions) and ring opened materials such as N-formylkynurenine, 3-hydroxykynurenine, kynurenine, and further oxidation products [68–70]. Some of these reactions occur via the initial formation of a neutral indolyl radical (itself formed via the ring radical-cation, or addition/elimination reactions [68,70]), and subsequent reaction with O_2. The complexity of these products, and their propensity for undergoing further reaction, make these materials difficult to use as quantitative markers. Furthermore, a number of these materials can be generated via enzymatic reactions (e.g., the indoleamine pathway [71]) and such processes may give misleading quantitative measurements. Of the compounds identified to date, the ring hydroxylated materials are probably the most reliable indicators of oxidant-mediated Trp degradation (and have been used as markers [72]), though these represent only a small proportion of the total Trp lost. The loss of the parent amino acid itself has often been suggested, and used, as a marker of protein oxidation as this can be readily examined, at least in vitro (e.g. [73]). Other oxidants also generate Trp oxidation products, though many of these processes have not been fully elucidated. Thus, $^1O_2$, direct photo-oxidation, HOCl, peroxynitrite, and $SO_3^{•-}$ also oxidize the indole ring [12,38,74–77]. The products of these reactions need to be studied further before they can be used as markers.

The oxidation products of His are poorly characterized [78]. A number of materials have been isolated by HPLC as a result of metal-ion catalyzed or other oxidation processes, and some have been characterized; these include asparagine, aspartic acid, and 2-oxo-histidine [79–83]. The last of these is the only one which has been used as a marker of His oxidation [80–83]. It is unclear whether this product is also generated by other oxidants, and whether it is specific for oxyradical chemistry; the former appears more likely as this species may be a result of ring oxidation and hydration reactions (i.e., might be generated by any potent oxidizing agent). The products of reaction of other oxidants, such as HOCl, peroxynitrite, and $^1O_2$ with His are poorly characterized, though it has been suggested that each reacts with this residue [76,84,85].

METHODOLOGY FOR THE MEASUREMENT OF SIDE CHAIN OXIDATION PRODUCTS

Protein carbonyls as generic markers can be assayed by use of a number of different methods; these have been extensively reviewed elsewhere and will not be covered further here (e.g., [2]). Three major methods have been developed for the measurement of specific oxidation products. These include measurement by immunologic methods on tissue or isolated protein samples, GC/MS, and HPLC with various different detectors. The first of these methods has only been used with a few oxidation products—chiefly carbonyl groups (e.g., [2]), 3-nitroTyr (e.g. [63]), and HOCl-modified proteins [86]; the exact epitope in the latter case is unknown. Though the antibodies used in these studies can be specific for the oxidation products under study, the quantification of tissue staining/binding obtained using this methodology is more difficult to achieve. This method also does not readily lend itself to the determination of multiple products present in a single sample, unlike either the GC or HPLC methods. Such measurements are of great use in that they can both provide measurement of the parent amino acid (and, hence, allow differences in the total protein present to be readily compensated for) and also allow some information to be obtained as to the nature of the oxidative process(es) that has generated the observed species (e.g. [26]).

Both the HPLC and GC/MS methods require the isolation, purification, and hydrolysis of the proteins under study before analysis of levels of particular oxidized side-chain lesions. Developments in mass spectrometric techniques should ultimately allow the measurement of these specific lesions on intact proteins; at present, protein samples require digestion before analysis can be carried out (e.g., [72,87]). As this is usually done enzymatically (e.g., using trypsin), this may have major

Table 2. Preparation of Biologic Samples for Analysis of Oxidized Amino-acid Side-chains

Plasma and plasma LDL	Atherosclerotic plaques	Brain tissue	Cataract lens
Plasma prepared by centrifugation of freshly collected human blood. Plasma LDL prepared by density-gradient ultra-centrifugation of plasma.	Samples collected in Chelex-treated and argon-flushed phosphate-buffered saline (PBS) containing 100-μM butylated hydroxy toluene (BHT) and 1-mM ethylenediamine-tetraacetate (EDTA)	Samples are collected in Chelex-treated and argon-flushed PBS containing 100-μM BHT and 1-mM EDTA.	Lenses stored at $-20°C$ immediately after surgical removal.
	↓		
	Intimas dissected from adventitia and media.		Cortices and nuclei dissected after thawing, and powdered in liquid nitrogen.
↓	↓		↓
	Material homogenised at 4°C in collection buffer.		After freeze-drying, powder (~4 mg) transfered into a 1mL brown glass autosampler vial (Alltech).
↓	↓	↓	↓

Samples treated with sodium borohydride (1 mg/mL). Homogenates or plasma samples (700 μL, protein concentration 1-10 mg per mL) transfered into a 1 mL brown glass autosampler vial (Alltech) and mixed with 0.3% sodium deoxycholate (50 μL) and 50% (w/v) trichloroacetic acid (100 μL). After centrifugation, the protein pellets are resuspended and washed twice with cold acetone and once with diethyl ether and freeze-dried.

↓

Vials placed in a Pico-Tag reaction vessel (Alltech, Baulkham Hills, New South Wales, Australia), containing 1 mL of 6-M HCl, 1% (v/v) phenol and 50 μL mercaptoacetic acid. After evacuation of air, the reaction vessels are incubated at 110°C for 16 h. Hydrolysate subsequently freeze-dried and redissolved in water for amino acid analysis.

advantages over the high temperature acid hydrolysis techniques currently used in many GC/MS and HPLC methods.

The methods used in the authors' laboratories for isolating oxidized proteinaceous materials are outlined in Table 2 [17,19,26,27,57,88,89]. Care must be taken in the storage, handling, and processing of the samples to avoid artifactual oxidation of the samples. The extent of further oxidation induced by hydroperoxides present in the samples can be minimized via the use of a reductive step (usually involving sodium borohydride) early in the isolation and purification procedure [17,19,88]. Effective delipidation of the samples is a prerequisite for successful protein hydrolysis in the final stage of the processing; the presence of lipid at this stage results in frothing of the sample in the hydrolysis vial and consequent extensive sample loss. Incomplete recovery of protein during these (necessary) steps can be compensated for by measurement of both the levels of the oxidative lesion and the parent amino acid from which it is generated, and expressing the concentration of the oxidized material with respect to the parent; this assumes, of course, that both the parent and oxidized material are lost equally, but there is no evidence to date that this assumption is incorrect.

The methods used to quantify the oxidized side chains are outlined in Table 3 [17,19,26,27,57,88,89]. These utilize HPLC separation and a number of different de-tection methods depending on the oxidized molecules under study; these are primarily fluorescence or electro-chemical in nature due to the sensitivity required. Absolute identification of the materials present in the various HPLC peaks can be obtained by fraction collection and mass spectroscopy (e.g., [27]). Concurrent UV/visible measurements are used, in most cases, to measure levels of the parent amino acid in the samples undergoing analysis in order to allow expression of the levels of the oxidized amino acid lesion to be made relative to the parent. Studies with isolated proteins and well-defined quantifiable radical sources (e.g. γ-irradiation) have shown that the yield of the majority of the oxidative lesions studied increases in a linear fashion with the concentration of initial attacking radical, and that this also correlates with the loss of the parent amino acid [19]; in no cases, however, has a complete balance of products formed, compared with parent lost, been achieved. The recovery of many of the oxidized stable markers, taken through such isolation and preparation procedures is good (usually > 70%) [17,19,30]. The only cases where the recovery is poor are with 4-hydroxyVal and 4-hydroxyLeu, where internal cyclization to give the γ-lactone is believed to occur under the acidic conditions used in the hydrolysis step [19]. The accuracy of determinations of both the oxidized and parent amino acids obtained using this HPLC methodology has been assessed; the interassay variation for the oxidized amino

Table 3. HPLC Quantification of Amino Acids in Protein Hydrolysates Isolated from Biologic Samples

	Derivatization	HPLC column	Mobile phase	Detector used
Parent amino acids	OPA	Zorbax ODS column (250 × 4.6 mm, 5 μm) Hewlett Packard	Gradient system 1, 1 mL/min	Fluorescence Ex 340nm Em 440nm
DOPA, o-Tyr m-Tyr }	None	Zorbax ODS column (250 × 4.6 mm, 5 μm) Hewlett Packard	Gradient system 2, 1 mL/min	Fluorescence Ex 280nm Em 320nm
Di-Tyr	None	Zorbax ODS column (250 × 4.6 mm, 5 μm) Hewlett Packard	Gradient system 2, 1 mL/min	Fluorescence Ex 280nm Em 410nm
3-ChloroTyr 3,5-DichloroTyr }	None	Zorbax ODS column (250 × 4.6 mm, 5 μm) Hewlett Packard	Gradient system 2, 1 mL/min	UV 280nm, and electrochemical at +600 mV
3-NitroTyr	None	Zorbax ODS column	Gradient system 2, 1 mL/min	UV 280nm, and electrochemical at +850 mV
3-HydroxyVal 5-HydroxyLeu }				
1st HPLC step for purification	None	LC-NH₂ column (250 × 4.6 mm, 5 μm) Supelco	20% 10-mM sodium phosphate pH 4.3 in acetonitrile, 1.5 mL/min	UV 210 nm
2nd HPLC step for quantification	OPA	Zorbax ODS column (250 × 4.6 mm, 5 μm) Hewlett Packard	Gradient system 1, 1 mL/min	Fluorescence Ex 340nm Em 440nm

Gradient system 1: solvent A, 20-mM sodium acetate pH 5.2/tetrahydrofuran/methanol (31:1:8, by vol.); solvent B, same as A (7:1:32, by vol.). 0% B to 25% B in 8 min; isocratic elution at 25% B for 5 min; then to 40% B in 10 min; then to 50% B in 2 min; isocratic elution at 50% B for 6 min, then to 100% B in 4 min; isocratic elution at 100% B for 3 min; then re-equilibration at 0% B for 12 min before next analysis.

Gradient system 2: solvent A, 100-mM sodium perchlorate in 10 mM sodium phosphate pH 2.5; solvent B, 80% methanol in water (v/v). Isocratic elution with 0% B for 12 min; then to 20% B in 8 min; further elution at 20% B for 3 min before changing to 50% B in 3 min; isocratic elution at 50% B for a further 3 min; then re-equilibration with 100% A for 10 min.

acids tested (DOPA, di-tyrosine, o- and m-Tyr, 3-hydroxyvaline, and 5-hydroxyleucine) are in the range 9–15%, whereas those of the parent amino acids (Tyr, Phe, Val, Leu) are 2–5% [27].

The use of GC/MS, and particularly the use of isotope dilution mass spectroscopy, as a method for determining some of the above oxidation products is covered in another article in this Forum (by J. Heinecke; see also [90,91]); the reader is referred to these articles for further details.

COMPARISON BETWEEN METHODOLOGIES

As indicated already, the possibility of artifactual generation of oxidized amino acids needs to be considered. A few methods of analysis can be performed on nonhydrolyzed protein samples. These include: the immunologic methods, in which specificity is of greatest concern; assay of protein-bound carbonyls; and assay of protein-bound DOPA [51]. The former is commonly used even on complex tissue samples, but in the authors' experience it is difficult to use for exact quantification. The carbonyl assay, and also that of protein-bound DOPA, are most useful in experiments with purified proteins, and the protein-bound DOPA assay has also been developed to be applicable to LDL (Armstrong et al., unpublished). However, the utility of the protein-

bound DOPA assay on complex tissue samples remains to be fully investigated. It has already been noted that the yield of fluorescence per mole of protein-DOPA varies between proteins by a factor of at least 4, but this is less than the range of variation in the color yield in the Lowry assay between different proteins, and may be acceptable.

Most of the established methods involve protein hydrolysis as a first step, and this is a key step in which oxidation can definitely occur. Artifactual formation of oxidized amino acid during such hydrolysis is probably one of the most serious possibilities, and it is necessary to quantitate this for every analyte and every different type of biological sample under study. This can be done by several complementary approaches. First, add free parent amino acids to the samples before hydrolysis, and determine their conversion to oxidized products: a disadvantage of this procedure is that the free amino acids are often more reactive than the same species within a protein. Second, add an unoxidized protein to the hydrolysis vials, in the presence or absence of sample material, and determine the extent of oxidation of the added protein. Only when the limits of artifact formation are established by these means can one securely interpret measured levels as indicative of unoxidized or oxidized samples, and determine whether the oxidation products are present under physiologic, or only under pathophysiologic, conditions. Complementary procedures to assess and control the extent of oxidation during tissue processing have been developed in our work with atherosclerotic plaque samples, utilizing the addition and subsequent quantitation of isoascorbate and tocotrienol [92].

HPLC procedures normally involve little further risk of artifact, and these can readily be controlled and assessed. On the other hand, mass spectroscopic methods frequently involve a further derivatization step that may itself generate oxidation products, as well as a thermal volatilization step, which can also be chemically damaging. A common approach to this problem is to add a stable isotope-labeled internal standard, whose recovery is assumed to represent the recovery of the analyte. The stable isotope standard may be chemically identical to the analyte, or it can sometimes be the parent. Particularly in the latter case it is important, but often neglected, to determine to what degree the isotopomeric form of the product is generated artifactually, as an index of the effect of the mass spectroscopic manipulations. The separate determination of the artifactual impact of the mass spectroscopic stages is critical, as this seems to have lead to some overestimated values for oxidized amino acids in published literature.

A number of products have been measured using both GC and HPLC techniques, which has allowed some crossreferencing of the values obtained using these two different methods. Unfortunately there has been little direct comparison between the different methods using identical samples; it is, therefore, difficult to provide absolute reference values for these specific markers in any tissues apart from fresh human plasma. Some of the values reported in early studies, particularly those for *o*- and *m*-Tyr in fresh human plasma detected by GC/MS, are higher than those measured by HPLC; this may arise from artifactual oxidation in the preparation/handling of the samples in the GC/MS experiments. In other cases, however, comparable values have been obtained. Thus, measurements of 3-chloroTyr in atherosclerotic plaque samples appear to be consistent between different experimental techniques (GC/MS vs HPLC; [26,93]), as are the levels of *o*- and *m*-Tyr in normal arterial tissue [26,93].

DETECTION OF PROTEIN SIDE-CHAIN OXIDATION PRODUCTS IN HUMANS

A number of the above oxidation products of amino acid side chains in proteins have been used as markers of oxidative damage to proteins in vivo; selected examples of such usage are given in Table 4, together with the levels detected in tissue or fluids from "normal" subjects as well as from a number of pathologic conditions. As can be readily seen from Table 4, most of the conditions

under which elevated levels of protein oxidation products have been detected to date are chronic in nature, with the highest levels of oxidized products seen in tissues/proteins that would be predicted, or are known to have slow rates of protein turnover (i.e., long half-lives), and, hence, the greatest likelihood of accumulating sufficient damage for ready detection. A few examples of elevated levels of oxidized amino acid side-chains detected in scenarios where acute damage is known or is likely to occur have also recently been reported.

CONFOUNDING FACTORS IN THE MEASUREMENT OF PROTEIN SIDE-CHAIN OXIDATION PRODUCTS

Little is known regarding the presence of the above mentioned amino acid side chain oxidation products in food stuffs. Therefore, it is difficult to discern what role dietary intake may play in determining the basal levels of oxidized amino acids detected in the studies undertaken to date. Such data is of major importance in determining what proportion of the materials detected in "normal" subjects arise from endogenous metabolic processes and what is from dietary intake. Little information is also available regarding the uptake and handling of these materials by the body, though it is reasonably well established at least under some circumstances, that oxidation of proteins in vivo can result in enhanced catabolism relative to the unoxidized parent (reviewed in [6,25,94–96]). This is, however, by no means universal, and a number of studies have shown that heavily oxidized materials are catabolised at slower rates than the unoxidized parent [97,98]. The subsequent tissue distribution, metabolism (see also above, for comments about further oxidation), and excretion of the oxidized free amino acids arising from either catabolism of oxidized proteins, or as a result of dietary intake, have not been studied in depth [99–101]. It has been established in cell culture experiments that oxidized amino acids can be taken up via normal amino acid transport systems (Rodgers et al., unpublished), and that this is a competitive process (i.e., high levels of oxidized materials can limit the uptake of normal amino acids using the same transport system). It is, therefore, possible that high levels of these oxidized materials could have metabolic effects, but under most conditions it would be predicted that the removal of the oxidized materials would limit any possible down regulation. It has also been shown in cell culture studies that there can be reincorporation of oxidized amino acids into newly synthesized proteins (Fu et al., unpublished), though this is inefficient in the presence of the parent (unoxidized) amino acid. The significance of these processes remains to be established, though this approach is a promising method for the study of the cellular turnover

Table 4. Oxidized Amino Acids in Normal and Pathologic Samples of Human Tissues and Fluids

Product	Physiologic levels	Pathologic levels	Source
DOPA	85-pmol/mg LDL protein (6/10,000 Tyr). Similar levels in total plasma protein		[24,25]
	9-pmol/mg wet weight intimal tissue in normal human artery.	14-pmol/mg wet weight intimal tissue (ca. 410-pmol/mg protein, 14/10,000 Tyr) in advanced human atherosclerotic plaques.	[24]
	51/10,000 Tyr in normal human lens	Stage dependent increase in levels with severity of human cataract. 598/10,000 Tyr in type IV cataract	[25]
o-, m-Tyr	62- and 35-pmol/mg LDL protein (5 and 3/10,000 Phe) respectively. Similar levels in total plasma protein.		[24,25]
	0.3 and 4/10,000 Phe respectively in fresh human plasma proteins.		[103]
	6-pmol o-Tyr and 1-pmol m-Tyr/mg wet weight intimal tissue in normal human artery.	105- and 175-pmol/mg wet weight intimal tissue (105 and 175-pmol/mg protein, 3.5 and 6/10,000 Phe respectively), in advanced human atherosclerotic plaques.	[24]
		No increase in atherosclerotic aorta samples compared with normal.	[93]
	5/10,000 Phe in human lenses of any age.	Unchanged with age.	[104]
	18 and 15/10,000 Phe respectively in normal human lenses of any age.	Stage dependent increase in levels with severity of human cataract. 216 and 143/10,000 Phe respectively in type IV cataract.	[25]
Di-tyrosine	0.2-pmol/mg LDL protein (0.02/10,000 Tyr). Similar levels in total plasma protein		[24,25]
	0.6 pmol/mg wet weight intimal tissue in normal human artery.	4.7-pmol/mg wet weight intimal tissue (150 pmol/mg protein, 5/10,000 Tyr) in advanced human atherosclerotic plaques.	[24]
		10-fold elevated in aortic lesions in comparison with normal aortic samples.	[93]
	0.01/10,000 Tyr in human lens proteins from young people.	0.03/10,000 Tyr in lens proteins from old people.	[104]
	0.6/10,000 Tyr in normal human lenses.	Stage dependent increase in levels with severity of human cataract. 6/10,000 Tyr in type IV cataract.	[25]
5-Hydroxy Leu	7-pmol/mg LDL protein (0.1/10,000 Leu). 14-pmol/mg total plasma protein (0.1/10,000 Leu).		[24,25]
	0.6-pmol/mg wet weight intimal tissue in normal human artery.	4-pmol/mg wet weight intimal tissue (2-20-pmol/mg protein, 0.2/10,000 Leu) in advanced human atherosclerotic plaques	[24]
	8-pmol/mg freeze-dried protein (2/10,000 Leu) in normal human lenses of any age.	Stage dependent increase in levels with severity of human cataract. 96 pmol/mg freeze dried protein (18/10,000 Leu) in type IV cataract.	[19,25]
3-Hydroxy Val	1.5-pmol/mg LDL protein (0.1/10,000 Val). Similar levels in total plasma protein		[24,25]
	1.0-pmol/mg wet weight intimal tissue in normal human artery.	2-pmol/mg wet weight intimal tissue (1-10-pmol/mg protein, 0.1/10,000 Val) in advanced human atherosclerotic plaques.	[24]
	5/10,000 Val in normal human lenses of any age.	Stage dependent increase in levels with severity of human cataract. 20/10,000 Val in type IV cataract	[25]
3-ChloroTyr	0.8/10,000 Tyr in normal human aorta.	4/10,000 Tyr in atherosclerotic aorta.	[93]
	Below detection limit (< 4/10,000 Tyr) in normal human lenses of any age.	Below detection limit (< 4/10,000 Tyr) in Type IV human cataract	[25]
3-NitroTyr	< 10-pmol/mg LDL protein (< 1/10,000 Tyr)	< 10-pmol/mg protein (< 4/10,000 Tyr) in advanced human atherosclerotic plaques.	[24]
		100-fold elevated in aortic lesion LDL in comparison with normal plasma LDL (8/10,000 Tyr vs 0.09/10,000 Tyr).	[93]
	Below detection limit (< 4/10,000 Tyr) in normal human lenses of any age.	Below detection limit (< 4/10,000 Tyr) in Type IV human cataract	[25]
Aminomalonic acid	0.04–0.3/1000 total amino acids in two Escherichia coli strains.	0.2/1000 Gly in post-mortem human plaque	[105]
5-Hydroxy-2-aminovaleric acid	150-pmol/mg protein in 100,000 g supernatants from young mouse livers.	Unchanged in old mouse livers. Elevated by hyperoxic exposure.	[32]
Protein carbonyls	Approx. 1-nmol/mg protein in many physiological tissue samples	Up to 8-nmol/mg protein in diseased brain samples	[2,106]

of oxidized proteins containing specific and well-defined levels of oxidized amino acids.

An important interpretive issue is whether proportional increments or factors of change are both informative when comparing tissue levels of oxidation products in different physiologic or pathologic conditions. For example, if muscle levels of an analyte per parent increase by 500 pmol per mole and this represents a 2-fold increment, how significant is a change in another analyte of +50 pmol per mole, which corresponds to a 10-fold increment?

The limited data available to date from both human and animal studies are consistent with the rapid excretion of oxidized amino acids via urine [100,101]. Thus, oxidized amino acids, presumably resulting from in vivo generation of oxidized proteins, have been detected in the urine of exercised or aging rats and the measurement of these materials has been suggested as a noninvasive methods of measuring oxidative stress in vivo [100,101]. The trends in the levels of these materials detected in rat urine have been reported to mirror those detected in skeletal muscle [100,101] suggesting that this might be a valid approach, though it is unclear whether other factors might confound such measurements. Much remains to be determined about the validity of such measurements, about what markers most accurately reflect tissue damage, and which are least affected by metabolic or clearance pathways. One interesting observation is that in fresh human plasma LDL DOPA levels are roughly 500-μmol/mol Tyr, while cholesterol-18:2-OOH:cholesterol-18:2 is roughly 0.4-μmol/mol. Even in HDL, in which the majority of normal plasma cholesterol ester hydroperoxides reside, the cholesterol-18:2-OOH:cholesterol-18:2 is only 11 μmol/mol [102]. That a reactive product of protein oxidation is present at higher substitution levels than one of lipid peroxidation, even though lipids are often thought to be more readily peroxidized than proteins, indicates that turnover and/or transport may dictate the levels observed in tissue samples, and highlights the need for caution in interpreting experimental data.

Acknowledgements — Work in the authors' laboratories is funded in part by the Australian Research Council, the National Health and Medical Research Council, the Wellcome Trust, the Clive and Vera Ramaciotti Foundations, the Diabetes Australia Research Trust, and the Juvenile Diabetes Foundation International.

REFERENCES

[1] Stadtman, E. R. Protein modification in aging. *J. Gerontol.* **43:**B112–B120; 1988.

[2] Levine, R. L.; Williams, J. A.; Stadtman, E. R.; Shacter, E. Carbonyl assays for determination of oxidatively modified proteins. *Meth. Enzymol.* **233:**346–357; 1994.

[3] Stadtman, E. R.; Berlett, B. S. Reactive oxygen-mediated protein oxidation in aging and disease. *Drug Metab. Rev.* **30:**225–243; 1998.

[4] Gebicki, J. M. Protein hydroperoxides as new reactive oxygen species. *Redox Report* **3:**99–110; 1997.

[5] Davies, M. J.; Dean, R. T. *Radical-mediated protein oxidation: from chemistry to medicine.* Oxford: Oxford University Press; 1997.

[6] Dean, R. T.; Fu, S.; Stocker, R.; Davies, M. J. The biochemistry and pathology of radical-mediated protein oxidation. *Biochem. J.* **324:**1–18; 1997.

[7] Davies, K. J.; Delsignore, M. E.; Lin, S. W. Protein damage and degradation by oxygen radicals. II. Modification of amino acids. *J. Biol. Chem.* **262:**9902–9907; 1987.

[8] Garrison, W. M. Reaction mechanisms in the radiolysis of peptides, polypeptides, and proteins. *Chem. Rev.* **87:**381–398; 1987.

[9] Stadtman, E. R.; Berlett, B. S. Fenton chemistry revisited: amino acid oxidation. *Basic Life Sci.* **49:**131–136; 1988.

[10] Stadtman, E. R. Metal ion-catalyzed oxidation of proteins: biochemical mechanism and biological consequences. *Free Radic. Biol. Med.* **9:**315–325; 1990.

[11] Stadtman, E. R. Role of oxidized amino acids in protein breakdown and stability. *Meth. Enzymol* **258:**379–393; 1995.

[12] Berlett, B. S.; Stadtman, E. R. Protein oxidation in aging, disease, and oxidative stress. *J. Biol. Chem.* **272:**20313–20316; 1997.

[13] von Sonntag, C. The chemical basis of radiation biology. London: Taylor and Francis; 1987.

[14] Trelstad, R. L.; Lawley, K. R.; Holmes, L. B. Nonenzymatic hydroxylations of proline and lysine by reduced oxygen derivatives. *Nature* **289:**310–322; 1981.

[15] Stadtman, E. R.; Berlett, B. S. Fenton chemistry. Amino acid oxidation. *J. Biol. Chem.* **266:**17201–17211; 1991.

[16] Simpson, J. A.; Narita, S.; Gieseg, S.; Gebicki, S.; Gebicki, J. M.; Dean, R. T. Long-lived reactive species on free-radical-damaged proteins. *Biochem. J.* **282:**621–624; 1992.

[17] Fu, S.; Hick, L. A.; Sheil, M. M.; Dean, R. T. Structural identification of valine hydroperoxides and hydroxides on radical-damaged amino acid, peptide, and protein molecules. *Free Radic. Biol. Med.* **19:**281–292; 1995.

[18] Ayala, A.; Cutler, R. G. The utilization of 5-hydroxyl-2-amino valeric acid as a specific marker of oxidized arginine and proline residues in proteins. *Free Radic. Biol. Med.* **21:**65–80; 1996.

[19] Fu, S. L.; Dean, R. T. Structural characterization of the products of hydroxyl-radical damage to leucine and their detection on proteins. *Biochem. J.* **324:**41–48; 1997.

[20] Kopoldova, J.; Liebster, J.; Babicky, A. The mechanism of the radiation chemical degradation of amino acids–V. Radiolysis of norleucine, leucine and isoleucine in aqueous solution. *Int. J. Appl. Radiat. Isot.* **14:**493–498; 1963.

[21] Kopoldova, J.; Liebster, J.; Babicky, A. The mechanism of the radiation chemical degradation of amino acids–IV. Radiolysis of valine in aqueous oxygenated and oxygen-free solutions. *Int. J. Appl. Radiat. Isot.* **14:**489–492; 1963.

[22] Gebicki, S.; Gebicki, J. M. Formation of peroxides in amino acids and proteins exposed to oxygen free radicals. *Biochem. J.* **289:**743–749; 1993.

[23] Fu, S.; Gebicki, S.; Jessup, W.; Gebicki, J. M.; Dean, R. T. Biological fate of amino acid, peptide and protein hydroperoxides. *Biochem. J.* **311:**821–827; 1995.

[24] Fu, S.; Davies, M. J.; Stocker, R.; Dean, R. T. Evidence for roles of radicals in protein oxidation in advanced human atherosclerotic plaque. *Biochem. J.* **333:**519–525; 1998.

[25] Fu, S.; Dean, R.; Southan, M.; Truscott, R. The hydroxyl radical in lens nuclear cataractogenesis. *J. Biol. Chem.* **273:**28603–28609; 1998.

[26] Stadtman, E. R. Oxidation of free amino acids and amino acid residues in proteins by radiolysis and by metal-catalysed reactions. *Annu. Rev. Biochem.* **62:**797–821; 1993.

[27] Stadtman, E. R. Metal ion-catalyzed oxidation of proteins: biochemical mechanism and biological consequences. *Free Radic. Biol. and Med.* **9:**315–325; 1990.

[28] Taborsky, G. Oxidative modification of proteins in the presence of ferrous ion and air. Effect of ionic constituents of the reaction medium on the nature of the oxidation products. *Biochemistry* **12:**1341–1348; 1973.

[29] Hammer, T.; Bode, R. Enzymatic production of α-aminoadipate-δ-semialdehyde and related compounds by lysine ε-dehydrogenase from *Candida albicans*. *Zentralbl. Microbiol.* **147:**65–70; 1992.

[30] Morin, B.; Bubb, W. A.; Davies, M. J.; Dean, R. T.; Fu, S. 3-Hydroxylysine, a potential marker for studying radical-induced protein oxidation. *Chem. Res. Toxicol.* **11:**1265–1273; 1998.

[31] Davies, M. J.; Fu, S.; Dean, R. T. Protein hydroperoxides can give rise to reactive free radicals. *Biochem. J.* **305:**643–649; 1995.

[32] Ayala, A.; Cutler, R. G. Comparison of 5-hydroxy-2-aminovaleric acid with carbonyl group content as a marker of oxidized protein in human and mouse liver tissues. *Free Radic. Biol. Med.* **21:**551–558; 1996.

[33] Amici, A.; Levine, R. L.; Tsai, L.; Stadtman, E. R. Conversion of amino acid residues in proteins and amino acid homopolymers to carbonyl derivatives by metal-catalyzed oxidation reactions. *J. Biol. Chem.* **264:**3341–3346; 1989.

[34] Climent, I.; Tsai, L.; Levine, R. L. Derivatization of gamma-glutamyl semialdehyde residues in oxidized proteins by fluoresceinamine. *Anal. Biochem.* **182:**226–232; 1989.

[35] Wardman, P.; von Sonntag, C. Kinetic factors that control the fate of thiyl radicals in cells. *Meth. Enzymol.* **251:**31–45; 1995.

[36] Swallow, A. J. *Radiation chemistry of organic compounds.* London: Pergamon; 1960.

[37] Brot, N.; Weissbach, H. Biochemistry and physiological role of methionine sulfoxide residues in proteins. *Arch. Biochem. Biophys.* **223:**271–281; 1983.

[38] Bensasson, R. V.; Land, E. J.; Truscott, T. G. *Excited states and free radicals in biology and medicine.* Oxford: Oxford University Press; 1993.

[39] Levine, R. L.; Mosoni, L.; Berlett, B. S.; Stadtman, E. Methionine residues as endogenous antioxidants in proteins. *Proc. Natl. Acad. Sci. USA* **93:**15036–15040; 1996.

[40] Yao, Y.; Yin, D.; Jas, G. S.; Kuczer, K.; Williams, T. D.; Schoneich, C.; Squier, T. C. Oxidative modification of a carboxyl-terminal vicinal methionine in calmodulin by hydrogen peroxide inhibits calmodulin-dependent activation of the plasma membrane Ca-ATPase. *Biochemistry* **35:**2767–2787; 1996.

[41] Garner, B.; Waldeck, A. R.; Witting, P. K.; Rye, K.-A.; Stocker, R. Oxidation of high density lipoproteins. II. Evidence for direct reduction of lipid hydroperoxides by methionine residues of apolipoproteins AI and AII. *J. Biol. Chem.* **273:**6088–6095; 1998.

[42] Moskovitz, J.; Rahman, M. A.; Strassman, J.; Yancey, S. O.; Kushner, S. R.; Brot, N.; Weissbach, H. *Escherichia coli* peptide methionine sulfoxide reductase gene: regulation of expression and role in protecting against oxidative damage. *J. Bacteriol.* **177:**502–507; 1995.

[43] Moskovitz, J.; Berlett, B. S.; Poston, J. M.; Stadtman, E. R. Methionine sulfoxide reductase in antioxidant defense. *Meth. Enzymol.* **300:**239–244; 1999.

[44] Brot, N.; Weissbach, L.; Werth, J.; Weissbach, H. Enzymatic reduction of protein-bound methionine sulfoxide. *Proc. Natl. Acad. Sci. USA* **78:**2155–2158; 1981.

[45] Minetti, G.; Balduini, C.; Brovelli, A. reduction of DABS-L-methionine-dl-sulfoxide by protein methionine sulfoxide reductase from polymorphonuclear leukocytes: stereospecificity towards the l-sulfoxide. *Ital. J. Biochem.* **43:**273–283; 1994.

[46] Moskovitz, J.; Jenkins, N. A.; Gilbert, D. J.; Copeland, N. G.; Jursky, F.; Weissbach, H.; Brot, N. Chromosomal localization of the mammalian peptide-methionine sulfoxide reductase gene and its differential expression in various tissues. *Proc. Natl. Acad. Sci. USA* **93:**3205–3208; 1996.

[47] Dizdaroglu, M.; Simic, M. G. Radiation induced conversion of phenylalanine to tyrosines. *Radiat. Res.* **83:**437; 1980.

[48] Huggins, T. G.; Wells-Knecht, M. C.; Detorie, N. A.; Baynes, J. W.; Thorpe, S. R. Formation of o-tyrosine and dityrosine in proteins during radiolytic and metal-catalyzed oxidation. *J. Biol. Chem.* **268:**12341–12347; 1993.

[49] Gieseg, S. P.; Simpson, J. A.; Charlton, T. S.; Duncan, M. W.; Dean, R. T. Protein-bound 3,4-dihydroxy phenylalanine is a major reductant formed during hydroxyl radical damage to proteins. *Biochemistry* **32:**4780–4786; 1993.

[50] Heinecke, J. W.; Li, W.; Daehnke, H. D.; Goldstein, J. A. Dityrosine, a specific marker of oxidation, is synthesized by the myeloperoxidase-hydrogen peroxide system of human neutrophils and macrophages. *J. Biol. Chem.* **268:**4069–4077; 1993.

[51] Armstrong, S. G.; Dean, R. T. A sensitive fluorometric assay for protein-bound DOPA and related products of radical-mediated protein oxidation. *Redox Report* **1:**291–298; 1995.

[52] Morin, B.; Davies, M. J.; Dean, R. T. The protein oxidation product 3,4-dihydroxyphenylalanine (DOPA) mediates oxidative DNA damage. *Biochem. J.* **330:**1059–1067; 1998.

[53] Ito, S.; Kato, T.; Shinpo, K., Fujita, K. Oxidation of tyrosine residues by tyrosinase—formation of protein-bound 3,4-dihydroxyphenylalanine and 5-S-cysteinyl-3,4-dihydroxyphenylalanine. *Biochem. J.* **222:**407–411; 1984.

[54] Ito, S.; Kato, T.; Fujita, K. Covalent binding of catechols to protein through the sulphydyrl group. *Biochem. Pharmacol.* **37:**1707–1710; 1988.

[55] Domigan, N. M.; Charlton, T. S.; Duncan, M. W.; Winterbourn, C. C.; Kettle, A. J. Chlorination of tyrosyl residues in peptides by myeloperoxidase and human neutrophils. *J. Biol. Chem.* **270:**16542–16548; 1995.

[56] Kettle, A. J. Neutrophils convert tyrosyl residues in albumin to chlorotyrosine. *FEBS Letts* **379:**103–106; 1996.

[57] Fu, S.; Wang, H.; Davies, M. J.; Dean, R. T. Reaction of hypochlorous acid with tyrosine and peptidyl-tyrosyl residues gives dichlorinated and aldehydic products in addition to 3-chlorotyrosine. *J. Biol. Chem.* submitted; 1999.

[58] Hazen, S. L.; Hsu, F. F.; Mueller, D. M.; Crowley, J. R.; Heinecke, J. W. Human neutrophils employ chlorine gas as an oxidant during phagocytosis. *J. Clin. Invest.* **98:**1283–1289; 1996.

[59] Hazen, S. L.; Hsu, F. F.; Heinecke, J. W. p-Hydroxyphenylacetaldehyde is the major product of L-tyrosine oxidation by activated human phagocytes. A chloride-dependent mechanism for the conversion of free amino acids into reactive aldehydes by myeloperoxidase. *J. Biol. Chem.* **271:**1861–1867; 1996.

[60] Wu, W.; Chen, Y.; d'Avignon, A.; Hazen, S. L. 3-Bromotyrosine and 3,5-dibromotyrosine are major products of protein oxidation by eosinophil peroxidase: potential markers for eosinophil-dependent tissue injury in vivo. *Biochemistry* **38:**3538–3548; 1999.

[61] Ischiropoulos, H.; Zhu, L.; Chen, J.; Tsai, M.; Martin, J. C.; Smith, C. D.; Beckman, J. S. Peroxynitrite-mediated tyrosine nitration catalyzed by superoxide dismutase. *Arch. Biochem. Biophys.* **298:**431–437; 1992.

[62] van der Vliet, A., O'Neill, C. A.; Halliwell, B.; Cross, C. E.; Kaur, H. Aromatic hydroxylation and nitration of phenylalanine and tyrosine by peroxynitrite. Evidence for hydroxyl radical production from peroxynitrite. *FEBS Lett.* **339:**89–92; 1994.

[63] Ischiropoulos, H. Biological tyrosine nitration: a pathophysiological function of nitric oxide and reactive oxygen species. *Arch. Biochem. Biophys.* **356:**1–11; 1998.

[64] Yi, D.; Smythe, G. A.; Blount, B. C.; Duncan, M. W. Peroxynitrite-mediated nitration of peptides: characterization of the products by electrospray and combined gas chromatography-mass spectrometry. *Arch. Biochem. Biophys.* **344:**253–259; 1997.

[65] Kettle, A. J.; van Dalen, C. J.; Winterbourn, C. C. Peroxynitrite and myeloperoxidase leave the same footprint in protein nitration. *Redox Report* **3:**257–258; 1997.

[66] Sampson, J. B.; Ye, Y.-Z.; Rosen, H.; Beckman, J. S. Myeloperoxidase and horseradish peroxidase catalyze tyrosine nitration in proteins from nitrite and hydrogen peroxide. *Arch. Biochem. Biophys.* **356:**207–213; 1998.

[67] Crowley, J. R.; Yarasheski, K.; Leeuwenburgh, C.; Turk, J.; Heinecke, J. W. Isotope dilution mass spectrometric quantification of 3-nitrotyrosine in proteins and tissues is facilitated by reduction to 3-aminotyrosine. *Anal. Biochem.* **259:**127–135; 1998.

[68] Jovanovic, S. V., Steenken, S., Simic, M. G. Kinetics and energetics of one-electron-transfer reactions involving tryptophan neutral and cation radicals. *J. Phys. Chem.* **95:**684–687; 1991.

[69] Maskos, Z.; Rush, J. D.; Koppenol, W. H. The hydroxylation of phenylalanine and tyrosine: a comparison with salicylate and tryptophan. *Arch. Biochem. Biophys.* **296:**521–529; 1992.

[70] Josimovic, L. J.; Jankovic, I.; Jovanovic, S. V. Radiation induced decomposition of tryptophan in the presence of oxygen. *Radiat. Phys. Chem.* **41:**835–841; 1993.

[71] Christen, S.; Peterhans, E.; Stocker, R. Antioxidant activities of some tryptophan metabolites: possible implication for inflammatory diseases. *Proc. Natl. Acad. Sci. USA* **87:**2506–2510; 1990.

[72] Finley, E. L.; Dillon, J.; Crouch, R. K.; Schey, K. L. Radiolysis-induced oxidation of bovine alpha-crystallin. *Photochem. Photobiol.* **68:**9–15; 1998.

[73] Giessauf, A.; Steiner, E.; Esterbauer, H. Early destruction of tryptophan residues of apolipoprotein B is a vitamin E-independent process during copper-mediated oxidation of LDL. *Biochim. Biophys. Acta* **1256:**221–232; 1995.

[74] Weil, L. On the mechanism of the photo-oxidation of amino acids sensitized by methylene blue. *Arch. Biochem. Biophys.* **110:**57–68; 1965.

[75] Yang, S. F. Destruction of tryptophan during the aerobic oxidation of sulfite ions. *Environ. Res.* **6:**395–402; 1973.

[76] Hazell, L. J.; van den Berg, J. J.; Stocker, R. Oxidation of low-density lipoprotein by hypochlorite causes aggregation that is mediated by modification of lysine residues rather than lipid oxidation. *Biochem. J.* **302:**421–428; 1994.

[77] Ischiropoulos, H.; al-Mehdi, A. B. Peroxynitrite-mediated oxidative protein modifications. *FEBS Letts.* **364:**279–282; 1995.

[78] Kopoldova, J.; Hrneir, S. Gamma-radiolysis of aqueous solution of histidine. *Z. Naturforsch. [C]* **32c:**482–487; 1977.

[79] Dean, R. T.; Wolff, S. P.; McElligott, M. A. Histidine and proline are important sites of free radical damage to proteins. *Free Rad. Res. Commun.* **7:**97–103; 1989.

[80] Uchida, K.; Kawakishi, S. Ascorbate-mediated specific oxidation of the imidazole ring in a histidine derivative. *Bioorg. Chem.* **17:**330–343; 1989.

[81] Uchida, K.; Kawakishi, S. Selective oxidation of imidazole ring in histidine residues by the ascorbic acid-copper ion system. *Biochem. Biophys. Res. Commun.* **138:**659–665; 1986.

[82] Uchida, K.; Kawakishi, S. 2-Oxo-histidine as a novel biological marker for oxidatively modified proteins. *FEBS Letts.* **332:**208–210; 1993.

[83] Lewisch, S. A.; Levine, R. L. Determination of 2-oxohistidine by amino acid analysis. *Meth. Enzymol.* **300:**120–124; 1999.

[84] Michaeli, A.; Feitelson, J. Reactivity of singlet oxygen toward amino acids and peptides. *Photochem. Photobiol.* **59:**284–289; 1994.

[85] Ischiropoulos, H.; al Mehdi, A. B. Peroxynitrite-mediated oxidative protein modifications. *FEBS Lett.* **364:**279–282; 1995.

[86] Hazell, L. J.; Arnold, L.; Flowers, D.; Waeg, G.; Malle, E.; Stocker, R. Presence of hypochlorite-modified proteins in human atherosclerotic lesions. *J. Clin. Invest.* **97:**1535–1544; 1996.

[87] Chowdhury, S. K.; Eshraghi, J.; Wolfe, H.; Forde, D.; Hlavac, A. G.; Johnston, D. Mass spectrometric identification of amino acid transformations during oxidation of peptides and proteins: modifications of methionine and tyrosine. *Anal. Chem.* **67:**390–398; 1995.

[88] Dean, R. T.; Fu, S.; Gieseg, S.; Armstrong, S. G. Protein hydroperoxides, protein hydroxides, and protein-bound DOPA. In: Punchard, N.; Kelly, F. J., eds. *Free radicals: a practical approach.* Oxford: Oxford University Press; 1996:171–183.

[89] Fu, S.; Fu, M.-X.; Baynes, J. W.; Thorpe, S. R.; Dean, R. T. Presence of dopa and amino acid hydroperoxides in proteins modified with advanced glycation end products: amino acid oxidation products as possible source of oxidative stress induced by age proteins. *Biochem. J.* **330:**233–239; 1998.

[90] Heinecke, J. W.; Hsu, F. F.; Crowley, J. R.; Hazen, S. L.; Leeuwenburgh, C.; Mueller, D. M.; Rasmussen, J. E.; Turk, J. Detecting oxidative modification of biomolecules with isotope dilution mass spectrometry: sensitive and quantitative assays for oxidized amino acids in proteins and tissues. *Meth. Enzymol.* **300:**124–144; 1999.

[91] Heinecke, J. W. Mass spectrometric quantification of amino acid oxidation products in proteins: insights into pathways that promote LDL oxidation in the human artery wall. *FASEB J.* **13:**1113–1120; 1999.

[92] Suarna, C.; Dean, R. T.; May, J.; Stocker, R. Human atherosclerotic plaque contains both oxidized lipids and relatively large amounts of alpha-tocopherol and ascorbate. *Arterioscler. Thromb. Vasc. Biol.* **15:**1616–1624; 1995.

[93] Hazen, S. L.; Heinecke, J. W. 3-Chlorotyrosine, a specific marker of myeloperoxidase-catalysed oxidation, is markedly elevated in low density lipoprotein isolated from human atherosclerotic intima. *J. Clin. Invest.* **99:**2075–2081; 1997.

[94] Stadtman, E. R. Protein oxidation and aging. *Science* **257:**1220–1224; 1992.

[95] Dean, R. T.; Armstrong, S. G.; Fu, S.; Jessup, W. Oxidised proteins and their enzymatic proteolysis in eucaryotic cells: a critical appraisal. In: Nohl, H.; Esterbauer, H.; Rice-Evans, C., eds. *Free radicals in the environment, medicine and toxicology.* London: Richelieu Press; 1994:47–79.

[96] Grune, T.; Reinheckel, T.; Davies, K. J. Degradation of oxidized proteins in mammalian cells. *FASEB J.* **11:**526–534; 1997.

[97] Grant, A. J.; Jessup, W.; Dean, R. T. Accelerated endocytosis and incomplete catabolism of radical-damaged protein. *Biochim. Biophys. Acta* **1134:**203–209; 1992.

[98] Grant, A. J.; Jessup, W.; Dean, R. T. Inefficient degradation of oxidized regions of protein molecules. *Free Rad. Res. Commun.* **18:**259–267; 1993.

[99] Leeuwenburg C.; Hansen, P.; Shaish, A.; Holloszy, J. O.; Heinecke, J. W. Markers of protein oxidation by hydroxyl radical and reactive nitrogen species in tissues of aging rats. *Am. J. Physiol.* **274:**R453–R461; 1998.

[100] Leeuwenburgh, C.; Hansen, P. A.; Holloszy, J. O.; Heinecke, J. W. Hydroxyl radical generation during exercise increases mitochondrial protein oxidation and levels of urinary dityrosine. *Free Radic. Biol. Med.* **27:**186–192; 1999.

[101] Leeuwenburgh, C.; Hansen, P. A.; Holloszy, J. O.; Heinecke, J. W. Oxidized amino acids in the urine of aging rats: potential markers for assessing oxidative stress in vivo. *Am. J. Physiol.* **276:**R128–R135; 1999.

[102] Bowry, V. W.; Stanley, K. K.; Stocker, R. High density lipoprotein is the major carrier of lipid hydroperoxides in human blood plasma from fasting donors. *Proc. Natl. Acad. Sci. USA* **89:**10316–10320; 1992.

[103] Blount, B. C.; Duncan, M. W. Trace quantification of the oxidative damage products, meta- and ortho-tyrosine, in biological samples by gas chromatography-electron capture negative ionization mass spectrometry. *Anal. Biochem.* **244:**270–276; 1997.

[104] Wells-Knecht, M. C.; Huggins, T. G.; Dyer, D. G.; Thorpe, S. R.; Baynes, J. W. Oxidized amino acids in lens protein with age. Measurement of o-tyrosine and dityrosine in the aging human lens. *J. Biol. Chem.* **268:**12348–12352; 1993.

[105] Van Buskirk, J. J.; Kirsch, W. M.; Kleyer, D. L.; Darklen, R. M.; Koch, T. H. Aminomalonic acid: identification in *E. coli* and atherosclerotic plaque. *Proc. Natl. Acad. Sci. USA* **81:**722–725; 1984.

[106] Lyras, L.; Evans, P. J.; Shaw, P. J.; Ince, P. G.; Halliwell, B. Oxidative damage and motor neurone disease. Difficulties in the measurement of protein carbonyls in human brain tissue. *Free Rad. Res.* **24:**397–406; 1996.

ABBREVIATIONS

3-ChloroTyr—3-chlorotyrosine
Cholesterol-18:2—cholesterol linoleate
Cholesterol-18:2-OOH—cholesterol linoleate hydroperoxides
3,5-DichloroTyr—3,5-dichlorotyrosine
3,5-DinitroTyr—3,5-dinitrotyrosine
Di-Tyr—dityrosine
DOPA—3,4-dihydroxyphenylalanine
GC/MS—gas chromatography/mass spectroscopy

4-HydroxyLeu—4-hydroxyleucine
5-HydroxyLeu—5-hydroxyleucine
3-HydroxyVal—3-Hydroxyvaline
4-HydroxyVal—4-hydroxyvaline
LDL—low-density lipoproteins
MPO—myeloperoxidase
3-NitroTyr—3-nitrotyrosine
m-Tyr—meta-tyrosine (3-hydroxyphenylalanine)
o-Tyr—ortho-tyrosine (2-hydroxyphenylalanine)

Bioassays for Oxidative Stress Status (BOSS). Edited by W.A. Pryor

MEASUREMENT OF OXIDANT-INDUCED SIGNAL TRANSDUCTION PROTEINS USING CELL IMAGING

Matthew E. Poynter, Yvonne M. W. Janssen-Heininger, Sylke Buder-Hoffmann, Douglas J. Taatjes,
and Brooke T. Mossman

Department of Pathology, University of Vermont, Burlington, VT, USA

(Received 2 September 1999; Accepted 3 September 1999)

Abstract—In addition to their capacity to damage macromolecules, oxidants play important roles in initiation of a number of signal transduction pathways. These include phosphorylation and dephosphorylation of members of the extracellular-regulated kinase (ERK) family of the mitogen-activated protein kinase (MAPK) cascade and events leading to activation of the transcription factor nuclear factor-kappaB (NF-κB). These cascades are key to transcriptional upregulation of genes important for cell survival, apoptosis, proliferation, transformation, and inflammation. To complement biochemical assays, cell-imaging approaches are necessary to detect the phosphorylated proteins of these cascades and their nuclear translocation, i.e., activation in cells. Protocols for these studies are presented, and the advantages of in situ microscopy-based techniques to detect oxidant-induced signaling pathways are discussed. © 1999 Elsevier Science Inc.

Keywords—Free radical, Cell imaging, Nuclear factor-κB, Extracellular-regulated kinase (ERK), Mitogen-activated protein kinase (MAPK), Cell signaling, Epithelial cells

INTRODUCTION

Cellular exposure to exogenously or endogenously generated oxidants causes macromolecular damage including protein oxidation, lipid peroxidation, and nucleic acid instability and mutation [1,2]. Oxidative damage of cellular constituents has been associated with increased incidence of a number of diseases, and is likely to be an important contributor to inflammation, carcinogenesis, and the pathophysiology of aging and a number of other diseases [3]. Despite an abundance of literature demonstrating oxidant-mediated damage to cellular macromol-

ecules, only recently has the role of oxidants in the cellular processes of signal transduction been addressed. Two distinct signal transduction pathways that are activated in response to oxidative stress (e.g., asbestos mineral fibers) are the mitogen activated protein kinase cascade that includes extracellular regulated kinase 1 and 2 (ERK1/2), and signaling pathways that mediate activation of nuclear factor-kappaB (NF-κB).

The ERK family are typically activated in a series of protein phosphorylation after phosphorylation of cell surface receptors (i.e., the epidermal growth factor receptor [EGFR]) or other extracellular signals. Phosphorylated members of the ERK family (ERK1, ERK2, ERK3, ERK4, and ERK5) then function to transcriptionally regulate specific subsets of genes. For example, phosphorylated ERK2 translocates to the nucleus to phosphorylate ternary complex factor (TCF). TCF then binds to the serum response element of c-fos, which then elicits transcriptional activation (Fig. 1), including early response protooncogenes such as c-fos.

NF-κB represents a family of transcription factors that become activated in response to inflammatory cytokines, mitogens, microbial products, and physical and oxidative stress [4]. NF-κB is composed of protein dimers, the

Address correspondence to: Dr. Brooke T. Mossman, University of Vermont, Department of Pathology, Burlington, VT 05405, USA; Tel: (802) 656-0382; Fax: (802) 656-8892; E-Mail: bmossman@zoo.uvm.edu.

Matthew Poynter, Ph.D., is a postdoctoral fellow in the Environmental Pathology Training Program (National Institute of Environmental ealth Sciences T32 ES07122-16) at the University of Vermont. Yvonne Janssen-Heininger, Ph.D., is an Assistant Professor in the Department of Pathology, University of Vermont College of Medicine. Sylke Buder-Hoffmann, Ph.D., is a postdoctoral fellow in the Department of Pathology supported by a grant from the Federal Republic of Germany. Douglas Taatjes, Ph.D., is an Associate Professor of Pathology and Director of the Cell Imaging Facility at the University of Vermont. Brooke Mossman, Ph.D, is a Professor of Pathology and Director of the Environmental Pathology Training Program at the University of Vermont.

Reactive Oxygen or Reactive Nitrogen Species

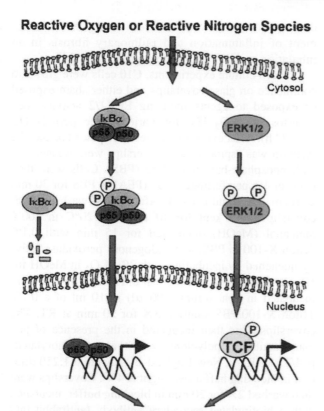

Transcription of genes regulating cell survival, apoptosis, proliferation, transformation, and inflammation.

Fig. 1. NF-κB and ERK1/2 signal transduction pathways are activated in response to oxidants. Exposure to oxidants causes activation of the transcription factor NF-κB (illustrated as p50/p65 heterodimers), often through the proteolytic degradation of phosphorylated IκBα, translocation of NF-κB to the nucleus, and binding to DNA. Oxidants induce phosphorylation of the signal transduction molecules, ERK1/2, through MAPK-dependent pathways. Phospho-ERK translocates to the nucleus where it phosphorylates ternary complex factor (TCF).

subunits of which contain Rel homology domains, and include the classic transcription-activating heterodimer consisting of p50 and p65 (RelA) subunits. NF-κB activity is controlled by members of a family of inhibitory proteins containing repeating ankyrin domains, known as IκB (inhibitor of κB). The mammalian IκB family consists of IκBα, IκBβ, and IκBε, which bind directly to NF-κB dimers in the cytosol, thereby preventing the nuclear localization of NF-κB and ensuring low basal levels of transcriptional activity. In addition, IκBγ (p105) and IκBδ (p100) contain both ankyrin repeats and Rel homology domains, thereby imparting upon them both inhibitor and transcription-activation domains. Upon cellular stimulation, IκB (or the similar motifs of the p105 or p100 molecules) becomes phosphorylated at specific serine residues (serines 32 and 36 in the case of the best-studied family member, IκBα). Phosphorylation causes the dissociation of IκB from the NF-κB dimer,

ubiquitination of IκB, and the immediate degradation of IκB by the 26S proteasome. Dissociation of NF-κB from IκB exposes the NF-κB nuclear localization sequence, which allows entry of the transcription factor into the nucleus, allowing an interaction between NF-κB and DNA (Fig. 1) at specific sites termed κB motifs (consensus GGGRNNT(Y)CC). DNA binding by NF-κB leads to the recruitment of essential components of the basal transcriptional machinery and the enhanced expression of genes proximal to the κB motif. Genes regulated by NF-κB encode inflammatory cytokines and enzymes, growth factors, chemotactic factors, cell adhesion molecules, and acute phase proteins, in addition to numerous others [5]. Furthermore, NF-κB activation is associated with initiation of an anti-apoptotic program, thereby inhibiting programmed cell death. Evidence supporting the capacity of oxidants to activate NF-κB comes from numerous studies in which H_2O_2 itself activates the transcription factor. In addition, chemical antioxidants such as pyrollidone dithiocarbamate, dithiothreitol, and N-acetyl cysteine, endogenous antioxidant molecules such as α-lipoic acid, α-tocopherol, and glutathione, as well as enzymatic antioxidants, including catalase, are capable of inhibiting NF-κB activation [6].

A number of effective strategies are routinely used to demonstrate activation of ERK1/2 or NF-κB in normal cells from in vitro and in vivo studies, including Western blotting, electrophoretic mobility shift assays (EMSA), and microscopy-based cell imaging, the latter of which is the focus of this article. Western blotting is commonly used to demonstrate reductions in IκB levels from whole cell extracts in response to a stimulant, including oxidant stress. While this approach can yield important information, difficulties arise because NF-κB transcriptionally regulates the synthesis of its inhibitor, resulting in higher levels of IκBα at later time points. In addition, NF-κB can become activated in the absence of IκBα degradation in response to oxidants [7]. Western blotting coupled to immunoprecipitation is also used to evaluate levels of phosphorylated ERK1/2 in whole cell extracts [8].

Western blotting, however, does not provide evidence that the transcription factor is actually capable of binding DNA, a step critical for the initiation of transcription. In order to demonstrate DNA binding activity by transcription factors, EMSA is routinely used. In the EMSA protocol used by our laboratories to determine NF-κB binding to DNA [9], an oligonucleotide containing sequences from known promoters or consensus motifs to which activated NF-κB is able to bind is radiolabeled in an in vitro kinase reaction and purified. The radiolabeled oligonucleotide is incubated with a few micrograms of extracted nuclear protein to facilitate the formation of specific protein-DNA complexes, which are then resolved in nondenaturing acrylamide gels and visualized by autoradiography or phosphorimaging. Free probe migrates rapidly through the gel while complexes exhibit a "mobility shift," migrating more slowly. The specificity of the oligonucleotide sequence allows identification of the transcription factor binding to it. In the case of NF-κB, however, it is important to establish which of the multiple family members comprise the dimer. To further identify components of the DNA-binding complex, antibodies recognizing epitopes specific to a single component are incubated in combination with the nuclear extract and the radiolabeled oligonucleotide. When resolved on a gel, binding of an antibody to a component of the DNA-binding complex further slows migration, thereby causing a "supershift" from which the components present in the DNA-binding complex may be identified. In this manner it is possible to determine, for instance, whether the NF-κB present in a nuclear extract is composed of transcriptionally active p50/p65 heterodimers or transcriptionally inactive p50/p50 homodimers, which both exhibit similar DNA-binding activities but dramatically different transcription-initiating activities.

Our laboratories routinely use a variety of biochemical techniques, some of which are described above for the measurement of activated signaling molecules and binding of transcription factors to DNA. However, these methods do not allow assessment of the cellular events occurring upon activation of these pathways. Herein, we present protocols used by our laboratories for the measurement of oxidant-induced signaling cascades both in vitro and in vivo using cell-imaging techniques.

EXPERIMENTAL PROCEDURES

Immunoperoxidase techniques

An advantage of immunoperoxidase techniques for detection of proteins in situ is that cells or tissues are examined on a light, as opposed to a fluorescence, microscope. In addition, coverslips of cells or tissue sections can be stored at room temperature for prolonged periods of time. Described below are studies using an antibody for phosphorylated ERK proteins that reacts with both ERK1 and ERK2 (New England Biolabs, Beverly, MA, USA). The specificity of this antibody was confirmed by Western blot analysis in that two bands were observed at 44 and 42 kDa, consistent with localization of ERK1 and ERK2. In brief, the objective of this work was to determine in an alveolar type II epithelial cell line (C10) [10], a target cell type of oxidant stresses such as hyperoxia or asbestos, whether H_2O_2 and asbestos caused distinct patterns of ERK1/2 localization. Moreover, we were interested in whether phosphorylation of ERK1/2 protein, an indi-

cator of its activation, occurred during the development of inflammation and pulmonary fibrosis in an inhalation model of lung disease.

For cell culture experiments, C10 cells were grown to confluence on glass coverslips and either sham exposed or exposed to agents inducing ERK1/2 activity, i.e., asbestos or H_2O_2 [8], for various time periods (15 min–24 h). After exposure, dishes were placed on ice, the medium was aspirated, and coverslips were washed 2× with phosphate-buffered saline (PBS). Cells were then fixed in 4% paraformaldehyde (PFA) in PBS for 30 min at room temperature (RT). After a 1× wash in PBS, coverslips were kept for 10 min at −20°C in 100% methanol (MeOH), incubated for 15 min with 0.1% Triton X-100 in PBS, and endogenous peroxidase activity quenched by incubating with 3% H_2O_2 in MeOH for 10 min at RT. After a 1× wash in PBS, coverslips were incubated in goat serum (200 μl) in 10 ml of a 0.1% Triton X-100/PBS solution 3× for 20 min at RT. The coverslips were then incubated in the presence of primary antibody (polyclonal rabbit antiphosphorylated-p44/42 ERK1/2; New England Biolabs) at a 1:250 dilution in blocking buffer overnight at 4°C. Coverslips were then washed 2× for 20 min in blocking buffer, incubated with a biotinylated secondary antibody (antirabbit-IgG from the Vectastain ABC Elite kit; Vector Laboratories, Burlingame, CA, USA) (1 drop per 10-ml blocking buffer) for 1 h at RT. After a 1× wash in PBS for 5 min, coverslips were incubated with the ABC reagent (Vector Laboratories) for 60 min according to the manufacturer's protocol. Coverslips were then washed 2× for 5 min in PBS before incubation in the chromagen 3,3'-diaminobenzidine (DAB) according to the manufacturer's protocol (Vector Laboratories). After a 1× wash in H_2O, coverslips were counterstained with hematoxylin and mounted on slides using 70% glycerin in H_2O. Slides were then observed by light microscopy. Controls consisted of coverslips stained in the absence of primary antibody.

As shown in Fig. 2A, phospho-ERK1/2 was not observed in confluent control cultures, but increased in the cytoplasm of asbestos-treated cells at 2 h (Fig. 2B). In contrast, both more intense cytoplasmic and nuclear staining, indicating translocation of phospho-ERKs, were seen in asbestos-exposed cells at 4 h (Fig. 2C). The time frame of increased cytoplasmic and nuclear localization of phospho-ERKs by H_2O_2 was more rapid, occurring within a 30-min period, and corresponded with patterns in Western blots at these time periods (data not shown). These results correlate with the patterns of ERK1/2 activity in epithelial and mesothelial cells in which the soluble agent, H_2O_2, causes earlier and transient increases whereas asbestos fibers are associated with protracted and prolonged ERK activity [8].

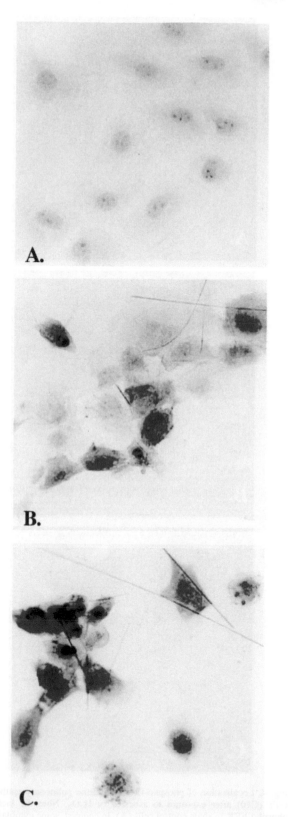

Fig. 2. Exposure to asbestos causes nuclear translocation of phospho-ERK in murine pulmonary epithelial (C10) cells as shown by immunoperoxidase technique: (A) sham controls; (B) asbestos-exposed cells.

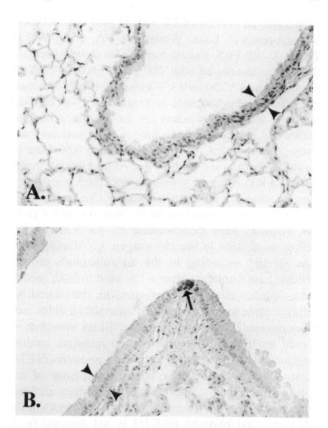

Fig. 3. Phospho-ERK in the bronchiolar epithelium (arrowheads) of the lung by immunoperoxidase staining as shown: (A) sham mouse lung; (B) asbestos-exposed lung at 20 d after initial exposure to asbestos. Note the focal cytoplasmic and nuclear localization of phospho-ERK (arrows).

In Fig. 3, we show an example of localization of phospho-ERK1/2 in the lung section of a sham control mouse and one exposed to chrysotile asbestos (7–10 mg/m^3 air for 30 d). In these experiments, groups of C57B1/6 mice ($n = 5$ per group/time point) were exposed by inhalation and killed using a lethal injection of pentobarbital before their chest cavities were opened and their lungs were perfused with PBS via the right ventricle. The left lung lobes were then fixed by intratracheal instillation of 4% PFA at a constant pressure of 14-cm H$_2$O and immersed in fixative overnight before embedding in paraffin. Lung sections were cut at 5-μm thickness for immunocytochemistry using the phospho-ERK antibody described above. Sections were deparaffinized in xylene 3 × for 5 min, rehydrated through a series of ethanols, and equilibrated in PBS. They were then permeabilized in 100% MeOH at -20°C for 10 min and washed with 0.1% Triton X-100/PBS before blocking of endogenous peroxidase activity by incubation in 3% H$_2$O$_2$ in MeOH for 10 min followed by 10 min in PBS. Sections were encircled with a hydrophobic film (PAP Pen; Electron Microscopy Services, Ft. Washington, PA, USA), and nonspecific protein binding was obstructed with a 200-μl overlay of 2% normal goat serum (Jackson Immunoresearch Labs, Westgrove, PA, USA) in 0.1% Triton X-100/PBS. Excess buffer was absorbed and sections were overlayed with 200 μl of phospho-ERK antibody diluted 1:250 in 0.1% Triton X-100/PBS containing 2% normal goat serum overnight at 4°C in a humidor. Sections were then washed 2× for 20 min in blocking buffer and incubated with a biotinylated secondary antibody (antirabbit-IgG from the Vectastain ABC Elite kit; Vector Laboratories) (1 drop per 10-ml blocking buffer) for 1 h at RT. After a 1× wash in PBS for 5 min, sections were incubated with the ABC reagent (Vector Laboratories) for 60 min according to the manufacturer's protocol. Sections were then washed 2× for 5 min in PBS before incubation in the chromagen 3,3′-diaminobenzidine (DAB) according to the manufacturer's protocol (Vector Laboratories). After a 1× wash in H$_2$O, sections were counterstained with hematoxylin, dehydrated in a series of ethanol and xylene, and mounted on slides using Vectamount (Vector Laboratories). Slides were then observed by light microscopy. Mouse intestinal sections, which contain a high proportion of phospho-ERK1/2-positive cells, and negative controls consisting of lung sections incubated in the absence of primary antibody, were run in parallel for each experiment. Results in Fig. 3A show that phospho-ERK1/2 is not detected in the normal mouse lung. However, prominent cytoplasmic and nuclear translocation in foci of bronchiolar epithelial cells is noted in asbestos-exposed lungs (Fig. 3B). A more comprehensive study shows that phospho-ERKs are prominent in epithelial cells of the bronchiole and alveolar duct region in asbestos-exposed mice. These are sites of initial deposition of asbestos fibers as well as sites of development of asbestos-associated epithelial cell hyperplasia and fibrotic lesions [8].

Multifluorescence techniques and confocal scanning laser microscopy (CSLM)

These approaches may be used on cells or tissues to determine the colocalization of multiple proteins using various fluorescence labels. This approach is generally more sensitive than an immunoperoxidase technique. Furthermore, when combined with methodology for in situ hybridization, one may determine whether genes transcriptionally activated by ERK1/2 or NF-κB dependent pathways are simultaneously expressed in cells that proliferate or undergo apoptosis, i.e., programmed cell death [11]. In addition, we have used these approaches in combination with mitochondrial protein markers and stains selective for DNA to colocalize DNA repair enzymes to the mitochondria and nucleus [12].

In Fig. 4, we show an example of C10 cells in culture stained with propidium iodide (PI) to localize DNA (red)

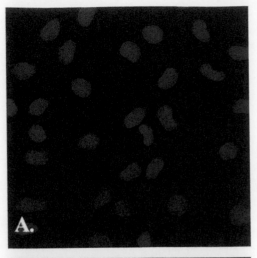

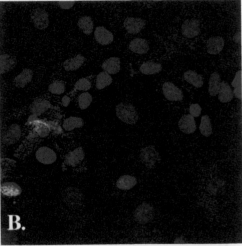

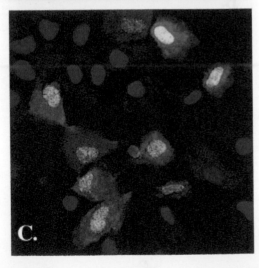

Fig. 4. Localization of phospho-ERK in murine pulmonary epithelial cells (C10) after exposure to asbestos or H$_2$O$_2$. Note the lack of phospho-ERK in sham control cells (A). In contrast, some cytoplasmic fluorescence (green) is seen in cells exposed to H$_2$O$_2$ (200 μM) for 2 h (B), and in asbestos-exposed cells (5 μg/cm^2) for 2 h (C). Note also the nuclear colocalization of phospho-ERK (green) and PI (red) to give a yellow signal.

and with the phospho-ERK1/2 antibody described above (green). These studies were identical in terms of design to the immunoperoxidase studies presented in Fig. 2. In this approach, coverslips of cells were prepared as described above for the immunoperoxidase studies. After the initial incubation in 0.1% Triton X-100/PBS, they were washed 1× in PBS before incubation in 2% defatted milk powder in 0.1% Triton-X-100/PBS for 30 min at RT. Coverslips were then washed 1× in PBS, and washed in 1% bovine serum albumin (BSA) before incubation with the phospho-ERK1/2 antibody in 1% BSA/PBS overnight at 4°C. After 2× incubations for 20 min in 1% BSA, coverslips were incubated in a secondary antibody solution (Alexa 488-conjugated goat anti-rabbit IgG; 1:200 in 1% BSA) for 1 h at RT. After 1× wash in PBS, coverslips were then incubated in a PI solution consisting of 0.1% Triton X-100, 20-μg/ml PI, 0.2 mg/ml RNase A, and 0.5-mM ethylenediaminetetraacetate, pH 8.0 for 25 min at RT. After a 1× wash each in PBS and H_2O, they were mounted on slides in an antifade mounting medium. Slides were then examined on a BioRad MRC1024ES Confocal Scanning Laser Microscope (Hercules, CA, USA), and separate confocal images were collected in the fluorescence modes followed by electronic merging of the images. These studies support the results of immunoperoxidase experiments presented in Fig. 2 in that control cells show little phospho-ERK1/2 (Fig. 4A). After exposure to asbestos, the numbers of cells expressing nuclear phospho-ERK1/2 increase at subsequent time points (Buder-Hoffmann et al., manuscript in preparation). The yellow color indicates colocalization of the phospho-ERK1/2 and PI in the nucleus. Increased phospho-ERK1/2 proteins are observed in H_2O_2-exposed cells at 2 h (Fig. 4B), but are more intense at earlier time periods (data not shown). In contrast, cytoplasmic and nuclear localization of phopho-ERK1/2 proteins is most dramatic in asbestos-exposed cells at this time point (Fig. 4C).

Another line of investigation in our laboratories has focused on the pathways of NF-κB activation in models of oxidative stress. NF-κB activation is associated with the translocation of p65 into the nucleus of cells in which it is transcriptionally active. This nuclear translocation can be observed by immunofluorescence techniques using an antibody recognizing p65, and the colocalization with DNA labeled with a variety of nuclear stains such as YOYO-1 iodide or PI. Fig. 5 demonstrates an example of nuclear translocation of p65 in response to lipopolysaccharide (LPS) or asbestos. In this study [9], rat lung epithelial (RLE) [13] cells were grown to confluence on coverslips, as described above for C10 cells. The protocol for fluorescent labeling and visualization was as described above, with the exception that the nuclear stain was YOYO-1 iodide (Molecular Probes, Eugene, OR,

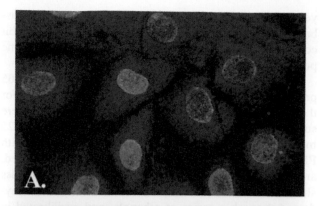

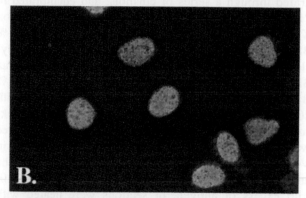

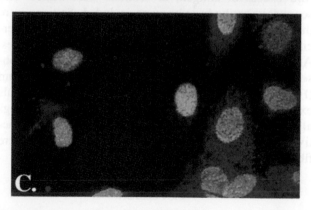

Fig. 5. Exposure to lipopolysaccharide (LPS) or asbestos causes activation of NF-κB in rat lung epithelial (RLE) cells. RLE cells grown to confluence on coverslips were sham-exposed (A), exposed to LPS for 2 h (B), or exposed to chrysotile asbestos for 8 h (C). Cells were fluorescently labeled and visualized using confocal microscopy as described in the Experimental Procedures. Exposure to LPS or asbestos causes cytoplasmic p65 (red) to translocate to the nucleus where it colocalizes with YOYO-1-stained DNA (green), producing a yellow color.

USA), which was collected in the green channel, whereas anti-p65 (Santa Cruz Biotechnology, Santa Cruz, CA, USA) was the primary antibody and collected in the red channel. As is seen in Fig. 5A, control cells exhibit immunoreactive p65 (red) exclusively in the cytoplasm. After exposure to LPS for 2 h, however, nearly all p65 protein translocates to the nucleus where it is colocalized with YOYO-1 iodide (green) to produce a

yellow color (Fig. 5B). Similarly, after exposure to asbestos for 8 h, many cells in the selected field exhibit
colocalization of p65 to the nucleus, although at this time
point some p65 remains in the cytoplasm (Fig. 5C).

We have also evaluated the expression of the p65
protein in rat lungs after exposure to asbestos [9]. For
these studies, rats ($n = 3–4$ per group/time point) were
subjected to sham or chrysotile asbestos exposure (8.25
mg/m^3 air for 5 d) and killed. Lungs were perfused with
PBS, removed, fixed in 70% ethanol, paraffin embedded,
and 5-μm sections were prepared and placed onto glass
slides. Sections were deparaffinized in xylene, rehydrated through a series of ethanols, and equilibrated in
PBS. Sections were then permeabilized in 1% Triton
X-100/PBS for 30 min and blocked 2× for 30 min in 1%
BSA/PBS. Sections were incubated with 2.5 μg/ml polyclonal rabbit anti-p65 antibody (SC-372, Santa Cruz
Biotechnology) for 1 h. After 3× incubation in 1%
BSA/PBS for 20 min each, sections were incubated with
20 μg/ml of a Cy5-conjugated secondary antibody (Jackson ImmunoResearch Laboratories) for 1 h. After three
washes in PBS for 5 min, each section was mounted in
Vectashield (Vector Laboratories). Controls in which the
anti-p65 antibody was incubated overnight with a 10-
fold excess of antigen, or sections were treated with
secondary antibody only, were included. Sections were
then examined on a confocal microscope (Bio Rad, Hercules, CA, USA) with identical settings for all samples.
As is seen in the representative field presented, we observed marked fluorescence in bronchiolar epithelium
(Fig. 6A), which was increased dramatically in animals
exposed to asbestos for 5 days (Fig. 6B). These findings
allowed us to conclude that increased p65 immunoreactivity is observed at early time points at sites where
fibrotic lesions develop.

DISCUSSION

Work in recent years has implicated an important role
for oxidants in multiple signal transduction cascades.
Two major pathways sensitive to oxidative stress are the
MAP kinases that include ERK1/2 and the cascade that
mediates induction of NF-κB. The oxidant-mediated activation of these cascades may have multiple consequences including, apoptosis, cell proliferation, differentiation, transformation, and the production of
inflammatory and immune modulatory molecules (Fig.
1). These outcomes likely depend on the model of oxidant stress, including the dose and kinetics of exposure,
as well as the cell or tissue under investigation.

Biochemical and molecular assays are available to
quantitatively measure the activation of signal transduction proteins or transcription factors in vitro. These approaches, however, are less informative when investigat-

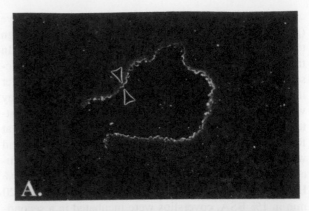

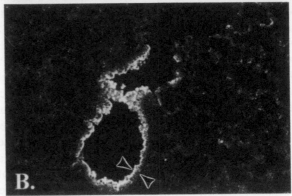

Fig. 6. Elevated immunoreactivity of p65 in the lungs of asbestos-
exposed mice. Mice were sham exposed (A) or exposed to aerosolized
chrysotile asbestos (B) for 5 d [9]. Lung sections were prepared and
visualized by confocal microscopy as described in the Experimental
Procedures. After asbestos exposure, increased p65 immunoreactivity
was seen in the bronchioles (arrowheads) of this section (B) as well as
in the alveolar duct epithelium but not in vessels (data not shown).
Reprinted with permission from Janssen et al. [9] (© 1997 the American Society for Investigative Pathology).

ing the localization of these proteins in intact cells,
especially in animal models of disease or in tissue sections where multiple cell types or specific locations of
injury are affected. Therefore, novel cell imaging approaches have become available that allow the localization of activated forms of signaling intermediates in
tissues or in subcellular compartments. These microscopy-based approaches facilitate the use of small amounts
of tissues or biopsies as well as archival tissues that may
not be amenable to biochemical analyses. With the incorporation of multiple fluorescent labels, investigators
have the ability to convincingly identify the cell type in
which transcription factors become activated on the basis
of the molecular phenotype in addition to architectural
localization within a tissue section. Furthermore, it is
possible to determine whether a signaling pathway is
active in a dividing cell or in a cell undergoing apoptosis
by the simultaneous incorporation of markers of apoptosis or proliferation. The ability to document these events

in situ may provide investigators with a better understanding of cellular responses and disease processes.

Our laboratories have focused on the pathways that mediate apoptosis, proliferation, or inflammation in models of pulmonary injury. Past research has demonstrated that the mineral fiber asbestos, a documented oxidative stress, induces the ERK pathway and downstream expression of proto-oncogenes [8]. Using specific chemical inhibitors of the ERK pathway, a decrease in apoptosis was observed in response to asbestos, demonstrating a causal link between ERK activation and apoptosis after exposure of mesothelial cells to asbestos [8]. To demonstrate the localization of activated phosphorylated ERK in cells or tissues, we utilized the phospho-specific ERK antibody and demonstrated that it is found in subpopulations of cells or tissues exposed to asbestos at sites of developing fibrotic lesions [14]. Our data described here suggest that the ERK cascade is activated in cells within bronchiolar epithelium and at bifurcations in terminal bronchioles where it may play a role in tissue remodeling, proliferation, or apoptosis [15]. However, a causal role of this protein awaits the evaluation of transgenic mice that express a dominant negative MEK, the upstream activator of ERK, at these sites.

The selective localization of p65 in bronchiolar epithelium in lung using an antibody to p65 and fluorescence microscopy may point to an important role of NF-κB in regulating inflammatory responses within bronchiolar epithelium. One problem of these studies was an apparent lack of nuclear localization under conditions of inflammation where nuclear presence of NF-κB can be expected [9]. Unpublished results from our laboratory have indicated that rapid fixation and staining is required for the preservation of nuclear localization that may explain some of our findings in this study. Importantly, a different antibody has become available that recognizes the nuclear localization signal of p65 (clone 12H11; Boehringer Mannheim, Indianapolis, IN, USA). In its latent form, NF-κB is complexed to IκB, which masks this nuclear localization signal and, therefore, abolishes binding of the antibody to p65. This new antibody provides an approach for the detection of active NF-κB complexes under conditions where fixation was insufficient to retain nuclear localization.

In summary, the recent availability of a battery of antibodies that recognize activated forms of signaling molecules have enabled investigators to examine these pathways in vitro and in vivo in response to oxidant injury. These microscopy-based approaches have lead to the identification of critical sites of injury and proliferation in the lung and enhance our understanding of mechanisms of oxidant-induced pulmonary diseases. Ideally, these approaches in cell culture or in tissues should be combined with biochemical and molecular studies to quantitate binding of transcription factors to DNA and transcriptional activation of NF-κB or ERK1/2 dependent gene expression.

Acknowledgements — This research was supported by National Institute of Environment Health Sciences T32ES07122, grants RO1 ES/HL09213 and RO1 HL39469 (B.T.M.), and grant RO1 HL60014 (Y.M.W.J.-H.).

REFERENCES

[1] Janssen, Y. M. W.; Borm, P. J. A.; Van Houten, B.; Mossman, B. T. Cell and tissue responses to oxidative damage. *Lab. Invest.* **69:**261–274; 1993.
[2] Halliwell, B. H.; Gutteridge, J. M. C. *Free radicals in biology and medicine.* Oxford: Oxford University Press; 1998.
[3] Beckman, K. B.; Ames, B. N. The free radical theory of aging matures. *Physiol. Rev.* **78:**547–581; 1988.
[4] Ghosh, S.; May, M. J.; Kopp, E. B. NF-κB and rel proteins: evolutionarily conserved mediators of immune responses. *Ann. Rev. Immunol.* **16:**225–260; 1998.
[5] Baeuerle, P. A.; Henkel, T. Function and activation of NF-κB in the immune system. *Ann. Rev. Immunol.* **12:**141–179; 1994.
[6] Schreck, R.; Albermann, K.; Baeuerle, P. A. Nuclear factor-kappa B: an oxidative stress-responsive transcription factor of eurkaryotic cells. *Free Radic. Res. Commun.* **365:**182–185; 1992.
[7] Imbert, V.; Rupec, R. A.; Livolsi, A.; Pahl, H. L.; Traenckner, B. M.; Mueller-Dieckmann, C.; Farahifar, D.; Rossi, B.; Augberger, P.; Baeuerle, P. A. Tyrosine phosphorylation of IκB activates NF-κB without proteolytic degradation of IκB-α. *Cell* **86:**787–798; 1996.
[8] Jimenez, L. A.; Zanella, C.; Fung, H.; Janssen, Y. M. W.; Vacek, P.; Charland, C.; Goldberg, J.; Mossman, B. T. Role of extracellular signal-regulated protein kinases in apoptosis by asbestos and H$_2$O$_2$. *Am. J. Physiol.* **273:**L1029–L1035; 1997.
[9] Janssen, Y. M. W.; Driscoll, K. E.; Howard, B.; Quinlan, T. R.; Treadwell, M.; Barchowsky, A.; Mossman, B. T. Asbestos causes translocation of p65 protein and increases NF-κB DNA binding activity in rat lung epithelial and pleural mesothelial cells. *Am. J. Pathol.* **151:**389–401; 1997.
[10] Malkinson, A. M.; Dwyer-Nield, L. D.; Rice, P. L.; Dinsdale, D. Mouse lung epithelial cell lines—tools for the study of differentiation and the neoplastic phenotype. *Toxicology* **123:**53–100; 1997.
[11] BeruBe, K. A.; Quinlan, T. R.; Moulton, G.; Hemenway, D.; O'Shaughnessy, P.; Vacek, P.; Mossman, B. T. Comparative proliferative and histopathologic changes in rat lungs after inhalation of chrysotile and crocidolite asbestos. *Toxicol. Appl. Pharmacol.* **137:**67–74; 1996.
[12] Fung, H.; Kow, Y. W.; Van Houten, B.; Taatjes, D. J.; Hatahet, Z.; Janssen, Y. M. W.; Vacek, P.; Faux, S. P.; Mossman, B. T. Asbestos increases mammalian AP-endonuclease gene expression, protein levels, and enzyme activity in mesothelial cells. *Cancer Res.* **58:**189–194; 1998.
[13] Driscoll, K. E.; Carter, J. M.; Iype, P. T.; Kumari, H. L.; Crosby, L. L.; Aardema, M. J.; Isfort, R. J.; Cody, D.; Chestnut, M. H.; Burns, J. L.; LeBoeuf, R. A. Establishment of immortalized alveolar type II epithelial cell lines from adult rats. *In Vitro Cell. Dev. Biol. Anim.* **31:**516–527; 1995.
[14] Zanella, C. L.; Posada, J.; Tritton, T. R.; Mossmann, B. T. Asbestos causes stimulation of the extracellular signal-regulated kinase 1 mitogen-activated protein kinase cascade after phosphorylation of the epidermal growth factor receptor. *Cancer Res.* **56:**5334–5338; 1996.
[15] Robledo, R. F.; Buder-Hoffmann, S. A.; Cummings, A. B.; Walsh, E. S.; Taatjes, D. J.; Mossman, B. T. Increased phosphorylated ERK immunoreactivity associated with proliferative and morphologic alterations following chrysotile asbestos inhalation in mice. *Am. J. Pathol.* In revision.

ABBREVIATIONS

NF-κB—nuclear factor-kappaB
ERK—extracellular-regulated kinase
MAPK—mitogen-activated protein kinase
EMSA—electrophoretic mobility shift assay
TCF—ternary complex factor
IκB—inhibitor of κB

PBS—phosphate buffered saline
BSA—bovine serum albumin
MeOH—methanol
DAB—3,3'-diaminobenzamide
RT—room temperature
PI—propidium iodide
LPS—lipopolysaccharide

Bioassays for Oxidative Stress Status (BOSS). Edited by W.A. Pryor
© 2001 Elsevier Science B.V. All rights reserved.

IN VIVO TOTAL ANTIOXIDANT CAPACITY: COMPARISON OF DIFFERENT ANALYTICAL METHODS[1]

Ronald L. Prior and Guohua Cao

USDA Jean Mayer USDA Human Nutrition Research Center on Aging at Tufts University, Boston, MA, USA

(Received 3 September 1999; Accepted 7 September 1999)

Abstract—Several methods have been developed to measure the total antioxidant capacity of a biological sample. The use of peroxyl or hydroxyl radicals as pro-oxidants in the oxygen radical absorbance capacity (ORAC) assay makes it different and unique from the assays that involve oxidants that are not necessarily pro-oxidants. An improvement in quantitation is achieved in the ORAC assay by taking the reaction between substrate and free radicals to completion and using an area-under-curve technique for quantitation compared to the assays that measure a lag phase. The interpretation of the changes in plasma or serum antioxidant capacity becomes complicated by the different methods used in detecting these changes. The interpretation also depends upon the conditions under which the antioxidant capacity is determined because the measurement reflects outcomes in a dynamic system. An increased antioxidant capacity in plasma or serum may not necessarily be a desirable condition if it reflects a response to increased oxidative stress. Similarly, a decrease in plasma or serum antioxidant capacity may not necessarily be an undesirable condition if the measurement reflects decreased production of reactive species. Because of these complications, no single measurement of antioxidant status is going to be sufficient, but a "battery" of measurements, many of which will be described in Forum articles, will be necessary to adequately assess oxidative stress in biological systems.   © 1999 Elsevier Science Inc.

Keywords—Antioxidant, Free radical, ORAC, Oxidant, Pro-oxidant, Reductant, Total antioxidant capacity

INTRODUCTION

Living organisms have developed a complex antioxidant network to counteract reactive species that are detrimental to human life. Nonenzymatic antioxidants, such as albumin, GSH, ascorbic acid, α-tocopherol, β-carotene, uric acid, bilirubin, and flavonoids, constitute an important aspect of the network. It is essential to measure these antioxidants in assessing in vivo antioxidant status. However, the number of different antioxidants in plasma, serum, urine, or other biological samples makes it difficult to measure each antioxidant separately. The possible interaction among different antioxidants in vivo could also make the measurement of any individual antioxidant less representative of the overall antioxidant status. Therefore, several methods including our oxygen radical absorbance capacity (ORAC) assay, have been developed and used to determine the total antioxidant capacities of various biologic samples [1–17]. This review will discuss the principles, characteristics, and applications of these methods along with the comparison of the ORAC assay to other methods in assessing in vivo antioxidant status.

ANTIOXIDANT AND REDUCTANT: SEVERAL CONCEPTS RELATED TO ANTIOXIDANT CAPACITY ASSAYS

Methods developed for measuring total antioxidant capacity of a biological sample have been classified as

[1]Mention of a trade name, proprietary product, or specific equipment does not constitute a guarantee by the U.S. Department of Agriculture and does not imply its approval to the exclusion of other products that may be suitable.

Address correspondence to: R. L. Prior, Ph.D., USDA, ARS, HNRCA, 711 Washington St., Boston, MA 02111, USA; Tel: (617) 556-3311; Fax: (617) 556-3222; E-Mail: prior@hnrc.tufts.edu

Dr. Ronald Prior is a Nutritionist and Laboratory Chief of the Phytochemical Laboratory at the Jean Mayer USDA Human Nutrition Research Center on Aging (HNRCA) at Tufts University, Boston, MA. Dr. Prior received his B.S. degree with honors from the University of Nebraska and he received his Ph.D. in Nutrition and Biochemistry from Cornell University in 1972. Dr. Prior has worked with the USDA for more than 20 years. During the past 12 years at the HNRCA, he has been Scientific Program Officer and has directed research activities dealing with the role of flavonoid and other phenolic food components on antioxidant status, their metabolism, and relationships to diseases of aging.

Guohua Cao, M.D., Ph.D., is currently a Scientist II at HNRCA. Dr. Cao studied medicine in Nantong Medical College in 1979 and at Nanjing Medical University in 1984. He obtained his Ph.D. in nutritional biochemistry from Beijing Medical University in 1990. Dr. Cao came to the United States in 1991 and worked at NIH where he was instrumental in developing the ORAC method.

inhibition methods involving reactive species [6]. Benzie and Strain [15] stated recently that all these inhibition methods were *indirect* tests of total antioxidant power. They claimed further that their noninhibition method, the FRAP assay, (which was defined as "ferric reducing ability of plasma" [9] and now has been defined to mean "ferric reducing/antioxidant power" [15]) was a *direct* test of the total antioxidant power [15]. These statements about the inhibition methods and FRAP assay may be misleading.

Before these statements can be clarified, it is necessary to discuss the concepts of a reductant, an oxidant, an antioxidant, and a pro-oxidant. This discussion will also help with the comparison of different analytical methods for assessing total antioxidant capacity of a biological sample.

Reductant and oxidant

Reduction of a chemical is defined as a gain of electrons. Oxidation is defined as a loss of electrons. A reductant or a reducing agent is a substance that donates electrons and, thereby, causes another reactant to be reduced. An oxidant or an oxidizing agent is a substance that accepts electrons and causes another reactant to be oxidized. An oxidation is impossible without a reduction elsewhere in the system. For example, when a sodium atom (Na) transfers an electron to a chlorine atom (Cl) to form NaCl (Na + Cl → NaCl), Na is oxidized to Na^+ ($Na → Na^+ + e^-$; oxidation) and Cl is reduced to Cl^- ($Cl + e^- → Cl^-$; reduction). Thus, in this reaction, Na is a reductant and Cl is an oxidant. When reduction and oxidation characterize a chemical reaction, it is called a redox reaction. Redox reactions are at the heart of biological oxidation, the chain of chemical reactions whereby we use oxygen from air to oxidize chemicals from the breakdown of food to provide energy for living.

Antioxidant and pro-oxidant

Reductant and *oxidant* are chemical terms, whereas *antioxidant* and *pro-oxidant* have meaning in the context of a biological system. An antioxidant can be defined as a substance that, when present at low concentrations compared with those of an oxidizable substrate, significantly prevents or delays a pro-oxidant initiated oxidation of the substrate. A pro-oxidant is a toxic substance that can cause oxidative damage to lipids, proteins and nucleic acids, resulting in various pathologic events and/or diseases. Pro-oxidant is a synonym for reactive species. Thus, chemically, a pro-oxidant is an oxidant of pathologic importance. An antioxidant can efficiently reduce a pro-oxidant with the formed products having no or low toxicity.

An antioxidant is a reductant, but a reductant is not necessarily an antioxidant

Most inhibition methods including our ORAC assay involve a pro-oxidant, which is usually a free radical, and an oxidizable substrate. The pro-oxidant induces oxidative damage (to the substrate), which is inhibited in the presence of an antioxidant. This inhibition is measured and related to antioxidant capacity of the antioxidant. The measured antioxidant capacity may have physiologic importance, because the pro-oxidants used in these systems are pathologically important. It is the existence of various harmful pro-oxidants or reactive species (i.e., $O_2^{-•}$, H_2O_2, $ROO^•$, and $OH^•$) in vivo that makes antioxidants essential to a healthy life. Therefore, the compounds measured using these inhibition methods are reductants that are able to reduce pro-oxidants and protect oxidizable substrates. These compounds or reductants are certainly antioxidants.

But, not all reductants defined in a chemical reaction are antioxidants. Otherwise, Na, which we used earlier to illustrate a redox reaction, glucose, ethanol, and many others should also be called antioxidants. Also, the reducing capacity of a reductant, in a mixture of reductants, to reduce an oxidant is not necessarily equal or proportional to the antioxidant capacity of the reductant to reduce a pro-oxidant. When the FRAP assay is regarded as a direct test of "total antioxidant power," a chemically defined reductant and a biologically defined antioxidant are basically treated as equal. The FRAP assay involves neither a pro-oxidant nor an oxidizable substrate. It depends upon the reduction of a ferric tripyridyltriazine (Fe^{3+}-TPTZ) complex to the ferrous tripyridyltriazine (Fe^{2+}-TPTZ) by a reductant at low pH. Fe^{2+}-TPTZ has an intensive blue color and can be monitored at 593 nm [9,15]. What it really measures is the ability of a compound to reduce Fe^{3+} to produce Fe^{2+}. Fe^{2+} is a well-known "pro-oxidant." It can react with H_2O_2 to produce $OH^•$, the most harmful free radical found in vivo. Then, why is the ability of a compound to produce Fe^{2+} from Fe^{3+} defined as "antioxidant power" in the FRAP assay? The answer is probably that some antioxidants, such as ascorbic acid and uric acid, can reduce both reactive species and Fe^{3+}, and their ability in reducing Fe^{3+} may reflect their ability in reducing reactive species. But, not all reductants that are able to reduce Fe^{3+} are antioxidants. In addition, *an antioxidant that can effectively reduce pro-oxidants may not be able to efficiently reduce Fe^{3+}*. For example, the FRAP assay does not measure GSH, an important antioxidant in vivo.

Nevertheless, the ferric-reducing ability measured for a biological sample may indirectly reflect the total antioxidant power of the sample. We observed a significant but weak linear correlation between serum FRAP and serum ORAC [18]. The important concept to emphasize here is not the advantage or disadvantage of the FRAP assay, but the *indirectness* of the assay in measuring "total antioxidant power" when no pro-oxidants are involved in the measurement. In other words, the FRAP assay is an indirect test of total antioxidant power.

ANTIOXIDANT CAPACITY ASSAYS INVOLVING OXIDANTS THAT ARE NOT NECESSARILY PRO-OXIDANTS

FRAP assay

As we discussed above, FRAP assay [9,15] involves an oxidant, i.e., Fe^{3+}. But, Fe^{3+} is not necessarily a pro-oxidant. Fe^{2+}, which is produced from the reduction of Fe^{3+} in the FRAP assay, could be a pro-oxidant because of its reaction with H_2O_2. However, neither Fe^{2+} nor Fe^{3+} is able to directly cause oxidative damage to lipids, proteins or nucleic acids.

TEAC assay

The Trolox equivalent antioxidant capacity (TEAC) assay was reported first by Miller et al. [5,6] and then modified by Re et al. [17]. The TEAC assay is based on the inhibition by antioxidants of the absorbance of the radical cation of 2,2′-azinobis(3-ethylbenzothiazoline 6-sulfonate) (ABTS), which has a characteristic long-wavelength absorption spectrum showing maxima at 660, 734, and 820 nm. The ABTS radical cation in the original version, which has been commercialized by Randox Laboratories (San Francisco, CA, USA), is formed by the interaction of ABTS with the ferrylmyoglobin radical species, generated by the activation of metmyoglobin with H_2O_2. This original TEAC assay measures the ability of a compound in reducing ABTS radical, although the compound under analysis can also reduce ferrylmyoglobin radicals. The ABTS radical is not necessarily a pro-oxidant because it is quite stable [17]. The intra- and interassay CV of the original TEAC assay were reported to be 0.54–1.59% and 3.6–6.1%, respectively [6]. The modified or improved TEAC assay uses ABTS radicals preformed by oxidation of ABTS with potassium persulfate. This preformed ABTS radical is stable for at least 2 d when stored in the dark at room temperature [17]. The original TEAC assay has produced useful information regarding the antioxidant activities of phytochemicals [19–22], although it has certain short-comings [23,24]. By using the original TEAC assay, plasma antioxidant capacity was found to be increased in patients with acute myocardial infarction [25], and decreased in premature infants [5,6]. However, the lower plasma TEAC found in premature infants was mainly due to the lower protein content in the plasma. The plasma TEAC in premature infants will be significantly higher, not lower, than that in term infants or adults if the data are based on the amount of total protein contained in the plasma [26].

Cyclic voltammetry method

The cyclic voltammetry procedure reported recently by Kohen et al. [27,28] evaluates the overall reducing power of low molecular weight antioxidants in a biological fluid or tissue homogenate. Following preparation, the sample is introduced into a well in which three electrodes are placed: the working electrode (e.g., glassy carbon), the reference electrode (Ag/AgCl), and the auxiliary electrode (platinum wire). The potential is applied to the working electrode at a constant rate (100 mV/sec) either toward the positive potential (evaluation of reducing equivalent) or toward the negative potential (evaluation of oxidizing species). During operation of the cyclic voltammetry, a potential current curve is recorded (cyclic voltammogram). The reducing power of a sample is composed of two parameters: the peak potential $[E_{p(a)}]$ and the anodic current (AC). The $E_{p(a)}$ is measured at the half increase of the current at each anodic wave (AW) and is referred as $E_{1/2}$. The $E_{1/2}$ correlates with the type of reductant: the lower the $E_{1/2}$, the higher the ability of the tested compounds to donate electrons to the working electrode. The AC is measured from the y axis of each AW and correlates with the overall concentration of the reductants. However, not all the common antioxidants donate their electrons to the glass carbon electrode at a sufficient rate. For example, thiol compounds such as glutathione will not be detected by the glass carbon electrode (other electrodes such as an Au/Hg electrode are needed for glutathione measurement) [28]. The sensitivity of this CV procedure is relatively low (1–10 μM) [28].

ANTIOXIDANT CAPACITY ASSAYS INVOLVING OXIDANTS THAT ARE PRO-OXIDANTS

TRAP assay

Antioxidant activity is an ability of a compound to reduce pro-oxidants or reactive species of pathologic significance. Most methods that have been defined as inhibition methods [6] involve reactive species, which are usually free radicals. The total radical trapping parameter (TRAP) assay of Wayner et al. [1] was the most widely used method for measuring total antioxidant capacity of plasma or serum during the last decade. The TRAP assay uses peroxyl radicals generated from 2,2′-azobis(2-amidinopropane) dihydrochloride (AAPH) and peroxidizable materials contained in plasma or other biological fluids. After adding AAPH to the plasma, the oxidation of the oxidizable materials is monitored by measuring the oxygen consumed during the reaction. During an induction period, this oxidation is inhibited by the antioxidants in the plasma. The length of the induction period (lag phase) is compared to that of an internal standard, Trolox (Aldrich, Milwaukee, WI, USA) (6-hydroxyl-2,5,7,8,-tetramethylchroman-2-carboxylic acid), and then quantitatively related to the antioxidant capacity of the plasma. One modification made by Wayner et al. [29] was the addition of linoleic acid in the plasma sample before the oxidation was initiated by peroxyl radicals. However, the major problem with the original TRAP assay lies in the oxygen electrode end point; an oxygen electrode will not maintain its stability over the period of time required [6]. Using this original TRAP assay, Mulholland and Strain [30] reported that serum TRAP experimental values were significantly lower in patients with acute myocardial infarction when compared to the sex- and age-matched controls. The plasma TRAP experimental values were also reported to decline significantly by about 40% during chemotherapy (16-mg/kg busulfan, 30–45 mg VP-16, and 120-mg/kg cyclophosphamide) in patients with various haematologic malignancies [31].

Luminol-based assays

Metsä-Ketelä et al. first developed and reported a chemiluminescence-based TRAP assay in 1991. The method was recently described in detail by Alho and Leinonen [14]. The principle of this method is that peroxyl radicals produced from AAPH oxidize luminol, leading to the formation of luminol radicals that emit light. The emitted light can be detected by a luminometer. Antioxidants in the sample inhibit this chemiluminescence for a time that is directly proportional to the total antioxidant potential of the sample. This potential of the sample is compared with that of Trolox and related quantitatively to the antioxidant capacity of the sample. It was claimed that both intra- and interassay coefficient of variation (CV) of the method were 2% ($n = 10$) [32]. By using this method, Aejmelaeus et al. [32] also found that in healthy females, TRAP increased significantly during the life span. In healthy males, TRAP increased until age 51–74, and then significantly decreased. The increase in TRAP in females and the decrease in TRAP in males were mainly due to unidentified antioxidants [32]. By using an analytical system very similar to this method, Pascual et al. [33] found that plasma antioxidant capacity was lower in septic patients but higher in septic shock patients, as compared with controls. Bilirubin was the greatest contributor to the increase with shock, fol-

lowed by uric acid. Slavíková et al. [34] reported that serum TRAP, measured using this luminol-based assay, increased in early but not later after intestinal ischemia in rats.

Whitehead et al. [3] reported an enhanced chemiluminescent assay for antioxidant capacity in biological fluids in 1992. The principle of this method is based on the oxidation of luminol by hydrogen peroxide or perborate in a reaction catalyzed by horseradish peroxidase, which emits light. Under normal circumstances this reaction produces low intensity light emission that may decay rapidly. The characteristics of the reaction can be altered substantially by the addition of the enhancer para-iodophenol giving a more intense, prolonged, and stable light emission. The light emission is sensitive to interference by antioxidants, but will be restored when all the added antioxidants have been consumed in the reaction. Using this method, Whitehead et al. [35] showed that the consumption of red wine or ascorbic acid, but not white wine, significantly increased the serum antioxidant capacity. However, the original procedure used the commercial reagent kits as ECL Antioxidant Detection Pack NK8989 from Amersham International (Buckingshamshire, UK).

Dichlorofluorescin-diacetate (DCFH-DA) based assay

Another method that also measures TRAP is the spectrophotometric assay reported recently by Valkonen and Kuusi [10]. This assay uses AAPH to generate peroxyl radicals and DCFH-DA as the oxidizable substrate for the peroxyl radicals. The oxidation of DCFH-DA by peroxyl radicals converts DCFH-DA to dichlorofluorescein (DCF). DCF is highly fluorescent (Ex 480 nm, Em 526 nm) and also has absorbance at 504 nm. Therefore, the produced DCF can be monitored either fluorometrically or spectrophotometrically. In this assay, DCF fluorescence or absorbance formation contains four phases. The first lag phase is due to the antioxidants in the sample. After the consumption of antioxidants by peroxyl radicals, the reaction proceeds to the first propagation phase. The second lag phase, which interrupts the first propagation, is due to Trolox (the internal standard) added during the first propagation phase. The second propagation of the reaction follows the consumption of Trolox. The first lag phase is compared to the second lag phase and, thus, related to the antioxidant capacity of the sample. This TRAP assay, also involving the determination of lag phases, was reported to have intra- and interassay CV of 3.4% and 4.6%, respectively. The measured TRAP was significantly increased in 11 human subjects after supplementation with 300-mg/day α-tocopherol for 1 week [10].

TOSC assay

Winston and his coworkers [11] recently reported an assay called total oxyradical scavenging capacity (TOSC) assay. It is based on the oxidation of α-keto-γ-methiolbutyric acid (KMBA) to ethylene by peroxyl radicals produced from AAPH. The ethylene formation, which is partially inhibited in the presence of antioxidants, is monitored by gas chromatographic analysis of head space from the reaction vessel. The TOSC is calculated according the equation: TOSC = $100 - (\int SA / \int CA \times 100)$, where $\int SA$ and $\int CA$ are the integrated area from the curve defining the sample and control reactions, respectively. The antioxidant or TOSC value is from 0 to 100. The intra- and interassay CV of the TOSC assay were reported to be 2% ($n = 30$) and 6% ($n = 30$), respectively. Obviously, this is a "open" system in terms of the area integration, because the production of ethylene from KMBA should increase continuously after the consumption of antioxidants if the available KMBA is not a limiting factor. There have been no TOSC data reported yet regarding its application in assessing in vivo antioxidant status.

Crocin based assays

A competition kinetics procedure for measuring plasma antioxidant capacity was recently described by Tubaro et al. [12]. This procedure is based on the oxidation (bleaching) of crocin by peroxyl radicals produced from AAPH. The bleaching rate of crocin by peroxyl radicals in the absence (V_0) and presence (V) of antioxidants were recorded for 10 min. The slope of the linear regression of the plot of [A]/[C] vs. V_0/V is theoretically the ratio between k_a and k_c (k_a/k_c), and, therefore, indicates the relative capacity of the compound under analysis to interact with peroxyl radicals (where k_a = the rate constant for the reaction between antioxidant and peroxyl radicals, k_c = the rate constant for the reaction between crocin and peroxyl radicals, [A] = concentration of antioxidant, and [C] = concentration of crocin). By dividing the k_a/k_c of a given antioxidant by the k_a/k_c of Trolox, the result is then expressed as Trolox equivalent. The intraassay CV was reported to be <8%. Tubaro et al. [12] believe this kinetic approach provides a more precise evaluation of the efficiency of antioxidant defense. However, this method measures the ability of an antioxidant in competing against another antioxidant, i.e., crocin for peroxyl radicals. In addition, the concentration of crocin used in the assay is fixed (10 μM). However, several concentrations of an analyzed compound are required in the assay. It is not clear whether a lag phase, or a period of maximal protection of crocin from oxidation by peroxyl radicals, can be seen in the presence of some antioxidants, such as Trolox, ascorbic acid, or uric acid. These antioxidants usually show a lag phase in the assay systems that use lipids or proteins as oxidizable substrate for peroxyl radicals. The antioxidant capacity of ascorbic acid analyzed using this kinetics method was reported to be 7.7 Trolox equivalent [12], much higher than those obtained with any other methods. Using this method, Tubaro et al. [12] also reported that plasma antioxidant capacity is influenced by the consumption of wine in humans. A microplate-based antioxidant capacity assay also using peroxyl radicals and crocin was reported recently [16]. This assay uses inhibition percentage instead of the kinetics.

Phycoerythrin (PE) based assays

The TRAP assay reported by Ghiselli et al. [7] in 1994 is basically a duplicate of that part of Glazer's method [2] that uses peroxyl radicals. Glazer's work [2] was published in 1990. It uses AAPH to generate peroxyl radicals or Cu^{2+}-ascorbate to produce hydroxyl radicals. The oxidizable substrate used in the method is B- or R-PE. Glazer assumes that the decrease in fluorescence of B- or R-PE in the presence of peroxyl radicals or hydroxyl radicals is linear with time. A period of complete protection (the length of the lag phase) of B- or R-PE by a sample against peroxyl or hydroxyl radicals is compared to Trolox and then quantitatively related to antioxidant capacity of the sample. However, the kinetics of PE fluorescence quenching are not linear in the presence of peroxyl or hydroxyl radicals. The lag phase is usually difficult to determine when a plasma or serum sample is analyzed.

The procedure of Ghiselli et al. [7] was published as a new method but did not cite Glazer's work [2]. Ghiselli et al. [7] showed (particularly Figs. 1 and 3 in [7]) that the decrease in R-PE fluorescence was linear with time until about 80% of the R-PE fluorescence was quenched, which could not be reproduced in our hands [4,8]. They indicated that their method had a high reproducibility, or a very low CV (0.96%). Ghiselli et al. [7] initially reported their TRAP assay to measure total plasma antioxidant capacity. However, when they validated their work they used artificial plasma samples instead of real plasma samples. The artificial plasma samples were the plasma samples to which an antioxidant cocktail had been added after complete exhaustion of endogenous antioxidants. The cocktail consisted of 400-μmol/L glutathione, 300-μmol/L uric acid, and 30-μmol/L ascorbic acid with a final dilution of 1:250. The measured TRAP value in this way was 1251 $\pm$ 12 μmol/L (CV = 0.96%). Ghiselli et al. [7] gave neither the procedure for exhausting the endogenous antioxidants nor any reasons for using the artificial plasma in the validation. An obvious explanation is that the artificial plasma will likely produce a much better lag phase than the real plasma. However, when the difficulty that people usually have in measuring a lag phase is considered, a CV of 0.96% obtained by using authentic artificial plasma is still unusually low.

OARC assay [4,8,13]

The ORAC assay is based largely on the work reported by Glazer's laboratory [2]. It uses PE as an oxidizable protein substrate and AAPH as a peroxyl radical generator or Cu^{2+}-H_2O_2 as a hydroxyl radical generator. To date, it is the only method that takes free radical action to completion and uses an area-under-curve (AUC) technique for quantitation, and, thus, combines both inhibition percentage and the length of inhibition time of the free radical action by antioxidants into a single quantity. The ORAC assay has been used by different laboratories [36–41] and has provided significant information regarding the antioxidant capacity of various biological samples from pure compounds such as melatonin [32], dopamine [34], and flavonoids [39,42, 43], to complex matrices such as tea [45], fruits [44], vegetables [45], student rasayana (an herbal mixture) [40], and animal tissues [28,37,41,46]. We found in 36 healthy nonsmokers that daily intake of total antioxidant capacity, measured as ORAC, from fruits and vegetables was significantly correlated in a curvilinear manner with the fasting plasma antioxidant capacity also measured as ORAC [47]. Increasing the consumption of fruits and vegetables from the usual 5 servings/day to the experimental 10 servings/day resulted in a significant increase of plasma ORAC [47]. Increased plasma ORAC was also seen in humans after the consumption of strawberries, spinach, red wine, and ascorbic acid [48].

THE COMPARISON OF ORAC ASSAY WITH OTHER ANTIOXIDANT CAPACITY ASSAYS, PARTICULARLY THE FRAP AND TEAC ASSAYS

The use of peroxyl or hydroxyl radicals as pro-oxidants in the ORAC assay makes it different from the assays that involve oxidants that are not necessarily pro-oxidants. The use of a protein substrate (PE) in the ORAC assay makes it different from the assays that use luminol [3,14] or crocin [12,16] as substrate for oxidation. The oxidation of luminol produces luminol radicals that emit light. The antioxidant under analysis may reduce not only radicals produced from AAPH [14] or H_2O_2-horseradish peroxidase [3], but also the luminol radicals resulting from the oxidation of luminol. Assays that use crocin as a substrate may simply measure one antioxidant against another in reducing peroxyl radicals.

Taking the reaction between PE and free radicals to completion and using an AUC technique for quantitation make the ORAC different and unique from the assays that also use PE but measure a lag phase [2,7]. The reproducible measurement of the lag phase is difficult because of the nonzero order kinetics of the PE fluorescence quenching. The AUC technique considers the kinetics of an antioxidant action, which was also addressed recently by the TOSC assay [11] and the kinetics procedure using crocin [12]. The AUC technique should be superior to all those assays that use either an inhibition percentage at a fixed time or a length of inhibition time at a fixed inhibition percentage(s) (i.e., a lag phase). AAPH undergoes spontaneous decomposition and produces peroxyl radicals with a rate primarily determined by temperature. Because of the very high molar ratio (more than 2000) of AAPH to antioxidants used in the ORAC assay, this procedure has high specificity and, thus, measures the capacity of antioxidants to *directly quench free radicals*. By utilizing different extraction techniques in the ORAC assay, one can remove serum proteins and also make some gross differentiation between aqueous and lipid-soluble antioxidants. Due to the use of the fluorescence technique, a dilute emulsion of an oil sample containing antioxidants can also be measured in the ORAC assay. However, the ORAC assay requires about 70 min to quantitate results.

The ORAC assay was compared to the TEAC assay, commercialized by Randox Laboratories, and the FRAP assay in a human study [18]. It was found that there was a weak but significant linear correlation between serum ORAC and serum FRAP. There was no correlation either between serum ORAC and serum TEAC or between serum FRAP and serum TEAC. The effect of dilution on the serum TEAC value and the use of inhibition percentage at a fixed time, without considering the length of inhibition time in the quantitation of results, adversely affected the Randox-TEAC assay. The FRAP assay is simple and inexpensive, but does not measure the SH-group containing antioxidants.

PLASMA OR SERUM ANTIOXIDANT CAPACITY AND IN VIVO ANTIOXIDANT STATUS: INTERPRETATION OF DATA

It appears that total antioxidant capacity in plasma or serum is tightly regulated. The interpretation of the changes in plasma or serum antioxidant capacity depends upon not only the method used in detecting these changes, but also the conditions under which the plasma or serum antioxidant capacity is determined, because the determined antioxidant capacity reflects outcomes in a dynamic system. The increase in total antioxidant capacity in plasma or serum after consumption of antioxidants should indicate an absorption of the antioxidants and an improved in vivo antioxidant defense status [47,48]. An increased plasma or serum antioxidant capacity could also be an adaptation to an increased oxidative stress at an early stage. In addition, an increased antioxidant capacity in plasma or serum may not necessarily be a desirable condition. For example, total antioxidant capacity of serum was significantly increased in patients with chronic renal failure when compared with control subjects [49]. The increase was almost entirely due to the relatively high serum urate; both serum antioxidant capacity and urate decreased after hemodialysis [49]. Thus, in this case, an increased urate is unrelated to oxidative stress. An increased serum antioxidant capacity, measured as ORAC, was also observed in rats exposed to hyperoxia because of a hyperoxia-induced increase in capillary permeability and the redistribution of antioxidants between tissues that followed [50]. Similarly, a decrease in plasma or serum antioxidant capacity is not necessarily an undesirable condition when the production of reactive species decreases. Serum ORAC decreased significantly in rats consuming a diet restricted by 40% in calories [51]; the dietary restriction was known to reduce the production of reactive species and has been recognized as the most effective manipulation by which to extend life span and retard the aging process in laboratory rodents and other short-lived species. The increased production of reactive species may result in a decrease in total antioxidant capacity in vivo. But, as a measure of the amount of systemic oxidative stress plasma or serum total antioxidant capacity may not be the tool of choice. The production of reactive species would probably have to be very extensive to disturb the system's steady state level of antioxidants. Because of these complications, no single measurement of antioxidant status is going to be sufficient, but a "battery" of measurements, many of which will be described in these Forum articles, will be necessary to adequately assess oxidative stress in biologic systems.

CONCLUSION

Antioxidant capacity can be defined as the ability of a compound to reduce pro-oxidants. Pro-oxidants are oxidants of pathologic importance. *Antioxidant* and *pro-oxidant* can be treated as biological terms, while *reductant* and *oxidant* are chemical terms. The use peroxyl or hydroxyl radicals as pro-oxidants, and a protein (PE) as oxidizable substrate in the ORAC assay makes it different from the assays that involve oxidants that are not necessarily pro-oxidants, and the assays that use luminol or crocin as a substrate for oxidation. Taking the reaction between PE and free radicals to completion and using an AUC technique for quantitation in the ORAC assay

improved the quantitation compared with the assays that also use PE but measure a lag phase. A significant linear but weak correlation was found between serum ORAC and serum FRAP, but neither between serum ORAC and serum TEAC, nor between serum FRAP and serum TEAC. The interpretation of the changes in plasma or serum antioxidant capacity depends upon the conditions under which the plasma or serum antioxidant capacity is determined. An increased antioxidant capacity in plasma or serum is not necessarily a desirable condition if it is due to an adaptive response to increased oxidative stress at an early stage. Similarly, the decrease in plasma or serum antioxidant capacity is not necessarily an undesirable condition when the production of reactive species decreases. Increased production of reactive species may result in a decrease in total antioxidant capacity in vivo. But, as a measure of the amount of systemic oxidative stress plasma or serum total antioxidant capacity may not be the tool of choice. A "battery" of measurements are necessary to adequately assess oxidative stress in biological systems.

REFERENCES

[1] Wayner, D. D. M.; Burton, G. W.; Ingold, K. U.; Locke, S. Quantitative measurement of the total, peroxyl radical-trapping antioxidant capacity of human blood plasma by controlled peroxidation. *FEBS Lett.* **187**:33–37; 1985.

[2] Glazer, A. N. Phycoerythrin fluorescence-based assay for reactive oxygen species. *Meth. Enzymol.* **186**:161–168; 1990.

[3] Whitehead, T. P.; Thorpe, G. H. G.; Maxwell, S. R. J. Enhanced chemiluminescent assay for antioxidant capacity in biological fluids. *Anal. Chim. Acta* **266**:265–277; 1992.

[4] Cao, G.; Alessio, H. M.; Cutler, R. G. Oxygen-radical absorbance capacity assay for antioxidants. *Free Radic. Biol. Med.* **14**:303–311; 1993.

[5] Miller, N. J.; Rice-Evans, C.; Davies, M. J.; Gopinathan, V.; Milner, A. A novel method for measuring antioxidant capacity and its application to monitoring the antioxidant status in premature neonates. *Clin. Sci.* **84**:407–412; 1993.

[6] Rice-Evans, C.; Miller, N. J. Total antioxidant status in plasma and body fluids. *Meth. Enzymol.* **234**:279–293; 1994.

[7] Ghiselli, A.; Serafini, M.; Maiani, G.; Assini, E.; Ferro-Luzzi, A. A fluorescence-based method for measuring total plasma antioxidant capability. *Free Radic. Biol. Med.* **18**:29–36; 1994.

[8] Cao, G.; Verdon, C. P.; Wu, A. H. B.; Wang, H.; Prior, R. L. Automated oxygen radical absorbance capacity assay using the COBAS FARA II. *Clin. Chem.* **41**:1738–1744; 1995.

[9] Benzie, I. F. F.; Strain, J. J. The ferric reducing ability of plasma (FRAP) as a measure of "antioxidant power": the FRAP assay. *Anal. Biochem.* **239**:70–76; 1996.

[10] Valkonen, M.; Kuusi, T. Spectrophotometric assay for total peroxyl radical-trapping antioxidant potential in human serum. *J. Lipid Res.* **38**:823–833; 1997.

[11] Winston, G. W.; Regoli, F.; Dugas, A. J., Jr.; Fong, J. H.; Blanchard, K. A. A rapid gas chromatographic assay for determining oxyradical scavenging capacity of antioxidants and biological fluids. *Free Radic. Biol. Med.* **24**:480–493; 1998.

[12] Tubaro, F.; Ghiselli, A.; Papuzzi, P.; Maiorino, M.; Ursini, F. Analysis of plasma antioxidant capacity by competition kinetics. *Free Radic. Biol. Med.* **24**:1228–1234; 1998.

[13] Cao, G., Prior, R. L. The measurement of oxygen radical absorbance capacity in biological samples. *Meth. Enzymol.* **299**:50–62; 1999.

[14] Alho, H.; Leinonen, J. Total antioxidant activity measured by chemiluminescence methods. *Meth. Enzymol.* **299**:3–14; 1999.

[15] Benzie, I. F. F.; Strain, J. J. Ferric reducing/antioxidant power assay: direct measure of total antioxidant activity of biological fluids and modified version for simultaneous measurement of total antioxidant power and ascorbic acid concentration. *Meth. Enzymol.* **299**:15–27; 1999.

[16] Lussignoli, S.; Fraccaroli, M.; Andrioli, G.; Brocco, G.; Bellavite, P. A microplate-based colorimetric assay of the total peroxyl radical trapping capability of human plasma. *Anal. Biochem.* **269**:38–44; 1999.

[17] Re, R.; Pellegrini, N.; Proteggente, A.; Pannala, A.; Yang, M.; Rice-Evans, C. Antioxidant activity applying an improved ABTS radical cation decolorization assay. *Free Radic. Biol. Med.* **26**:1231–1237; 1999.

[18] Cao, G.; Prior, R. L. Comparison of different analytical methods for assessing total antioxidant capacity of human serum. *Clin. Chem.* **44**:1309–1315; 1998.

[19] Rice-Evans, C.; Miller, N. J.; Bolwell, P. G.; Bramley, P. M.; Pridham, J. B. The relative antioxidant activities of plant-derived polyphenolic flavonoids. *Free Radic. Res.* **22**:375–383; 1995.

[20] Salah, N.; Miller, N. J.; Paganga, G.; Tijburg, L.; Bolwell, G. P.; Rice-Evans, C. Polyphenolic flavanols as scavengers of aqueous phase radicals and as chain-breaking antioxidants. *Arch. Biochem. Biophys.* **322**:339–346; 1995.

[21] Miller, N. J.; Sampson, J.; Candeias, L. P.; Bramley, P. M.; Rice-Evans, C. A. Antioxidant activities of carotenes and xanthophylls. *FEBS Lett.* **384**:240–242; 1996.

[22] Miller, N. J.; Castelluccio, C.; Tijburg, L.; Rice-Evans, C. The antioxidant properties of theaflavins and their gallate esters—radical scavengers or metal chelators? *FEBS Lett.* **392**:40–44; 1996.

[23] Strube, M.; Haenen, G. R. M. M.; Van Den Berg, H.; Bast, A. Pitfalls in a method for assessment of total antioxidant capacity. *Free Radic. Res.* **26**:515–521; 1997.

[24] Schofield, D.; Braganza, J. M. Shortcomings of an automated assay for total antioxidant status in biological fluids. *Clin. Chem.* **42**:1712–1714; 1996.

[25] Güler, K.; Palanduz, S.; Ademoğlu, E.; Sahnayenli, N.; Gökkusu, C.; Vatansever, S. Total antioxidant status, lipid parameters, lipid peroxidation and glutathione levels in patients with acute myocardial infarction. *Med. Sci. Res.* **26**:105–106; 1998.

[26] Cao, G.; Giovanoni, M.; Prior, R. L. Antioxidant capacity decreases during growth but not aging in rat serum and brain. *Arch. Gerontol. Geriatr.* **22**:27–37; 1996.

[27] Chevion, S.; Berry, E. M.; Kitrossky, N. K.; Kohen, R. Evaluation of plasma low molecular weight antioxidant capacity by cyclic voltammetry. *Free Radic. Biol. Med.* **22**:411–421; 1997.

[28] Kohen, R.; Beit-Yannai, E.; Berry, E. M.; Tirosh, O. Overall low molecular weight antioxidant activity of biological fluids and tissues by cyclic voltammetry. *Meth. Enzymol.* **300**:285–296; 1999.

[29] Wayner, D. D. M.; Burton, G. W.; Ingold, K. U. The antioxidant efficiency of vitamin C is concentration-dependent. *Biochem. Biophys. Acta* **884**:119–123; 1986.

[30] Mulholland, C. W.; Strain, J. J. Serum total free radical trapping ability in acute myocardial infarction. *Clin. Bichem.* **24**:437–441; 1991.

[31] Dürken, M.; Agbenu, J.; Finckh, B.; Hübner, C.; Pichlmeier, U.; Zeller, W.; Winkler, K.; Zander, A.; Kohlschütter, A. Deteriorating free radical-trapping capacity and antioxidant status in plasma during bone marrow transplantation. *Bone Marrow Transplant.* **15**:757–762; 1995.

[32] Aejmelaeus, R.; Holm, P.; Kaukinen, U.; Metsä-Ketelä, T. J. A.; Laippala, P.; Hervonen, A. J. L.; Alho, H. E. R. Age-related changes in the peroxyl radical scavenging capacity of human plasma. *Free Radic. Biol. Med.* **23**:69–75; 1997.

[33] Pascual, C.; Karzai, W.; Meier-Hellmann, A.; Oberhoffer, M.; Horn, A.; Bredle, D.; Reinhart, K. Total plasma antioxidant capacity is not always decreased in sepsis. *Crit. Care Med.* **26**:705–709; 1998.

[34] Slavíková, H.; Lojek, A.; Hamar, J.; Dušková, M.; Kubala, L.; Vondráček, J.; Číž, M. Total antioxidant capacity of serum increased in early but not late period after intestinal ischemia in rats. *Free Radic. Biol. Med.* **25**:9–18; 1998.

[35] Whitehead, T. P.; Robinson, D.; Allaway, S.; Syms, J.; Hale, A. Effect of red wine ingestion on the antioxidant capacity of serum. *Clin. Chem.* **41**:32–35; 1995.

[36] Pieri, C.; Maurizio, M.; Fausto, M.; Recchioni, R.; Marcheselli, F. Melatonin: a peroxyl radical scavenger more effective than vitamin E. *Life Sci.* **55**:PL271–PL276; 1994.

[37] Testa, R.; Testa, I.; Manfrini, S.; Bonfigli, A. R.; Piantanelli, L.; Marra, M.; Pieri, C. Glycosylated hemoglobin and fructosamines: does their determination really reflect the glycemic control in diabetic patients? *Life Sci.* **59**:43–49; 1996.

[38] Miller, J. W.; Selhub, J.; Joseph, J. A. Oxidative damage caused by free radicals produced during catecholamine autoxidation: protective effects of O-methylation and melatonin. *Free Radic. Biol. Med.* **21**:241–249; 1996.

[39] Lin, Y. L.; Juan, I. M.; Chen, Y. L.; Liang, Y. C.; Lin, J. K. Composition of polyphenols in fresh tea leaves and associations of their oxygen-radical-absorbing capacity with antiproliferative actions in fibroblast cells. *J. Agric. Food Chem.* **44**:1387–1394; 1996.

[40] Sharma, H. M.; Hanna, A. N.; Kauffman, E. M.; Newman, H. A. I. Effect of herbal mixture *student rasayana* on lipoxygenase activity and lipid peroxidation. *Free Radic. Biol. Med.* **18**:687–697; 1995.

[41] Yu, J.; Fox, J. G.; Blanco, M. C.; Yan, L.; Correa, P.; Russell, R. M. Long-term supplementation of canthaxanthin does not inhibit gastric epithelial cell proliferation in *helicobacter mustelae*-infected ferrets. *J. Nutr.* **125**:2493–2500; 1995.

[42] Cao, G.; Sofic, E.; Prior, R. L. Antioxidant and pro-oxidant behavior of flavonoids: structure-activity relationships. *Free Radic. Biol. Med.* **22**:749–760; 1997.

[43] Wang, H.; Cao, G.; Prior, R. L. The oxygen radical absorbing capacity of anthocyanins. *J. Agric. Food Chem.* **45**:304–309; 1997.

[44] Wang, H.; Cao, G.; Prior, R. L. Total antioxidant capacity of fruits. *J. Agric. Food Chem.* **44**:701–705; 1996.

[45] Cao, G.; Sofic, E.; Prior, R. L. Antioxidant capacity of tea and common vegetables. *J. Agric. Food Chem.* **44**:3426–3431; 1996.

[46] Cao, G.; Giovanoni, M.; Prior, R. L. Antioxidant capacity in different tissues of young and old rats. *Proc. Soc. Exp. Biol. Med.* **211**:359–365; 1996.

[47] Cao, G.; Booth, S. L.; Sadowski, J. A.; Prior, R. L. Increases in human plasma antioxidant capacity following consumption of controlled diets high in fruits and vegetables. *Am. J. Clin. Nutr.* **68**:1081–1087; 1998.

[48] Cao, G.; Russell, R. M.; Lischner, N.; Prior, R. L. Serum antioxidant capacity is increased by consumption of strawberries, spinach, red wine or vitamin C in elderly women. *J. Nutr.* **128**:2383–2390; 1998.

[49] Jackson, P.; Loughrey, C. M.; Lightbody, J. H.; McNamee, P. T.; Young, I. S. Effect of hemodialysis on total antioxidant capacity and serum antioxidants in patients with chronic renal failure. *Clin. Chem.* **41**:1135–1138; 1995.

[50] Cao, G.; Shukitt-Hale, B.; Bickford, P. C.; Joseph, J. A.; McEwen, J.; Prior, R. L. Hyperoxia-induced changes in antioxidant capacity and the effect of dietary antioxidants. *J. Appl. Physiol.* **86**:1817–1822; 1999.

[51] Cao, G.; Prior, R. L.; Cutler, R. G.; Yu, B. P. Effect of dietary restriction on serum antioxidant capacity in rats. *Arch. Gerontol. Geriatr.* **25**:245–253; 1997.

ABBREVIATIONS

AAPH—2,2′-azobis(2-amidinopropane) dihydrochloride

ABTS—2,2′-azinobis(3-ethylbenzothiazoline 6-sulfonate)

AC—anodic current

AUC—area-under-curve
AW—anodic wave
CV—coefficient of variation
DCF—dichlorofluorescein
DCFH-DA—dichlorofluorescin-diacetate
FRAP—ferric reducing/antioxidant power
KMBA—α-keto-γ-methiolbutyric acid

ORAC—oxygen radical absorbance capacity
PE—phycoerythrin
TEAC—Trolox equivalent antioxidant capacity
TOSC—total oxyradical scavenging capacity
TRAP—total radical trapping parameter
Trolox—6-hydroxyl-2,5,7,8,-tetramethylchroman-2-carboxylic acid

Bioassays for Oxidative Stress Status (BOSS). Edited by W.A. Pryor

CLINICAL APPLICATION OF BREATH BIOMARKERS OF OXIDATIVE STRESS STATUS

TERENCE H. RISBY and SHELLEY S. SEHNERT

Department of Environmental Health Sciences, The Johns Hopkins University School of Hygiene and Public Health, Baltimore, MD USA

(Received 29 September 1999; Accepted 1 October 1999)

Abstract—Isolation and quantification of volatile breath biomarkers indicative of relevant alterations in clinical status has required development of new techniques and applications of existing analytical chemical methods. The most significant obstacles to successful application of this type of sample have been reduction in required sample volume permitting replicate analysis (an absolute requirement for all clinical studies), separation of the analyte(s) of interest from background molecules, water vapor and other molecules with similar physical properties, introduction of automation in analysis and the use of selective detection systems (electron impact mass spectrometry, flame photometric, thermionic detectors), and automated sample collection from the human subject. Advances in adsorption technology and trace gas analysis have permitted rapid progress in this area of clinical chemistry. © 1999 Elsevier Science Inc.

Keywords—Free radical, Reactive oxygen species, Antioxidant vitamins, Smoking, Ionizing radiation, Reperfusion injury, Breath ethane

INTRODUCTION

In 1990 our research group at Johns Hopkins began the first in a decade-long series of clinical and basic investigations into the use of the isolation and quantification of breath biomarkers of oxidant stress status in normal and diseased humans. These studies have expanded from our early investigations concerning vitamin E status in children to encompass reperfusion injury during organ transplant, the effect of oxidant stress status on organ viability, diet and longevity, and environmental prooxidants, to mention just a few areas to which we have applied this versatile technique.

Essential to all of these investigations was the need to improve the method of breath collection and analysis, making the method reproducible, accurate, and rapid. Optimization of the collection and separation of volatile breath constituents led to enhanced techniques to separate and accurately quantify successively smaller quantities of the analytes of interest. Of course, many other investigators explored areas of breath research simultaneously, and we are fortunate to have enjoyed a productive interchange of information with our colleagues globally regarding this new tool in clinical research.

In this article we review the state of the art of human breath analysis as we have contributed to the field, hoping that this information will inspire others to use this flexible, noninvasive technique to explore problems that will lead to enhanced approaches for diagnosis and disease prevention and amelioration of suffering in humans.

Address correspondence to: Terence H. Risby, Ph.D., Department of Environmental Health Sciences, School of Hygiene & Public Health, The Johns Hopkins University, 615 North Wolfe Street, Baltimore, MD 21205, USA.

Terence H. Risby, Ph.D., a Professor of Environmental Health Sciences, Pathology and International Health at the Johns Hopkins University Medical Institutions, received his doctorate in 1970 from Imperial College of Science, Technology and Medicine, London University. Shelley S. Sehnert, Ph.D., an Instructor of Environmental Health Sciences at the Johns Hopkins University School of Hygiene and Public Health, received her doctorate in 1988 from Johns Hopkins University School of Hygiene and Public Health. These researchers have worked together for more than 15 years using modern analytical chemistry methods to make physiologically relevant measurements.

BREATH COLLECTION AND ANALYSIS

Collection of breath

Humans breathing spontaneously. The primary function of the lung is to exchange oxygen for carbon dioxide by diffusion across the alveolar-capillary membrane. When gas is inspired into the lungs most of the inhaled gas reaches the alveolar-capillary membrane (alveolar ventilation) and the remainder fills the conducting airway (anatomic dead space ventilation). In resting humans, the extent of alveolar ventilation is controlled by the rate of carbon dioxide production by aerobic cellular metabolism. Therefore, the rate of alveolar ventilation is indirectly related to the rate of perfusion in the pulmonary vasculature with the result that the rate of carbon dioxide excretion reaches a steady state. Changes in acid/base balance and systemic metabolism will alter the levels of carbon dioxide in blood and the arterial concentration of carbon dioxide is involved in the regulation of alveolar ventilation and pulmonary perfusion. In resting healthy subjects, 80% of the alveolar gas is exchanged; the remainder is referred to as physiologic dead space ventilation. The composition of expired breath will reflect normal physiologic processes and varies during the process of expiration. If the molecules of interest are derived from blood and their excretion is perfusion limited, their concentrations will increase at the end of expiration and correlate to the highest concentration of expired carbon dioxide (end-tidal carbon dioxide concentration). The maximum exhalation of a molecule, whose excretion is diffusion controlled, may not correlate with end-tidal carbon dioxide concentration. Molecules such as nitric oxide and carbon monoxide are also produced by cells in the conducting airway and, therefore, their exhalation profiles will reflect these additional contributions.

The collection of breath for analysis must consider normal pulmonary physiology because composition of expired breath changes during expiration (anatomic and instrumental dead space ventilation vs. alveolar ventilation). Breath collection is "noninvasive," although the human subject is cognizant of the presence of the mouthpiece during breath collection and is aware of their pulmonary ventilation. Hypo- or hyperventilation will produce changes in the composition of expired breath. During breath collection, an adolescent or adult in a seated position, is asked to inhale and exhale through a disposable mouthpiece and respiratory membrane filter into a two-way nonrebreathing valve, and exhaled breath is collected from the expiratory arm of this valve. For neonates, a silicone infant respiratory mask is used in place of the mouthpiece. Two techniques for breath collection have been investigated: (i) collection of end-tidal breath, and (ii) collection of total breath for a defined period of time (usually 1 min).

1.) Collection of end-tidal breath requires that the subject is breathing at rest and that the breath is collected when the concentration of carbon dioxide reaches a plateau (end-tidal). This approach requires that this portion of breath is representative of all previous and subsequent breaths during resting or relaxed breathing, i.e., steady state, and will produce breath samples with the highest concentration of molecules. These molecules are derived from exchange at the alveolar-capillary membrane and require that analysis be performed on a volume of breath that is a fraction (< 30%) of the tidal volume of an individual breath. This method of breath collection is ideally suited to monitors that can determine the concentration of analyte molecules directly without concentration in real time. Currently, no analytical methods are available to detect breath ethane in real time without concentration.

2.) Collection of total breath for a defined period of time has been the method that has been used most extensively for the determination of breath ethane. This approach has been used by our laboratory to collect breath from premature neonates to normal and diseased adults. The implicit assumption of this approach is that if breath is collected over a period of time, concentrated, and analyzed, then the analytical result reflects the average concentration of breath ethane in exhaled breath over the period of sampling. By definition, the concentration of breath ethane in total collected breath is lower than the end-tidal concentration of breath ethane because it has been diluted by the physiologic and anatomic dead space ventilation. However, the concentration represents the average or integration of all the breaths collected during the sampling period and is less susceptible to spurious breaths. Based upon our experience, the coefficient of variation for the analysis of successive timed breath samples for a given individual is < 3%. We have recently developed a breath sampling device for adolescent and adult subjects that continuously measures and records the instantaneous gas flow, concentration of carbon dioxide, and mouth pressure as a function of time during breath collection. After collection, the tidal volume, breathing frequency, end-tidal concentration of carbon dioxide for each breath, the total minute ventilation, and average concentration of exhaled carbon dioxide are calculated for the period of breath collection. This technique permits complete characterization of the collected breath sample. Studies are in progress that will provide comparable information for the rate of perfusion in the pulmonary vasculature during breath collection. The combination of all this information will enable breath analysis to be related to normal physiology, which is a prerequisite for an understanding of the excretion of endogenously produced analyte molecules.

Humans breathing with mechanical assistance. The collection of breath from intubated humans whose breathing (frequency and tidal volume) is supported by ventilators or anesthesia machines has been performed successfully during our research. Breath can be collected at the end of the endotracheal tube or on the expiratory side of the breathing circuit. The only difference between humans supported by ventilators and anesthesia machines is that the latter involves closed circuit rebreathing to reduce the amounts of anesthetic gases that are used. Closed circuit breathing involves recirculation of the expiratory air through a cartridge containing soda lime to remove carbon dioxide. Additional air or oxygen (fresh gas supply > 1 l/min) is supplied in the closed circuit to maintain the required level of inspiratory oxygen. The gas collected from intubated humans who are supported by ventilators or anesthesia machines is always diluted with the supply gases because a portion of the gas will flow to the expiratory arm bypassing the endotracheal tube. For our studies, the fresh gas flow was maintained at 1 l/min to reduce dilution. The major difficulty experienced with breath collection from intubated humans is that the supply air or oxygen used in ventilators or anesthesia machines often contains quantifiable levels of the analyte molecules.

Background levels of analyte molecules in inspiratory air. Background correction for the concentration of analyte molecules found in inspiratory air is a problem that has no easy solution. From previous discussion it should be apparent that exhaled breath is diluted with inspiratory gas from the anatomic and physiologic dead spaces and, therefore, background correction for the contributions from this inspiratory gas should be performed. Two methods to correct for background contamination have been used in our laboratory. The first involves collection of room air samples (or inspiratory air for intubated humans) concomitantly to the collection of breath samples for subsequent background correction. Alternatively, humans breathing spontaneously can breathe purified air from high-pressure gas cylinders. Four minutes of prebreathing with purified air before breath collection was found to be sufficient because we have demonstrated by the nitrogen wash-out technique that residual gas will be removed from the lungs of normal adult subjects after a period of approximately 2 min of relaxed breathing. However, if the analyte molecule of interest is lipid soluble, a more extensive period of washout is necessary before any total body storage of the analyte molecule is reduced to the concentration of the inspiratory air. The time for total body wash-out has not been determined although based upon our studies with humans supported by anesthesia machines, total body equilibration to the inspiratory gas occurs after a period of about 30 min. In our research both methods to correct for background

concentrations of analyte molecules in inspiratory air have been used successfully.

Breath storage devices. Because real-time monitors for breath biomarkers are currently unavailable, exhaled breath must be collected and carried to the laboratory for concentration and analysis. Most published studies have collected total exhaled breath in inert gas sampling bags. Analysis is performed as soon as possible after collection. The requirements for gas sampling bags are that the bags must be inert and not adsorb analyte molecules. These bags should have a volume that corresponds to at least twice the amount of breath collected thereby minimizing any back pressure from the collected gas that could restrict the exhalation of the study subject. Similarly, the inlet to the gas sampling bag should have sufficient dimensions to minimize any pressure drop across the inlet to the gas sampling bag. Recently, evacuated polished stainless steel cylinders (SUMMA) or thermal desorption tubes packed with adsorbents have been used to collect expiratory gases. These devices collect known volumes of the exhaled breath at controlled flow rates. The advantages of these latter devices are that the subject experiences no resistance to breathing and the collected breath samples can be transported from the sampling site to the laboratory.

Concentration of breath

Off-line concentration. In our initial studies we adopted the published method for breath concentration [1] in which the entire 1-min breath sample was concentrated by cryogenic adsorption in a metal coil packed with silica gel (Porasil-C) (Alltech Associates, Deerfield, IL, USA) maintained at $-180°C$. After concentration, the adsorbent was warmed to $-120°C$, a gas-tight syringe was connected to the outlet of the coil containing the silica gel, and the adsorbent was heated to 100°C. The coil was subsequently flushed with 10 ml of hydrocarbon-free nitrogen gas and the concentrated breath sample was injected onto the gas chromatographic column using a gas sampling valve. Separation was achieved using gas solid chromatography.

On-line concentration. Our preliminary studies using the external concentration of breath were successful but it was found that it was easy to lose an entire breath sample. Therefore, we investigated alternate approaches that would allow only a small volume of the collected breath to be analyzed. We examined a number of adsorbents to trap the molecules of interest in breath at various sampling temperatures: activated charcoal, silica gel, bonded silica gel, alumina, graphitized carbon, and porous polymers. All these adsorbents were found to adsorb breath analyte molecules cryogenically, however, thermal desorption of the adsorbed breath molecules proved to be problematic because some of the adsorbents had to be ramped rapidly to high temperatures to produce analyte peaks with reasonable efficiencies. The porous polymer 2,6-diphenyl-p-phenylene oxide (Tenax GC, [Alltech Associates] 60/80 mesh) was found to trap all analyte breath molecules that were of interest to us at $-121°C$ and to quantitatively thermally desorb them at 135°C. Various designs for the cryogenic trap were investigated and the most successful was a stainless steel "U" tube (15 cm × 1.65 mm o.d.; 1.19 mm i.d.) packed with the porous polymeric adsorbent. This collection tube was connected in place of the standard sample loop of a six-port gas sampling valve. This collection tube was cooled cryogenically with a liquid nitrogen/ethanol slush bath and allowed to equilibrate for 6 min. After this time, a 60-ml aliquot of collected breath was drawn through the adsorbent bed. After collection, the valve was rotated and the cryogenic bath was replaced immediately with specially designed heating block maintained at 135°C and the concentrated breath molecules were swept onto the gas chromatographic column (gas solid chromatographic column or wide-bore capillary gas chromatographic column). This was the method of breath concentration used by our laboratory for most of our published studies. Using this method for on-line concentration, each sample of breath could be analyzed in triplicate with a coefficient of variation for replicate samples < 3%. The major difficulty for this method of concentration was that the method is labor intensive and not amenable to automation.

The limitations of using gas sampling bags, the need to develop a method of breath concentration at room temperature, and availability of an analytical method that could be automated led us to evaluate commercially available thermal desorption systems. We selected an automated two-stage thermal desorption system (Perkin Elmer Corporation, Norwalk, CT, USA) for evaluation and examined all the available adsorbents and mixed beds of adsorbents that were commercially available for the analysis of gases and vapors. These adsorbents were examined for their ability to collect and concentrate exhaled breath. None of these systems were found to be acceptable for breath collection and concentration. Our criteria in these studies were the ability to collect breath ethane quantitatively at ambient temperatures in the presence of water vapor and carbon dioxide. Subsequently, we were asked to evaluate some experimental adsorbents that were manufactured for research purposes (Supelco Corporation, Bellefonte, PA, USA). A number of these adsorbents were examined and a glass thermal desorption tube packed with equal amounts of a graphitized carbon (Carbopack X) and a carbon molecular sieve

(Carboxen-1018) separated by glass wool met our criteria. This graphitized carbon has a nitrogen surface area of 240 m^2/g, density of 0.41 g/ml, and a pore volume restricted to the meso range of 0.62 ml/g. The carbon molecular sieve has a nitrogen surface area of 700 m^2/g, density of 0.6 g/ml, and a pore volume restricted to micro range of 0.35 ml/g. Breath is sampled through the bed of graphitized carbon into the bed of carbon molecular sieve. Thermal desorption is performed in the opposite direction to the sampling flow. Quantitative concentration of the molecules of interest in breath was achieved at room temperature and quantitative thermal desorption of the adsorbed molecules was produced at around 300°C. The temperature difference between adsorption and desorption is approximately 270°C, which is similar to the temperature difference we had shown to be optimum for our earlier studies with cryogenic adsorption and thermal desorption using the polymeric adsorbent. Breath analysis was performed with a minimum volume of 60 ml of collected breath. Room temperature concentration of exhaled breath using this thermal desorption tube is considerably easier than collection in gas sampling bags, and results in significantly smaller amounts of water and carbon dioxide being trapped.

Analysis

Packed column gas chromatography. In our initial studies we adopted a modification of a published method for the analysis of breath ethane using gas solid chromatography [1]. Separation was performed using a column packed with porous divinylbenzene-styrene copolymer at an isothermal temperature (40°C). After ethane had eluted, the column was back-flushed to remove all the less volatile molecules found in breath. Because this method of analysis was time consuming and we were interested in quantifying other molecules in exhaled breath, we examined alternate column temperature programs and gas solid adsorbents (silica gel, bonded silica gels, alumina, graphitized carbon, and porous polymers). Major problems were experienced with silica gel and alumina due to the water vapor that was trapped by cryogenic concentration that produced a variation in retention data as a function of the number of samples run. Water vapor was deactivating these columns and retention volumes decreased. In addition, it was difficult to separate ethylene from ethane. Ethylene is found in the breath of humans and may be another molecule involved in molecular signaling. The optimum column (1.83 m × 3.2 mm o.d., 2.16 mm i.d.) packing material was found to be octadecyl silyl bonded silica (Chemisorb C18 80/100 mesh) and separation was achieved using the following temperature program: isothermal 35°C for 5 min, linear temperature program 35–215°C at 10°C/min, and iso-

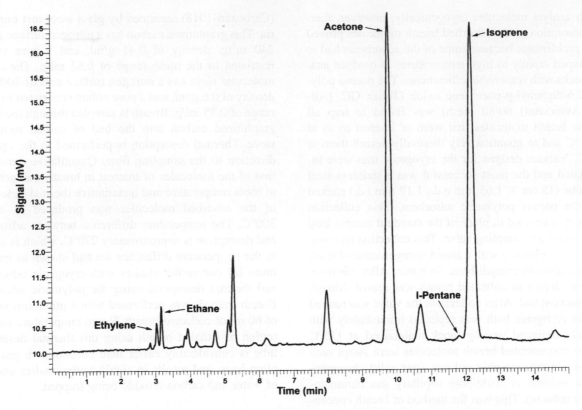

Fig. 1. Capillary gas chromatogram of a representative breath sample collected and analyzed using a thermal desorption tube.

thermal at 215°C for 13 min. These conditions produced resolution of ethylene and ethane and of 1-pentane and isoprene but not to baseline. This protocol was used for many of our earlier studies that quantified ethane in a variety of breath samples. However, the gas chromatographic column had to be replaced regularly and, therefore, we investigated alternate columns for the separation of exhaled breath.

Capillary gas chromatography. Wide-bore (0.53 mm i.d.) fused silica columns were examined because regular bore columns are not suitable for samples introduced by gas sampling valves. Gas sampling valves require higher flow rates in order to minimize sample dispersion. Various columns were investigated including porous layer open tubular (PLOT) columns (polydivinylbenzene and alumina) and various bonded methyl silicones. The polydivinylbenzene PLOT column was found to be excellent for the separation of molecules in the range of C1–C3, but had significant baseline drift when the column was temperature programmed. The alumina PLOT column was easily deactivated by the presence of water vapor in concentrated breath. A fused silica open tubular column wall coated with a thick film (7 μm) of cross-linked bonded dimethyl silicone (60 m) was found to produce base-

line resolution of target molecules with the following temperature protocol: isothermal at 25°C for 5 min, 25–200°C at 5°C/min, and isothermal at 200°C for 5 min. One additional change was to use pressure regulation to control the gas flow since the rotation of the gas sampling valve produced significant flow variations when flow controllers were used. This system was reproducible and was used successfully until the conversion to two-stage thermal desorption was made. Currently, a fused silica open tubular column (0.32 mm) wall coated with a thick film (5 μm) of cross-linked bonded dimethyl silicone (60 m) is used to separate molecules found in exhaled breath (Fig. 1), with the following temperature protocol: isothermal at 35°C for 10 min, 35–200°C at 5°C/min, and isothermal at 200°C for 10 min.

Expression of data

The response factors for each of the breath molecules were obtained by injections of known volumes of standard gas mixtures from calibration curves. For all our studies the concentrations of the analyte molecules fell within the linear range. The minimum detectable signal for ethane corresponded to approximately 0.1 parts per billion (ppb). Initially, the data

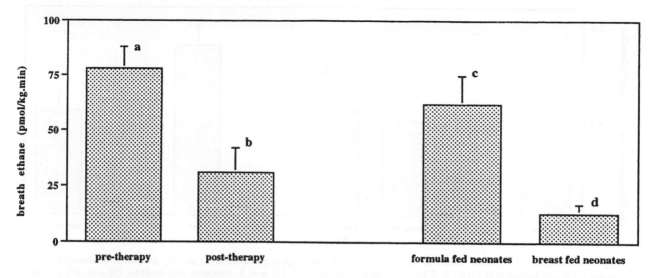

Fig. 2. Effects of nutrition on breath ethane: (a) and (b) data collected from pediatric patients with vitamin E deficiency, (a) $p < .001$, $n = 8$, compared with (b); (c) and (d) data collected from neonates, (c) $p < .04$, $n = 26$, compared with (d), $n = 10$. Statistical analyses were performed with ANOVA, data are means $\pm$ SE.

were expressed in concentration units (ppb or pmol/l) but this information does not permit intersubject comparisons to be made. Therefore, the concentration data were converted to generation rates by the measurement of the minute ventilation (l/min) and subsequently corrected for body weight (pmol/[kg · min]). For intubated subjects the data were corrected for body surface area (nmol/[m^2 · min]) because these are the units that are used clinically by most anesthesiologists. More recently, an additional correction has been made for anatomic, physiologic, and instrumental dead space based upon the concentration of exhaled carbon dioxide. During collection, the end-tidal concentration and average concentration of carbon dioxide over each breath are quantified and recorded. After the breath sample has been collected, the average values for these measurements during breath collection can be made. The average end-tidal concentration of carbon dioxide and the average concentration of carbon dioxide during exhalation (mixed expired) can be used to determine the volume that corresponds to the anatomic, physiologic, and instrumental dead space by the following calculation: {[end-tidal concentration of carbon dioxide in Torr] − [average concentration of carbon dioxide during exhalation in Torr]}/{[end-tidal concentration of carbon dioxide in Torr] × [average tidal volume of each breath]}. This calculation assumes that the subject is relaxed breathing at steady state. The concentration of any analyte molecule that is measured in exhaled breath is decreased by this factor and, therefore, the actual concentration should be corrected by this factor.

CLINICAL APPLICATIONS

Antioxidant vitamin status in children

In 1988, Lemoyne and associates [1] published a clinical report that demonstrated the levels of breath pentane correlated inversely with vitamin E status in adult study subjects. Our investigations [2] extended these studies to include pediatric study subjects, and used levels of breath ethane instead of breath pentane to monitor vitamin E status in a study population with severe chronic liver disease accompanied by decreased metabolic capacity. Measuring levels of markers that correlate with mechanisms that focus on pro- and antioxidant status in these children led us to support Lemoyne et al.'s [1] conclusions that reduced levels of antioxidants selenium and vitamin E synergistically promote tissue lipid peroxidation and elevated levels of breath ethane. Therapeutic administration of parenteral vitamin E at high levels produced both clinical improvement and decreased levels of breath ethane for most of the study subjects (Fig. 2). However, for one study subject the level of breath ethane remained high after this vitamin E therapy and normal levels of breath ethane were achieved after additional therapy with oral selenium. The extension of breath analysis to include pediatric study subjects, and subsequently to premature neonates (discussed later), required that methods of analysis for breath hydrocarbons based upon small volumes of collected breath were to be developed.

Studies in animal models by both our group [3] and others led us to study molecular mechanisms that induce

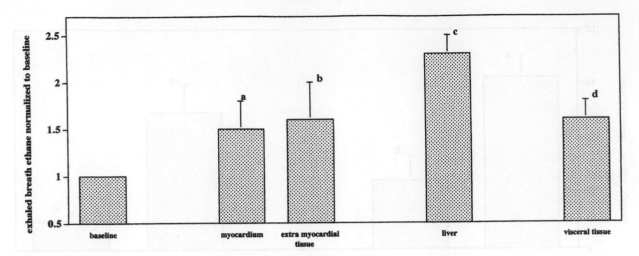

Fig. 3. Reperfusion of ischemic tissues in humans during different surgical procedures: (a) and (b) data collected during human cardiopulmonary bypass time points 0–5 min after reperfusion, (a) $p < .05$, $n = 8$, compared with baseline, (b) $p < .02$, $n = 8$, compared with baseline; (c) data collected during human liver transplantation time point 1–3 min after portal vein opened, (c) $p < .001$, $n = 8$, compared with baseline; (d) data collected during human supraceliac aortic cross clamping time points 16–25 min after reperfusion, (d) $p < .05$, $n = 6$, compared with baseline. Statistical analyses were performed with ANOVA, data are means ± SE.

elevated levels of breath hydrocarbons as a result of changes in oxidative stress status.

Origins of breath hydrocarbons: hepatic studies

We focused our clinical research program on the liver because it is the site of intense metabolic and synthetic activity, and is well perfused by blood. We postulated that pro-oxidant injury to this distal organ may produce rapid increases in the levels of exhaled breath hydrocarbons. Peroxidation of the polyunsaturated fatty acids in cell membranes became the focus of our energies. We selected the surgical setting because it minimized the effects of extrinsic factors such as dietary pro-oxidants, antioxidants, and fats. Moreover, by design, each subject acted as his/her own control. We limited our evaluation of pro-oxidant sources to reactive oxygen species, and the situation we sought to dissect was organ reperfusion, specifically, reperfusion of the liver during human orthotopic transplantation [4].

Granger and collaborators (for review see [5]) have proposed that reperfusion injury occurs during organ transplantation by the following mechanism: during the anoxic period between harvest and implantation of the donor graft, catabolism of ATP is interrupted and hypoxanthine and xanthine accumulate in the ischemic tissue. Additionally, ischemia may mediate the conversion of xanthine dehydrogenase to xanthine oxidase. When reperfusion of the transplanted graft occurs, the oxidation of the purine metabolites by xanthine oxidase generates superoxide initiating a free radical mediated chain reaction that peroxidizes cell membrane lipids, leading to

cellular and organ damage (for review see [5]). Preservation of organs in cold solutions containing antioxidants seeks to reduce this contribution to injury. In the intrasurgical setting we used breath ethane, a stable endproduct of the free radical peroxidation of lipids, to determine whether it was possible to quantify reperfusion injury and thereby separate this injury from ischemic damage. Our group at Johns Hopkins successfully conducted these studies, and demonstrated clinically the observation that reperfusion injury occurs during human orthotopic liver transplantation (Fig. 3).

During the course of these studies many technical difficulties presented themselves to us, not the least of which was the issue of collection of breath intraoperatively from an anesthetized patient whose ventilation is artificially supported. Unfortunately, it was not possible to collect simultaneous collection of tissue samples for analysis of complementary markers of oxidative stress during this procedure. Blood samples were collected but no definitive results were obtained because many the transplant recipients received significant quantities of blood and blood products during surgery.

Of course, in this setting, we could not provide the definitive element that would indisputably correlate breath ethane to pro-oxidant status via the mechanisms that have been proposed. For example, the reversal of pro-oxygenation by the administration of an exogenous substrate that quenches the free radical reaction, namely, superoxide dismutase. However, we subsequently successfully performed this experiment many times in an in vivo animal model, and demonstrated also that ethane levels correlated to other biomarkers of increased/de-

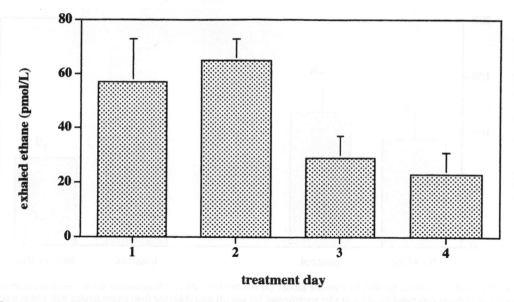

Fig. 4. Pretreatment levels of breath ethane on each of the 4 treatment days of total body irradiation. Data are means ± SE, $n = 4$, data do not reach statistical significance.

creased oxidative stress status including reperfusion injury, Factor VII, and HSP 72.

Sources of oxidative injury: radiotherapy and ethane evolution

In order to confirm that the production of ethane was an oxygen free radical mediated event, we used a clinical radiotherapeutic approach that is well-established to function via oxygen free radical mediated intracellular reactions. In collaboration with a team of radiation oncologists from the Johns Hopkins Hospital, we used, to our advantage, therapeutic doses of radiation administered during clinical radiotherapy that are quantifiable (whereas, during organ reperfusion, the dose of "free radical generating agent" is not readily quantifiable). Of course, levels of enzymes involved in the reactions leading to hydrocarbon generation can be quantified and related to levels of blood, tissue, and breath biomarkers of oxidative stress.

Using total body irradiation (TBI) before bone marrow transplant for treatment of hemopoietic malignancies as our clinical setting [6], we demonstrated that oxygen free radicals induced by this form of ionizing radiation initiated a chain of cellular events that lead to elevated breath ethane levels. Samples were collected over a 4-d TBI treatment protocol with breath sampled before and after each treatment interval.

This study indicated that levels of breath ethane correlated positively with the clinical radiotherapeutic dose events. Moreover, antioxidant defenses evidenced by the daily breath sample collected before treatment appeared to be increased by continued radiation treatment. This induction of antioxidant defenses, over the four treatment protocol, has been confirmed by subsequent studies with additional study patients undergoing TBI before bone marrow transplantation (Fig. 4). Unfortunately, blood and tissue samples that could have confirmed this increased in antioxidant defenses are not normally collected during radiotherapy.

Charting reperfusion injury in the absence of cold ischemia

Human liver transplantation is a complex surgical undertaking, complicated by many factors. For example, the recipient's biochemistry is severely compromised by terminal liver disease. In addition, variability induced by storage and preservation of organs between harvest and transplantation, donor status, and intrasurgical complications all contribute to difficulties in assessing and quantifying the oxygen stress status of transplanted organs. To minimize these variables, we studied patients in a surgical setting that involved a warm ischemic phase for a relatively brief and controlled interval without a preservation scenario, because transplantation was not involved. Crossclamping of the supraceliac aorta for repair of complex aneurysms was selected for study, and we monitored levels of breath ethane throughout this surgical procedure. Reperfusion injury of the viscera in this surgical setting produces a familiar pattern of increased oxidative stress characterized by increased breath ethane levels. Interestingly, in this setting, reperfusion injury

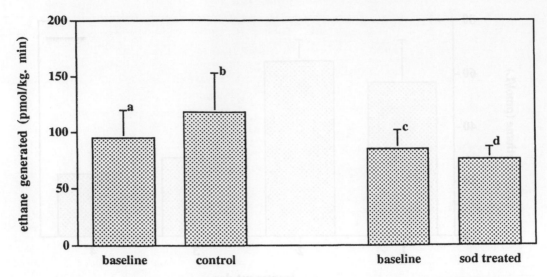

Fig. 5. Reperfusion of warm ischemic liver in a porcine model: (a) and (b) data collected from control swine, (a) ethane generated prior to reperfusion, (b) ethane generated 10–15 min after reperfusion; (c) and (d) data collected from swine treated with sod at reperfusion, (c) ethane generated prior to reperfusion, (d) ethane generated 10–15 min after reperfusion [$p < .05$, $n = 4$, compared with (b)]. Statistical analyses were performed with ANOVA, data are means ± SE.

occurred more slowly than described above in the hepatic transplantation setting (Fig. 3 [7]).

These results were explored in a porcine model of warm ischemia of hepatic tissue that was developed to include all aspects of liver transplantation without retrieval and transplantation. Veno-venous bypass of the liver for 2 h was used because it is typical of the warm ischemic time required for retrieval and implantation during human liver transplantation. Samples of blood, liver tissue, and breath were collected consecutively through the procedure. Correlations between several different tissue biomarkers of oxygen free radical mediated injury were observed (generation of breath ethane and conjugated dienes) and tissue injury (levels of liver enzymes [serum aspartate aminotransferase (AST) and serum alanine aminotransferase (ALT)], expression of heat shock protein (HPS)-72 [8], and markers of liver function, Factor VII, and clearance [indocyanine green]). Of particular significance, ablation of tissue injury demonstrated by the levels of AST, ALT, HSP-72, Factor VII, and indocyanine clearance was achieved by the administration of the free radical scavenger superoxide dismutase at reperfusion. Tissue protection correlated significantly to the decline in the levels of breath ethane (Fig. 3 [9]).

Reperfusion injury during cardiopulmonary bypass

Cardiac surgery is a clinical setting in which two reperfusion events occur that are temporally separated. An initial reperfusion event occurs when the aortic cross-clamp is removed and coronary blood flow is re-estab-

lished, and this is followed by the second event when the arrested heart and lungs are restored to normal function. This surgical procedure was evaluated for evidence of reperfusion injury because it involved two events in different tissue masses. Moreover, this surgical procedure provides unique clinical information, such as cardiac output, that is not normally measured during other surgeries. The levels of breath ethane produced after the reperfusion of the myocardium correlated to indices of cardiac function. The levels of breath ethane after reperfusion of extra-myocardial organs correlated to improved hemodynamics and better oxygen carrying capacity (Fig. 3 [10]). These results suggest that breath ethane levels may be predictive of successful interventions aimed at reducing reperfusion injury and improving clinical outcome.

Oxidative stress status: measuring the effect of dietary antioxidants

In the preceding discussion we presented several different types of acute pro-oxidant tissue injury that provided significant evidence that breath ethane is a sensitive, reproducible biomarker of oxidative stress status in humans. We have also mechanistically accounted for the pathways that lead to ethane production in these pro-oxidant situations, and have correlated increases in the levels of breath ethane under different conditions to levels of blood and tissue biomarkers of oxidative injury. Moreover, we have biochemically quenched the pro-oxidant response and shown that this effect was paralleled by the changes in breath ethane. We expanded on

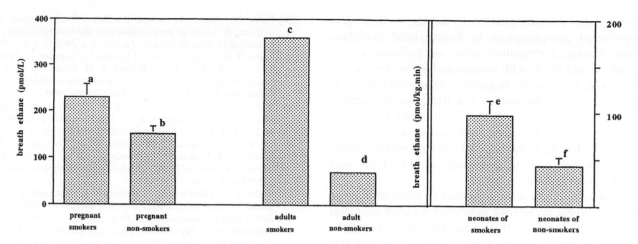

Fig. 6. Effects of smoking on breath ethane. Data collected from pregnant women [(a) and (b)]: (a) $p < .03$, $n = 19$, compared with (b), $n = 25$. Statistical analyses were performed with ANOVA, data are means $\pm$ SE. Data collected from adult population [(c) and (d)]: (c) $p < .05$, $n = 14$, compared with (d), $n = 107$. Statistical analyses were performed with Wilcoxon two-sample test, data are medians. Data collected from neonates [(e) and (f) data expressed as pmol/kg $\cdot$ min]: (e) $p < .03$, $n = 19$, compared with (f), $n = 26$. Statistical analyses were performed with ANOVA, data are means $\pm$ SE.

these results in collaboration with Kathleen Schwartz, a pediatric gastroenterologist specializing in antioxidant supplementation in compromised children, by characterizing the effect of chronic pro-oxidants and dietary antioxidants on breath ethane levels [11]. We found that current dietary standards for adult antioxidant sufficiency do not provide protection from the pro-oxidant effects of maternal tobacco use during pregnancy. We observed a direct correlation between the number of cigarettes smoked and breath ethane levels (Fig. 5). No correlation was observed between breath ethane and vitamin E supplementation in smokers, which may be an indicator that increased vitamin E prophylaxis at the doses studied was insufficient to compensate for the strongly pro-oxidant effect of maternal smoking [11].

We also explored the effect of dietary vitamin A and E in newborn formula on the relationship between the pro-oxidant toxicity of maternal smoking and oxidative stress status in neonates as measured by breath ethane. As expected, maternal smoking was strongly pro-oxidant to the neonate; and, interestingly, vitamin A and E supplemented formula was found to be pro-oxidant as compared with breast milk [12] (Fig. 6). Thus, maternal smoking appears to compromise the neonate irrespective of all other factors. Serum vitamin A and E levels did not correlate to infant breath ethane.

We were interested to evaluate further whether vitamin A, which has receptor-mediated effects and may be a more potent antioxidant than vitamin E, could be protective. In premature neonates deficient in vitamin A, we found supplementation produced positive therapeutic effect that correlated inversely to breath ethane levels [13].

Similarly case-controlled studies designed to better

dissect the relationship between dietary antioxidant status and breath ethane levels in nonpregnant adults showed that there was a direct correlation between cigarette smoking and several biomarkers of increased oxidative stress, including increased levels of breath ethane (Fig. 6 [14,15]).

We explored breath ethane as a biomarker of the dietarily-induced pro-oxidant processes that induce atherogenesis in normal adults. During a controlled human feeding study that was designed to limit dietary fat intake and increase antioxidant consumption (via increased fruit and vegetable consumption), we found that dietary modification in ways that increase healthy lifestyle factors significantly affect serum antioxidant capacity. Importantly, this response was protective against oxygen free radical lipid peroxidation [16].

Applicability of breath analysis to human diagnostic requires further study to understand the effects of endogenous and exogenous factors on analytes (clinically relevant) in breath. Obviously, certain medications, diet, and, most obviously, smoking, are lifestyle factors that seriously compromise the validity of the breath analysis.

Breath isoprene and ozone exposure

Of course, ethane is not the only analyte of interest in breath, which contains as many as 400 different compounds, and only about 30 of which have been identified. We have initiated studies aimed at identifying the pathophysiologic origins of other breath biomarkers including additional stable products of lipid peroxidation. Isoprene, a major hydrocarbon found in breath, is related to the biosynthesis of cholesterol and is postulated to be produced by an acid catalyzed elimination reaction from dimethylallyl pyrophosphate or by the Mg++ dependent isopentenyl pyrophosphate to dimethylallyl pyrophosphate. Ozone, a ubiquitous urban air pollutant, causes oxidation and reacts with unsaturated hydrocarbons. We postulated that breath isoprene levels would decrease during exposure to ozone, and thus, breath isoprene might be useful as a noninvasive biomarker of ozone exposure. Ozone pollution is linked to several health effects, including increased incidence of asthma attacks and decreased ease of breathing in individuals with chronic lung disease.

Healthy subjects were subjected to laboratory ozone exposure using previously published standard protocols [17]. Results indicated that ozone exposure initiated the induction of antioxidant defenses. We postulated that ozone induced the release of reactive oxygen species that damage cellular membranes and initiate oxidation of the cholesterol dependent secretions, as for example, in lung surfactant. In 24-h postexposure, significant increases in breath isoprene were found and suggest breath isoprene utility as a delayed marker of the lung response to an ambient pollutant.

CONCLUSIONS

Breath ethane is currently being used as a biomarker of oxidative stress to study the pathogenesis of a number of other disease states such as Downs syndrome, Alzheimer's disease, diabetes, infection, and systemic inflammation. Ethane generation has been utilized to investigate the generation of free radicals during dialysis for kidney failure. Moreover, breath biomarkers of additional biochemical pathways are receiving active investigation by our group. The development of real-time monitors for specific breath analytes is an area of research that will continue to expand with the potential for exhaled biomarkers to evolve as diagnostic tools and clinical indicators of homeostasis.

Acknowledgements — We wish to acknowledge our many clinical colleagues who have collaborated with us during the course of these studies, for without their help, advice, and support these studies would not have been possible. The generous gift of automatic thermal desorption systems, capillary gas chromatograph and data system by the Perkin Elmer Corporation is gratefully acknowledged. This work was partially supported by funds from the US Air Force Office of Scientific Research (F49620-95-1-0270, F49620-98-1-0403), and the National Institutes of Health (PO1HL56091).

REFERENCES

[1] Lemoyne, M.; VanGossum, A.; Kurian R.; JeeJeebhoy, K. N. Plasma vitamin E and selenium and breath pentane in home parenteral nutrition patients. *Am. J. Clin. Nutr.* **48**:1310–1315; 1988.

[2] Refat, M.; Moore, T.; Kazui, M.; Risby, T. H.; Perman, J.; Schwarz, K. Utility of breath ethane as a non-invasive vitamin E biomarker of status in children. *Pediatr. Res.* **30**:396–403; 1991.

[3] Risby, T. H.; Jiang, L.; Stoll, S.; Ingram, D.; Spangler, E.; Heim, J.; Cutler, R.; Roth, G. S.; Rifkind, J. M. Breath ethane as a marker of hypoxia and oxidant stress in dietarily restricted female Fisher 344 rats. *J. Appl. Physiol.* **86**:617–622; 1999.

[4] Risby, T. H.; Maley, W.; Scott, R. P. W.; Bulkley, G. B.; Kazui, M.; Sehnert, S. S.; Schwarz, K. B.; Potter, J.; Mezey, E.; Klein, A. S.; Colombani, P.; Fair, J.; Merritt, W. T.; Beattie, C.; Mitchell, M. C.; Williams, G. M.; Perler, B. A.; Donham, R. T.; Burdick, J. F. Evidence for free radical-mediated lipid peroxidation at reperfusion of human orthotopic liver transplants. *Surgery* **115**:94–101; 1994.

[5] Reilly, P. M.; Schiller, H. J.; Bulkley, G.B. Pharmacologic approach to tissue injury mediated by free radicals and other reactive oxygen metabolites. *Am. J. Surg.* **161**:488–503; 1991.

[6] Arterbery, V. E.; Pryor, W. A.; Jiang, L.; Sehnert, S. S.; Foster, W. M.; Abrams, R. A.; Williams, J. R.; Wharam, M. D. Jr.; Risby, T. H. Breath ethane generation during clinical total body irradiation as a marker of oxygen-free-radical-mediated lipid peroxidation: a case study. *Free Radic. Biol. Med.* **17**: 569–576; 1994.

[7] Kazui, M.; Andreoni, K. A.; William, G. M.; Perler, B. A.; Bulkley, G.; Beattie, C.; Donham, R.; Sehnert, S.; Burdick, J.; Risby, T. Viceral lipid peroxidation occurs at reperfusion after supraceliac aortic crossclamping. *J. Vasc. Surg.* **19**:473–477; 1994.

[8] Schoeniger, L. O.; Andreoni, K. A.; Ott, G. R.; Risby, T. H.; Bulkley, G. B.; Udelsman, R.; Burdick, J. F.; Buchman, T. G. Induction of heat shock gene expression in the post-schemic liver is dependent upon superoxide generation at reperfusion. *Gastroent.* **106**:177–184; 1994.

[9] Kazui, M.; Andreoni, K. A.; Norris, E. J.; Klein, A. S.; Burdick, J. F.; Beattie, C.; Sehnert, S. S.; Bell, W. R.; Bulkley, G. B.; Risby, T. Breath ethane: a specific indicator of free radical-mediated lipid peroxidation following reperfusion of the ischemic liver. *Free Radic. Biol. Med.* **13**:509–515; 1992.

[10] Andreoni, K. A.; Kazui, M.; Cameron, D. E.; Nyham, D.; Sehnert, S. S.; Rohde, C. A.; Risby, T. H. Ethane: a marker of lipid peroxidation during cardiopulmonary bypass in humans. *Free Radic. Biol. Med.* **26**:439–455; 1999.

[11] Schwarz, K. B.; Cox, J.; Sharma, S.; Witter, F.; Clement, L.; Sehnert, S.; Risby, T. H. Cigarette smoking is pro-oxidant in pregnant women regardless of antioxidant nutrient intake. *J. Nutrit. Environ. Med.* **5**:225–234; 1995.

[12] Schwarz, K. B.; Cox, J. M.; Sharma, S.; Clement, L.; Witter, F.; Abbey, H.; Sehnert, S.; Risby, T. Pro-oxidant effects of maternal cigarette smoking in formula-fed newborns. *J. Pediat. Gastroent. Nutr.* **24**:68–74; 1997.

[13] Schwarz, K. B.; Cox, J. M.; Sharma, S.; Clement, L.; Humphrey, J.; Gleason, C.; Abbey, H.; Sehnert, S.; Risby, T. Possible antioxidant effect of vitamin A supplementation in premature infants. *J. Pediat. Gastroent. Nutr.* **25**:408–414; 1997.

[14] Miller, E. R.; Appel, L. J.; Jiang, L.; Risby, T. H. The impact of cigarette smoking on measures of oxidative damage. *Circulation* **94**:I143; 1996.

[15] Miller, E. R.; Appel, L. J.; Jiang, L.; Risby, T. H. The association of cigarette smoking and lipid peroxidation in a controlled feeding study. *Circulation* **96**:1097–1101; 1997.

[16] Miller, E. R.; Appel, L. J.; Jiang, L.; Risby, T. H. The effect of dietary patterns on measures of lipid peroxidation: results from a randomized clinical trial. *Circulation* **98**:2390–2395; 1998.

[17] Foster, W. M.; Jiang, L.; Stetkiewicz, P.; Risby, T. H. Breath isoprene: temporal changes in respiratory output after exposure to ozone. *J. Appl. Phys.* **80**:706–710; 1996.

Bioassays for Oxidative Stress Status (BOSS). Edited by W.A. Pryor

ANALYSIS OF OXIDIZED HEME PROTEINS AND ITS APPLICATION TO MULTIPLE ANTIOXIDANT PROTECTION

AL L. TAPPEL

Department of Food Science and Technology, University of California, Davis, CA, USA

(Received 25 August 1999; Accepted 22 September 1999)

Abstract—Oxidation in tissues and homogenates can be determined by the analysis of oxidized heme proteins. Oxidation of heme proteins can be measured by spectral changes and the deconvolution of the spectra of mixtures of heme proteins by a spreadsheet heme spectra analysis program (HSAP), incorporating the spectra of the individual pure heme proteins. HSAP also is used to analyze the spectra of mixtures of heme proteins found in the literature. HSAP is applied in measuring the protective effects in rats of multiple antioxidants suitable for use in humans for protection against diseases. © 1999 Elsevier Science Inc.

Keywords—Heme proteins, Spectra, Tissues, Liver, Heart, Vitamin E, Selenium, Antioxidants, Free radicals

ANALYSIS OF OXIDIZED HEME PROTEINS

Because new methods are needed to measure tissue oxidation and the oxidative processes in animals, we have developed spectral techniques and a heme spectra analysis program (HSAP) [1,2] as new methods of measuring heme protein oxidation in tissues. Existing analyses of oxidation in biological samples quantitate primarily irreversible oxidative products. HSAP measures reversible and irreversible damage to heme proteins. We have used rat heart, liver, kidney, and spleen in studies of heme protein oxidation in homogenates and tissue slices.

HSAP uses the known spectral absorption of all of the major heme proteins in the calculation we have devel-

oped to measure the oxidized and reduced products. The experimental absorbance spectra of tissue homogenates were obtained with a Beckman DU-50 spectrophotometer (Beckman Instruments, Inc., Fullerton, CA, USA). For tissue homogenates, a light path of 10 mm was used. Four layers of parafilm, representing turbidity, were used as a background to subtract some of the absorbance caused by turbidity inherent in tissue homogenates. Samples were scanned from 390 to 450 nm (γ region) at a $10 \times$ dilution, and from 500 to 640 nm (α and β regions). Absorbance vs. wavelength at 2-nm intervals was automatically recorded by a scan program in the spectrophotometer. Spectral data obtained for fresh homogenates and for incubant samples were transferred to the Excel spreadsheet computer program (Microsoft Corp., Redmond, WA, USA).

Heme proteins in tissue slices were measured after incubation of the tissue slices. The absorbance spectrum of each sample was obtained with a Beckman DU-50 spectrophotometer. A 50-mg sample was put into a small cylindrical spectrophotometer cell of 5.5-mm diameter and a light path of 2.0 mm. In the airtight cell, the heme proteins came to a redox equilibrium determined by the physiologic conditions of the tissues. Parafilm was again used to correct for some of the turbidity. The cell was sealed by a microscope cover glass and mounted at the center of the window of the spectrophotometer as close as possible to the photoreceptor to reduce light scattering

Address correspondence to: Professor Al L. Tappel, Department of Food Science and Technology, University of California, Davis, CA 95616, USA; Tel: (530) 752-1488; Fax: (530) 752-4759.
Al L. Tappel is Professor Emeritus at the University of California, Davis. He received a Ph.D. in Biochemistry from the University of Minnesota. He has done considerable research on the topic of methods for measuring oxidative damage in biologic components. He has taught analytical methods at the undergraduate and graduate levels. Areas of research include in vivo and in vitro lipid peroxidation and oxidative damage to proteins including heme proteins. His work includes: research on vitamin E as a biological lipid antioxidant, selenium glutathione peroxidase as a protective enzyme, and the use of multiple antioxidants in protecting against in vivo oxidative stress. He has published over 450 articles. He has received a total of seven research awards from The Oxygen Club, American Society for Nutritional Sciences, American Chemical Society, American Oil Chemists' Society, and a Guggenheim Fellowship.

Reprinted from: *Free Radical Biology & Medicine*, Vol. 27, Nos. 11/12, pp. 1193-1196, 1999

caused by the tissue. The sample was scanned from 390
to 450 nm and from 500 to 640 nm. The absorbance
versus wavelength at 2-nm intervals was automatically
recorded by the scan program in the spectrophotometer.
Spectral data were transferred to the Excel spreadsheet
computer program.

To deconvolute the spectra obtained in the DU-50
spectrophotometer, HSAP uses the multiple-regression
capabilities of an Excel spreadsheet. Micromolar extinction coefficients at 2-nm intervals from 390 to 450 and
500 to 640 nm were preprogrammed into the spreadsheet
for the literature standard spectra of hemoglobin, oxyhemoglobin, methemoglobin, hemochrome, hemichrome,
reduced and oxidized cytochromes of the mitochondria,
and reduced and oxidized cytochromes of the microsomes. Because all these pigments are present in tissue,
for example, heart tissue, it is a major problem to deconvolute the spectra of the combined absorption of all
pigments into the spectral absorption of each component
and thus determine the quantity of each heme pigment.
We have done this. The spectra of rat heart homogenate
shows in the visible region the major peak of oxyhemoglobin and in the near-ultraviolet region the corresponding γ peak of the heme proteins (Fig. 1) [3]. Oxyhemoglobin from blood dominates the spectra of nonperfused
heart. After the homogenate was allowed to oxidize for
up to 4 h, the spectrum indicated principally methemoglobin with the corresponding γ peak in the near ultraviolet region. The spectrum also incorporates a correction for turbidity absorption that must be subtracted from
the absorption of the homogenate to get the absorption of
heme proteins. In HSAP, experimental absorbance values at the same wavelengths are inserted into a reserved
column in the spreadsheet. Multiple regression is performed to determine heme protein concentrations using
the extinction coefficients of the heme proteins as independent variables and the experimental absorbance values as the dependent variables.

Precise quantitation of the amount of heme proteins
present in a homogenate sample provided accurate assessment of the oxidized heme proteins calculated by
HSAP. This quantitation was achieved through modification of existing pyridine hemochrome methods. Input
into HSAP of the total heme protein content via the
pyridine hemochrome value generated reproducible values for oxidized heme proteins.

The program has broad potential as a multicomponent
analysis tool. Modification of HSAP led to the development of a difference spectra analysis program (DSAP),
which was used to quantitate the type and amount of
heme proteins observed in mitochondrial difference
spectra. HSAP and DSAP provide methods for interpreting complex spectral information of multicomponent biological samples that undergo oxidation [1].

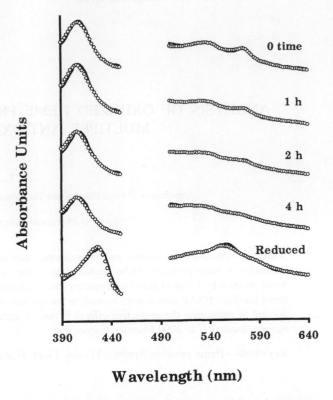

Wavelength (nm)

Fig. 1. Visible spectra of heme proteins in rat heart homogenates
incubated at 37°C for various amounts of time and analyzed with
HSAP: OOO = experimental spectra; — = calculated spectra. The
reduced spectrum was obtained from the 0 time sample after reduction
with sodium dithionite. The spectra are stacked, and the Soret bands
(below 450 nm) are plotted at approximately one-fifth of the original
absorbancies. Calculated spectra closely match and are hidden behind
the experimental spectra. Reprinted from North and Tappel [3].

APPLICATION ON HSAP TO LITERATURE SPECTRA OF HEME PROTEINS

We have used rat kidney, heart, liver, and spleen
homogenates and liver and heart slices. The amount of
oxidized heme proteins obtained by HSAP correlates
with other measurements of tissue oxidation, thiobarbituric acid reacting products [3], and fluorescent lipid
peroxidation products [4].

We have applied HSAP to spectra of heme proteins
from the literature that are otherwise difficult for quantitative analysis [1,2]. As an example, Fig. 2 shows
absorbance values from a spectrum of isolated adult rat
heart cells under anaerobic conditions and the spectrum
calculated with HSAP. Because the published spectrum
provided only relative measurement of absorbance,
HSAP provided only qualitative information. HSAP
matched the experimental spectrum using deoxymyoglo-
bin and reduced cytochromes of the mitochondria as the
heme proteins responsible for the spectral data in the
proportions expected in rat heart, approximately 38 and
62%, respectively. The results shown in Fig. 2 demon-

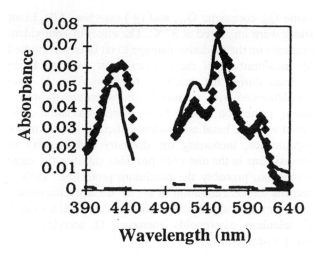

Fig. 2. HSAP analysis of absorbance spectrum of isolated adult rat heart cells under anaerobic conditions: ◆ = values determined from spectra from the literature; — = model spectrum calculated with HSAP. The (—) line under the spectra is the absorbance calculated for turbidity with HSAP. Reprinted from North et al. [1] with permission (Copyright © 1996 Academic Press, Inc.).

strate the ability of HSAP to supply good calculated fits to the experimental data. For a number of published spectra, HSAP approximations support the qualitative description of the spectra and are in accord with the expected biochemistry. The published spectra in the literature are usually calculated to contain only one or two hemoglobin species, whereas tissue and homogenate samples represent complex spectra containing multiple heme protein mixtures [1,2].

Table 1. Addition of Antioxidant Nutrients to Basal Diet

	Amount of antioxidant (mg/kg) diet group		
Antioxidant	B	E + Se	All
Vitamin E	0	25	25
Selenium		0.3	0.3
β-Carotene	0	0	45
Coenzyme Q_{10}	0	0	30
Ascorbic acid 6-Palmitate	0	0	100
Canthaxanthin	0	0	45
Trolox C	0	0	50
Acetylcysteine	0	0	200
Coenzyme Q_0	0	0	100
(+)-Catechin	0	0	100

In addition to its applications to animal tissue slices and homogenates, HSAP can have applications to human studies. The passage of light through tissue is the critical requirement. We have analyzed spectra from the literature with HSAP to determine the relative amounts of heme proteins in a reflectance spectrum of human skin. Figure 3 shows values for the logarithm of the inverse of reflectance of white skin and the model spectrum calculated with HSAP. HSAP matched the experimental spectrum using oxyhemoglobin, deoxyhemoglobin, and a slanted baseline to approximate the absorbance from melanin ($r^2 = 0.95$). Because the logarithm of the inverse of reflectance only approximates absorbance, HSAP could not be used quantitatively in this example.

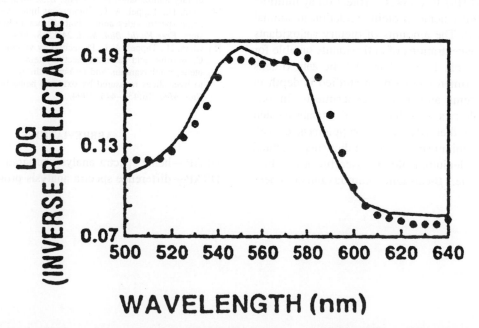

Fig. 3. Inverse of reflectance of caucasian skin and the spectrum calculated with HSAP: ●●● = values determined from the experimental data; —— = model spectrum calculated with HSAP. Reprinted with permission from Boyle et al. [2] (Copyright © 1994 American Chemical Society).

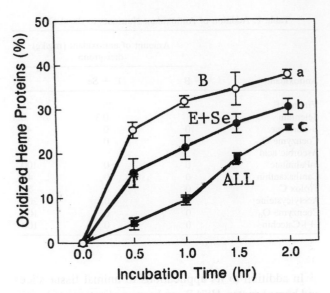

Fig. 4. Production of oxidized heme proteins during the spontaneous oxidative reaction in liver slices. For composition of each diet see Table 1. The values are expressed as mean ± SD for three rats. The curves marked with different lower case letters are significantly different from each other at a 95% confidence level. Reprinted from Inform. *Production of oxidized heme proteins during the spontaneous oxidative reaction in liver slices.*, (vol. 6). 1995: 778–783. with permission (Copyright © 1995 AOCS Press).

The results support the published qualitative explanation of the experimental spectrum.

HSAP IN MULTIPLE ANTIOXIDANT STUDIES

We have used HSAP to assess protection by multiple antioxidants against heme protein oxidation in animal dietary studies [5]. The amounts of dietary antioxidants fed to rats and concentrations of antioxidants suitable for human intake, (Table 1) correlate. All the antioxidants used in our experiments have been studied in depth in biochemical systems and animal experiments. In one investigation male rats were fed a basal vitamin E– and selenium-deficient diet (B), a diet supplemented with vitamin E and selenium (E + Se) or a diet supplemented with vitamin E, selenium, trolox C, ascorbic acid palmitate, acetylcysteine, β-carotene, canthaxanthin, coen-

zyme Q_0, coenzyme Q_{10}, and (+)-catechin (All). Liver slices were incubated at 37°C. The effect of antioxidant nutrients on the oxidative damage to rat liver was studied by measurement of the production of oxidized heme proteins during the oxidative reactions (Fig. 4). Diet supplemented with vitamin E and selenium showed a strong protection against heme protein oxidation compared with the basal antioxidant-deficient diet. Of major significance, increasing the diversity and quantity of antioxidants in the diet (All) provided significantly more protection, probably the maximum protection. Table 1 records the concentration of all the dietary antioxidants used for maximal protection. Humans have taken vitamin E, selenium, carotenoids, coenzyme Q, acetylcysteine, and (+)-catechins.

SUMMARY

HSAP is useful in resolving spectra of mixtures of heme proteins and in analyzing oxidation of heme proteins with a major application of determining protection given by multiple antioxidants.

REFERENCES

[1] North, J. A.; Rein, D.; Tappel, A. L. Multicomponent analysis of heme protein spectra in biological materials. *Anal. Biochem.* **233:** 115–123; 1996.
[2] Boyle, R. C.; Tappel, A. L.; Tappel, A. A.; Chen, H.; Andersen, H. J. Quantitation of heme proteins from spectra of mixtures. *J. Agric. Food Chem.* **42:**100–104; 1994.
[3] North, J. A.; Tappel, A. L. Measuring the oxidation of heme compounds in heart homogenates from rats supplemented with dietary antioxidants. *Free Radic. Biol. Med.* **22:**175–184; 1997.
[4] Rein, D.; Tappel, A. L. Fluorescent lipid oxidation products and heme spectra index antioxidant efficacy in kidney tissue of hamsters. *Free Radic. Biol. Med.* **24:**1278–1284; 1998.
[5] Chen, H.; Tappel, A. L. Protection by vitamin E, selenium, trolox C, ascorbic acid palmitate, acetylcysteine, coenzyme Q, beta-carotene, canthaxanthin, and (+)-catechin against oxidative damage to liver slices measured by oxidized heme proteins. *Free Radic. Biol. Med.* **16:**437–444; 1994.

ABBREVIATIONS

HSAP—heme spectra analysis program
DSAP—difference spectra analysis program

Bioassays for Oxidative Stress Status (BOSS). Edited by W.A. Pryor

OXIDATIVE STRESS STATUS—THE SECOND SET

WILLIAM A. PRYOR

Biodynamics Institute, Louisiana State University, Baton Rouge, LA, USA

In the December 1999 issue of *FRBM* (volume 27, numbers 11/12), we presented the first group of seven papers in a Forum on Oxidative Stress Status (OSS). In my introduction, I explained that this OSS Forum eventually will include a substantial number of papers, which will be published in batches as they are completed and pass through the refereeing process.

In this issue, *FRBM* is pleased to present the second group of four papers that are a part of this OSS Forum. This Forum leads off with a contribution from the Vanderbilt group that has developed methodology for detecting isoprostanes. I had predicted in 1975–1976 that isoprostanes would be formed, and that some would have biological activity [1,2], but it took the inspiration, the marvelous applications of separation science, and the development of exquisite analytical techniques by the group of Roberts, Morrow, and their associates to bring this technique to life [3]. It now appears that levels of isoprostanes may be the most reliable method of measuring OSS. In the first article in this Forum, a precis of this exciting development is presented by its discoverers.

The second article reviews the technique of measuring ethane and pentane in expired breath. This is an extremely promising method, because it is noninvasive and can be done as a function of time. It suffers, however, from the relatively small differences between elevated values and control values, as these hydrocarbon gases are ubiquitous. The analytical techniques required, therefore, are exacting. The review of this technology by Knutson and his associates presents some new insights into this method.

The years preceding our entrance into the 21st century were enlivened by the discovery that reactive nitrogen species (RNS) are as commonplace and important in cells as are our somewhat older friends, the reactive oxygen species (ROS). This discovery stems, of course, from the ground-breaking concept that nitric oxide is a signal molecule, a discovery that lead to the awarding of the Nobel Prize to Louis Ignarro, Ferid Murad, and Robert Furchgott. Shortly after our attention was centered on the new kid on the free radical block, •NO, Beckman and his associates pointed out that •NO and superoxide were both radicals, are often produced by the same cells, and combine at nearly the diffusion limit.

Thus, the biochemistry of the combination product of nitric oxide and superoxide, peroxynitrite (O=N-O-O$^-$), sprang on the scene. The pioneer explorers Irwin Fridovich and Joe McCord set sail and discovered the large and important "continent" in free radical biology called superoxide, and years of exciting trading with the newly discovered "natives" followed. It appears that the "brother continent" of RNS will produce even more exciting discoveries. A good bit of attention in recent years has centered on developing a biomarker for peroxynitrite activity, and the one most often studied is nitrotyrosine. In the third contribution in this Forum, Kenneth Hensley and his associates at Oklahoma present analytical methods for the detection of both nitrotyrosine and nitro-γ-tocopherol.

The final contribution in this set of OSS articles is by H. M. Shen and C. N. Ong, who discuss the detection of oxidized DNA, including 8-oxodeoxyguanosine, in human sperm. Preserving viable sperm has become increasingly important in recent years, as techniques for aided conception have advanced.

The editors of *FRBM* hope you enjoy this second set of our OSS Forum. Another, considerably larger group of papers will appear in the coming few months—even though this is the second set, we are still a good distance from the end of the match.

Address correspondence to: Dr. William A. Pryor, Biodynamics Institute, 711 Choppin Hall, Louisiana State University, Baton Rouge, LA 70803, USA.

REFERENCES

[1] Pryor, W. A.; Stanley, J. P. A suggested mechanism for the production of malonaldehyde during the autoxidation of polyunsaturated fatty acids. Nonenzymatic production of prostaglandin endoperoxides during autoxidation. *J. Org. Chem.* 40:3615–3617; 1975.

[2] Pryor, W. A.; Stanley, J. P.; Blair, E. Autoxidation of polyunsat-

Reprinted from: *Free Radical Biology & Medicine, Vol. 28, No. 4, pp. 503-504, 2000*

urated fatty acids. Part II. A suggested mechanism for the formation of TBA-reactive materials from prostaglandin-like endoperoxides. *Lipids* 11:370–379; 1976.

[3] Morrow, J. D.; Hill, K. E.; Burk, R. F.; Nammour, T. M.; Badr,
K. F.; Roberts, L. J., II. A series of prostaglandin F$_2$-like compounds are produced *in vivo* in humans by a non-cyclooxygenase,
free radical-catalyzed mechanism. *Proc. Natl. Acad. Sci. USA* 87:
9383–9387; 1990.

Bioassays for Oxidative Stress Status (BOSS). Edited by W.A. Pryor

MEASUREMENT OF F$_2$-ISOPROSTANES AS AN INDEX OF OXIDATIVE STRESS IN VIVO

L. Jackson Roberts, II and Jason D. Morrow

Departments of Pharmacology and Medicine, Vanderbilt University, Nashville, TN, USA

(Received 3 December 1999; Accepted 7 December 1999)

Abstract—In 1990 we discovered the formation of prostaglandin F$_2$-like compounds, F$_2$-isoprostanes (F$_2$-IsoPs), in vivo by nonenzymatic free radical–induced peroxidation of arachidonic acid. F$_2$-IsoPs are initially formed esterified to phospholipids and then released in free form. There are several favorable attributes that make measurement of F$_2$-IsoPs attractive as a reliable indicator of oxidative stress in vivo: (i) F$_2$-IsoPs are specific products of lipid peroxidation; (ii) they are stable compounds; (iii) levels are present in detectable quantities in all normal biological fluids and tissues, allowing the definition of a normal range; (iv) their formation increases dramatically in vivo in a number of animal models of oxidant injury; (v) their formation is modulated by antioxidant status; and (vi) their levels are not effected by lipid content of the diet. Measurement of F$_2$-IsoPs in plasma can be utilized to assess total endogenous production of F$_2$-IsoPs whereas measurement of levels esterified in phospholipids can be used to determine the extent of lipid peroxidation in target sites of interest. Recently, we developed an assay for a urinary metabolite of F$_2$-IsoPs, which should provide a valuable noninvasive integrated approach to assess total endogenous production of F$_2$-IsoPs in large clinical studies. © 2000 Elsevier Science Inc.

Keywords—Free radical, Isoprostanes, Lipid peroxidation, Oxidant stress

BIOCHEMISTRY OF THE FORMATION OF ISOPROSTANES

One of the greatest needs in the field of free radical research has been the availability of a reliable noninvasive approach to assess oxidative stress status in humans

Address correspondence to: L. Jackson Roberts, II, MD, Department of Pharmacology, 522 MRB-1, Vanderbilt University, Nashville, TN 37232, USA; Tel: (615) 343-1816; Fax: (615) 343-9446; E-Mail: jack.roberts@mcmail.vanderbilt.edu

L. Jackson Roberts, II, received his M.D. degree from the University of Iowa and was elected to Alpha Omega Alpha. He did an Internal Medicine residency at Washington University, a fellowship in Clinical Pharmacology at Vanderbilt University, and was a recipient of a Burroughs Wellcome Scholar in Clinical Pharmacology Award. He is currently Professor of Pharmacology and Medicine at Vanderbilt University.

Jason D. Morrow received his M.D. degree from Washington University. He did an Internal Medicine residency at Vanderbilt University, an Infectious Disease fellowship at Washington University, and a Clinical Pharmacology fellowship at Vanderbilt University. He is currently a Professor of Medicine and Pharmacology at Vanderbilt University. He was a Howard Hughes Medical Institute Physician Research Fellow and the recipient of an NIH Physician Scientist Award, a International Life Sciences Institute Career Development Award, and a Burroughs Wellcome Fund Clinical Scientist Award in Translational Research.

[1]. It has long been recognized that methods previously developed for this purpose lack specificity, sensitivity, or are too invasive for human investigation [2]. In 1990, we reported the discovery of the formation of prostaglandin F$_2$-like compounds in vivo in humans by nonenzymatic free radical–induced peroxidation of arachidonic acid [3]. The mechanism by which these compounds are formed is depicted in Fig. 1. Initially, three initial arachidonoyl radicals are formed which then undergo endocylization to form four prostaglandin H$_2$-like bicyclic endoperoxide intermediate regioisomers which are reduced to four F-ring regioisomers. Each regioisomer is comprised of eight racemic diastereomers. Because these compounds are isomeric to prostaglandin F$_{2\alpha}$ formed by the cyclooxygenase, they have been termed F$_2$-isoprostanes (F$_2$-IsoPs). The regioisomers are designated as different series according to the carbon number on which the side chain hydroxyl is located. This is in based on the nomenclature system for IsoPs approved by the Eicosanoid Nomenclature Committee, sanctioned by the Joint Commission on Biochemical Nomenclature (JCBN) of the International Union of Pure and Applied Chemistry (IUPAC) [4]. The vast majority of arachidonic acid is

Reprinted from: *Free Radical Biology & Medicine, Vol. 28, No. 4, pp. 505-513, 2000*

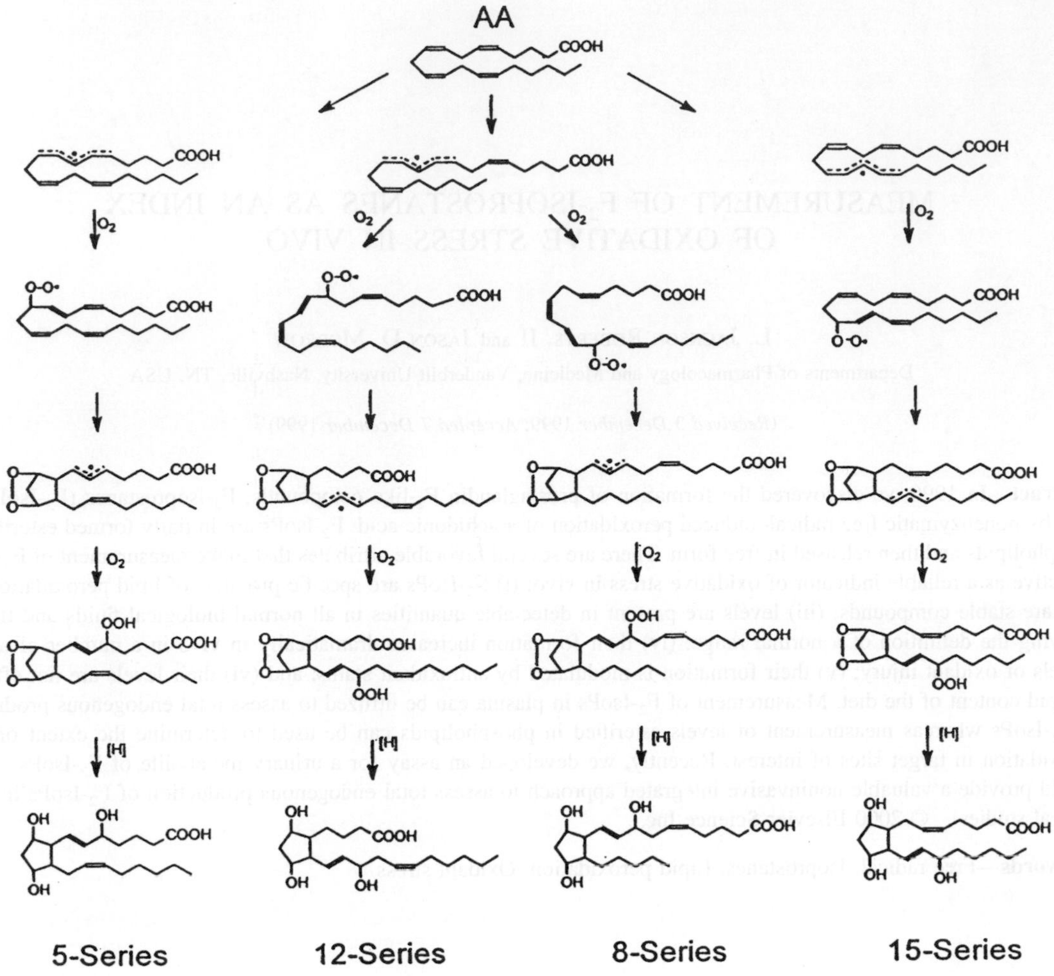

Fig. 1. Mechanism of formation of F$_2$-IsoPs.

present esterified in phospholipids rather than in free form. In this regard, we have found that F$_2$-IsoPs are initially formed esterified on phospholipids and then released in free form by a phospholipase(s) [5].

Central in the pathway of formation of IsoPs are the endoperoxide intermediates which are reduced to F-ring IsoPs. We recently demonstrated that glutathione is an important effector of this reduction [6]. However, we have shown that this reduction is not completely efficient and that the H$_2$-IsoP endoperoxides rearrange in vivo to form E-ring, D-ring, thromboxane-ring, A-ring, and J-ring IsoPs [7–9]. More recently, we have shown that they also undergo rearrangement to form highly reactive acyclic γ-ketoaldehydes, termed *isolevuglandins* [10]. In addition, several IsoPs have been found to exert potent and interesting bioactivity. This involves both receptor-mediated actions, e.g., vasoconstriction, in the case of F$_2$-IsoPs and E$_2$-IsoPs and biological effects due to inherent chemical reactivity, e.g., adduct formation, in the case of A$_2$-IsoPs, J$_2$-IsoPs, and isolevuglandins [9–12].

While the products of the IsoP pathway produced in addition to the F$_2$-IsoPs are of interest, they are not as stable as F$_2$-IsoPs and thus are not ideal candidates for measurement as a marker of lipid peroxidation. For this forum, therefore, we will focus on discussion of the utility of measuring F$_2$-IsoPs as an index of oxidative stress status.

METHODOLOGY FOR MEASUREMENT OF F$_2$-ISOPROSTANES

Some discussion about the methodology used for measurement of F$_2$-IsoPs is important. The initial discovery of F$_2$-IsoPs was made possible by the use of mass spectrometric analysis. We, and others, continue to use stable isotope dilution gas chromatography negative ion chemical ionization mass spectrometry for measurement of F$_2$-IsoPs [13]. Although mass spectrometric methodology is expensive and time consuming, it is highly specific and sensitive. The accuracy of our method of assay

Table 1. Favorable Characteristics of Measurement of F$_2$-IsoPs as an Index Oxidative Stress Status In Vivo

Stable compounds
Specific products of lipid peroxidation
Present in detectable quantities in all normal biological tissues and fluids, thus allowing the definition of a normal range
Levels increase substantially in animal models of oxidant injury
Levels are modulated by antioxidant status
Levels are unaffected by lipid content in the diet

is 96% and the precision is ± 5%. Enzyme-linked immunosorbent assay (ELISA) kits are now available for measurement of F$_2$-IsoPs by several commercial vendors. However, immunoassays for F$_2$-IsoPs are associated with the same potential shortcomings that have been recognized with immunoassays for prostaglandins for over 3 decades [14]. These problems are primarily related to substances in biological fluids that interfere with the immunoassay. In this regard, the immunoassays can usually be demonstrated to work very well in buffer systems that do not contain large amounts of biological substances but in complex biological fluids and tissues, interfering substances are frequently encountered. More often than not, biological samples must be purified to some extent before performing the immunoassay. Simple partial purification procedures such as extraction may actually concentrate interfering substances, thus requiring more extensive purification by thin-layer chromatography or high-performance liquid chromatography. A recent article by Proudfoot and colleagues compared measurement of F$_2$-IsoPs in urine by ELISA and mass spectrometry and found considerable inconsistencies [15]. Thus, at present, measurement of F$_2$-IsoPs by mass spectrometry remains the method of choice.

FAVORABLE CHARACTERISTICS OF F$_2$-ISOPROSTANES AS A MEASURE OF OXIDATIVE STRESS IN VIVO

The discovery of the formation of F$_2$-IsoPs was initially of biochemical interest but evolving from subsequent studies is the notion that measurement of these compounds likely represents one of the most reliable approaches to assess oxidative stress status in vivo. There are a number of favorable attributes that imply that measurement of F$_2$-IsoPs may provide a reliable marker of lipid peroxidation in vivo (Table 1). First, these are stable compounds. F$_2$-IsoPs are also specific products of free radical–induced lipid peroxidation. F$_2$-IsoPs have been found to be present in detectable quantities esterified in all normal biological tissues and in free form in all normal biological fluids. This is important because it allows the definition of a normal range such that small increases in their formation can be detected in situations of mild oxidant stress. It should be mentioned that IsoPs are not a major product of lipid peroxidation vide infra

but levels are present in vivo that can be readily detected by currently available methodology. Importantly, the formation of F$_2$-IsoPs has been shown to increase dramatically in well-established animal models of oxidant injury, e.g., administration of CCl$_4$ to normal rats and administration of diquat to Se-deficient rats [3]. Further, levels are modulated by endogenous antioxidant status, e.g., vitamin E and Se, and by exogenous administration of antioxidant agents [16–18]. Finally, there may be a concern that levels of F$_2$-IsoPs may be influenced by lipid content in the diet which can contain IsoPs as a result of oxidation of dietary arachidonic acid. However, we had previously shown urinary levels of F$_2$-IsoPs in subjects ingesting a normal diet were unchanged after 4 d of a diet consisting of only glucose polymers [3]. In a recent study by Richelle and colleagues, it was also confirmed that levels of IsoPs are unaffected by lipid content of the diet [19].

DIRECT COMPARISON OF MEASURING F$_2$-ISOPS WITH OTHER MEASURES OF LIPID PEROXIDATION TO ASSESS OXIDANT INJURY IN VITRO AND IN VIVO

To explore whether the value of measuring F$_2$-IsoPs surpasses that of some routinely used measures of lipid peroxidation in vivo, we directly compared measurements of F$_2$-IsoPs with malondialdehyde (MDA) and lipid hydroperoxides, commonly used measures of lipid peroxidation, both in vitro and in vivo. The results obtained were most revealing.

Comparison between the formation of malondialdehyde and F$_2$-IsoPs

Initially, we compared the time-course of formation of F$_2$-IsoPs with MDA in vitro during oxidation of rat liver microsomes with iron/ADP/ascorbate [20]. MDA was measured by the thiobarbituric acid reacting substances (TBARS) assay [21]. As shown in Fig. 2A, the time-course of formation of both MDA and F$_2$-IsoPs were highly correlated. Of note, however, are the different scales on the two y-axes indicating that there was approximately 25,000 times more MDA generated than F$_2$-IsoPs. We then compared the amounts of MDA and esterified F$_2$-IsoPs formed in the liver of rats after ad-

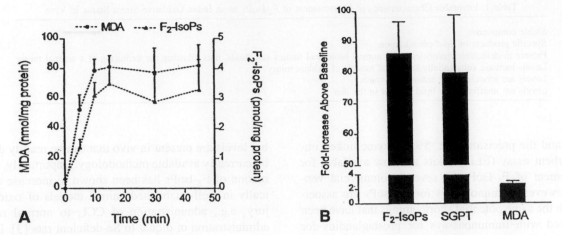

Fig. 2. (A) Time-course of formation of MDA and F₂-IsoPs during oxidation of rat liver microsomes. (B) Levels of F₂-IsoPs esterified in liver lipids, levels of MDA in liver, and plasma concentrations of SGPT (serum glutamic pyruvic transaminase) following administration of CCl₄ to rats.

ministration of CCl₄ to induce a severe oxidant injury to the liver (Fig. 2B). As a index of the severity of liver injury, plasma concentrations of serum glutamic pyruvic transaminase (SGPT) were also measured. Levels of esterified F₂-IsoPs in the liver increased strikingly by approximately 85-fold. This was accompanied by a comparable increase in plasma concentrations of SGPT, indicating a good correlation between the magnitude F₂-IsoP production and severity of hepatocellular injury. In contrast, levels of MDA increased only less than 3-fold. Thus, whereas the relative increases in the formation of MDA and F₂-IsoPs were found to correlate highly during oxidation of microsomal lipids in vitro, the relative increase in levels of F₂-IsoPs detected in the setting of CCl₄-induced oxidant injury in vivo far surpassed the increase in levels of MDA detected. The reason why the relative increase in MDA after administration of CCl₄ in vivo was much less compared with the increase in F₂-IsoPs remains speculative but may be due to rapid metabolic clearance [22]. This demonstration of the greater utility of measurement of F₂-IsoPs compared with MDA in vivo as an index of oxidant injury is even further enhanced by the recognition that the TBARS assay is not specific for MDA nor is MDA a specific marker of lipid peroxidation [2,23].

Comparison between the formation of lipid hydroperoxides and F₂-IsoPs

We also carried out separate studies comparing the generation of lipid hydroperoxides with F₂-IsoPs both in vitro and in vivo. In collaboration with Frei and colleagues [24], we compared the formation of cholesterol ester lipid hydroperoxides and F₂-IsoPs during Cu⁺²-induced oxidation of human low-density lipoprotein

(LDL). As shown in Fig. 3A, the relative increases and time-courses of formation of cholesterol ester lipid hydroperoxides, measured by chemiluminescence assay, and F₂-IsoPs were highly correlated. Of note again, however, is that the levels of cholesterol ester lipid hydroperoxides are plotted on a micromole scale whereas the levels of F₂-IsoPs are plotted as nanomoles, indicating that approximately 1000-times higher amounts of cholesterol ester lipid hydroperoxides were formed compared with F₂-IsoPs. The decline in levels of both during later times of incubation of LDL with Cu⁺² were due to hydrolysis of the lipid hydroperoxides and F₂-IsoPs. Although not shown, this was confirmed by demonstrating that the decline in levels of esterified F₂-IsoPs was mirrored by an increase in levels of free F₂-IsoPs in the incubation buffer.

In collaboration with Matthews and colleagues [22], we also undertook studies in which we compared the relative amounts of formation of arachidonic acid–derived lipid hydroperoxides and F₂-IsoPs in vivo esterified in liver lipids and free in the circulation during administration of CCl₄ to rats [17]. In these experiments, lipid hydroperoxides were measured by a highly sensitive and accurate stable isotope dilution negative ion chemical ionization gas chromatography/mass spectrometry (GC/MS) method reduction of the lipid hydroperoxides to stable alcohols (hydroxyeicoastetraenoic acids [HETEs]). As shown in Fig. 3B, the fold-increase in levels of F₂-IsoPs esterified in liver of rats treated with CCl₄ compared to levels measured in untreated rats greatly exceeded that of HETEs. Also of note is that F₂-IsoPs could be detected in the circulation of untreated rats and levels increased more than 20-fold following administration of CCl₄, whereas HETEs could not be detected in the circulation, even in this setting of severe

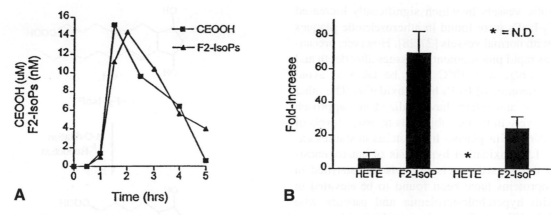

Fig. 3. (A) Time-course of formation of cholesterol ester lipid hydroperoxides and esterified F$_2$-IsoPs during Cu^{+2}-induced oxidation of LDL. (B) Levels of lipid hydroperoxides and F$_2$-IsoPs esterified in liver and circulating concentrations in plasma following administration of CCl$_4$ to rats. Lipid hydroperoxides were measured following reduction to alcohols (HETE's). N.D. = not detected.

CCL$_4$-induced oxidant injury using a highly sensitive method of assay. The results of this study again highlights the greater utility of measuring F$_2$-IsoPs as an index of free radical–induced lipid peroxidation compared with another measure of lipid peroxidation, namely measurement of lipid hydroperoxides.

VARIOUS APPROACHES TO ASSESS ENDOGENOUS PRODUCTION OF F$_2$-ISOPS

There are several approaches that can be utilized to assess endogenous production of F$_2$-IsoPs, each of which has certain advantages and/or drawbacks.

Measurement of free unmetabolized IsoPs

Sampling of biological fluids for measurement of free unmetabolized F$_2$-IsoPs usually involves plasma or urine although other biological fluids can also be used in special situations, e.g., cerebrospinal fluid. Plasma sampling has the potential problem of artifactual generation of IsoPs by autoxidation of plasma arachidonic acid if the sample is not processed and stored properly. Plasma samples cannot be stored at −20°C because autoxidation can occur during storage at this temperature [25]. However, we have found that plasma can be stored at −70°C for at least 6 months without the occurrence of generation of F$_2$-IsoPs by autoxidation. Autoxidation is not a problem with urine [3], which is understandable because urine does not have a high lipid content. Conceptually, concentrations of F$_2$-IsoPs in plasma can provide a useful index of total endogenous production of IsoPs because levels in plasma presumably derive from all tissues in the body. A potential contribution of local formation of F$_2$-IsoPs in the kidney may confound interpretation of urinary levels of unmetabolized F$_2$-IsoPs [11]. As men-

tioned, IsoPs are initially formed esterified in lipids and then released in free form. This can be a dynamic process [4]. Depending on the experimental situation, timing of sampling of blood for determination of F$_2$-IsoPs may not be critical. Because the elimination half-life of IsoPs in the circulation is relatively short, less than 20 min [26], measurement of F$_2$-IsoPs in a single sample of blood only provides an index of IsoP formation during a relatively short period of time. In chronic disease states in which there may be a relatively steady rate of formation and release of IsoPs from phospholipids into the circulation, timing of sampling of blood is not critical. However, in some chronic disease states there may be significant intraday fluctuations in the formation of IsoPs. In dynamic situations in which there is an oxidant insult for only a relatively short period of time, e.g., ischemia/reperfusion injury, multiple sequential sampling of blood is necessary to assess the full magnitude of the increase in IsoP generation during rapidly changing rates of production over time [16].

Measurement of esterified F$_2$-IsoPs

Because IsoPs are initially formed esterified to phospholipids, this can be utilized to assess overproduction of F$_2$-IsoPs in specific target sites of interest. With this approach, phospholipids and neutral lipids are subjected to Folch lipid extraction and then hydrolyzed with base after which liberated free F$_2$-IsoPs are quantified [13]. Although sampling of tissues for analysis is primarily limited to studies in experimental animals, the sensitivity of the mass spectrometric assay is sufficient to allow determination of levels of F$_2$-IsoPs in small biopsy specimens from human tissues. This approach has been utilized to explore levels of esterified F$_2$-IsoPs in postmortem samples obtained from humans, e.g., normal and atherosclerotic vessels in which significantly increased levels of F_2-IsoPs were found in atherosclerotic plaques compared with normal vessels [27,28]. However, precautions such as rapid procurement of tissues after death and immediate storage at $-70°C$ must be taken to avoid artifactual generation of IsoPs by autoxidation. The other area where we and others have utilized this approach rather extensively in living subjects is to assess levels of F_2-IsoPs esterified in plasma lipoproteins in studies exploring the LDL oxidation hypothesis of atherogenesis [16,29]. In this regard, levels of F_2-IsoPs esterified in plasma lipoproteins have been found to be elevated in patients with hypercholesterolemia and patients who smoke [16,29,30], conditions that are high risk factors for atheroslerosis. In an equal volume of plasma, F_2-IsoPs esterified in plasma lipoproteins are present at levels approximately 4-fold higher than levels of free F_2-IsoPs. Thus, for such studies, only small amounts of plasma need to be obtained for analysis (~0.5 ml).

Measurement of a urinary metabolite of F_2-IsoPs

Obtaining plasma for determination of F_2-IsoPs, although only minimally invasive, is frequently not suitable for large clinical trials because of the logistics of drawing blood and the requirement that the plasma be rapidly isolated and stored at $-70°C$. Although collection of urine is feasible in large clinical trials, the interpretation of measurement of unmetabolized F_2-IsoPs in urine as an index of total endogenous IsoP production can be confounded by the potential contribution of local IsoP production in the kidney. Therefore, we recently carried out a study to identify a urinary metabolite of an F_2-IsoP to circumvent these problems. It has been well established in the prostaglandin field that measurement of the urinary excretion of metabolites of prostaglandins represents the most reliable approach to assess total endogenous production of prostanoids [31]. Thus, measurement of the urinary excretion of an F_2-IsoP metabolite should also afford an accurate measure of endogenous production of IsoPs. This has the additional advantage in that blood does not have to be obtained and also measurement of the level of a metabolite in urine collected over many hours can provide an integrated index of IsoP production over time.

One of the F_2-IsoPs, which we have shown is produced in vivo, is 15-F_{2t}-IsoP (8-iso-prostaglandin $F_{2\alpha}$) [32]. Metabolism of prostaglandins in most instances has been found to produce a plethora of metabolites. However, we found that a single metabolite predominated in the profile of derivatives produced from metabolism of 15-F_{2t}-IsoP. This metabolite was identified by mass spectrometric analysis as 2,3-dinor-5,6-dihydro-15-F_{2t}-IsoP [33] (Fig. 4). This metabolite was synthesized and

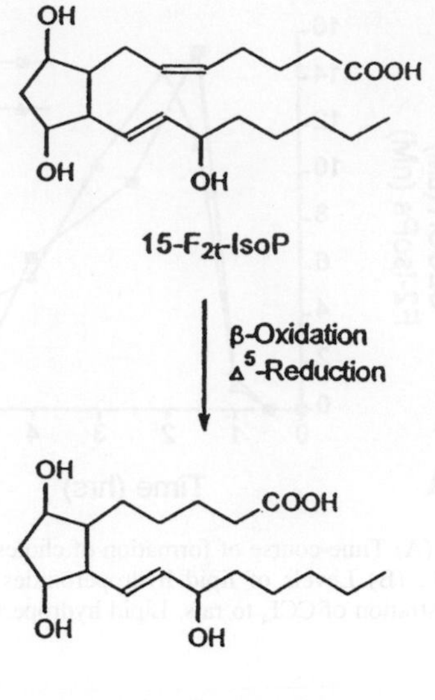

Fig. 4. Formation of the major urinary metabolite of 15-F_{2t}-IsoP by processes of one step of β-oxidation and reduction of the Δ^5-double bond.

converted to an $[^{18}O_4]$ derivative for use as an internal standard and recently we developed a stable isotope dilution negative ion chemical ionization GC/MS method for its analysis [34]. Levels of this metabolite in normal individuals were found to be 0.39 ± 0.18 ng/mg creatinine (mean $\pm$ 2 SD). The levels of this metabolite increased a mean of 24-fold compared with baseline in urine collected over 12 h after administration of CCl_4 to rats. Additionally, the excretion of this metabolite was found to be increased a mean of 2.5-fold in patients with polygenic hypercholesterolemia and these increases were suppressed by a mean of 54% following 8 weeks of treatment with a combination of vitamin E, vitamin C, and β carotene. These data suggest that quantification of the urinary excretion of 2,3-dinor-5,6-dihydro-15-F_{2t}-IsoP will contribute importantly to our ability to reliably assess free radical–induced lipid peroxidation in vivo and provide an approach that should be applicable to large clinical studies. Development of an ELISA assay for this metabolite is currently under commercial development. Assuming that the accuracy of the ELISA assay for the metabolite in urine can be validated by GC/MS, this may eventuate in the wide availability of this measure of lipid peroxidation by investigators in the free radial field.

SUMMARY AND CONCLUSIONS

In summary, initially the discovery of F_2-IsoPs was primarily of biochemical interest. This discovery, how-

Table 2. Disorders in Which Measurements of F$_2$-IsoPs Has Implicated a Role for Free Radicals in the Disease Process

Smoking [30,35]	Rhabdomyolysis renal injury [50,51]
Alzheimer's disease [36–38]	Acute cholestasis [52,53]
Huntington's disease [39]	Adult respiratory distress syndrome [54]
Hypercholesterolemia and atherosclerosis [16,27–30,40]	Halothane hepatoxicity [55]
Hyperhomocysteinemia [41]	Acetaminophen poisoning [11]
Hepatorenal syndrome [42]	Ischemia/reperfusion injury [16,56–59]
Scleroderma [43]	Cr (IV) poisoning [60]
Age-related decline in renal function [44]	Diquat poisoning [3,61]
Se deficiency [18]	Cisplatin-induced renal dysfunction [62]
Vitamin E deficiency [18]	Transplant organ injury during cold preservation [63]
Retinopathy of prematurity [45]	Chronic obstructive lung disease [64]
Alcohol-induced liver injury [46,47]	Interstitial lung disease [65]
Allergic Asthma [48]	Organophosphate poisoning [66]
Diabetes [49]	CCl$_4$-induced hepatotoxicity [3,26]

ever, has evolved over the last several years in a number of areas. These include the discovery of a number of additional classes of compounds that are generated as products of the IsoP pathway and the findings that several of these compounds can exert potent biological activity either through receptor-mediated actions or in the case of A$_2$/J$_2$-IsoPs and isolevuglandins because of their inherent chemical reactivity. However, an important aspect of the discovery of IsoPs is that evidence continues to accumulate suggesting that measurement of F$_2$-IsoPs represents one of the most reliable approaches to assess oxidative stress status in vivo. Utilizing measurements of F$_2$-IsoPs has implicated strongly, often for the first time, a role for free radicals in the pathogenesis of a large number of diseases (Table 2). In most of the studies listed in Table 2, F$_2$-IsoPs were measured by mass spectrometry which is accurate but requires expensive instrumentation. In this regard, the major limitation for the wide-spread use of measurements of F$_2$-IsoPs by investigators in free radical research has been the issue of reliability and accuracy of immunoassays for F$_2$-IsoPs. Hopefully, these potential problems regarding the accuracy of immunoassays for F$_2$-IsoPs can be overcome in the future which would undoubtedly lead to a great expansion in the use of measurement of F$_2$-IsoPs to assess oxidative stress status in vivo.

Acknowledgements — Supported by grants GM42056, GM15431, DK48831, CA77839, DK26657, and CA68485 from the National Institutes of Health. J. D. M. is a recipient of a Burroughs Wellcome Fund Clinical Scientist Award in Translational Research.

REFERENCES

[1] Pryor, W. A. On the detection of lipid hydroperoxides in biological samples. *Free Radic. Biol. Med.* **7**:177–178; 1989.

[2] Halliwell, B.; Grootveld, M. The measurement of free radical reactions in humans. *FEBS Lett.* **213**:9–14; 1987.

[3] Morrow, J. D.; Hill, K. E.; Burk, R. F.; Nammour, T. M.; Badr, K. F.; Roberts, L. J., II. A series of prostaglandin F$_2$-like compounds are produced in vivo in humans by a non-cyclooxygenase

free radical catalyzed mechanism. *Proc. Natl. Acad. Sci. USA* **87**:9383–9387; 1990.

[4] Taber, D. F.; Morrow, J. D.; Roberts, L. J., II. A nomenclature system for the isoprostanes. *Prostaglandins* **53**:63–67; 1997.

[5] Morrow, J. D.; Awad, J. A.; Boss, H. J.; Blair, I. A.; Roberts, L. J., II. Non-cyclooxygenase derived prostanoids (F$_2$-isoprostanes) are formed in situ on phospholipids. *Proc. Natl. Acad. Sci. USA* **89**:10721–10725; 1992.

[6] Morrow, J. D.; Roberts, L. J., II; Daniel, V. C.; Mirotchnechenko, O.; Swift, L.; Burk, R. F. Comparison of the formation of D$_2$/E$_2$-isoprostanes to F$_2$-isoprostanes in vitro and in vivo: effect of oxygen tension and glutathione. *Arch. Biochem. Biophys.* **353**:160–171; 1998.

[7] Morrow, J. D.; Minton, T. A.; Mukundan, C. R.; Campbell, M. D.; Zackert, W. E.; Daniel, V. C.; Badr, K. F.; Blair, I. A.; Roberts, L. J., II. Free radical induced generation of isoprostanes in vivo: Evidence for the formation of D-ring and E-ring isoprostanes. *J. Biol. Chem.* **269**:4317–4326; 1994.

[8] Morrow, J. D.; Awad, J. A.; Wu, A.; Zackert, W. E.; Daniel, V. C.; Roberts, L. J., II. Free radical generation of thromboxane-like compounds (isothromboxanes) in vivo. *J. Biol. Chem.* **271**:23185–23190; 1996.

[9] Chen, Y.; Morrow, J. D.; Roberts, L. J., II. Formation of reactive cyclopentenone compounds in vivo as products of the isoprostane pathway. *J. Biol. Chem.* **274**:10863–10868; 1999.

[10] Brame, C. J.; Salomon, R. G.; Morrow, J. D.; Roberts, L. J., II. Identification of extremely reactive γ-ketoaldehydes (isolevuglandins) as products of the isoprostane pathway and characterization of their lysyl protein adducts. *J. Biol. Chem.* **274**:13139–13146; 1999.

[11] Roberts, L. J., II; Morrow, J. D. The generation and actions of isoprostanes. *Biochim. Biophys. Acta* **1345**:121–135; 1997.

[12] Morrow, J. D.; Roberts, L. J., II. The isoprostanes: Unique bioactive products of lipid peroxidation. *Prog. Lipid Res.* **36**:1–21; 1997.

[13] Morrow, J. D.; Roberts, L. J., II. Mass spectrometric quantification of F$_2$-isoprostanes in biological fluids and tissues as a measure of oxidant stress. *Methods Enzymol.* **300**:3–12; 1998.

[14] Granstrom, E.; Kindahl, H. Radioimmunoassay of prostaglandins and thromboxanes. *Adv. Prostaglandin Thromboxane Leukot. Res.* **5**:119–210; 1978.

[15] Proudfoot, J.; Barden, A.; Mori, T. A.; Burke, V.; Croft, K. D.; Beilin, L. J.; Puddey, I. B. Measurement of urinary F(2)-isoprostanes as markers of in vivo lipid peroxidation—a comparison of enzyme immunoassay with gas chromatography/mass spectrometry. *Anal. Biochem.* **272**:209–215; 1999.

[16] Richelle, M.; Turini, M. E.; Guidoux, R.; Tavazzi, I.; Metairon, S.; Fay, L. B. Urinary isoprostane excretion is not confounded by the lipid content of the diet. *FEBS Lett.* **459**:259–262; 1999.

[17] Longmire, A. W.; Swift, L. L.; Roberts, L. J., II; Awad, J. A.; Burk, R. F.; Morrow, J. D. Effect of oxygen tension on the

generation of F_2-isoprostanes and malondialdehyde in peroxidizing rat liver microsomes. *Biochem. Pharmacol.* **47:**1173–1177; 1994.

[18] Burk, R. F. Glutathione-dependent protection by rat liver microsomal protein against lipid peroxidation. *Biochim. Biophys. Acta* **757:**21–28; 1983.

[19] Marnett, L. J.; Buck, J.; Tuttle, M. A.; Bull, A. W. Distribution and oxidation of malondialdehyde in mice. *Prostaglandins* **30:** 241–254; 1985.

[20] Janero, D. R. Malondialdehyde and thiobarbituric acid-reactivity as diagnostic indices of lipid peroxidation and peroxidative tissue injury. *Free Radic. Biol. Med.* **9:**515–540; 1990.

[21] Lynch, S. M.; Morrow, J. D.; Roberts, L. J.; Frei, B. Formation of non-cyclooxygenase derived prostanoids (F_2-isoprostanes) in human plasma and isolated low density lipoproteins exposed to metal ion-dependent and -independent oxidative stress. *J. Clin. Invest.* **93:**998–1004; 1994.

[22] Matthews, W. R.; McKenna, R.; Guido, D. M.; Petre, T. W.; Jolly, R. A.; Morrow, J. D.; Roberts, L. J. A comparison of gas chromatography-mass spectrometry assays for in vivo lipid peroxidation. *Proceedings of 41st ASMS Conference Mass Spectrom. Allied Topics* 1993:865A–865B.

[23] Morrow, J. D.; Harris, T. M.; Roberts, L. J. Non-cyclooxygenase oxidative formation of a series of novel prostaglandins; Analytical ramifications for measurement of eicosanoids. *Anal. Biochem.* **184:**1–10; 1990.

[24] Morrow, J. D.; Awad, J. A.; Kato, T.; Takahashi, K.; Badr, B. F.; Roberts, L. J., II; Burk, R. F. Formation of novel non-cyclooxygenase derived prostanoids (F_2-isoprostanes) in carbon tetrachloride hepatotoxicity, an animal model of lipid peroxidation. *J. Clin. Invest.* **90:**2502–2507; 1992.

[25] Roberts, L. J.; Morrow, J. D. Isoprostanes as markers of lipid peroxidation in atherosclerosis. In: Serhan, C. N.; Ward, P. A., eds. *Molecular biology of inflammation.* Totowa: Humana Press; 1999:141–163.

[26] Gniwotta, C.; Morrow, J. D.; Roberts, L. J., II; Kuhn, H. Prostaglandin F_2-like compounds, F_2-isoprostanes, are present in increased amounts in human atherosclerotic lesions. *J. Arteriosclerosis Thromb. Vasc. Biol.* **17:**2975–2981; 1997.

[27] Pratico, D.; Iuliano, L.; Mauriello, A.; Spagnoli, L.; Lawson, J. A.; Rokach, J.; Maclouf, J.; Violi, F.; FitzGerald, G. A. Localization of distinct F_2-isoprostanes in human atherosclerotic lesions. *J. Clin. Invest.* **100:**2028–2034; 1997.

[28] Reilly, M. P.; Pratico, D.; Delanty, N.; DiMinno, G.; Ttremoli, E.; Rader, D.; Kapoor, S.; Kapoor, S.; Lawson, J.; FitzGerald, G. A. Increased formation of distinct F_2-isoprostanes in hypercholesterolemia. *Circulation* **98:**2822–2828; 1998.

[29] Morrow, J. D.; Frei, B.; Longmire, A. W.; Gaziano, J. M.; Lynch, S. M.; Stauss, W. E.; Oates, J. A.; Roberts, L. J., II. Increase in circulating products of lipid peroxidation (F_2-isoprostanes) in smokers: Smoking as a cause of oxidative damage. *N. Engl. J. Med.* **332:**1198–1203; 1995.

[30] Roberts, L. J., II. Comparative metabolism and fate of the eicosanoids. In: Willis, A. L., ed. *CRC handbook of eicosanoids: prostaglandins and related lipids, vol. 1, part A.* Boca Raton: CRC Press; 1987:239–244.

[31] Morrow, J. D.; Badr, K. F.; Roberts, L. J., II. Evidence that the F_2-isoprostane, 8-epi-PGF$_{2\alpha}$, is formed in vivo. *Biochim Biophys. Acta* **1210:**244–248; 1994.

[32] Roberts, L. J., II; Moore, K. P.; Zackert, W. E.; Oates, J. A.; Morrow, J. D. Identification of the major urinary metabolite of the F_2-isoprostane, 8-iso-prostaglandin $F_{2\alpha}$, in humans. *J. Biol. Chem.* **271:**20617–20620; 1996.

[33] Morrow, J. D.; Zackert, W. E.; Yang, J. P.; Kurhts, E. H.; Callawaert, D.; Kanai, K.; Taber, D. F.; Moore, K. P.; Oates, J. A.; Roberts, L. J., II. Quantification of the major urinary metabolite of the isoprostane 15-F_{2t}-isoprostane (8-iso-PGF$_{2\alpha}$) by stable isotope dilution mass spectrometric assay. *Anal. Biochem.* **269:**326–331; 1999.

[34] Reilly, M.; Delanty, N.; Lawson, J. A.; FitzGerald, G. A. Mod-

ulation of oxidant stress in vivo in chronic cigarette smokers. *Circulation* **94:**19–25; 1996.

[35] Montine, T. J.; Markesbery, W. R.; Morrow, J. D.; Roberts, L. J., II. Cerebrospinal fluid F_2-isoprostane levels are increased in patients with Alzheimer's disease. *Ann. Neurol.* **44:**410–413; 1998.

[36] Montine, T. J.; Beal, M. F.; Cudkowicz, M. D.; O'Donnel, H.; Margolin, R. A.; McFarland, L.; Cachrach, A. F.; Zacker, W. E.; Roberts, L. J., II; Morrow, J. D. Increased CSF F_2-isoprostane concentration in probable AD. *Neurology* **52:**562–565; 1999.

[37] Pratico, D.; Lee, M. Y.; Trojanowski, J. Q.; Rokach, J.; FitzGerald, G. A. Increased F_2-isoprostanes in Alzheimer's disease: evidence for enhanced lipid peroxidation in vivo. *FASEB J.* **12:** 1777–1783; 1998.

[38] Montine, T. J.; Beal, M. F.; Roberts, D.; Cudkowicz, M. E.; Brown, R. H.; O'Donnel, H.; Zackert, W. E.; Roberts, L. J., II; Morrow, J. D. Cerebrospinal levels of F_2-isoprostanes, specific markers of lipid peroxidation, are elevated in Huntington's disease patients. *Neurology* **52:**1104–1105; 1999.

[39] Pratico, D.; Tangirala, R. K.; Rader, D. J.; Rokach, J.; FitzGerald, G. A. Vitamin E suppresses isoprostane generation in vivo and reduces atherosclerosis in ApoE-deficient mice. *Nat. Med.* **4:**1189–1192; 1998.

[40] Voutilainen, S.; Morrow, J. D.; Roberts, L. J., II; Alfthan, G.; Alho, H.; Nyssonen, K.; Salonen, J. T. Enhanced in vivo lipid peroxidation at elevated plasma homocysteine levels. *Arteriosclerosis Thromb. Vasc. Biol.* **19:**1263–1266; 1999.

[41] Morrow, J. D.; Moore, K. P.; Awad, J. A.; Ravenscraft, M. D.; Marini, G.; Badr, K. F.; Williams, R.; Roberts, L. J., II. Marked overproduction of non-cyclooxygenase derived prostanoids (F_2-isoprostanes) in the hepatorenal syndrome. *J. Lipid. Mediators* **6:**417–420; 1993.

[42] Stein, C. M.; Awad, J. A.; Tanner, S. B.; Roberts, L. J., II; Morrow, J. D. Evidence for free radical mediated injury (isoprostane overproduction) in scleroderma. *Arthritis Rheumat.* **39:**1146–1150; 1996.

[43] Reckelhoff, J. F.; Kanjii, V.; Racusen, L.; Schmidt, A. M.; Yan, S. D.; Morrow, J. D.; Roberts, L. J., II; Salahudeen, A. K. Vitamin E ameliorates enhanced renal lipid peroxidation and accumulation of F_2-isoprostanes in aging kidneys. *Am. J. Physiol.* **274:**R767–R774; 1998.

[44] Awad, J. A.; Morrow, J. D.; Hill, K. E.; Roberts, L. J., II; Burk, R. F. Detection and localization of lipid peroxidation in vitamin E and selenium deficient rats using F_2-isoprostanes. *J. Nutr.* **124:** 810–864; 1994.

[45] Lahaie, I.; Hardy, P.; Hou, X.; Hassessian, H.; Asselin, P.; Lachapelle, P.; Almazan, G.; Varma, D. R.; Morrow, J. D.; Roberts, L. J., II; Chemtob, S. A novel mechanism for vasoconstrictor action of 8-isoprostaglandin $F_{2\alpha}$, on retinal vessels. *Am. J. Physiol.* **274:**R1406–R1416; 1998.

[46] Nanji, A. A.; Khwaja, S.; Tahan, S. R.; Sadrzadeh, S. M. Plasma levels of a novel non-cyclooxygenase derived prostanoid (8-isoprostane) correlate with severity of liver injury in experimental alcoholic liver disease. *J. Pharmacol. Exp. Therap.* **269:**1280–1285; 1994.

[47] Alehnik, S. I.; Leo, M. A.; Aleynik, M. K.; Lieber, C. S. Increased circulating products of lipid peroxidation in patient with alcoholic liver disease. *Alcohol. Clin. Exp. Res.* **22:**192–196; 1998.

[48] Dworski, R.; Murray, J. J.; Roberts, L. J., II; Oates, J. A.; Morrow, J. D.; Fisher, L.; Sheller, J. R. Allergen-induced synthesis of F_2-isoprostanes in atopic asthmatics: evidence for oxidative stress. *Am J. Resp. Crit. Care Med.* **160:**1947–1951; 1999.

[49] Davi, G.; Ciabattoni, G.; Consoli, A.; Mezzetti, A.; Falco, A.; Santarone, S.; Pennese, E.; Vitacolonna, E.; Bucciarelli, T.; Constantini, F.; Capani, F.; Patrono, C. In vivo formation of 8-isoprostaglandin $F_{2\alpha}$ and platelet activation in diabetes mellitus: effects of improved metabolic control and vitamin E supplementation. *Circulation* **99:**224–229; 1999.

[50] Moore, K.; Patel, R.; Darley-Usmar, V.; Holt, S.; Zackert, W. E.; Clozel, M.; Anand, R.; Wilson, M.; Harvey, S.; Morrow, J. D.; Roberts, L. J., II. A causative role for redox cycling of myoglobin and its inhibition by alkalinzation in the pathogenesis and treat-

ment of rhabdomyolysis-induced renal failure. *J. Biol. Chem.* **273**:31731–31737; 1998.

[51] Holt, S.; Reeder, B.; Wilson, M.; Harvey, S.; Morrow, J. D.; Roberts, L. J., II; Moore, K. Increased lipid peroxidation in patients with rhabdomyolysis. *Lancet* **353**:1241; 1999.

[52] Leo, M. A.; Aleynik, S. I.; Siegel, J. H.; Kasmin, F. E.; Aleynik, M. K.; Lieber, C. S. F$_2$-isoprostane and 4-hydroxynonenal excretion in human bile of patients with biliary tract and pancreatic disorders. *Am. J. Gastroenterol.* **92**:2069–2072; 1997.

[53] Holt, S.; Marley, R.; Fernando, B.; Harry, D.; Anand, R.; Goodier, D.; Moore, K. Acute cholestasis-induced renal failure: effects of antioxidants and ligands for the thromboxane A$_2$ receptor. *Kidney Int.* **55**:271–277; 1999.

[54] Carpenter, C. T.; Price, P. V.; Christman, B. W. Exhaled breath condensate isoprostanes are elevated in patients with acute lung injury and ARDS. *Chest* **114**:1653–1659; 1998.

[55] Awad, J. A.; Horn, J. I.; Roberts, L. J., II; Franks, J. J. Demonstration of halothane-induced hepatic lipid peroxidation in rats using F$_2$-isoprostanes. *Anesthesiology* **84**:910–916; 1996.

[56] Mobert, J.; Becker, B. G. Cyclooxygenase inhibition aggravates ischemia-reperfusion injury in the perfused guinea pig heart: involvement of isoprostanes. *J. Am. Coll. Cardiol.* **31**:1687–1694; 1998.

[57] Reilly, M. P.; Delanty, N.; Roy, L.; Rokach, J.; Callaghan, P. O.; Crean, P.; Lawson, J. A.; FitzGerald, G. A. Increased formation of isoprostanes IPF2α-I and 8-epi-prostaglandin F$_{2\alpha}$ in acute coronary angioplasty: evidence for oxidant stress during coronary reperfusion in humans. *Circulation* **96**:3314–3320; 1997.

[58] Mathews, W. R.; Guido, D. M.; Fisher, M. A.; Jaeschke, H. Lipid peroxidation as a molecular mechanism of liver cell injury during reperfusion after ischemia. *Free Radic. Biol. Med.* **16**:763–770; 1994.

[59] Takahashi, K.; Nammour, T. K.; Fukunaga, M.; Ebert, J.; Morrow, J. D.; Roberts, L. J., II; Badr, K. F. Glomerular actions of a free radical generated novel prostaglandin, 8-epi-prostaglandin F$_{2\alpha}$, in the rat. *J. Clin. Invest.* **90**:136–141; 1992.

[60] Kadiiska, M. B.; Morrow, J. D.; Awad, J. A.; Roberts, L. J., II; Mason, R. P. Enhanced formation of free radicals and F$_2$-isoprostanes in vivo by acute Cr (IV) poisoning. *Chem. Res. Toxicol.* **11**:1516–1520; 1998.

[61] Awad, J. A.; Burk, R. F.; Roberts, L. J., II. Effect of selenium

deficiency and glutathione modulating agents on diquat toxicity and lipid peroxidation. *J. Pharmac. Exp. Therap.* **270**:858–864; 1994.

[62] Salahudeen, A.; Wilson, P.; Pande, R.; Poovala, V.; Kanji, V.; Ansari, N.; Morrow, J. D., Roberts, L. J., II. Cisplatin induces N-acetyl cysteine suppressible F$_2$-isoprostane production and injury in renal tubular epithelial cells. *J Am. Soc. Nephrol.* **9**:1448–1455; 1998.

[63] Salahudeen, A.; Nawaz, M.; Poovala, V.; Kanjii, V.; Wang, C.; Morrow, J. D.; Roberts, L. J., II. Cold induces time-dependent F$_2$-isoprostane formation in renal tubular cells and rat kidneys stored in University of Wisconsin solution: implications for immediate post-transplant renal vasoconstriction. *Kidney Int.* **55**:1759–1762; 1999.

[64] Pratico, D.; Basili, S.; Vieri, M.; Cordova, C.; Violi, V.; Fitzgerald, G. A. Chronic obstructive pulmonary disease is associated with an increase in urinary levels of isoprostane F2alpha-II, an index of oxidant stress. *Am. J. Resp. Crit. Care Med.* **158**:1709–1714; 1998.

[65] Montuschi, P.; Ciabattoni, G.; Paredi, P.; Pantelidis, P.; DuBois, R. M.; Kharitoniv, S. A.; Barnes, P. J. 8-isoprostane as a biomarker of oxidative stress in interstitial lung diseases. *Am J. Resp. Crit. Care Med.* **158**:1524–1527; 1998.

[66] Yang, Z. P.; Wu, A.; Morrow, J. D.; Roberts, L. J., II; Dettbarn, W-D. Increases in malondialdehyde-thiobarbituric acid complex (MDA-TBA) and F$_2$-isoprostanes in diisopropylfluorophosphate induced muscle hyperactivity. *Biochem. Pharmacol.* **52**:357–361; 1996.

ABBREVIATIONS

F$_2$-IsoPs—F$_2$-isoprostanes
MDA—malondialdehyde
SGPT—serum glutamic pyruvic transaminase
TBARS—thiobarbituric acid reacting substances
LDL—low-density lipoprotein
HETE—hydroxyeicosatetraenoic acid

Bioassays for Oxidative Stress Status (BOSS). Edited by W.A. Pryor

METHODS FOR MEASURING ETHANE AND PENTANE IN EXPIRED AIR FROM RATS AND HUMANS

MITCHELL D. KNUTSON,* GARRY J. HANDELMAN,[†] and FERNANDO E. VITERI[‡]

*Department of Nutrition, Harvard School of Public Health, Boston, MA, [†]Department of Health and Clinical Science, University of Massachusetts, Lowell, MA, and [‡]Department of Nutritional Sciences, University of California, Berkeley, CA, USA

(Received 5 October 1999; Accepted 19 October 1999)

Abstract—Numerous studies in animals and humans provide evidence that ethane and pentane in expired air are useful markers of in vivo lipid peroxidation. The measurement of breath hydrocarbons, being noninvasive, is well suited for routine use in research and clinical settings. However, the lack of standardized methods for collecting, processing, and analyzing expired air has resulted in the use of a wide variety of different methods that have yielded highly disparate results among investigators. This review outlines the methods that we have developed and validated for measuring ethane and pentane in expired air from rats and humans. We describe the advantages of these methods, their performance, as well as potential errors that can be introduced during sample collection, concentration, and analysis. A main source of error involves contamination with ambient-air ethane and pentane, the concentrations of which are usually much greater and more variable than those in expired air. Thus, it appears that the effective removal of ambient-air hydrocarbons from the subject's lungs before collection is an important step in standardizing the collection procedure. Also discussed is whether ethane or pentane is a better marker of in vivo lipid peroxidation. © 2000 Elsevier Science Inc.

Keywords—Lipid peroxidation, Oxidative stress, Free radical, Hydrocarbons, Alkanes, Breath tests, Iron

INTRODUCTION

Ethane and pentane are the main volatile hydrocarbons formed during the breakdown of peroxidized n-3 and n-6 fatty acids, respectively [1]. The formation of ethane and pentane in the process of lipid peroxidation represents

Address correspondence to: Dr. M. D. Knutson, Harvard School of Public Health, Department of Nutrition, Building II, Room 257, 665 Huntington Avenue, Boston, MA 02115, USA; Tel: (617) 432-2533; Fax: (617) 432-2435; E-Mail: mknutson@hsph.harvard.edu

Mitchell Knutson obtained his Ph.D. degree in 1998 at the University of California at Berkeley. His doctoral research focused on the effects of daily and intermittent iron supplements on lipid peroxidation in rats and humans. He now studies the molecular mechanisms of iron transport on a postdoctoral fellowship at Harvard. Garry Handelman obtained his Ph.D. in biochemistry from Tufts University in 1991, focusing on the metabolism and oxidative degradation of carotenoids. His current studies include role of carotenoids in vision, and development of new markers for oxidative stress in humans. Fernando Viteri, M.D. (University of San Carlos, Guatemala), Sc.D. (Med-Physiol, University of Cincinnati, Ohio) is Professor of Nutritional Sciences, University of California at Berkeley. He has worked on the functional consequences of nutritional deficiencies and especially on the effective and safe prevention/correction of micronutrient deficiencies, with emphasis on iron.

only a secondary pathway [2], but is of special importance because these hydrocarbons are exhaled through the breath and can thus be studied noninvasively.

The first documentation of breath ethane as a marker of in vivo lipid peroxidation dates from 1974, when Reily et al. [3] reported that mice exhaled large amounts of ethane after exposure to carbon tetrachloride, a potent catalyst of in vitro lipid peroxidation. In 1977, Tappel's group extended these findings by demonstrating the presence of markedly increased amounts of not only ethane, but also pentane in the breath of vitamin E–deficient rats [4]. Since these initial reports, a number of compelling studies have shown increased exhalation of ethane and/or pentane in response to various oxidant stresses. In animals, for example, substantial increases have been observed after exposure to hyperbaric oxygen [5], iron excess [6], and phenylhydrazine [7]. In humans, increases in ethane and/or pentane have been observed subsequent to hyperbaric oxygen [8], cigarette smoking [9], total body irradiation [10], and acute aerobic exercise [11]. Consistent with the relationship between ex-

Reprinted from: *Free Radical Biology & Medicine,* Vol. 28, No. 4, pp. 514-519, 2000

haled ethane and pentane and in vivo lipid peroxidation is the observation that various antioxidants have decreased the exhalation of these hydrocarbons in studies of animals [12,13] and humans [14–16]. Taken together, these studies provide substantial evidence that expired-air ethane and pentane are representative and valid markers of in vivo lipid peroxidation.

Yet despite the growing number of reports on breath ethane and pentane there exist no widely accepted methods for collecting and analyzing expired air. As a result, most investigators have had to develop their own techniques. However, descriptions of these techniques often provide limited methodological detail and limited documentation of method validation and performance; various investigators have noted this fact [17–20]. The lack of extensive validation and the manifold differences between different techniques are very likely responsible for much of the striking variability in expired-air ethane and pentane values reported by different investigators. The large variability in normal adult values has been reviewed elsewhere [20–22].

Our interests in methodological detail and method performance motivated us to develop and validate techniques for the collection and analysis of expired air from laboratory rats [23] and humans [20]. The purpose of this review is to outline these methods, and to focus on their advantages as well as the potential errors and difficulties that can be introduced during sample collection, storage, concentration, and analysis. Most of these potential errors and difficulties are relevant to other methods as well. Also discussed are comparisons of ethane and pentane as sensitive and reliable markers of in vivo lipid peroxidation. For a comprehensive treatment of technical and physiological aspects of measuring breath ethane and pentane, the reader is referred to the excellent review by Kneepkens et al. [21].

Measurement of ethane and pentane in expired air from laboratory rats

We use an open-flow respiratory chamber system to collect expired air from rats. This involves placing a rat in a hermetically sealed glass jar through which hydrocarbon-free air (HCFA) is passed for an initial 20-min washout period. The washout period serves to flush ambient-air ethane and pentane from the chamber and the rat's lungs. During the next 40 min, two consecutive 4 l air-breath samples are collected into gas sampling bags and stored until analysis for ethane and pentane. An advantage of the collection system is that it is constructed from simple and inexpensive equipment; this enables multiple collection systems to be set up and used concurrently.

Because ethane and pentane concentrations in expired air are usually below the detection limit of most chromatographic methods, samples are concentrated before analysis. To do this, we have developed a trap-and-purge technique that concentrates samples at least 80-fold. Briefly, a 4 l air-breath sample is passed through a precolumn of Indicating Drierite (W. A. Hammond Drierite, Co., Ltd., Xenia, OH, USA) and soda lime (to remove water vapor and carbon dioxide, respectively) and then through a cooled loop containing adsorbants that trap alkanes, but very little nitrogen and oxygen. Next, the alkanes are desorbed with heat and analyzed by gas chromatography. We find that adult rats typically exhale ethane and pentane at rates of about 1 pmol/100 g body wt/min.

We feel that the primary advantage of this method is that it traps highly volatile ethane (b.p. $-88.6°C$) very efficiently. Indeed, our motivation for developing our trap-and-purge technique was because we were unable to trap ethane effectively with techniques that use adsorbants cooled to $-130°C$ in an ethanol-liquid nitrogen slurry [4]. Other investigators have reported similar difficulties [24]. Thus to trap ethane, we cool the adsorbants in liquid nitrogen (b.p. $-196°C$); at this temperature, trapping efficiency is essentially 100%, even over a range that far exceeds ethane concentrations found in expired air from rats (and humans). Moreover, the use of liquid nitrogen is far less cumbersome and subject to less temperature variations than is an ethanol-liquid nitrogen slurry.

Measurement of ethane and pentane in expired air from humans

We collect expired-air samples from humans using an apparatus we have designed. Briefly, a seated subject breathes hydrocarbon-free air from a 100-l bag. The subject wears a noseclip and breathes through a mouthpiece to prevent inhaling ambient air. After 6 min, the subject's lungs are flushed of ambient-air ethane and pentane, and two consecutive 3-l expired-air samples are collected into gas sampling bags. Concentrations of ethane and pentane are then determined using the same trap-and-purge and analysis procedures as described above. The total volume of the collected expired air and the time required for the collection of each sample are used for calculations of concentration (pmol alkane/l) and exhalation rate (pmol alkane/kg body weight [b.wt.]/min). We find that normal adults typically exhale ethane and pentane at rates of about 2 pmol/kg b.wt./min.

One advantage of the method we have developed is that we offer a detailed description of the components of the breath collection apparatus, along with explicit instructions for the collection and analysis of samples [20]. Methodological details are of paramount importance to the reliability of the measurements, mainly because contamination from ambient air may render the result unreliable, as discussed below.

Various features of the method make it readily adaptable for routine use in the clinic. First, the collection apparatus can be made portable. Second, the 10-min collection procedure, which is well tolerated by subjects, makes the test ideal for repeated measurements of the same subject. Third, the expired-air samples can be collected and stored (or transported) for later analysis. Finally, but perhaps most important, is that the method is both reliable and practicable.

Evaluation of method performance

The performance of analytical methods should be judged on two factors: (i) reliability and (ii) practicability [25]. The reliability is assessed on the basis of the magnitude of analytical variation, the specificity, and the limit of detection. The practicability of a method depends on numerous factors such as speed, dependability, required technical skill, cost, and safety.

As for the reliability of our methods, we have determined that our trap-and-purge technique has an acceptably low analytical variation of 6% (expressed as % coefficient of variation, % CV). The 6% CV represents an average of three values that span the usual range of ethane and pentane concentrations in human expired air. We have evaluated the specificity of our method by demonstrating adequate gas chromatographic separation between significant components in breath that could possibly interfere with ethane and pentane, i.e., ethylene, propane, butane, isopentane, hexane, and isoprene [20,23]. However, the use of mass spectrometry is still needed for unequivocal peak identification. The limit of detection for ethane and pentane are 0.22 and 0.77 pmol, respectively. These limits are about $30\times$ and $10\times$ less, respectively, than the ethane and pentane signals obtained from the analysis of a typical sample of human expired air.

One feature of the practicability of our method is demonstrated by its speed (and thus high throughput): it takes 10 min to collect two 3 l expired-air samples, 5 min for the trap-and-purge procedure, and less than 10 min for gas chromatographic analysis. Another feature is its dependability. Method failures, from collection to analysis, are rare. Indeed we have used our hydrocarbon adsorbent trap for the analysis of well over a thousand samples with no detectable changes in quantitation.

Potential errors introduced during the collection and storage of samples

A potential error introduced during the collection of expired air involves the failure to distinguish between endogenous and exogenous hydrocarbons. Exogenous ethane and pentane, i.e., those in ambient air, are usually found in much greater concentrations (Table 1) than those in washed-out human breath. In their comprehensive review of breath ethane and pentane, Kneepkens et al. [21] conclude that according to the majority of studies that use a washout period, ethane and pentane are exhaled at rates not exceeding about 5 pmol/kg b.wt./min, or about 50 pmol/l (assuming a b.wt. of 70 kg and a minute ventilation of 7 l/min). We usually obtain mean values of about 25 pmol/l for ethane and about 18 pmol/l for pentane. Therefore, considering that ambient-air ethane and pentane concentrations can sometimes be $15-30\times$ greater than those in expired air, it seems that the accurate measurement of endogenous ethane and pentane production would benefit by eliminating the atmospheric component.

A relatively easy way to do this is to have subjects breathe HCFA for a period of time in order to flush ambient hydrocarbons from the lungs before collecting expired air. Perhaps it is feasible that the ambient hydrocarbons could be measured simultaneously and then subtracted out; but, this introduces more error because ambient levels are not only much higher than those in expired air, they are also much more variable. Indeed, ambient-air ethane can fluctuate "very quickly from moment to moment" [29], particularly if the air intake vent is near a source of pollution, such as a parking lot. Thus, the inclusion of a washout period to remove ambient-air ethane and pentane seems desirable as a means of establishing methodological standardization. In healthy subjects, we demonstrated that 4 min of breathing HCFA was adequate to flush high levels of ethane and pentane from the lungs, and that washout times up to 30 min resulted in no further reductions in breath hydrocarbons [20]. However, longer washout times are probably necessary in patients with compromised lung function. For example, Habib et al. found that an 8-min washout including multiple vital capacity breaths was needed to flush high levels of ambient ethane from the lungs of patients with chronic obstructive pulmonary disease [29].

Table 1. Range of Ambient-Air Ethane and Pentane Concentrations Reported by Various Investigators

Ref.	n	Ethane (pmol/l)	Pentane (pmol/l)
26	7[a]	70–220	110–340
20	23	68–224	2–76
27	•••[b]	73–726	51–209
28	5[c]	800 ± 390	30 ± 60

[a] Values represent the range of values from seven urban centers. Chromatographic peak assignments were verified using mass spectrometry.
[b] Not available.
[c] Values are mean ± SD.

Errors may also be introduced during the storage of expired-air samples. Drury et al. [30] found that pentane concentrations in 1 1 Tedlar (Norton Performance Plastics, Akron, OH, USA) gas sampling bags did not change significantly after 24 or 48 h of storage, but did increase by 30% (range 22–108%, $n = 5$) after 72 h. Using 3 1 Tedlar bags, we found that ethane and pentane concentrations can increase appreciably after only 24 h of storage. The increases are likely due to a slow diffusion of ambient hydrocarbons into the gas sampling bags. In the study by Drury et al. [30], the large variability in pentane accumulation after storage might reflect differences between individual bags. We have noted that Tedlar bags invariably develop leaks after extensive use. We therefore test the bags periodically by filling them with nitrogen and leaving them overnight; bags with small leaks will be partially deflated the next morning. The finding of large differences in ethane and pentane values between consecutive breath samples can sometimes also help identify a bag with leaks.

Potential errors introduced during sample concentration and analysis

A potential problem introduced during sample concentration involves the trapping of oxygen while an expired-air sample is passed through the cold loop. Although our adsorbent-containing loop does not trap oxygen while immersed in liquid nitrogen, subsequent loops we have built trap oxygen to various degrees. The packing characteristics seem critical. Incidentally, we have found that the trapped oxygen can be completely vented without affecting the recovery of ethane and pentane. This is done by taking the loop out of the liquid nitrogen after an expired-air sample has been passed through; as the loop starts to warm, the trapped oxygen expands rapidly and can be vented into a syringe. The loop is then heated, and the trapped hydrocarbons are purged with a small amount of nitrogen into another syringe. Venting the trapped oxygen has the advantage of allowing the samples to be concentrated into volumes as small as 10–15 ml (an overall 200- to 300-fold concentration).

Contamination with ethane and pentane from system components can be another source of error. For example, we have found that Drierite and soda lime (Fisher Scientific, Santa Clara, CA, USA) are consistently contaminated with high levels of hydrocarbons, especially pentane. These materials can only be used after being adequately purged with a stream of nitrogen gas [20].

During sample analysis, the presence of methane in expired air can sometimes pose a problem. We have identified some subjects who consistently exhale unusually large amounts of methane. When these breath samples are concentrated, the methane yields a large chromatographic peak that can run into that of ethane. We have since determined that better separation between methane and ethane can be achieved using a PoraPak Q, 100/120 mesh (Waters Corp., Milford, MA, USA) column (2 m × 3.2 mm outer diameter) at 40°C. This column however is not optimized for pentane.

General technical aspects that promote optimal results

Despite the lack of a widely accepted standardized method for measuring ethane and/or pentane in expired air, we feel that optimal results can be obtained by any method provided it includes several technical aspects. First, the method should include a means of effectively dealing with the high and very variable concentrations of ambient-air ethane and pentane. We feel that this is best achieved by using a washout period, and by scrupulously avoiding ambient-air contamination during sample collection, storage, concentration, and analysis. Contamination with even small amounts of ambient air may render the results unreliable. Second, the concentration procedure must be reliable; the documentation of recovery, linearity, within- and between-day variability, and long-term stability of measurements is desirable. Assessment of these aspects will particularly enhance the reliability of detecting changes by repeated measurements of the same subject. Third, the chromatography should adequately resolve significant breath constituents that may coelute with ethane or pentane—namely, methane, ethylene, isopentane, and isoprene. If these aspects are incorporated, then it seems that the researcher or clinician should obtain accurate and reproducible results, even without between-laboratory standardization of methods.

Which is the better marker of in vivo lipid peroxidation: ethane or pentane?

In our studies of iron-related lipid peroxidation in rats [31] and humans [32,33], we found that exhaled ethane was a more sensitive indicator than pentane. Others have also reported the greater response of ethane than pentane to iron excess [6,34].

From a methodological point of view, we have determined that a single ethane measurement is more reliable than one of pentane. This is based on the finding that the between-day biological variation of ethane in expired air was less than that of pentane (24% CV vs. 38% CV) [20]. The greater biological variation in pentane exhalation rates may reflect changes in rates of pentane metabolism. Hydrocarbons are metabolized by hepatic mono-oxygenases at rates that correlate with molecular weight [35]. Accordingly, Wade and van Rij, using human subjects who breathed through a rebreathing circuit, measured pentane metabolism to be considerably faster ($T_{1/2}$ = 52 min) than that of ethane ($T_{1/2}$ = 4.1 h) [36]. They subsequently showed that pretreatment with a blocker of hepatic mono-oxygenases increased ethane exhalation by 70% and pentane by 1250% [37]. These studies demonstrate that pentane exhalation is influenced more by hepatic metabolism, and that under some circumstances, increases in breath pentane may represent altered hepatic metabolism rather than increased in vivo lipid peroxidation.

It has been proposed that the high solubility of pentane in adipose may further complicate interpretations of the measurement of breath pentane [22]. Springfield and Levitt tested the hypothesis that adipose tissue would provide an enormous pentane sink and found that obese rats exposed to a high pentane environment excreted pentane more slowly than did normal rats [22]. Based on these findings, they predicted that many hours of breathing HCFA would be required to wash environmental pentane that is solubilized in the adipose of even a nonobese human. Although this prediction may be true, data from two studies of various washout times lead to the conclusion that if pentane measured in the breath represents a slow washout from the adipose, then its contribution to total breath pentane output must be very small. We studied washout times of 4, 10, 20, and 30 min in nine healthy subjects and found no significant differences between pentane exhalation rates after washouts of 4 and 30 min [20]. Indeed, the mean exhalation rate of pentane in these subjects was 1.1 pmol/kg b.wt./min—a value 2.5 times less than the mean exhalation rate of ethane. Morita et al. studied even longer washout times in 15 healthy volunteers [8]. They found that after subjects breathed HCFA for 0, 30, 60, and 90 min, mean pentane exhalation rates, in pmol/kg b.wt./min ($\pm$ SEM), were 10.2 ($\pm$ 1.5), 1.6 ($\pm$ 0.2), 1.2 ($\pm$ 0.9), 1.3 ($\pm$ 0.4), and 1.3 ($\pm$ 0.3), respectively. Thus, although it remains uncertain as to whether pentane measured in expired air represents endogenous production or a slow release of environmental pentane solubilized in body fat, the above studies demonstrate that relatively stable, low rates of pentane exhalation can be achieved with relatively short washout periods. This indicates that, at least in healthy volunteers, measured increases in exhaled pentane may represent true in vivo lipid peroxidation and not simply a release of pentane from the adipose. However, studies are still needed comparing pentane washout times from obese and normal human subjects after prior exposure (and equilibration) to a high pentane environment.

Overall, ethane seems to be a more reliable marker of in vivo lipid peroxidation: it is metabolized to a much lesser extent than pentane, it is poorly soluble in tissues, and it displays less day-to-day biological variation than does pentane. Nevertheless, several studies have shown the exhalation of pentane, but not ethane, to increase under some circumstances [38–40]. This may be because pentane is derived from n-6 fatty acids, the predominant lipid class in the body. It seems, therefore, that the measurement of both ethane and pentane is still desirable.

REFERENCES

[1] Dumelin, E. E.; Tappel, A. L. Hydrocarbon gases produced during in vitro peroxidation of polyunsaturated fatty acids and decomposition of preformed hydroperoxides. *Lipids* 12:894–900; 1977.

[2] Frankel, E. N. Volatile lipid oxidation products. *Prog. Lipid Res.* 22:1–33; 1983.

[3] Riely, C. A.; Cohen, G.; Lieberman, M. Ethane evolution: a new index of lipid peroxidation. *Science* 183:208–210; 1974.

[4] Dillard, C. J.; Dumelin, E. E.; Tappel, A. L. Effect of dietary vitamin E on expiration of pentane and ethane by the rat. *Lipids* 12:109–114; 1977.

[5] Habib, M. P.; Eskelson, C.; Katz, M. A. Ethane production rate in rats exposed to high oxygen concentration. *Am. Rev. Respir. Dis.* 137:341–344; 1988.

[6] Dillard, C. J.; Tappel, A. L. Volatile hydrocarbon and carbonyl products of lipid peroxidation: a comparison of pentane, ethane, hexanal, and acetone as in vivo indices. *Lipids* 14:989–995; 1979.

[7] Clemens, M. R.; Remmer, H.; Waller, H. D. Phenylhydrazine-induced lipid peroxidation of red blood cells in vitro and in vivo: monitoring by the production of volatile hydrocarbons. *Biochem. Pharmacol.* 33:1715–1718; 1984.

[8] Morita, S.; Snider, M. T.; Inada, Y. Increased N-pentane excretion in humans: a consequence of pulmonary oxygen exposure. *Anesthesiology* 64:730–733; 1986.

[9] Habib, M. P.; Clements, N. C.; Garewal, H. S. Cigarette smoking and ethane exhalation in humans. *Am. J. Respir. Crit. Care Med.* 151:1368–1372; 1995.

[10] Arterbery, V. E.; Pryor, W. A.; Jiang, L.; Sehnert, S. S.; Foster, W. M.; Abrams, R. A.; Williams, J. R.; Wharam, M. D., Jr.; Risby, T. H. Breath ethane generation during clinical total body irradiation as a marker of oxygen-free-radical-mediated lipid peroxidation: a case study. *Free Radic. Biol. Med.* 17:569–576; 1994.

[11] Leaf, D. A.; Kleinman, M. T.; Hamilton, M.; Barstow, T. J. The effect of exercise intensity on lipid peroxidation. *Med. Sci. Sports Exerc.* 29:1036–1039; 1997.

[12] Dillard, C. J.; Gavino, V. C.; Tappel, A. L. Relative antioxidant effectiveness of alpha-tocopherol and gamma-tocopherol in iron-loaded rats. *J. Nutr.* 113:2266–2273; 1983.

[13] Dillard, C. J.; Downey, J. E.; Tappel, A. L. Effect of antioxidants on lipid peroxidation in iron-loaded rats. *Lipids* 19:127–133; 1984.

[14] Do, B. K.; Garewal, H. S.; Clements, N. C., Jr.; Peng, Y. M.; Habib, M. P. Exhaled ethane and antioxidant vitamin supplements in active smokers. *Chest* 110:159–164; 1996.

[15] Kanter, M. M.; Nolte, L. A.; Holloszy, J. O. Effects of an antioxidant vitamin mixture on lipid peroxidation at rest and postexercise. *J. Appl. Physiol.* 74:965–969; 1993.

[16] Steinberg, F. M.; Chait, A. Antioxidant vitamin supplementation and lipid peroxidation in smokers. *Am. J. Clin. Nutr.* 68:319–327; 1998.

[17] Kohlmuller, D.; Kochen, W. Is n-pentane really an index of lipid peroxidation in humans and animals? A methodological reevaluation. *Anal. Biochem.* 210:268–276; 1993.

[18] Mendis, S.; Sobotka, P. A.; Euler, D. E. Pentane and isoprene in

expired air from humans: gas-chromatographic analysis of single breath. *Clin. Chem.* **40:**1485–1488; 1994.

[19] Seabra, L.; Braganza, J. M.; Jones, M. F. A system for the quantitative determination of hydrocarbons in human breath. *J. Pharm. Biomed. Anal.* **9:**693–697; 1991.

[20] Knutson, M. D.; Lim, A. K.; Viteri, F. E. A practical and reliable method for measuring ethane and pentane in expired air from humans. *Free Radic. Biol. Med.* **27:**560–571; 1999.

[21] Kneepkens, C. M.; Lepage, G.; Roy, C. C. The potential of the hydrocarbon breath test as a measure of lipid peroxidation (published erratum appears in *Free Radic. Biol. Med.* 1994 Dec; 17(6):609). *Free Radic. Biol. Med.* **17:**127–160; 1994.

[22] Springfield, J. R.; Levitt, M. D. Pitfalls in the use of breath pentane measurements to assess lipid peroxidation. *J. Lipid Res.* **35:**1497–1504; 1994.

[23] Knutson, M. D.; Viteri, F. E. Concentrating breath samples using liquid nitrogen: a reliable method for the simultaneous determination of ethane and pentane (published erratum appears in *Anal. Biochem.* 1999; 270:186). *Anal. Biochem.* **242:**129–135; 1996.

[24] Lawrence, G. D.; Cohen, G. Concentrating ethane from breath to monitor lipid peroxidation in vivo. *Methods Enzymol.* **105:**305–311; 1984.

[25] Statland, B. E.; Per Winkel, M. D. Sources of variation in laboratory measurements. In: Henry, J. B., ed. *Clinical diagnosis and management by laboratory methods* (16th edn.). Philadelphia: W. B. Saunders Co.; 1979:17–28.

[26] Sexton, K.; Westberg, H. Nonmethane hydrocarbon composition of urban and rural atmospheres. *Atmos. Environ.* **18:**1125–1132; 1984.

[27] Dumelin, E. E.; Dillard, C. J.; Tappel, A. L. Breath ethane and pentane as measures of vitamin E protection of *Macaca radiata* against 90 days of exposure to ozone. *Environ. Res.* **15:**38–41; 1978.

[28] Zarling, E. J.; Clapper, M. Technique for gas-chromatographic measurement of volatile alkanes from single-breath samples. *Clin. Chem.* **33:**140–141; 1987.

[29] Habib, M. P.; Tank, L. J.; Lane, L. C.; Garewal, H. S. Effect of vitamin E on exhaled ethane in cigarette smokers. *Chest* **115:** 684–690; 1999.

[30] Drury, J. A.; Nycyk, J. A.; Cooke, R. W. I. Pentane measurement in ventilated infants using a commericaly available system. *Free Radic. Biol. Med.* **22:**895–900; 1997.

[31] Knutson, M. D.; Walter, P. B.; Ames, B. N.; Viteri, F. E. Both iron deficiency and daily iron supplements increase lipid peroxidation in rats. *J. Nutr.* in press; 2000.

[32] Knutson, M. D.; Walter, P. B.; Mendoza, C.; Ames, B. N.; Viteri, F. E. Effects of daily and weekly oral iron supplements on iron status and lipid peroxidation in women. *FASEB J.* **13:**A698 abstr.; 1999.

[33] Mertz, S.; Donanagelo, C.; Knutson, M.; King, J. C.; Walter, P.; Ames, B.; Viteri, F. E. Breath ethane (Eth) excretion rate in women is increased by daily iron (Fe) but not by daily zinc (Zn) supplementation. *FASEB J.* **13:**A241 abstr.; 1999.

[34] Filser, J. G.; Bolt, H. M.; Muliawan, H.; Kappus, H. Quantitative evaluation of ethane and n-pentane as indicators of lipid peroxidation in vivo. *Arch. Toxicol.* **52:**135–147; 1983.

[35] Frank, H.; Hintze, T.; Bimboes, D.; Remmer, H. Monitoring lipid peroxidation by breath analysis: endogenous hydrocarbons and their metabolic elimination. *Toxicol. Appl. Pharmacol.* **56:**337–344; 1980.

[356] Wade, C. R.; van Rij, A. M. In vivo lipid peroxidation in man as measured by the respiratory excretion of ethane, pentane, and other low-molecular-weight hydrocarbons. *Anal. Biochem.* **150:** 1–7; 1985.

[37] Wade, C. R.; van Rij, A. M. The metabolism of low molecular weight hydrocarbon gases in man. *Free Radic. Res. Commun.* **4:**99–103; 1987.

[38] Allard, J. P.; Royall, D.; Kurian, R.; Muggli, R.; Jeejeebhoy, K. N. Effects of beta-carotene supplementation on lipid peroxidation in humans. *Am. J. Clin. Nutr.* **59:**884–890; 1994.

[39] Toshniwal, P. K.; Zarling, E. J. Evidence for increased lipid peroxidation in multiple sclerosis. *Neurochem. Res.* **17:**205–207; 1992.

[40] Bilton, D.; Maddison, J.; Webb, A. K.; Seabra, L.; Jones, M.; Braganza, J. M. Cystic fibrosis, breath pentane, and lipid peroxidation (letter; comment). *Lancet* **337:**1420; 1991.

ABBREVIATIONS

HCFA—hydrocarbon-free air
CV—coefficient of variation

MEASUREMENT OF 3-NITROTYROSINE AND 5-NITRO-γ-TOCOPHEROL BY HIGH-PERFORMANCE LIQUID CHROMATOGRAPHY WITH ELECTROCHEMICAL DETECTION

KENNETH HENSLEY, KELLY S. WILLIAMSON, and ROBERT A. FLOYD

Free Radical Biology and Aging Research Program, Oklahoma Medical Research Foundation, Oklahoma City, OK, USA

(Received 03 September 1999; Revised 09 December 1999; Accepted 28 December 1999)

Abstract—Nitric oxide (NO) is a lipophilic gaseous molecule synthesized by the enzymatic oxidation of L-arginine. During periods of inflammation, phagocytic cells generate copious quantities of NO and other reactive oxygen species. The combination of NO with other reactive oxygen species promotes nitration of ambient biomolecules, including protein tyrosine residues and membrane-localized γ-tocopherol. The oxidative chemistry of NO and derived redox congeners is reviewed. Techniques are described for the determination of 3-nitro-tyrosine and 5-nitro-γ-tocopherol in biological samples using high-performance liquid chromatography with electrochemical detection.   © 2000 Elsevier Science Inc.

Keywords—Free radical, Nitric oxide; Nitrotyrosine, Nitrotocopherol, HPLC, Electrochemical; Review

INTRODUCTION

Nitric oxide (nitrogen monoxide, ·NO or NO) is an ubiquitous gaseous free radical whose importance to cell biology has become manifestly evident since its discovery as a signaling substance in the late 1980s [1,2]. Originally described as endothelium-derived vasorelaxing factor [1], NO has since been recognized to possess other beneficial and malign biochemical capacities. Most of the signaling functions inherent to NO are manifest through its reaction with heme prosthetic groups or by reversible S-nitrosation of protein thiols. In contrast, the cytopathic nature of NO is evident under conditions where the reasonably innocuous NO converts to more reactive redox congeners. These derived species capriciously attack protein, lipid, and DNA with conse-

quences that have yet to be fully appreciated. In this review, the cytotoxic potential of NO is discussed with reference to recently developed techniques that allow the determination of protein and lipid nitration as biomarkers of NO-dependent oxidative tissue damage.

TWO FACES OF NO: INFLUENCE OF THE REDOX MILIEU

NO is derived from the reduced nicotinamide adenine dinucleotide phosphate (NADPH)-dependent oxidation of L-arginine by the enzyme NO synthase (NOS). Since the discovery of NOS, several isoforms of the enzyme have been described [3]. Constitutively expressed versions of NOS exist in endothelium and neurons. These isoforms are transiently activated by Ca^{2+} influx into the cytosolic compartment, and they synthesize NO in brief pulses of activity. This NO diffuses freely across cell membranes until finding a suitable target for binding. The best-studied molecular target for NO is the heme group of guanylate cyclase in vascular endothelial cells [4–6]. NO binding to guanylate cyclase results in enzyme activation and subsequent conversion of guanosine monophosphate (GMP) to cyclic GMP. The ultimate consequence of cyclic GMP synthesis is relaxation of vascular smooth muscle. Addition of NO to other heme

---

Kenneth Hensley holds B.Sc degrees in biology and chemistry from the University of Kentucky (1992) and a Ph.D. in chemistry from the University of Kentucky (1995). He has served as a research scientist at the Oklahoma Medical Research Foundation for the past 4 years. His current research investigates the relationship between oxidative stress and neuroinflammation in the aging human brain, with special emphasis on basic mechanisms of neurodegeneration in Alzheimer's disease.

Address correspondence to: Kenneth Hensley, Free Radical Biology and Aging Research Program, Oklahoma Medical Research Foundation, 825 N.E. 13th Street, Oklahoma City, OK 73104, USA; Tel: (405) 271-7580; Fax: (405) 271-1795;
E-Mail: kenneth-hensley@omrf.ouhsc.edu.

groups has been reported and may significantly influence the redox environment of the cell. For instance, Kanner and colleagues [7] have shown that the NO-myoglobin complex is incapable of catalyzing H_2O_2-dependent lipid peroxidation, although metmyoglobin and oxymyoglobin were able to do so.

More recently, NO has been recognized to form metastable S-nitrosothiol derivatives which may be relevant to signal transduction processes [8]. The mechanism of nitrosation is somewhat debatable. In one of the first demonstrations of NO reaction with thiols, Pryor et al. [9] showed that disulfides of glutathione could form in aqueous solution near physiological pH. These researchers postulated a nitrosothiol intermediate formed by direct attack of the thiolate anion on the NO free radical. Other mechanisms favor indirect pathways, where NO is first oxidized to the nitrosonium cation (NO^+) or, more likely, to dinitrogen trioxide (N_2O_3) before reaction with sulfhydryls [10–12]. NO oxidation to N_2O_2 requires molecular O_2, although other reactive oxygen species (ROS) are not involved [12]. The full significance of biological S-nitrosation is not clear at this point in time. One possible consequence of S-nitrosation is an increase in the effective diffusion distance for NO. As a case in point, S-nitrosation of hemoglobin may facilitate the delivery of NO to capillary beds as a means of perfusion control [13–15]. Another possibility is that S-nitrosation activates (or inhibits) regulatory elements involved in signal transduction cascades in a manner analogous to the reaction of NO with guanylate cyclase. In support of this latter paradigm, Lander and his colleagues [16–19] have described S-nitrosation of the monomeric guanosine triphosphate–binding protein and oncogene p21-ras. The S-nitrosation process occurs on a specific cysteine residue [18,19] and stimulates guanine nucleotide exchange by the p21-ras enzyme. Activation of p21-ras is expected to stimulate downstream protein kinases involved in cell cycle control, apoptosis, and inflammatory gene expression. As expected, NO activation of p42/p44 mitogen-activated protein kinase, c-Jun amino-terminal kinase, and p38-mitogen-actived protein kinase has been reported [20]. The chemistries described above pertain to NO synthesized in brief pulses at low concentrations in the near absence of other ROS. Quite different chemistries become pertinent under conditions where NO is synthesized in greater quantities, chronically, in the presence of other ROS. This situation occurs under certain inflammatory states when a specific inducible isoform of nitric oxide synthase (iNOS) is expressed [21–23]. Unlike endothelium NOS and brain NOS, iNOS is constitutively active and less well regulated. iNOS seems to function as part of the mammalian armamentarium against invading pathogens and also as part of the autoimmune defense against neoplasia [24,25]. Al-

though originally described in the context of macrophage biology, iNOS can be expressed in a variety of nonimmune cell types under appropriate conditions of inflammatory stimulation, where cytokine overproduction is evident [22,23,26].

The synthesis of NO by iNOS often occurs simultaneously with the generation of other, distinct ROS such as superoxide ($O_2^{\bullet-}$), hydrogen peroxide (H_2O_2), and hypochlorous acid. This is particularly true of iNOS produced by macrophages responding to a local inflammatory insult, where neutrophil degranulation and reduced nicotinamide adenine dinucleotide phosphate oxidase-dependent superoxide production are salient phenomena [27–31]. Reaction of NO with other ROS generates secondary species that are more reactive and cytotoxic than their precursor components [11,29,32–34]. Pertinent reactions of NO with other ROS are summarized in Fig. 1. When considering the chemistry outlined in Fig. 1, it is important to recognize that these reactions are not comprehensive and that they are quite dependent on the relative stoichiometry between NO and other ROS. For instance, peroxynitrite (oxoperoxonitrate [$ONOO^-$]) is formed readily when the ratio of $O_2^{\bullet-}$ to NO is equal to or greater than unity, although excess NO flux apparently inhibits $ONOO^-$ formation [35]. Similarly, NO may act as a potent radical chain terminator (i.e., an antioxidant) in lipid peroxidation processes under circumstances where $ONOO^-$ formation is not likely to occur [36].

Of the many possible NO-derived redox congeners, peroxynitrite ($ONOO^-$) has been the most studied. Peroxynitrite is formed at almost diffusion-limited rates (approximately 6×10^9 M^{-1}/s^{-1}) by the reaction between NO and $O_2^{\bullet-}$ [37]. Peroxynitrite is reasonably stable under conditions of extreme alkalinity and may be synthesized by the reaction of H_2O_2 with NO_2^- in acid solution, followed by a rapid quenching step in NaOH [38]. The pK_a of ONOOH is approximately 6.8, depending on the ionic composition of the buffer medium [39, 40]. Dilution of alkaline $ONOO^-$ into neutral solutions results in immediate decomposition of the anion and nitration of available targets, particularly phenolic substrates [41–44]. The exact mechanism of phenolic nitration is subject to speculation. Conceivably, ONOOH could decompose to $^{\bullet}NO_2$ and $^{\bullet}OH$, which would then react with aromatic rings. Although thermodynamic and kinetic arguments have been made against such homolytic scission [45,46], the presence of catalytic elements such as metal chelates could facilitate the reaction [41, 42]. In fact, the O-O bond within ONO-OH is quite weak (22 vs. 51 kcal/M in the case of H_2O_2), and equally valid thermodynamic arguments can be made in favor of homolytic scission of peroxynitrous acid [47]. Peroxynitrite-mediated phenolic nitration is enhanced by a num-

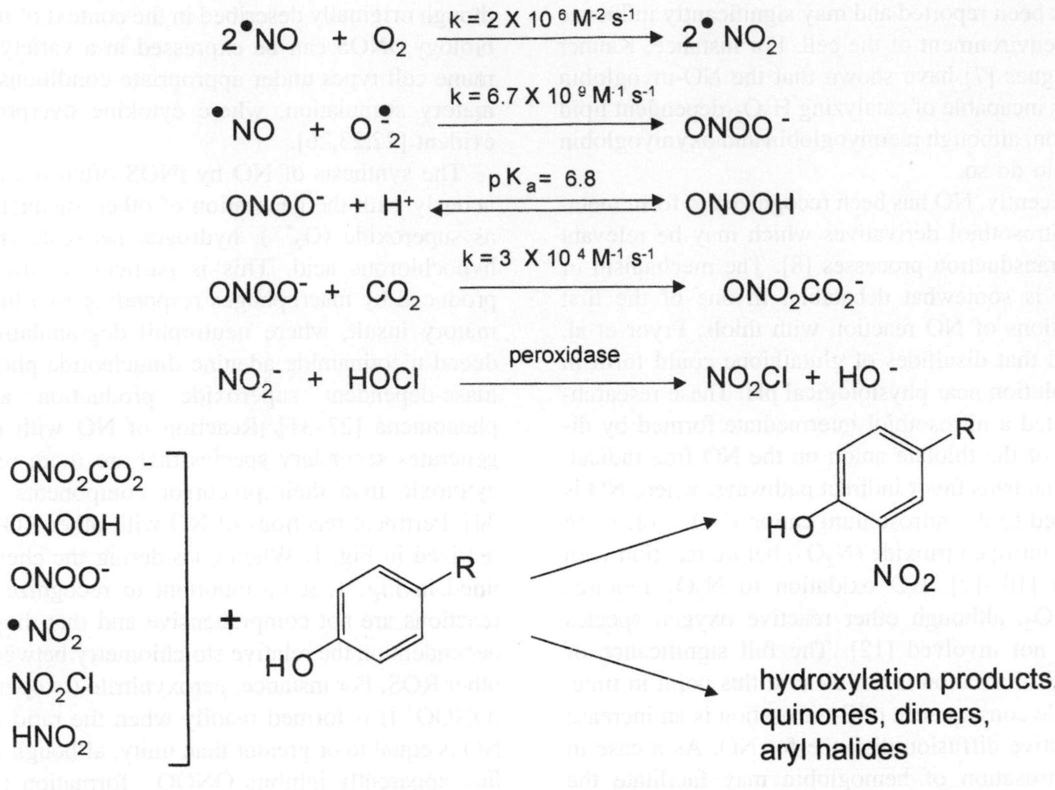

$$2\,^{\bullet}NO + O_2 \xrightarrow{k = 2 \times 10^6\ M^{-2}\ s^{-1}} 2\,^{\bullet}NO_2$$

$$^{\bullet}NO + O_2^{\bullet-} \xrightarrow{k = 6.7 \times 10^9\ M^{-1}\ s^{-1}} ONOO^-$$

$$ONOO^- + H^+ \xrightleftharpoons{pK_a = 6.8} ONOOH$$

$$ONOO^- + CO_2 \xrightarrow{k = 3 \times 10^4\ M^{-1}\ s^{-1}} ONO_2CO_2^-$$

$$NO_2^- + HOCl \xrightarrow{peroxidase} NO_2Cl + HO^-$$

Fig. 1. Several biochemically salient reactions of NO leading to the formation of nitrated phenolic biomarkers.

ber of substances, including CO_2 and, surprisingly, superoxide dismutase [40,41,48–50]. The reaction of $ONOO^-$ with CO_2 is rapid ($k = 3 \times 10^4\ M^{-1}/s^{-1}$) and the physiological concentration of carbonate is high (>1 mM); thus, $ONO_2CO_2^-$ may be the most physiologically relevant nitrating agent [44–50]. To complicate matters even further, NO-derived oxidants other than peroxynitrite may well be involved with inflammatory pathophysiologies. Nitryl chloride, for example, has recently been identified as a product of NO reaction with hypochlorous acid in the vicinity of activated neutrophils [51], and nitryl chloride is quite capable of nitrating tyrosine [51].

Regardless of the nitrating species and chemistry, phenolic nitration is a thoroughly documented phenomenon that occurs during NO liberation in a permissive milieu. The instrumental determination of nitrated aromatic species is therefore a promising strategy for assessment of NO-dependent tissue damage.

3-NITROTYROSINE AS AN INDICATOR OF NO-DEPENDENT PROTEIN OXIDATION

The study of NO in vivo has long been hindered by technical difficulties inherent to the noninvasive measurement of NO, $ONOO^-$, and related species. In an effort to circumvent this difficulty, biomarkers that specifically indi-

cate tissue nitration have been sought. Nitrotyrosine was the first obvious candidate for such a biomarker. Tyrosine is an abundant amino acid in both the free form and when it is incorporated into protein, and it is thus likely to react to a significant extent with ambient electrophiles. As discussed previously, peroxynitrite-derived electrophiles rapidly attack phenolic targets, yielding stable and unique products. Peroxynitrite reaction with tyrosine gives 3-nitrotyrosine (3-NO_2-Tyr) in high yield, with some production of 3,3'-dityrosine, 3,4-dihydroxyphenylalanine, and the corresponding quinone [41–43].

Early attempts to detect 3-NO_2-Tyr used a combination of high-performance liquid chromatography with ultraviolet detection (HPLC-UV) and immunochemical methods. Although both approaches met with some success, both were inherently limited. UV detection is sufficiently sensitive for determination of 3-NO_2-Tyr in vitro but generally not sensitive enough for the routine quantitation of 3-NO_2-Tyr in vivo. Salman-Tabcheh and colleagues [52] measured free 3-NO_2-Tyr in human plasma using reverse-phase HPLC with UV detection; however, few similar studies have been reported. Antibodies directed against 3-NO_2-Tyr provide some qualitative indication of protein nitration [53]. Unfortunately, quantitative immunochemical analysis is cumbersome and subject to variability in both antibody specificity and affinity. These caveats may become less important with the development of monoclonal anti-3-NO_2-Tyr antibodies and enzyme-linked immunosorbent assay techniques [54]. Mass spectrometric approaches have also been used to detect 3-NO_2-Tyr, but they require expensive and specialized equipment, extensive sample preparation, and isotopically labeled internal standards [55,56].

PRINCIPLES OF HPLC-ECD

HPLC with electrochemical detection (ECD) is emerging as the technique of choice for sensitive, selective, and facile instrumental determination of 3-NO_2-Tyr [43,44,57–60]. Electrochemical detectors have long been used to measure catecholamines, xenobiotics, and oxidized nucleotides [58], but only recently has ECD been brought to bear on protein and lipid analysis. The primary difference between ECD and other forms of HPLC detection is that the analyte undergoing electrochemical detection actually experiences a chemical transformation (oxidation or reduction) on the surface of the ECD. The current generated by this redox reaction depends on the structure of the analyte and is proportional to the analyte concentration. Phenolic compounds are ideal candidates for ECD as they typically oxidize at convenient potentials (200–800 mV) and are well-retained on reverse-phase HPLC columns.

Traditionally, ECD has been segregated into two variants: the amperometric and coulometric detectors. Amperometric detectors are little more than activated electrode surfaces over which the HPLC effluent is directed. Diffusion of analyte to the electrode surface allows a small percentage (typically <10 %) of the analyte to undergo electrochemical conversion. Thus, amperometric ECD is limited in sensitivity by mass transfer considerations. Coulometric detectors overcome this problem somewhat. A coulometric electrode detector resembles a porous sponge, typically of graphite, through which the analyte is forced at high pressure. As in the case of amperometric detectors, the analyte is electrochemically oxidized (or reduced) on the electrode surface. The increased surface area of the coulometric electrode allows for more efficient analyte conversion and lower detection limits. Coulometric detection is exquisitely sensitive, with detection limits often in the subpicomole range. Moreover, ECD offers an intrinsic selectivity function. Components of the analyte matrix that oxidize at potentials greater than the operating potential of the electrochemical cell are invisible to the ECD. Thus, electrochemically active species may be determined in a complex matrix without the need for extensive precolumn extraction and purification.

Placing several coulometric cells in series can further increase the selectivity of ECD. In this strategy, a series

of electrochemical cells are typically assigned incrementally increasing oxidation potentials. Thus, an electrochemical array is created, across which different components of the analyte mixture are sequentially detected. The addition of an electrochemical array produces a second dimension of chromatographic resolution that can be especially important during ultrasensitive detection of bioanalytes in complex matrices. Figure 2 illustrates this principle in operation during the separation of a mixture of tyrosine oxidation products, including 3-NO_2-Tyr.

DETERMINATION OF 3-NO_2-TYR BY HPLC-ECD

We have used HPLC-ECD array techniques to measure 3-NO_2-Tyr and other tyrosine oxidation products in cell culture [43], human brain tissue [59], and cerebral microvessels [58]. Both free and protein-bound tyrosine analogues can be determined, although sample preparation becomes more difficult when one is interested in protein-bound amino acids. For determination of free tyrosine congeners, a biological sample may be extracted with 5% trichloroacetic acid (TCA) and filtered before injection onto an HPLC-ECD instrument. This preparation is suitable for samples containing little or no nitrite; however, the analyst is advised that acid-catalyzed nitration of tyrosines may proceed under strong acid conditions and high concentrations of nitrite anion. Nitrite concentration can be readily determined by the Griess reaction or other techniques [61]; concentrations of NO_2^- less than 10 μM are tolerated by TCA without introduction of significant artifactual tyrosine nitration. If nitrite levels are extremely high, the sample may be dialyzed or otherwise preprocessed before HPLC analysis.

Determination of protein-bound 3-NO_2-Tyr requires more extensive sample preparation to liberate tyrosine congeners. Various techniques can be used to exhaustively digest proteins. Acid digestion in heated 6 N HCl is one possible strategy; however, the presence of lipids and undefined substances in biological samples often leads to "charring" of the protein under conditions of extreme acid hydrolysis, and acid-catalyzed nitration reactions become quite pronounced. Alternative enzymatic strategies for 3-NO_2-Tyr liberation have been described [42–44]. We have found that several bacterial proteases, particularly the "pronase" class of nonspecific protease isolated from *Staphylococcus griseus*, are most suitable for quantitative digestion of biological samples and adequate liberation of 3-NO_2-Tyr. The general protein digestion protocol is as follows. The biological sample is homogenized in hypotonic solution (10 mM of sodium acetate, with or without 0.1% triton-X-100, is an excellent choice of buffer for most samples; protease and phosphatase inhibitors and other electrophilic organic reagents should generally be excluded from the digestion

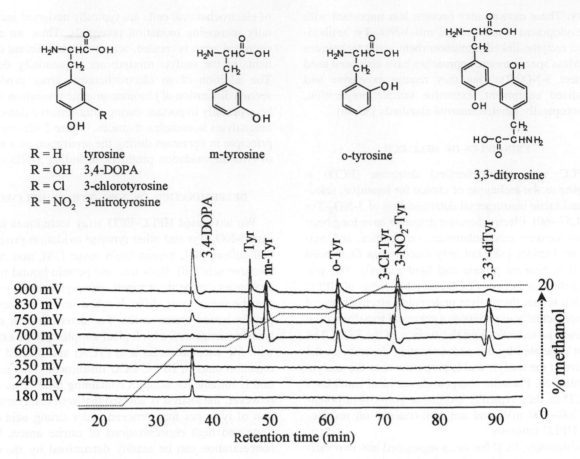

R = H   tyrosine   m-tyrosine   o-tyrosine
R = OH   3,4-DOPA
R = Cl   3-chlorotyrosine   3,3-dityrosine
R = NO₂   3-nitrotyrosine

Fig. 2. Separation and detection of tyrosine congeners by HPLC-ECD. Chromatography was performed as described in the text. Analytes were detected using a model 5600 CoulArray instrument (ESA, Chelmsford, MA, USA).

buffer). The protein concentration is determined by means of a Lowry or other assay and is adjusted to 5 mg/ml. A stock solution of *S. griseus* protease is made in sodium acetate at 10 mg/ml. The pronase is then added at a 1:10 dilution into the biological sample, which is heated 18 hours at 50°C in a water bath with gentle agitation. At the end of the digestion period, the sample is acidified by the addition of a 1:10 volume of 50% TCA. After centrifugation and filtration of the supernatant through a 0.4 μm polyvinylidene difluoride syringe filter, the clear supernatant is ready for injection onto the HPLC instrument.

The choice of column, mobile-phase, and instrument parameters must be determined by the individual investigator to suit his particular needs and sample matrix. As a general purpose column, we recommend using a Toso-Haas (Montgomeryville, PA, USA) ODS-80TM column (5 μm particle size, 4.6 × 250 mm) or a column of similar manufacture. Nitrotyrosine may be resolved under isocratic conditions from many matrices [43,58,59]. Nonetheless, better resolution of tyrosine oxidation products is obtained by employing a gradient elution strategy

with a 0 to 20% stepwise gradient using either methanol or acetonitrile as the organic component (Fig. 2). Ion-pairing agents or detergents may be included in the mobile phase; we have found that lithium dodecyl sulfate sometimes improves the resolution of tyrosine congeners in brain tissue homogenates [59]. It is critical for an ECD experiment that the mobile phase has sufficiently high ionic strength to facilitate electrochemical reactions at the electrode surface. For this reason, lithium acetate or lithium triphosphate (30 mM or higher) is recommended for use in the mobile phase in combination with an acetate or citrate buffer with a pH less than 5.0. Lithium salts are reasonably soluble in highly organic solution, resist the growth of bacteria, and are tolerably absent of redox-active metals. Transition metals such as iron and copper and small electrogenic compounds such as azide or cyanide must be assiduously avoided in preparing the mobile phase for an ECD experiment. Also, for maximum sensitivity, the mobile phase should be continuously sparged with helium or argon to maintain low levels of dissolved oxygen.

It should be noted that alternative strategies for ECD

of 3-NO$_2$-Tyr have been reported. These typically rely on precolumn chemical reduction of the analyte to 3-aminotyrosine by treatment with strong reducing agents such as sodium hydrosulfite [44]. Aminotyrosine is more easily oxidized than 3-NO$_2$-Tyr; hence, it is more readily detected than the 3-NO$_2$-Tyr precursor. Conversely, aminotyrosine analysis is hampered by inefficiencies in nitrotyrosine reduction, instability of the 3-amino-tyrosine product, and poor retention on reverse-phase columns. In an interesting recent variation, the nitrotyrosine→aminotyrosine conversion was catalyzed by passage of the sample through an ascorbate-treated platinum filter placed in line with the reverse-phase column and preceding the electrochemical detector [62]. This "on-line" reduction strategy circumvents many of the pitfalls of aminotyrosine determination and may have broad applicability to the ultrasensitive determination of protein nitration products.

ECD allows the simultaneous determination of multiple tyrosine congeners (Fig. 2); thus, it is possible to express 3-NO$_2$-Tyr concentration as a ratio to underivatized tyrosine in the same chromatographic run. As a note of caution, the analyst is advised that tyrosine concentrations are typically 1000 to 10,000 times the concentration of 3-NO$_2$-Tyr in a typical protein digest [43,58, 59]. The response sensitivity of most electrochemical cells saturates at analyte concentrations in excess of 10 μM. Thus, one may wish to insert a UV-visible or photodiode array detector in line with (but preceding) the electrochemical detector to accurately determine tyrosine content and use the extreme sensitivity and selectivity of ECD for measurement of tyrosine oxidation products.

NO REACTIONS IN THE LIPID PHASE: 5-NITRO-Γ-TOCOPHEROL AS A BIOMARKER OF OXIDATIVE DAMAGE

Although 3-NO$_2$-Tyr may prove to be an excellent biomarker for NO-dependent protein oxidation, measurement of protein-bound 3-NO$_2$-Tyr may not yield a complete and adequate representation of the NO-dependent tissue damage. NO is actually quite a lipophilic substance, and it has become appreciated that peroxynitrite flagrantly targets lipid-phase components [63–66]. Moreover, NO autoxidation to NO$_2$ and higher oxides is several hundred times more rapid in membranes than in the aqueous phase [67]. Lipid nitration products, particularly those formed from tocopherol nitration, may therefore prove to be better biomarkers for nitrative stress than 3-NO$_2$-Tyr. Mammalian cell membranes contain α-tocopherol, γ-tocopherol, and other forms of tocopherol in stoichiometries as high as 1:1000 with respect to phospholipids [68]. Interestingly, the γ-tocopherol concentration in

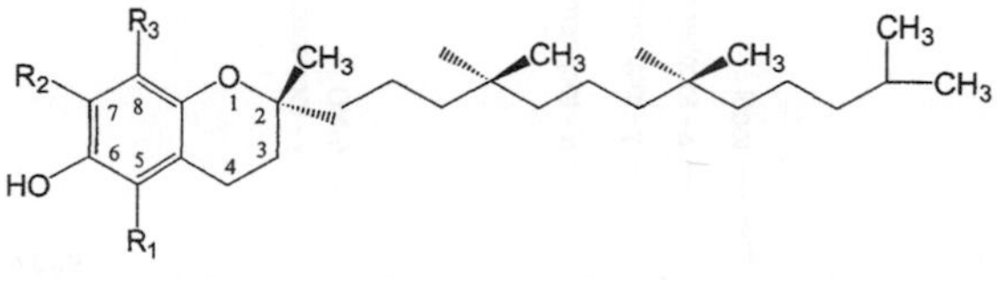

$R_1 = R_2 = R_3 = CH_3$	α-tocopherol
$R_1 = R_3 = CH_3$; $R_2 = H$	β-tocopherol
$R_1 = H$; $R_2 = R_3 = CH_3$	γ-tocopherol
$R_1 = R_2 = H$; $R_3 = CH_3$	δ-tocopherol
$R_1 = R_2 = R_3 = H$	tocol
$R_1 = NO_2$; $R_2 = R_3 = CH_3$	5-nitro-γ-tocopherol

Fig. 3. Structures of tocopherol variants and their nomenclature [69].

human plasma is up to 25% that of α-tocopherol, making γ-tocopherol a significant but frequently unappreciated component of the total tocopherol pool [68]. The chemistry of α-tocopherol as a free radical scavenger and chain-breaking antioxidant has been explored. To date, however, the reaction of tocopherol species with NO-derived oxidants has not been extensively investigated. Although α-tocopherol possesses a phenolic structure that renders this compound electrochemically detectable, there is no obvious space on the chromane ring system where an electrophilic nitration reaction might occur (Fig. 3). In contrast, γ-tocopherol lacks a methyl group in position 5 and hence might be anticipated to experience nitration in vivo. It has been reported that γ-tocopherol nitration proceeds during peroxynitrite-mediated lipid oxidation, although α-tocopherol is relatively spared from oxidative conversion [64]; however, this finding has been contradicted in a separate study [66]. Nonetheless, it seems clear that peroxynitrite reaction with tocopherol is favored over its reaction with tyrosine when tyrosine and α- or γ-tocopherol are simultaneously present as reactants [66].

Cooney et al. [63] used optical methods to detect 5-nitro-γ-tocopherol in pancreatic cells exposed to the inflammatory cytokine interleukin 1β [63]. Extending on this work, we have developed an HPLC-ECD method sufficient to separate 5-nitro-γ-tocopherol from other tocopherol species over the course of a 30 min chromatographic analysis (Fig. 4). The γ-tocopherol is not resolved from its structural isomer β-tocopherol under these conditions. HPLC-ECD is sensitive enough to measure extremely low tocopherol concentrations in endotoxin-stimulated primary rat glial cell culture (Fig. 5). The detection limit for 5-nitro-γ-tocopherol by HPLC-ECD is approximately 0.5 pmol on the column, which is well below the sensitivity of UV-visible detectors [58].

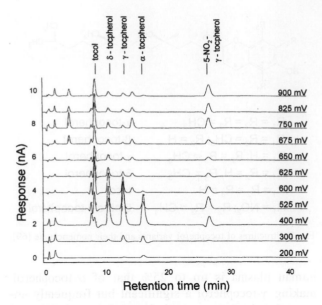

Fig. 4. Separation of tocopherol variants by HPLC-ECD. Chromatography was performed as described in the text. Analytes were detected using a model 5600 CoulArray instrument (ESA). A total of 500 pM of each analyte was loaded on the column.

CHROMATOGRAPHIC STRATEGY FOR 5-NITRO-Γ-TOCOPHEROL

Tocopherol chromatography is conducted in a manner similar to tyrosine chromatography, with some differences in sample preparation and mobile-phase composition. Samples are homogenized in 10 mM of sodium acetate (pH 7.0) containing 0.1% triton X-100 and adjusted to a 5 mg/ml protein concentration. Sam-

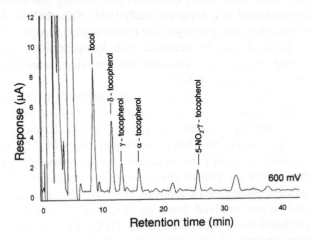

Fig. 5. HPLC-ECD identification of 5-nitro-γ-tocopherol in primary rat astrocytes stimulated with endotoxin (1 μg/ml of *Escherichia coli* lipopolysaccharide for 36 h). The 5-nitro-γ-tocopherol peak in this chromatogram represents 2.5 pmol of analyte on the column. The ratio of 5-nitro-γ-tocopherol to γ-tocopherol increased from a value of 0.023 ± 0.013 ($n = 5$) in unstimulated cells to a value of 0.151 ± 0.027 ($n = 5$) in lipopolysaccharide-stimulated cells.

ple volumes (1 ml) are centrifuged 10 min at 10,000 × g and 4°C. The resulting pellet is suspended in 1 ml of ethanol and thrice extracted in 5 ml of hexane, which is subsequently evaporated under a vacuum. Many published tocopherol analyses include chain-breaking antioxidants in the hexane to minimize artifactual loss of tocopherol during extraction. We have found that inclusion of butylated hydroxytoluene, a phenolic free radical scavenger, has no discernible effect on tocopherol or 5-nitro-γ-tocopherol concentrations measured in brain homogenates. After evaporation of the hexane, the resulting lipid film is redissolved in methanol and filtered before HPLC analysis. Chromatography is conducted using the same reverse-phase column as that described for nitrotyrosine determination. The isocratic mobile phase contains 83% acetonitrile, 12% methanol, 0.2% acetic acid, and 30 mM of lithium acetate. Using this mobile phase and a flow rate of 2 ml/ min, 5-nitro-γ-tocopherol elutes within 30 min (Figs. 4 and 5). Inclusion of an UV-visible or photodiode array detector in line with (preceding) the electrochemical array allows simultaneous determination of α-tocopheryl quinone as another index of lipid oxidation [58].

Currently, there are no commercially available sources for 5-nitro-γ-tocopherol standards. Fortunately, this compound is easily synthesized by exposing a hexane solution (10 mM in γ-tocopherol) to NO_2 (g) or auto-oxidized NO (g) [63]. This can be accomplished by filling a 15 ml glass tube two thirds full with tocopherol in hexane and then filling the head space with NO_2. After several minutes, a yellow color develops, indicating tocopherol nitration. The 5-nitro-γ-tocopherol is formed in high yield (>50%) by this reaction along with several other products, including γ-tocopheryl quinone [63]. Purification of 5-nitro-γ-tocopherol is readily accomplished by preparative thin-layer chromatography or column chromatography over silica with a mobile phase containing 99% dichloromethane plus 1% methanol. The 5-nitro-γ-tocopherol elutes near the solvent front as a yellow band, followed by one or more reddish bands ascribed to tocopheryl quinones ($R_f < 0.6$). The purified 5-nitro-γ-tocopherol can be washed from the silica matrix with ether, concentrated under a vacuum, and redissolved in 100% methanol. The chromatography may need to be repeated several times to obtain punctilious 5-nitro-γ-tocopherol. The molar extinction coefficient of 5-nitro-γ-tocopherol is 6750 M^{-1}/cm^{-1} at 302 nm and 1352 M^{-1}/cm^{-1} at 420 nm [63].

SUMMARY

Recent revelations concerning the toxicological importance of NO have necessitated new analytical strategies for determination of NO-dependent oxidative tissue damage. The interested researcher may find that nitrated phenolic components such as 3-NO$_2$-Tyr and 5-nitro-γ-tocopherol provide a suitable index for assessing NO-dependent pathophysiology. HPLC-ECD offers a convenient and sensitive instrumental strategy for measurement of these nitrated bioanalytes.

Acknowledgements — This work was supported in part by Grants from the National Institutes of Health (NS35747) and the Oklahoma Center for the Advancement of Science and Technology (OCAST H67-097). The authors thank R.V. Cooney for initial gifts of 5-nitro-γ-tocopherol and stimulating insights into nitrotocopherol chemistry.

REFERENCES

[1] Ignarro, L. J.; Buga, G. M.; Wood, K. S.; Byrns, R. E.; Chaudhuri, G. Endothelium-derived relaxing factor produced and released from artery and vein is nitric oxide. *Proc. Natl. Acad. Sci. USA* **84:** 9265–9269; 1987.

[2] Palmer, R. M.; Ferrige, A. G.; Moncada, S. Nitric oxide release accounts for the biological activity of endothelium-derived relaxing factor. *Nature* **327:**524–526; 1987.

[3] Fukuto, J. M.; Chaudhuri, G. Inhibition of constitutive and inducible nitric oxide synthase: potential selective inhibition. *Annu. Rev. Pharmacol. Toxicol.* **35:**165–194; 1995. [©4] Ignarro, L. J. Heme-dependent activation of soluble guanylate cyclase by nitric oxide: regulation of enzyme activity by porphyrins and metalloporphyrins. *Semin. Hematol.* **26:**63–76; 1989.

[5] Traylor, T. G; Sharma, V. S. Why NO? *Biochemistry* **31:**2847–2849; 1992.

[6] McDonald, L. J.; Murad, F. Nitric oxide and cGMP signaling. *Adv. Pharmacol.* **34:**263–275.

[7] Kanner, J.; Harel, S.; Granit, R. Nitric oxide as an antioxidant. *Arch. Biochem. Biophys.* **289:**130–136; 1991.

[8] Upchurch, G. R., Jr.; Welch, G. N.; Loscalzo, J. S. Nitrosothiols: chemistry, biochemistry, and biological actions. *Adv. Pharmacol.* **34:**343–349, 1995.

[9] Pryor, W. A.; Church, D. F.; Govinda, C. K.; Crank, G. Oxidation of thiols by nitric oxide and nitrogen dioxide: synthetic utility and toxicological implications. *J. Org. Chem.* **47:**159–161, 1982.

[10] Lipton, S. A.; Choi, Y. B.; Pan, Z. H.; Lei, S. Z.; Chen, H. S.; Sucher, N. J.; Loscalzo, J.; Singel, D. J.; Stamler, J. S. A redox-based mechanism for the neuroprotective and neurodestructive effects of nitric oxide and related nitroso-compounds. *Nature* **364:**626–632; 1993.

[11] Lipton, S. A.; Singel, D. J.; Stamler, J. S. Neuroprotective and neurodestructive effects of nitric oxide and redox congeners. *Ann. NY Acad. Sci.* **17:**382–387; 1994.

[12] Kharitonov, V. G.; Sunduist, A. R.; Sharma, V. S. Kinetics of nitrosation of thiols by nitric oxide in the presence of oxygen. *J. Biol. Chem.* **270:**28158–28164.

[13] Chan, N. L.; Rogers, P. H.; Arnone, A. Crystal structure of the S-nitroso form of liganded human hemoglobin. *Biochemistry* **37:**16459–16464; 1998.

[14] Ferranti, P.; Nalorni, A.; Mamone, G.; Sannolo, N.; Marino, G. Characterization of S-nitrosohaemoglobin by mass spectometry. *FEBS Lett.* **400:**19–24; 1997.

[15] Stamler, J. S.; Jia, L.; Eu, J. P.; McMahon, T. J.; Demchenko, I. T.; Bonaventura, J.; Gernert, K.; Piantadosi, C. A. Blood flow regulation by S-nitrosohemoglobin in the physiological oxygen gradient. *Science* **276:**2034–2037; 1997.

[16] Lander, H. M.; Hajjar, D. P.; Hempstead, B. L.; Mirza, U. A.; Chait, B. T.; Campbell, S.; Quilliam, L. A. A molecular switch on p21 (ras). Structural basis for the nitric oxide-p21 (ras) interaction. *J. Biol. Chem.* **272:**4323–4326; 1997.

[17] Lander, H. M.; Milbank, A. J.; Taurus, J. M.; Hajjar, D. P.; Hempstead, B. L.; Schwartz, G. D.; Kraemer, R. T.; Mirza, U. A.;

[18] Lander, H. M.; Ogiste, J. S.; Pearce, S. F.; Levi, R.; Novagrodsky, A. Nitric oxide stimulated guanine nucleotide exchange on p21 ras. *J. Biol. Chem.* **270:**7017–7020; 1995.

[19] Mirza, U. A.; Chait, B. T.; Lander, H. M. Monitoring reactions of nitric oxide with peptides and proteins by electrospray ionization–mass spectrometry. *J. Biol. Chem.* **270:**17185–17188; 1995.

[20] Lander, H. M.; Jacovina, A. T.; Davis, R. J.; Tauras, J. M. Differential activation of mitogen-activated protein kinases by nitric oxide related species. *J. Biol. Chem.* **271:**19705–19709; 1996.

[21] Xie, Q. W.; Cho, H. J.; Calaycay, J.; Mumford, R. A.; Swiderek, K. M.; Lee, T. D.; Ding, A.; Troso, T.; Nathan, C. Cloning and characterization of inducible nitric oxide synthase from mouse macrophages. *Science* **256:**226–228; 1992.

[22] Koide, M.; Kawahara, Y.; Tsuda, T.; Yokoyama, M. Cytokine-induced expression of an inducible type of nitric oxide synthase gene in cultured vascular smooth muscle cells. *FEBS Lett.* **18:** 213–217; 1993.

[23] Geller, D. A.; Lowenstein, C. J.; Shapiro, R. A.; Nussler, A. K.; Di Silvio, M.; Wang, S. C.; Nakayama, D. K.; Simmons, R. L.; Snyder, S. H.; Billiar, T. R. Molecular cloning and expression of inducible nitric oxide synthase from human hepatocytes. *Proc. Natl. Acad. Sci. USA* **90:**3491–3495; 1993.

[24] Xie, K.; Huang, S.; Dong, Z.; Gutman, M.; Fidler, I. J. Direct correlation between expression of endogenous inducible nitric oxide synthase and regression of M5076 reticulum cell sarcoma hepatic metastases in mice treated with liposomes containing lipopeptide CGP 31362. *Cancer Res.* **55:**3123–3131; 1995.

[25] Xie, K.; Huang, S.; Dong, Z.; Juang, S. H.; Gutman, M.; Xie, Q. W.; Nathan, C.; Fidler, I. J. Transfection with the inducible nitric oxide synthase gene suppresses tumorigenicity and abrogates metastasis by K-1735 murine melanoma cells. *J. Exp. Med.* **181:**1333–1343; 1995.

[26] Bhat, N. R.; Zhang, P.; Lee, J. C.; Hogan, E. L. Extracellular signal-regulated kinase and p38 subgroups of mitogen-activated protein kinases regulate inducible nitric oxide synthase and tumor necrosis factor alpha gene expression in endotoxin-stimulated primary glial cultures. *J. Neurosci.* **18:**1633–1641; 1998.

[27] Ischiropoulos, H.; Zhu, L.; Beckman, J. S. Peroxynitrite formation from macrophage-derived nitric oxide. *Arch. Biochem. Biophys.* **298:**446–451; 1992.

[28] Carreras, M. C.; Pargament, G. A.; Catz, S. D.; Poderoso, J. J.; Boveris, A. Kinetics of nitric oxide and hydrogen peroxide production and formation of peroxynitrite during the respiratory burst of human neutrophils. *FEBS Lett.* **341:**65–68; 1994.

[29] Crow, J. P.; Beckman, J. S. The importance of superoxide in nitric oxide-dependent toxicity: evidence for peroxynitrite-mediated injury. *Adv. Exp. Med. Biol.* **387:**147–161; 1996.

[30] Colton, C. A.; Snell, J.; Chernyshev, O.; Gilbert, D. L. Induction of superoxide anion and nitric oxide production in cultured microglia. *Ann. NY Acad. Sci.* **17:**54–63; 1994.

[31] Eiserich, J. P.; Hristova, M.; Cross, C. E.; Jones, A. D.; Freeman, B. A.; Halliwell, B.; van der Vliet, A. Formation of nitric oxide-derived inflammatory oxidants by myeloperoxidase in neutrophils. *Nature* **391:**393–397; 1998.

[32] Beckman, J. S.; Beckman, T. W.; Chen, J.; Marshall, P. A.; Freeman, B. A. Apparent hydroxyl radical production by peroxynitrite: implications for endothelial injury from nitric oxide and superoxide. *Proc. Natl. Acad. Sci. USA* **87:**1620–1624; 1990.

[33] Crow, J. P.; Beckman, J. S. Reactions between nitric oxide, superoxide and peroxynitrite: footprints of peroxynitrite in vivo. *Adv. Pharmacol.* **34:**17–43; 1995.

[34] Gutierrez, H.; Paler-Martinez, A.; Tarpey, M. M.; Rubbo, H. Oxygen radical-nitric oxide reactions in vascular diseases. *Adv. Pharmacol.* **34:** 45–69; 1995.

[35] Miles, A. M.; Bohle, D. S.; Glassbrenner, P. A.; Hansert, B.; Wink, D. A.; Grisham, M. B. Modulation of superoxide-dependent oxidation and hydroxylation reactions by nitric oxide. *J. Biol. Chem.* **271:**40–47; 1996.

[36] Rubbo, H.; Darley-Usmar, V.; Freeman, B. A. Nitric oxide regulation of tissue free radical injury. *Chem. Res. Toxicol.* **9:**809–820; 1996.

[37] Huie, R. E.; Padmaja, S. Reaction of NO with O_2^-. *Free Radic. Res. Commun.* **18:**195–199; 1993.

[38] Hensley, K.; Tabatabaie, T.; Stewart, C. A.; Pye, Q.; Floyd, R. A. Nitric oxide and derived species as toxic agents in stroke, AIDS dementia, and chronic neurodegenerative disorders. *Chem. Res. Toxicol.* **10:**527–532; 1997.

[39] Kissner, R.; Nauser, T.; Bugnon, P.; Lye, P. G.; Koppenol, W. H. Formation and properties of peroxynitrite studied by laser flash photolysis, high pressure stopped flow and pulse radiolysis. *Chem. Res. Toxicol.* **10:**1285–1292; 1997.

[40] Koppenol, W. H.; Moreno, J. J.; Pryor, W. A.; Ischiropoulos, H.; Beckman, J. S. Peroxynitrite, a cloaked oxidant formed by nitric oxide and superoxide. *Chem. Res. Toxicol.* **5:**834–842; 1992.

[41] Beckman, J. S.; Ischiropoulos, H.; Zhu, L.; van der Woerd, M.; Smith, C. D.; Chen, J.; Harrison, J.; Martin, J. C.; Tsai, M. Kinetics of superoxide dismutase and iron-catalyzed nitration of phenolics by peroxynitrite. *Arch. Biochem. Biophys.* **298:**438–445; 1992.

[42] Ramezanian, M. S.; Padmaja, S.; Koppenol, W. H. Hydroxylation and nitration of phenolic compounds by peroxynitrite. *Chem. Res. Toxicol.* **9:**232–240; 1996.

[43] Hensley, K.; Maidt, M. L.; Pye, Q. N.; Stewart, C. A.; Wack, M.; Tabatabaie, T.; Floyd, R. A. Quantitation of protein-bound 3-nitrotyrosine and 3,4-dihydroxyphenylalanine by high-performance liquid chromatography with electrochemical array detection. *Anal. Biochem.* **251:**187–195; 1997.

[44] Shigenaga, M. K.; Lee, H. H.; Blount, B. C.; Christen, S.; Shigeno, E. T.; Yip, H.; Ames, B. N. Inflammation and NO(X)-induced nitration: assay for 3-nitrotyrosine by HPLC with electrochemical detection. *Proc. Natl. Acad. Sci. USA* **94:**3211–3216; 1997.

[45] Koppenol, W. H. The basic chemistry of nitrogen monoxide and peroxynitrite. *Free Radic. Biol. Med.* **25:**385–391; 1998.

[46] Koppenol, W. H.; Kissner, R. Can O=NOOH undergo homolysis? *Chem. Res. Toxicol.* **11:**87–90; 1998.

[47] Gow, A.; Duran, D.; Thom, S. R.; Ischiropoulos, H. Carbon dioxide enhancement of peroxynitrite-mediated protein tyrosine nitration. *Arch. Biochem. Biophys.* **333:**42–48; 1996.

[48] Lymar, S. V.; Hurst, J. K. Radical nature of peroxynitrite reactivity. *Chem. Res. Toxicol.* **11:**714–715; 1998.

[49] Lymar, S. V.; Hurst, J. K. Rapid reaction between peroxynitrite ion and carbon dioxide: implications for biological activity. *J. Am. Chem. Soc.* **117:**8867–8868; 1995.

[50] Squadrito, G. L.; Pryor, W. A. Oxidative chemistry of nitric oxide: the roles of superoxide, peroxynitrite and carbon dioxide. *Free Radic. Biol. Med.* **25:**392–403, 1998.

[51] Eiserich, J. P.; Cross, C. E.; Jones, A. D.; Halliwell, B.; van der Vliet, A. Formation of nitrating and chlorinating species by reaction of nitrite with hypochlorous acid. A novel mechanism for nitric oxide-mediated protein modification. *J. Biol. Chem.* **271:**19199–19208; 1996.

[52] Salman-Tabcheh, S.; Guerin, M. C.; Torreilles, J. Nitration of tyrosyl residues from extra and intracellular proteins in human whole blood. *Free Radic. Biol. Med.* **19:**695–698; 1995.

[53] Viera, L.; Ye, Y. Z.; Estevez, A. G.; Beckman, J. S. Immunohistochemical methods to detect nitrotyrosine. *Methods Enzymol.* **301:**373–381; 1999.

[54] Ter Steege, J. C.; Koster-Kamphuis, L.; van Straaten, E. A.; Forget, P. P.; Buurman, W. A. Nitrotyrosine in plasma of celiac disease patients as detected by a new sandwich ELISA. *Free Radic. Biol. Med.* **25:**953–963; 1998.

[55] Crowley, J. R.; Yarasheski, K.; Leeuwenburgh, C.; Turk, J.; Heinecke, J. W. Isotope dilution mass spectrometric quantification of 3-nitrotyrosine in proteins and tissues is facilitated by reduction to 3-aminotyrosine. *Anal. Biochem.* **259:**127–135; 1998.

[56] Leeuwenburgh, C.; Hardy, M. M.; Hazen, S. L.; Wagner, P.; Oh-ishi, S.; Steinbrecher, U. P.; Heinecke, J. W. Reactive nitrogen intermediates promote low density lipoprotein oxidation in human atherosclerotic intima. *J. Biol. Chem.* **272:**1433–1436; 1997.

[57] Shigenaga, M. K. Quantitation of protein-bound 3-nitrotyrosine by high-performance liquid chromatography with electrochemical detection. *Methods Enzymol.* **301:**27–40; 1999.

[58] Hensley, K.; Williamson, K. S.; Maidt, M. L.; Gabbita, S. P.; Grammas, P.; Floyd, R. A. Determination of biological oxidative stress using high performance liquid chromatography with electrochemical detection (HPLC-ECD). *J. High Res. Chromatogr.* **22:**429–437; 1999.

[59] Hensley, K.; Maidt, M. L.; Yu, Z. Q.; Sang, H.; Markesbery, W. R.; Floyd, R. A. Electrochemical analysis of protein nitrotyrosine and diotyrosine in the Alzheimer brain reveals region-specific accumulation. *J. Neurosci.* **18:**8126–8132; 1998.

[60] Herce-Pagliai, C.; Kotecha, S.; Shuker, D. E. Analytical methods for 3-nitrotyrosine as a marker of exposure to reactive nitrogen species: a review. *Nitric Oxide* **2:**324–336; 1998.

[61] Ellis, G.; Adatia, I.; Yazdanpanah, M.; Makela, S. K. Nitrite and nitrate analyses: a clinical biochemistry perspective. *Clin. Biochem.* **31:**195–220; 1998.

[62] Ohshima, H.; Cellan, I.; Chazotte, L.; Pignatelli, B.; Mower, H. F. Analysis of 3-nitrotyrosine in biological fluids and protein hydrolyzates by high-performance liquid chromatography using a post-separation, on-line reduction column and electrochemical detection: results with various nitrating agents. *Nitric Oxide* **3:**132–141; 1999.

[63] Cooney, R. V.; Harwood, P. J.; Franke, A. A.; Narala, K.; Sundtrom, A. K.; Berggren, P. O.; Mordan, L. J. Products of gamma tocopherol reaction with NO_2 and their formation in rat insulinoma (RINm5F) cells. *Free Radic. Biol. Med.* **19:**259–260; 1995.

[64] Christen, S.; Woodall, A. A.; Shigenaga, M. K.; Southwell-Keely, P. T.; Duncan, M. W.; Ames, B. N. Gamma tocopherol traps mutagenic electrophiles such as NO(X) and complements alpha-tocopherol: physiological implications. *Proc. Natl. Acad. Sci. USA* **94:**3217–3222.

[65] O'Donnell, V. B.; Eiserich, J. P.; Chumley, P. H.; Jablonsky, M. J.; Krishna, N. R.; Kirk, M.; Barnes, S.; Darley-Usmar, V. M.; Freeman, B. A. Nitration of unsaturated fatty acids by nitric oxide-derived reactive nitrogen species peroxynitrite, nitrous acid, nitrogen dioxide, and nitronium ion. *Chem. Res. Toxicol.* **12:**83–92; 1999.

[66] Goss, S. P. A.; Hogg, N.; Kalyanaraman, B. The effect of α-tocopherol on the nitration of γ-tocopherol by peroxynitrite. *Arch. Biochem. Biophys.* **363:**333–340; 1999.

[67] Liu, X.; Miller, M. J. S.; Joshi, M. S.; Thomas, D. D.; Lancaster, J. R., Jr. Accelerated reaction of nitric oxide with O_2 within the hydrophobic interior of biological membranes. *Proc. Natl. Acad. Sci. USA* **95:**2175–2179; 1998.

[68] Vatassery, G. T.; Krezowski, A. M.; Eckfeldt, J. H. Vitamin E concentrations in human blood plasma and platelets. *Am. J. Clin. Nutr.* **37:**1020–1024; 1983.

[69] IUPAC-IUB Joint Commission on Biochemical Nomenclature (JCBN). Nomenclature of tocopherols and related compounds. Recommendations 1981. *Eur. J. Biochem.* **123:**473–475; 1982.

Bioassays for Oxidative Stress Status (BOSS). Edited by W.A. Pryor

DETECTION OF OXIDATIVE DNA DAMAGE IN HUMAN SPERM AND ITS ASSOCIATION WITH SPERM FUNCTION AND MALE INFERTILITY

HAN-MING SHEN and CHOON-NAM ONG

Department of Community, Occupational and Family Medicine, National University of Singapore, 16 Medical Drive,
Singapore, Republic of Singapore

(Received 26 August 1999; Revised 12 October 1999; Accepted 28 October 1999)

Abstract—The expanding research interest in the last two decades on reactive oxygen species (ROS), oxidative stress, and male infertility has led to the development of various techniques for evaluating oxidative DNA damage in human spermatozoa. Measurement of 8-hydroxydeoxyguanosine (8-OHdG) offers a specific and quantitative biomarker on the extent of oxidative DNA damage caused by ROS in human sperm. The close correlations of 8-OHdG level with male fertility, sperm function and routine seminal parameters indicate the potential diagnostic value of this technique in clinical applications. On the other hand, single cell gel electrophoresis (SCGE or comet assay) and terminal deoxynucleotidyl transferase (TdT) mediated dUTP nick end labeling (TUNEL) assay have also been demonstrated to be sensitive, and reliable methods for measuring DNA strand breaks in human spermatozoa. As certain technical limitations were inherent in each of these tests, it is believed that a combination of these assays will offer more comprehensive information for a better understanding of oxidative DNA damage and its biological significance in sperm function and male infertility. © 2000 Elsevier Science Inc.

Keywords—Sperm, Oxidative stress, 8-OHdG, Free radical, Comet, TUNEL, Biomarker

INTRODUCTION

In recent years, a number of studies have consistently indicated a worldwide decreasing trend of male fertility in terms of average sperm counts and sperm quality, especially in developed countries [1,2]. One of the significant developments in the last two decades in the study of human infertility has been the discovery that reactive oxygen species (ROS) and oxidative stress play a critical role in the etiology of defective sperm function and male infertility. Exten-

sive studies have been conducted to establish the link between oxidative stress and pathology of human spermatozoa: (i) direct detection of ROS formation from human spermatozoa and contaminating leukocytes in semen, (ii) the occurrence of lipid peroxidation and its vital role in the etiology of male infertility, (iii) measurement of oxidative DNA damage in sperm and its association with impaired sperm function and male fertility, and (iv) changes of the antioxidant defense system and the possible application of antioxidants in prevention of male infertility. A number of comprehensive review articles have well discussed these issues [3–8]. Oxidative DNA damage refers to the functional or structural alterations of DNA resulting from the insults of ROS, which have been widely implicated in many degenerative diseases including aging and cancer [9,10]. It is known that the chromatin in mammalian spermatozoa is constructed in a unique pattern that is extremely compact and stable and differs substantially from that of somatic cells [11]. Since the early 1990s, substantial studies have been conducted to analyze the integrity of DNA in human sperm and have yielded important information on the possible relationship of oxidative sperm DNA damage with defective sperm function and male infertility. The present review attempts to summarize and

Address Correspondence to: Dr. H.-M. Shen, National University of Singapore, Department of Community, Occupational and Family Medicine, Faculty of Medicine (MD3), 16 Medical Drive, Singapore 117597, Republic of Singapore; Tel: +65 874-4996; Fax: +65 779-1489; E-Mail: cofshm@nus.edu.sg.
Drs. Han-Ming Shen and Choon-Nam Ong are from the Department of Community Medicine, National University of Singapore (NUS).

H.M.S. received his Bachelor of Medicine and Master of Public Health degrees from Zhejiang Medical University, People's Republic of China. He has been working as a Research Scientist at the NUS after obtaining his Ph.D. in 1996.

C.-N.O. received his Ph.D. from the University of Manchester, England in 1977. He is currently the Director of the Centre for Environmental and Occupational Health at the NUS.

Both authors have been working in the area of reproductive toxicology, oxidative stress, and antioxidants in carcinogenesis and cancer chemoprevention, for the last 10 years.

Reprinted from: *Free Radical Biology & Medicine, Vol. 28, No. 4, pp. 529-536, 2000*

Table 1. Summary of Results of DNA Extraction From Human
Spermatozoa

Parameters	Geometric mean	95% Confidence interval
Sperm density ($\times 10^6$ cells/ml)	51.29[a]	40.04–65.69
Volume used for DNA extraction (ml)	1.45[a]	1.22–1.68
Number of sperm cells used for DNA extraction ($\times 10^6$)	68.23	53.88–86.40
Amount of DNA (μg) per sample	111.28	88.33–140.18
$A_{260/280}$ ratio	1.82[a]	1.80–1.84
DNA yield (μg/10^6 cells)	1.63	1.27–2.09
8-OHdG/10^5 dG	8.91	6.99–11.35

[a] Arithmetic mean.

evaluate such measurements and to establish the link between oxidative sperm DNA damage and sperm function and male infertility.

MEASUREMENT OF 8-HYDROXYDEOXYGUANOSINE IN HUMAN SPERM

8-Hydroxydeoxyguanosine (8-OHdG) is the most commonly studied biomarker for oxidative DNA damage. Among various oxidative DNA adducts, 8-OHdG has been selected as a representative of oxidative DNA damage due to its high specificity, potent mutagenicity, and relative abundance in DNA [12]. There are two main methods applied in 8-OHdG analysis in genomic DNA: (i) high-performance liquid chromatography equipped with electrochemical detection (HPLC-EC) and (ii) gas chromatography-mass spectrometry (GC/MS). Comparatively, HPLC-EC is a more popular method among many research groups.

Methodology of 8-OHdG analyses in human sperm

Compared with the method used in somatic cells, the main difference of 8-OHdG analysis in human sperm is at the stage of DNA extraction. In our laboratory, we conduct our 8-OHdG analysis in human spermatozoa according to the following procedures:

DNA extraction. The detailed protocol for DNA extraction from human sperm has been described earlier [13]. Briefly, sperm sample was first washed with sperm washing buffer (SWB, 10 mM Tris-HCl, 10 mM ethylenediaminetetra-acetate [EDTA], 1 M NaCl, pH 7.4), then incubated with dithiothreitol (DTT), proteinase K and SDS. Subsequently, sperm DNA was extracted with chloroform/isoamyl alcohol and digested with ribonuclease A. Finally, DNA was dissolved in 10 mM Tris-HCl for subsequent enzymatic DNA digestion. Table 1 sum-

marizes the results of DNA extraction using the above protocol from one typical experiment (30 semen samples from an infertility clinic). Good quality of DNA was obtained with an average $A_{260/280}$ ratio at about 1.82. The yield of DNA varied from 40 to 300 μg depending on the number of sperm cells contained in the semen, with a mean of 1.63 μg DNA/10^6 cells, which accounts for about 54% of the total DNA contained in human spermatozoa (human spermatozoa contain half the amount of DNA in somatic cells, equivalent to about 3 μg DNA/10^6 cells).

Enzymatic DNA digestion. DNA digestion was a critical step in 8-OHdG analysis. Similar to many other studies, a combination of enzymes including nuclease P1 and alkaline phosphatase were used to ensure an optimal condition [13,14]. In our study we found that complete DNA digestion was achievable only when bacterial source AP was used, and other source of AP such as bovine-born enzymes resulted in much poorer digestion (unpublished observations). The rate of DNA digestion in each sample, which is normally above 90%, could be calculated by the dG concentration using the conversion factor 0.62 nmol of dG = 1 μg DNA [14].

HPLC-EC analysis. The HPLC-EC analysis system consisted of a Gilson (Villers-le-Bel, France) pump, a Whatman (Singapore) partisphere 5 C18 column, an electrochemical detector (Ag/AgCl reference electrode, glassy carbon working electrode, potential 0.7 V, range 50 nA), an ultraviolet detector (254 nm) (Hewlett-Packard, Palo Alto, CA, USA), an autosampler, and an integrator. The mobile phase consisted of 20 mM $NH_4H_2PO_4$, 1 mM EDTA and 4% methanol (pH 4.7). The flow rate was 1 ml/min. The retention times for 8-OHdG and dG were 13 and 9 min, respectively. The results were presented as 8-OHdG/10^5 dG.

The detection limit, at a signal-to-noise ratio of three, was calculated to be 750 fmole. The batch-to-batch variation was measured using pooled semen samples, with a coefficient of variation (CV) of 10.8%. By treating the 8-OHdG standard with the same procedures for DNA digestion and HPLC analysis, the recovery rate was found to be more than 95%.

Although detection of 8-OHdG appears to be a sensitive, specific and quantitative method for studying oxidative DNA damage, there are obvious methodological problems: (i) the analytical procedure including DNA extraction and digestion may introduce artificial DNA oxidations; (ii) lack of standard protocols leads to substantial variation among different laboratories; and (iii) requirement of relatively larger amount (50 μg) of DNA hinders its wide application in clinical samples. A

Table 2. Determination of 8-OHdG Levels in Human Spermatozoa Using HPLC-EC

Subjects	Sample number	8-OHdG/ 10^5 dG	Reference
Free-living individuals	50	2.1	[14]
With vitamin C depletion	28	13.5[a]	
Smokers	19	2.0[a]	[21]
Nonsmokers	22	1.3	
Control subjects	17	1.0	[22]
Infertile patients	19	1.5	
Infertile patients with antioxidants supplement	14	1.1	
Smokers	22	6.2	[13]
Nonsmokers	32	3.9	
Control subjects	54	4.8	[24]
Infertile patients	60	10.0	

[a] Value calculated from fmole/μg DNA based on the conversion rate provided by the authors.

number of studies have well discussed the technical aspects of 8-OHdG measurement, attempting to standardize the procedure and control the artifacts inherited in the analysis [15–20].

With regards to the above discussion, the following precautions are recommended in ordered to obtain reliable and reproducible 8-OHdG data from human sperm: (i) incubation with DTT and proteinase K is essential to decondense and extract sperm DNA efficiently; (ii) avoid using phenol in DNA extraction to minimize the artifacts; (iii) exclude those samples with less than 30 μg of DNA extracted due to the detection limit of the system; (iv) use the same batch of chemicals and reagents including the 8-OHdG standard throughout the whole study; (v) always include a few pooled samples as internal quality controls in each analysis; and (vi) try to analyze the digested DNA sample as soon as possible after digestion (usually within the same day).

Biological significance of 8-OHdG analysis in human sperm

Fraga et al. [14] were the first to report the 8-OHdG level in human spermatozoa, and followed by several other laboratories including ours [13,21–24]. The results of these analyses are summarized in Table 2. Although precautions are always required when the results from different laboratories are compared, it can be noted that the mean values of 8-OHdG in human sperm are generally within the same magnitude and are also comparable to the 8-OHdG levels in many other human tissues or cells [16].

The most direct evidence suggesting the involvement of oxidative sperm DNA damage in male infertility is the finding that infertile patients contained higher level of 8-OHdG in sperm than control subjects. Kodama et al. [22] found that the levels of 8-OHdG in sperm DNA in

19 infertile patients were significantly higher than in the control (1.5 ± 0.2 vs. 1.0 ± 0.1 8-OHdG/10^5 dG). A recent study in our laboratory provided more convincing evidence suggesting the importance of oxidative sperm DNA in male infertility [24]. Based on a larger sample size (60 infertile patients and 54 healthy subjects), a nearly 110% increase of sperm 8-OHdG level in infertile patients was noted, compared with a 50% increase in the earlier report [22]. We have also analyzed the correlation of 8-OHdG levels in human sperm with seminal parameters in both infertile patients and normal controls [23, 24]. It was found that the 8-OHdG level in sperm DNA was positively correlated to the number of sperm cells with head defects, and inversely correlated to sperm density, total sperm number, sperm motility, and the number of sperm cells with normal morphology [24]. Therefore, data from all these studies strongly support the involvement of oxidative DNA damage in the pathological process of male infertility, and also suggest the potential diagnostic value of 8-OHdG in the evaluation of sperm function and male fertility.

On the other hand, oxidative DNA damage may also arise from a defeated antioxidant defense mechanism. The antioxidant system in semen and spermatozoa consists of enzymatic proteins such as superoxide dismutase, catalase and glutathione peroxidase, and small antioxidant molecules such as ascorbic acid, urate, and glutathione [6]. An earlier study by Fraga et al. [14] demonstrated two findings: (i) the 8-OHdG level in human sperm DNA was inversely related to seminal plasma ascorbic acid level and (ii) changes of dietary ascorbic acid intake affected the 8-OHdG level in human sperm. Subsequent studies confirmed the importance of antioxidants in protection against endogenous oxidative DNA damage in human sperm [21,22]. For instance, daily antioxidant supplements (ascorbic acid, α-tocopherol, and glutathione) in a group of 14 infertile patients for 2 months resulted in a significant reduction of 8-OHdG level in sperm DNA [22].

Cigarette smoking is a well-studied risk factor for sperm abnormalities and male infertility [25,26]. It is known that cigarette smoke contains a large amount of oxidants, hence studies on the possible effects of cigarette smoking on oxidative DNA damage in sperm will be of interest. Using seminal cotinine concentration as the biomarker of smoking exposure, a good correlation was found between seminal cotinine concentrations and sperm 8-OHdG levels in 28 healthy smokers [13]. As summarized in Table 2, smokers contained a significantly higher level of 8-OHdG in sperm DNA than nonsmokers based on a total of 60 healthy subjects [13], which is in good agreement with the data from an earlier report [21].

DETECTION OF DNA STRAND BREAKS IN HUMAN SPERMATOZOA

DNA base modification and DNA strand breaks are two of the major forms of oxidative DNA damage caused by ROS. In addition to 8-OHdG measurement as discussed above, numerous studies have attempted to explore DNA strand breaks in human spermatozoa and the implication of such damage in sperm function and male infertility [7,8]. Among various techniques used, single-cell gel electrophoresis (SCGE or comet assay) and terminal deoxynucleotidyl transferase (TdT) nick end-labeling assay (TUNEL) are the two popular methods in evaluation of DNA strand breaks or DNA fragmentation in human spermatozoa.

Methodology for comet assay in human spermatozoa

Comet assay is a visual fluorescent technique for measurement of DNA strand breaks in individual cells. Since its first introduction in the late 1970s, the assay has been widely used as a simple and sensitive method for assessing DNA damage in somatic cells [27–29]. This technique recently has been adapted for measuring DNA strand breaks in human sperm [30–38]. Although the detailed procedures may vary from laboratory to laboratory, principally it consists of the following steps: sperm suspension preparation, gel setting, cell lysis, DNA unwinding, electrophoresis, neutralization, DNA staining, and comet image analysis. The following method has been used in our laboratory based on an earlier report [38], with some modifications:

1) Sperm washing: Wash the sperm with phosphate-buffered saline (PBS) twice and resuspend in PBS at 2×10^6 cells/ml.
2) Slide preparation and gel setting:
   (i) Place 100 μl of 0.75% normal melting point agarose (NMPA) on a fully frosted slide, solidify the gel at room temperature for at least 5 min;
   (ii) Mix 8 μl of sperm suspension with 72 μl of 0.75% low melting point agarose (LMPA), and add the mixture on to the top of the first layer; and
   (iii) After solidifying the gel at 4°C for 8 min, add the third layer of 0.75% LMPA (100 μl) and keep the slide at 4°C for another 8 min.
3) Cell lysis and DNA decondensation:
   (i) Immerse the slides in a cold lysing solution (2.5 M NaCl, 100 mM Na$_2$ EDTA, 10 mM Tris, 1% sodium lauryl sarcosine, 1% Triton X-100, pH 10) at 4°C for 1 h;
   (ii) Transfer the slides into a solution for RNase treatment (of 2.5 M NaCl, 5 mM Tris, 0.05%

sodium lauryl sarcosine, pH 7.4, with 10 μg/ml RNase A) at 37°C for 4 h; and
   (iii) Treat the slides with proteinase K at for 15 h in a solution of 2.5 M NaCl, 5 mM Tris, 0.05% sodium lauryl sarcosine, pH 7.4, with 200 μg/ml proteinase K.
4) Electrophoresis: After equilibrating the slides in 300 mM sodium acetate and 100 mM Tris, pH 10 at 4°C in the electrophoresis tank for 20 min, run electrophoresis at 12 V (0.46V/cm) and 100 mA at 4°C for 1 h.
5) Neutralization: Neutralize the slides in 0.4 M Tris-HCl (pH 7.4) for at least 5 min.
6) Staining: Add 45 μl of ethidium bromide (15 μg/ml) onto each slide.
7) Microscopic image analysis: Commonly used parameters including percentage of cells with tail, percentage of DNA in the tail, tail length, and tail moment; a special imaging analysis software for comet assay (Komet system, Kinetic Imaging, UK) will offer a more objective and accurate measurement of above parameters.

The main differences of comet assay used for sperm from that of somatic cells were at the stage of cell lysis and DNA decondensation. The unique structure of sperm chromatin, plus a more resistant sperm head, makes the cell lysis and DNA decondensation in human sperm more difficult. Therefore, a prolonged incubation of sperm with proteinase K and detergents is required [31, 32,37,38]. In addition, a neutral condition (pH 9 or 10) is generally favored, which allows the detection of double strand breaks in spermatozoa. A stronger alkaline condition (pH 12 or 13) is likely to detect substantial amount of alkali labile sites existing in sperm DNA due to the extremely compact structure of sperm chromatin [37,38]. A few recent studies have discussed the methodology of the comet assay in human spermatozoa and it is generally believed that it is a simple, inexpensive, reliable, and reproducible technique for assessing DNA damage in sperm [32,35,37,38].

Methodology of TUNEL assay in human spermatozoa

TUNEL assay was originally designed for measuring DNA fragmentation occurring during apoptosis [39]. Recently this technique has been used for detecting DNA strand breaks in human sperm [34,40–45]. In our laboratory, we have successfully established the TUNEL assay in human spermatozoa based on its original application in a somatic cell line [46]. Unlike other protocols that either used biotin-dUTP/avidin or BrdUTP/anti-Br-dUTP-FITC system, we applied a FITC-labeled dUTP system (an in situ cell death detection kit from Boehr-

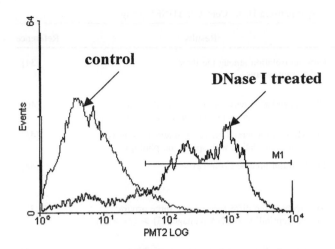

Fig. 1. Histograms of the TUNEL assay in human spermatozoa analyzed by flow cytometry. DNase I treatment was carried out as the positive control. After fixing and permeabilization, sperm cells were treated with 2.5 units DNase I in the reaction buffer (40 mM Tris.HCl, 10 mM NaCl, 6 mM MgCl$_2$, 10 mM CaCl$_2$, pH 7.9) for 30 min at room temperature, which resulted in a significant increase of TUNEL-positive cells (81%), compared with 4% in the control group.

inger Mannheim, Mannheim, Germany), which includes the following main steps:

1) Wash sperm samples and prepare sperm suspension as that for comet assay.
2) Fix sperm cells (about 3×10^6 spermatozoa) in 2% paraformaldehyde (30 min, RT).
3) Wash cells with PBS once, resuspend cells in 100 μl permeabilisation solution (0.1% Triton X-100, 0.1% sodium citrate) for 10 min on ice.
4) Wash with PBS and resuspend cells in 50 μl TdT reaction solution containing nucleotides and TdT enzyme. Keep one tube as a negative control without the addition of the enzyme.
5) Incubate the sample in a humidified chamber for 60 min at 37°C in the dark.
6) Analyze the sample using flow cytometry with an air-cooled argon 488 nm laser and a 550 nm dichroic mirror as detectors. Collect at least 10,000 cells in each group.
7) The obtained data were analyzed using WinMDI 2.7 (Scripps Inst., CA, USA) software for subtracting the histogram and calculating the percentage of TUNEL-positive cells.

Using DNase I as the positive control, we found that treatment of DNase I with sperm cells resulted in a significant increase of spermatozoa labeled with fluorescein isothiocynate (Fig. 1). Similarly, treatment with H$_2$O$_2$ was also able to increase the percentage of TUNEL-positive cells (data not shown).

Biological significance of comet and TUNEL assay in human spermatozoa

Technically, both the comet and TUNEL assay measure DNA strand breaks. A recent study by Aravindan et al. [34] applied these two tests in a group of infertile patients and found that results from these two assays were highly correlated. Moreover, it is also noted that results from the comet or TUNEL assay were also strongly correlated with the results from other techniques used for measuring DNA damage such as the sperm chromatin structure assay (SCSA) and in situ nick translation assay [34,40,41,47–49]. Therefore, all these findings suggest the usefulness of both techniques in the evaluation of DNA damage in human sperm.

Some of the recent studies using the comet and TUNEL assay for assessing DNA strand breaks in human spermatozoa are summarized in Table 3. Similar to the data of 8-OHdG analysis (Table 2), the prevalence of DNA strand breaks or DNA fragmentation in human sperm closely correlated with (i) sperm function such as sperm motility, sperm-oocyte fusion ability, in vitro fertilization or intracytoplasmic sperm injection fertilization rate [36,42,44]; and (ii) routine seminal parameters such as sperm morphology and concentration [42,44,45].

Two more recent studies by Host et al. [45,48] discovered that infertile patients contained higher level of DNA strand breaks than fertile subjects using the TUNEL assay, suggesting the diagnostic value of this technique in evaluation of sperm function and male fertility. However, other studies, using comet assay, failed to detect any significant difference of the baseline level of DNA strand breaks between normozoospermic fertile, normozoospermic infertile and asthenozoospermic infertile subjects [31,35]. As the number of subjects in each group was relatively small in those reports, more studies are certainly needed to evaluate the exact correlation of DNA strand breaks with male fertility.

On the other hand, it is interesting to note that spermatozoa from infertile patients are generally more susceptible to the effects of DNA damaging agents such as H$_2$O$_2$ and radiograph exposure [31,35]; and some antioxidants such as ascorbate and α-tocopherol were found to be able to provide significant protection against such damage induced by H$_2$O$_2$ in human spermatozoa measured by comet assay [50]. At present, two hypotheses have been proposed to explain the possible origin of sperm DNA damage: (i) incomplete maturation during spermiogenesis, and (ii) indicative of apoptosis [8]. The latter is supported by a very recent study by Sakkas et al. [51] showing that the presence of Fas death receptor in spermatozoa closely relates to abnormal sperm parameters and male fertility. It is believed that ROS and oxidative stress may be involved in both processes. There-

Table 3. Detection of DNA Strand Breaks in Human Spermatozoa Using Comet or TUNEL assay

Subjects	Number	Treatment	Methods	Results	Reference
Infertile patients	23		SCSA TUNEL comet	Close correlation among the three different assays	[34]
Infertile patients	25		SCSA TUNEL	Close correlation between the extent of DNA strand breaks and sensitivity to denaturation	[40]
Fertile donors Infertile patients	20 33		TUNEL	(i) More DNA strand breaks in infertile subjects (ii) Close correlation with sperm morphology	[45]
(1) Proven fertile donors (2) Unexplained infertile (3) Oligozoospermia infertile	20 39 74		TUNEL	DNA strand breaks: (3) > (2) > (1)	[48]
(1) Normozoospermic fertile (2) Normozoospermic infertile (3) Asthenozoospermic infertile	20/13 20/17 20/11	H_2O_2 Radiograph	Comet	(i) No baseline difference among three groups (ii) Susceptibility to DNA damage: (3) > (2) > (1)	[31,35]
Infertile patients	150	Radiograph Antioxidants	Comet	Vitamins C, E, and urate protect sperm DNA	[33]
Infertile patients	47	X/XO Antioxidants	TUNEL	(i) Induction of DNA damage by X/XO (ii) Protective effect of NAC and GSH	[43]
ICSI subjects	150		TUNEL	(i) Close correlation between DNA damage and motility and morphology (ii) Inverse correlation between DNA damage and ICSI fertilization rate	[44]
Infertile patients	298		TUNEL	(i) Inverse correlation between DNA damage and sperm motility, morphology and concentration (ii) Inverse correlation between DNA damage and IVF fertilization rate	[42]
Nonsmokers Smokers	35 35		SCSA TUNEL	Smokers contained more DNA damage than nonsmokers	[49]

GSH = glutathione; ICSI = intracytoplasmic sperm injection; IVF = in vitro fertilization; SCSA = sperm chromatin structure assay; X/XO = xanthine/xanthine oxidase.

fore, data from all these studies tend to suggest that DNA damage monitored by these comet or TUNEL assay may have important physiological and pathological relevance in terms of sperm quality and fertility.

SUMMARY

The expanding research interest in the last two decades on ROS, oxidative stress, and male infertility has

Table 4. Comparison Among Three Main Methods Used for Detection of Oxidative DNA Damage in Human Spermatozoa

Methods	No. of Cells Required	Time required	Forms of DNA damage	Main features	Main disadvantages
8-OHdG Analysis	3×10^7	2 d	Base modifications	(i) High specificity (ii) Quantitative (iii) High sensitivity (iv) Correlated with sperm function (v) Associated with fertility	(i) Large amount of samples required (ii) Introduction of artifacts (iii) Special equipment required (iv) Lack of standard protocols
Comet Assay	2×10^4	2 d	DNA double strand breaks	(i) Simple and inexpensive (ii) High sensitivity (iii) Correlated with seminal parameters (iv) Small number of cells required (v) Observation of individual cells	(i) Not specific to oxidative damage (ii) Subjectiveness in data acquire (iii) No evident correlation with fertility (iv) Lack of standard protocols
TUNEL assay	3×10^6	4–5 h	DNA single and double strand breaks	(i) Simple and fast (ii) High sensitivity (iii) Indicative of apoptosis (iv) Correlated with seminal parameters (v) Associated with fertility (vi) Availability of commercial kits	(i) Not specific to oxidative damage (ii) Special equipment required (iii) Expensive

led to the development of various techniques in evaluating oxidative DNA damage in human spermatozoa. Here we have discussed the methodology of three commonly used assays and their applications in the assessment of sperm function and male infertility. Table 4 summarizes the main features of these three methods that allow specific qualification and quantification of oxidative DNA damage in human spermatozoa. Due to the inherent technical limitations in each of these tests, it is believed that a combination of these assays will offer more complete information for a better understanding of oxidative DNA damage and its biological significance in male infertility.

Acknowledgements — We would like to acknowledge the following people for their valuable contributions to this work: Drs. S. E. Chia, D. X. Xu, S. Y. Dong, Z. Y. Ni, P. Bang, Mr. H. Y. Ong, and Ms. B. L. Lee. We also thank Prof. B. Halliwell for his valuable comments on the manuscript. The study was partially supported by the Centre for Environmental and Occupational Health Research at the National University of Singapore and the China Medical Board in New York.

REFERENCES

[1] Carlsen, E.; Giwercman, A.; Keiding, N.; Skakkebaek, N. E. Evidence for decreasing quality of semen during past 50 years. *Br. Med. J.* **305:**609–613; 1992.

[2] Auger, J.; Kuntsman, J. M.; Czyglik, F.; Jouannet, P. Decline in semen quality among fertile men in Paris during the past 20 years. *N. Engl. J. Med.* **332:**281–285; 1995.

[3] Cummins, J. M.; Jequier, A. M.; Kan, R. Molecular biology of human male infertility: links with aging, mitochondrial genetics, and oxidative stress? *Mol. Reprod. Dev.* **37:**345–362; 1994.

[4] Sharma, P. K.; Agarwal, A. Role of reactive oxygen species in male infertility. *Urology* **48:**835–850; 1996.

[5] Griveau, J. F.; Le Lannou, D. Reactive oxygen species and human spermatozoa: physiology and pathology. *Int. J. Androl.* **20:**61–69; 1997.

[6] Storey, B. T. Biochemistry of the induction and prevention of lipoperoxidative damage in human spermatozoa. *Mol. Hum. Reprod.* **3:**203–213; 1997.

[7] Aitken, R. J. The human spermatozoon—a cell in crisis? *J. Reprod. Fertil.* **115:**1–7; 1999.

[8] Sakkas, D.; Mariethoz, E.; Manicardi, G.; Bizzaro, D.; Bianchi, P. G.; Bianchi, U. Origin of DNA damage in ejaculated human spermatozoa. *Rev. Reprod.* **4:**31–37; 1999.

[9] Ames, B. N.; Shigenaga, M. K.; Hagen, T. M. Oxidants, antioxidants, and the degenerative disease and aging. *Proc. Natl. Acad. Sci. USA* **90:**7915–7922; 1993.

[10] Beckman, K. B.; Ames, B. N. Oxidative decay of DNA. *J. Biol. Chem.* **272:**19633–19636; 1997.

[11] Ward, W. S.; Coffey, D. S. DNA packaging and organization in mammalian spermatozoa: comparison with somatic cells. *Biol. Reprod.* **44:**569–574; 1991.

[12] Floyd, R. A. The role of 8-hydroguanine in carcinogenesis. *Carcinogenesis* **11:**1447–1450; 1990.

[13] Shen, H.-M.; Chia, S. E.; Ni, Z. Y.; New, A. L.; Lee, B. L.; Ong, C.-N. Detection of oxidative DNA damage in human sperm and its association with cigarette smoking. *Reprod. Toxicol.* **11:**675–680; 1997.

[14] Fraga, C. G.; Motchnik, P. A.; Shigenaga, M. K.; Helbock, H. J.; Jacob, R. A.; Ames, B. N. Ascorbic acid protects against endogenous oxidative DNA damage in human sperm. *Proc. Natl. Acad. Sci. USA* **88:**11003–11006; 1991.

[15] Collins, A.; Cadet, J.; Epe, B.; Gedik, C. Problems in the measurement of 8-oxoguanine in human DNA. Report of a workshop, DNA oxidation, held in Aberdeen, UK, January 19–21, 1997. *Carcinogenesis* **18:**1833–1836; 1997.

[16] Kasai, H. Analysis of a form of oxidative DNA damage, 8-hydroxy-2'-deoxyguanosine, as a marker of cellular oxidative stress during carcinogenesis. *Mutat. Res.* **387:**147–163; 1997.

[17] Helbock, H. J.; Beckman, K. B.; Shigenaga, M. K.; Walter, P. B.; Woodall, A. A.; Yeo, H. C.; Ames, B. N. DNA oxidation matters: the HPLC-electrochemical detection assay of 8-oxo-deoxyguanosine and 8-oxo-guanine. *Proc. Natl. Acad. Sci. USA* **95:**288–293; 1998.

[18] Cadet, J.; D'Ham, C.; Douki, T.; Pouget, J-P.; Ravanat, J-L.; Sauvaigo, S. Facts and artifacts in the measurement of oxidative base damage to DNA. *Free Radic. Res. Commun.* **29:**541–550; 1998.

[19] Halliwell, B. Can oxidative DNA damage be used as a biomarker of cancer risk in humans? Problems, resolutions and preliminary results from nutritional supplementation studies. *Free Radic. Res. Commun.* **29:**469–486; 1998.

[20] Moller, L.; Hofer, T.; Zeisig, M. Methological considerations and factor affecting 8-hydroxy-2'-deoxyguanosine analysis. *Free Radic. Res. Commun.* **29:**511–524; 1998.

[21] Fraga, C. G.; Motchnik, P. A.; Wyrobek, A. J.; Rempel, D. M.; Ames, B. N. Smoking and low antioxidant levels increase oxidative damage to sperm DNA. *Mutat. Res.* **351:**199–203; 1996.

[22] Kodama, H.; Yamaguchi, R.; Fukada, J.; Kasai, H.; Tanaka, T. Increased oxidative deoxyribonucleic acid damage in the spermatozoa of infertile male patients. *Fertil. Steril.* **68:**519–524; 1997.

[23] Ni, Z. Y.; Liu, Y. Q.; Shen, H.-M.; Chia, S. E.; Ong, C.-N. Does the increase of 8-hydroxyguanosine lead to poor sperm quality? *Mutat. Res.* **381:**77–82; 1997.

[24] Shen, H.-M.; Chia, S. E.; Ong, C.-N. Evaluation of oxidative DNA damage in human sperm and its association with male infertility. *J. Androl.* **20:**718–723; 1999.

[25] Stillman, R. J.; Rosenerg, M. J.; Sachs, B. P. Smoking and reproduction. *Fertil. Steril.* **46:**545–566; 1986.

[26] Hughes, E. G.; Brennan, B. G. Does cigarette smoking impair natural or assisted fecundity? *Fertil. Steril.* **66:**679–689; 1996.

[27] McKelvey-Martin, V. J.; Green, M. H. L.; Schmezer, P.; Pool-Zobel, B. L.; De Meo, M. P.; Collins, A. The single cell gel electrophoresis assay (comet assay): an European review. *Mutat. Res.* **288:**47–63; 1993.

[28] Fairbairn, D. W.; Olive, P. L.; O'Neill, K. L. The comet assay: a comprehensive review. *Mutat. Res.* **339:**37–59; 1995.

[29] Rojas, E.; Lopez, M. C.; Valverde, M. Single cell gel electrophoresis assay: methodology and applications. *J. Chromatogr. B* **722:**225–254; 1999.

[30] Singh, N. P.; Danner, D. B.; Tice, R.; McCoy, M. T.; Collins, G. D.; Schneider, E. L. Abundant alkaline-sensitive sites in DNA of human and mouse sperm. *Exp. Cell Res.* **184:**461–470; 1989.

[31] Hughes, C. M.; Lewis, S. E. M.; McKelvey-Martin, V. J.; Thompson, W. A comparison of baseline and induced DNA damage in human spermatozoa from fertile and infertile men, using a modified comet assay. *Mol. Hum. Reprod.* **2:**613–619; 1996.

[32] Hughes, C. M.; Lewis, S. E. M.; McKelvey-Martin, V. J.; Thompson, W. Reproducibility of human sperm DNA measurement alkaline single cell gel electrophoresis assay. *Mutat. Res.* **374:**261–268; 1997.

[33] Hughes, C. M.; Lewis, S. E. M.; McKelvey-Martin, V. J.; Thompson, W. The effects of antioxidant supplementation during Percoll preparation on human sperm DNA integrity. *Hum. Reprod.* **13:**1240–1247; 1998.

[34] Aravindan, G. R.; Bjordahl, J.; Jost, L. K.; Evenson, D. P. Susceptibility of human sperm to in situ DNA denaturation is strongly correlated with DNA strand breaks identified by single-cell electrophoresis. *Exp. Cell Res.* **236:**231–237; 1997.

[35] McKelvey-Martin, V. J.; Melia, N.; Walsh, I. K.; Johnston, S. R.; Hughes, C. M.; Lewis, S. E. M.; Thompson, W. Two potential clinical applications of the alkaline single-cell gel electrophoresis assay: (1) human bladder washings and transitional cell carcinoma of the bladder; and (2) human sperm and male infertility. *Mutat. Res.* **375:**93–104; 1997.

[36] Aitken, R. J.; Gordon, E.; Harkiss, D.; Twigg, J. P.; Miline, P.; Jennings, Z.; Irvine, D. S. Relative impact of oxidative stress on the functional competence and genomic integrity of human spermatozoa. *Biol. Reprod.* **59:**1037–1046; 1998.

[37] Haines, G.; Marples, B.; Daniel, P.; Morris, I. DNA damage in human and mouse spermatozoa after in vitro irradiation assessed by the comet assay. *Adv. Exp. Med. Biol.* **444:**79–91; 1998.

[38] Singh, N. P.; Stephens, R. E. X-ray-induced DNA double-strand breaks in human sperm. *Mutagenesis* **13:**75–79; 1998.

[39] Gavrieli, Y.; Sherman, Y.; Ben-Sasson, S. A. Identification of programmed cell death in situ via specific labeling of nuclear DNA fragmentation. *J. Cell. Biol.* **119:**493–501; 1992.

[40] Gorczyca, W.; Traganos, F.; Jesionowska, H.; Darzynkiewicz, Z. Presence of DNA strand breaks and increased sensitivity of DNA in situ to denaturation in abnormal human sperm cells: analogy to apoptosis of somatic cells. *Exp. Cell Res.* **207:**202–205; 1993.

[41] Sailer, B. L.; Jost, L. K.; Evenson, D. P. Mammalian sperm DNA susceptibility to in situ denaturation associated with the presence of DNA strand breaks as measured by the terminal deoxynucleotidyl transferase assay. *J. Androl.* **16:**80–87; 1995.

[42] Sun, J. G.; Jurisicova, A.; Casper, R. F. Detection of deoxyribonucleic acid fragmentation in human sperm: correlation with fertilization in vitro. *Biol. Reprod.* **56:**602–607; 1997.

[43] Lopes, S.; Jurisicova, A.; Sun, J. G.; Casper, R. F. Reactive oxygen species: potential cause for DNA fragmentation in human spermatozoa. *Hum. Reprod.*13:896–900; 1998.

[44] Lopes, S.; Sun, J. G.; Jurisicova, A.; Meriano, J.; Casper, R. F. Sperm deoxyribonucleic acid fragmentation is increased in poor-quality semen samples and correlates with failed fertilization in intracytoplasmic sperm injection. *Fertil. Steril.* **69:**528–532; 1998.

[45] Host, E.; Lindenberg, S.; Kahn, J. A.; Christensen, F. DNA strand breaks in human sperm cells: a comparison between men with normal and oligozoospermic sperm samples. *Acta Obstet. Gynecol. Scand.* **78:**336–339; 1999.

[46] Shen, H.-M.; Yang, C. F.; Ong, C.-N. Sodium selenite-induced oxidative stress and apoptosis in human hepatoma HepG$_2$ cells. *Int. J. Cancer* **81:**820–828; 1999.

[47] Manicardi, G. C.; Tombacco, A.; Bizzaro, D.; Bianchi, U.; Bianchi, P. G.; Sakkas, D. DNA strand breaks in ejaculated human spermatozoa: comparison of susceptibility to the nick translation and terminal transferase assays. *Histochem. J.* **30:**33–39; 1998.

[48] Host, E.; Linderberg, S.; Ernst, E.; Christensen, F. DNA strand breaks in human spermatozoa: a possible factor, to be considered in couples suffering from unexplained infertility. *Acta Obstet. Gynecol. Scand.* **78:**622–625; 1999.

[49] Potts, R. J.; Newburry, C. J.; Smith, G.; Notarianni, L. J.; Jefferies, T. M. Sperm chromatin damage associated with male smoking. *Mutat. Res.* **423:**103–111; 1999.

[50] Donnelly, E. T.; McClure N.; Lewis, S. E. The effect of ascorbate and alpha-tocopherol supplementation in vitro on DNA integrity and hydrogen peroxide–induced DNA damage in human spermatozoa. *Mutagenesis* **14:**505–512; 1999.

[51] Sakkas, D.; Mariethoz, E.; St. John, J. C. Abnormal sperm parameters in humans are indicative of an abortive apoptotic mechanisms linked to the fas-mediated pathways. *Exp. Cell Res.* **251:** 350–355; 1999.

OXIDATIVE STRESS STATUS—THE THIRD SET

WILLIAM A. PRYOR

Biodynamics Institute, Louisiana State University, Baton Rouge, LA, USA

This issue of *FRBM* presents the third group of articles in our ongoing Forum on Oxidative Stress Status (OSS). You can find the other sets in our earlier issues:

◆ The first set of seven in *FRBM* 27:11/12 (December 1999).
◆ The second set of four articles in *FRBM* 28:4 (February 2000).

And, yes, Dear And Constant Reader, *FRBM* will be publishing at least two more groups of articles in this OSS Forum in coming issues, so look for them. In fact, this Forum has grown like Topsy, since authors have suggested the inclusion of their methodologies in the Forum as they have seen the other articles appearing in earlier *FRBM* issues.

This third group of articles begins with a treatment of the Biomarkers of Oxidative Stress Status (BOSS) program, referred to in my introduction to the first Forum [1]. In an effort to confirm the reliability and usefulness of measures of OSS, the National Institutes of Health has organized a group of expert investigators to perform various measures of OSS on rats, each group using its own BOSS. This program is described in the lead-off article in the Forum, submitted by Ron Mason's group, entitled "Biomarkers of oxidative stress study: are plasma antioxidants markers of CCl_4 poisoning?". This article by Kadiiska et al., reports their search for noninvasive biomarkers of oxidative stress using plasma levels of antioxidants.

The Forum continues with two methods utilizing electron spin resonance to measure OSS: The first, by H. Togashi et al., is entitled "Analysis of hepatic oxidative stress status by electron spin resonance spectroscopy and imaging", and the second, by Y. Miura and T. Ozawa, is entitled, "Noninvasive study of radiation-induced oxidative damage using in vivo electron spin resonance".

The two articles that follow evaluate the use of cyclic voltammetry in OSS: the first, by S. Chevion et al., is entitled, "The use of cyclic voltammetry for the evaluation of antioxidant capacity" and the second, by R. Kohen et al., is entitled, "Quantification of the overall ROS scavenging capacity of biological fluids and tissues".

This set of OSS Forum articles concludes with an article by E. Aghdassi and J. P. Allard entitled, "Breath alkanes as a marker of oxidative stress in different clinical conditions". The measurement of exhaled gases is, potentially, an important method, and it was discussed by other researchers in both our first and second Forums. Each group of researchers, in this as well as the preceding sets, offers their own unique perspective on the various methods of measuring oxidative stress status, and thus contributes to painting the overall picture that represents today's state of the art.

REFERENCE

[1] Oxidative stress status: OSS, BOSS, and "Wild Bill" Donovan. *Free Radic. Biol. Med.* **27:**1135–1136; 1999.

Address correspondence to: Dr. William A. Pryor, Biodynamics Institute, 711 Choppin Hall, Louisiana State University, Baton Rouge, LA 70803, USA.

BIOMARKERS OF OXIDATIVE STRESS STUDY: ARE PLASMA ANTIOXIDANTS MARKERS OF CCl₄ POISONING?

Maria B. Kadiiska,* Beth C. Gladen,* Donna D. Baird,* Anna E. Dikalova,* Rajindar S. Sohal,[†]
Gary E. Hatch,[‡] Dean P. Jones,[§] Ronald P. Mason*, and J. Carl Barrett*

*National Institute of Environmental Health Sciences, National Institutes of Health, Research Triangle Park, NC; [†]Southern
Methodist University, Dallas, TX; [‡]National Health and Environmental Effects Research Laboratory, Health Science,
National Institutes of Health, United States Environmental Protection Agency, Research Triangle Park, NC;
and [§]Emory University, Atlanta, GA

(Received 20 July 1999; Revised 7 February 2000; Accepted 8 February 2000)

Abstract—Antioxidants in the blood plasma of rats were measured as part of a comprehensive, multilaboratory validation study searching for noninvasive biomarkers of oxidative stress. For this initial study an animal model of CCl₄ poisoning was studied. The time (2, 7, and 16 h) and dose (120 and 1200 mg/kg, intraperitoneally)-dependent effects of CCl₄ on plasma levels of α-tocopherol, coenzyme Q (CoQ), ascorbic acid, glutathione (GSH and GSSG), uric acid, and total antioxidant capacity were investigated to determine whether the oxidative effects of CCl₄ would result in losses of antioxidants from plasma. Concentrations of α-tocopherol and CoQ were decreased in CCl₄-treated rats. Because of concomitant decreases in cholesterol and triglycerides, it was impossible to dissociate oxidation of α-tocopherol and the loss of CoQ from generalized lipid changes, due to liver damage. Ascorbic acid levels were higher with treatment at the earliest time point; the ratio of GSH to GSSG generally declined, and uric acid remained unchanged. Total antioxidant capacity showed no significant change except for 16 h after the high dose, when it was increased. These results suggest that plasma changes caused by liver malfunction and rupture of liver cells together with a decrease in plasma lipids do not permit an unambiguous interpretation of the results and impede detection of any potential changes in the antioxidant status of the plasma. © 2000 Elsevier Science Inc.

Keywords—CCl₄, Rat, Plasma, α-Tocopherol, Coenzyme Q, Ascorbic acid, Glutathione, Uric acid, Total antioxidant capacity, Free radicals

INTRODUCTION

Links between oxidative stress and adverse health effects have been suggested for several groups of diseases, including cardiovascular, respiratory, and neurological [1–4] as well as for the general aging process [5,6]. Such adverse effects are mediated by free radical damage to lipids, proteins, and DNA. Protection from damage oc-

Disclaimer: The research described in this article has been reviewed by the Health Effects Research Laboratory, U.S. Environmental Protection Agency, and approved for publication. Approval does not signify that the contents necessarily reflect the views and policies of the Agency nor does mention of trade names or commercial products constitute endorsement or recommendation for use.

Maria B. Kadiiska, M.D., Ph.D., is a staff scientist at the National Institute of Environmental Health Sciences. She has organized and coordinated the Biomarkers of Oxidative Stress Study.

Beth C. Gladen, Ph.D., is a mathematical statistician at Biostatistics Branch, National Institute of Environmental Health Sciences.

Donna D. Baird, Ph.D., is an epidemiologist at Epidemiology Branch, National Institute of Environmental Health Sciences.

Anna E. Dikalova, Ph.D., is a visiting fellow at National Institute of Environmental Health Sciences.

Rajindar S. Sohal, Ph.D., is a University Distinguished Professor of Biological Sciences at the Southern Methodist University in Dallas

Gary E. Hatch, Ph.D., is a group leader in the Experimental Toxicology Division, U.S. EPA.

Dean P. Jones, Ph.D., is a professor in biochemistry at Emory University.

Ronald P. Mason, Ph.D. leads the Free Radical Metabolism Group at National Institute of Environmental Health Sciences.

J. Carl Barrett, Ph.D. is the Scientific Director of the Division of Intramural Research at the National Institute of Environmental Health Sciences. He has initiated and sponsored the nationwide validation study on the existing biomarkers for the assessment of oxidative stress.

Address correspondence to: Maria B. Kadiiska, NIEHS, P.O. Box 12233, MD F0-02, Research Triangle Park, NC 27709; Tel: 919-541-0201; Fax: 919-541-1043; E-Mail: kadiiska@niehs.nih.gov.

Reprinted from: *Free Radical Biology & Medicine*, Vol. 28, No. 6, pp. 838-845, 2000

curs through the action of multiple antioxidants, some endogenously produced, some provided through dietary intake [1,7]. It has been generally agreed among researchers who investigate free radicals that there is a need for validating sensitive and specific biomarkers for oxidative damage resulting from multiple types of oxidative insults in rodents, nonhuman primates, and humans. The National Institute of Environmental Health Sciences (NIEHS) has taken the lead in initiating the first comprehensive comparative study for determining which of the available biomarkers of oxidative stress are most specific, sensitive and selective. This article is the first report of the nationwide Biomarkers of Oxidative Stress Study (BOSS). Ideally, biomarkers of oxidative stress for human studies would be measurable in specimens that can be collected relatively easily such as blood or urine. In addition, the analysis procedure should be applicable to stored specimens.

Numerous markers of oxidative stress and antioxidant status have been tried [8], but few studies have compared the validity and sensitivity among several biomarkers [9]. The purpose of this research effort was to examine several proposed markers of oxidative stress in the same model system. To this end, we used a well-documented animal model of oxidative stress—carbon tetrachloride administered to rats [10–17]. We measured a series of biomarkers in blood plasma collected from rats treated under the same protocol. The present report describes the results for several antioxidant biomarkers: α-tocopherol, coenzyme Q, ascorbic acid, glutathione, uric acid, and an overall measure of antioxidant capacity.

It was hypothesized that plasma antioxidant levels would decline owing to the oxidative stress induced by CCl_4. Results suggest that the primary liver damage caused by CCl_4 acutely affected plasma constituents, making it difficult to unambiguously interpret changes in plasma antioxidant concentrations.

MATERIALS AND METHODS

Chemicals and reagents

Carbon tetrachloride and all other chemicals and reagents used in the study were obtained from Sigma-Aldrich Corporation (St. Louis, MO, USA).

Animals and treatment protocol

Male Fisher 344 rats (260–280 g) obtained from Charles River Laboratories (Raleigh, NC, USA) were used in all experiments. The animals were housed three to a cage. Autoclaved hardwood bedding was used in solid-bottom polycarbonate cages with filter tops. Animal rooms were maintained at 20–25°C with 35–70%

relative humidity with alternating 12 h light and dark cycles. The rats had free access to deionized, reverse-osmosis-treated water and received autoclaved National Institutes of Health (NIH) 31 rodent chow (Zeigler Bros, Gardners, PA, USA) ad libitum. For all experiments, rats were fasted overnight and then administered intraperitoneal injections of carbon tetrachloride in canola oil. Control rats received an equal volume of canola oil. Fasting continued through the experiment. Rats were anesthetized with Nembutal (Abbott, Chicago, IL, USA) (0.1 ml/100 g body weight[b.wt.]), and 5 ml of blood was removed from the dorsal aorta where it entered anterior to its distal bifurcation into the common iliac arteries. Rats were sacrificed after blood collection. Liver tissue was collected from a subsample of animals.

Study design

Animals from each of the three dose groups (canola oil, 120 mg/kg CCl_4, and 1200 mg/kg CCl_4) were killed at three time points (2, 7, and 16 h after CCl_4 injection). Each group consisted of five rats for each analysis. Animal treatment, sample preparation, liver histology, and serum enzyme activities were done at NIEHS (Research Triangle Park, NC, USA). Glutathione (GSH and GSSG) measurements were performed at Emory University (Atlanta, GA, USA). Coenzyme Q (CoQ) content was determined at Southern Methodist University (Dallas, TX, USA). All other parameters studied were measured at the U.S. Environmental Protection Agency (EPA) at Research Triangle Park, NC.

Each sample was marked with a code number so that those conducting the assays were not aware of the treatment status of the animals.

Electron spin resonance sample preparation and measurement

Two hours after administration of CCl_4 [120 mg/kg, intraperitoneally (i.p.)] and 1 h after the administration of phenyl N-tert-butylnitrone (PBN) (90 mg/kg, i.p.) rats were anesthetized with Nembutal (0.1 ml/100 g b.wt., i.p.) and euthanized by exsanguination. The livers were excised and homogenized in 30 mM 2,2'-dipyridyl and toluene. After centrifugation at 3000 rpm for 10 min, the toluene layer was isolated. ESR spectra were recorded immediately at room temperature using a 4 mm quartz tube and a Bruker (Billerica, MA, USA) EMX spectrometer with a Super High Q cavity. Spectra were recorded on an IBM-compatible computer interfaced to the spectrometer.

Specimen collection

Plasma preparation. Blood (5 ml) was drawn through single draw Vacutainer needles (21 gauge) into open Vacutainer blood collection tubes containing heparin. The tubes were gently inverted two to three times for mixing and immediately placed on ice. Blood was centrifuged (2000 rpm for 10 min at 4°C) no more than 30 min after collection.

Serum preparation. Blood (5 ml) was drawn and left on ice for 30 min. Before centrifugation, the clot was rimmed with a wooden applicator stick. The blood samples were centrifuged at 2000 rpm for 15 min. The serum was removed, frozen immediately on dry ice, and stored at −70°C.

Liver preparation for histopathology

Liver tissue from five animals per dose group was examined at 2, 7, and 16 h after exposure. Samples of liver tissue for histopathology were fixed by immersion in 10% neutral buffered formalin for 24 h. Slices (4 mm) of fixed tissue were processed, embedded in paraffin, sectioned to a thickness of 5 μm, mounted on glass slides, and stained with hematoxylin and eosin for histopathological evaluation.

Analysis of liver enzyme activities

Serum enzyme activities of lactic dehydrogenase (LDH), alanine aminotransferase (ALT), alkaline phosphatase (ALP), aspartate aminotransferase (AST), sorbitol dehydrogenase (SDH), 5′-nucleotidase (5′-NT), and the serum concentration of total bile acids (TBA) were measured according to published procedures [18–21].

Measurement of plasma cholesterol and triglycerides

These measurements were made using kits available from Sigma-Aldrich Corp. (St. Louis, MO, USA) adapted for use on the Cobas Fara II clinical analyzer (Roche Diagnostics, Branchburg, NJ, USA).

Analysis of antioxidant status

Alpha-tocopherol. Plasma samples were diluted with an equal volume of absolute ethanol containing 1 μg/ml butylated hydroxyanisole. One ml of heptane was added, and the sample was vortexed. This sample was centrifuged at 1000 rpm (5 min) at 4°C. The heptane layer was removed, and another 1 ml of fresh heptane added. The sample was again vortexed and centrifuged as before until three volumes of heptane had been added. These volumes were combined and dried under nitrogen. The residue was dissolved in 0.2 ml of methanol and 20 μl was injected onto a high-performance liquid chromatography (HPLC) column with colometric detection [22].

Coenzyme Q (CoQ$_n$). Plasma concentrations of total (oxidized and reduced) coenzyme Q (CoQ$_9$ and CoQ$_{10}$) were measured using an HPLC method described elsewhere [23].

Ascorbic acid and uric acid. Plasma (0.2 ml) was added to 0.6 ml of 4% perchloric acid, and the mixture was vortexed, frozen on dry ice, and stored at −70°C. Perchloric acid supernatants (27,000 $\times$ *g*, 20 min) of plasma were assayed for ascorbic acid and uric acid (the same isocratic run detects both substances) using amperometric detection [24].

Glutathione. For GSH and GSSG measurements, whole blood samples (0.5 ml) were added to a preservation solution (0.5 ml of 100 mM serine borate, pH 8.5, containing 0.5 mg of sodium heparin, 1 mg of bathophenanthroline, and 2 mg of iodoacetic acid per milliliter), which was designed to inhibit GSH degradation by glutamyltranspeptidase, clotting, and autoxidation [25]. Samples were mixed by inverting the microcentrifuge tube, then centrifuged to remove blood cells [25]. Aliquots (0.2 ml) of the supernatant were transferred to tubes containing a solution (0.2 ml of 10% (w/v) perchloric acid and 0.2 M boric acid) to precipitate protein and provide an internal standard (10 mM γ-glutamyl glutamate). Samples were derivatized with iodoacetic acid and dansyl chloride for analysis by HPLC with fluorescence detection [25]. GSH and GSSG were calculated relative to both the internal standard and external standards with comparable results.

Total antioxidant capacity. For plasma total antioxidant capacity measurement, an automated analyzer assay was adapted from the assay procedure of Miller et al. [26]. This assay expresses the total antioxidant capacity in millimolar "Trolox equivalents." Thus 1 mM of total antioxidant capacity would be equivalent to the antioxidant action of 1 mM of the water-soluble vitamin E derivative, Trolox (Sigma-Aldrich Corp.). It has been shown that urate and proteins account for at least 76% of the antioxidant activity detected by this assay in human plasma [26].

Statistical analysis

Statistical comparisons were done by analysis of variance. Comparisons were made between the canola oil

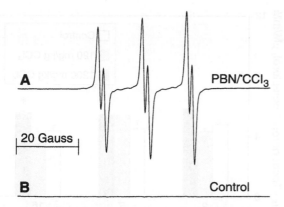

Fig. 1. (Spectrum A) ESR spectra of the PBN/trichloromethyl radical adduct detected in the toluene extract of rat liver. The sample was obtained 2 h after i.p. administration of 120 mg/kg CCl_4 in canola oil and 1 h after i.p. injection of 90 mg/kg spin-trapping agent PBN. The hyperfine coupling constants were $a^N = 13.99$ G and $a_{\beta}^H = 1.62$ G (linewidth, 0.72 G). (Spectrum B) Same as in spectrum A, but rat was given an injection of canola oil in place of CCl_4. Instrumental settings of Bruker EMX spectrometer: microwave power, 20 mW; modulation amplitude, 1 G; conversion time, 1.3 s; time constant, 2.6 s.

controls and each dose group at the same time point. When appropriate, analysis was done on the log scale. Values of $p < .05$ were considered statistically significant.

RESULTS

Trichloromethyl free radical formation

Carbon tetrachloride toxicity in rats was used as the model system, because it represents a well-documented injury that is oxidative in nature. The target organ is the liver. In this model, bioactivation to a reactive intermediate is required. The initial step is reduction of CCl_4 by the cytochrome P450 system to the trichloromethyl free radical ($^{\bullet}CCl_3$) [27] which, in the presence of oxygen, is subsequently converted into a peroxyl radical ($^{\bullet}OOCCl_3$) [28]. As shown in Fig. 1, spectrum A, when CCl_4 (120 mg/kg, i.p.) and PBN (90 mg/kg, i.p.) were administered to a rat, a strong six-line ESR spectra of PBN/$^{\bullet}CCl_3$ was detected. Treatment with PBN alone produced no detectable signal (Fig. 1, spectrum B).

Liver enzyme activities and histopathology

Table 1 shows substantial increases in plasma levels of ALT; differences are significant for the high dose at 2 h and for both doses at 7 and 16 h. The same pattern of significant increases was found for AST and SDH (data not shown). Alkaline phosphatase showed significant increases for both doses at 7 and 16 h (data not shown). The activities of LDH and 5′-nucleotidase showed sig-

nificant increases for the high dose at 7 and 16 h (data not shown). The histopathology review performed on liver tissue showed a dose-related increase in the incidence and severity of centrilobular hepatocellular cytoplasmic vacuolization and necrosis. In general, these changes were detectable by 2 h after administration in both low and high doses of CCl_4, but became more severe at each subsequent time point, particularly in the high dose group (data not shown). Hepatocellular necrosis varied from involvement of a few individual centrilobular hepatocytes in cases of minimal severity to approximately 50% of affected hepatic lobules in the most severe cases.

Plasma level of α-tocopherol

Compared with the controls at the same time point, statistically significant decreases in levels of α-tocopherol were found for the high dose at 7 h and for both doses at 16 h (Table 1). The remaining differences from the controls were not significant. Because α-tocopherol is present in the lipid fraction of blood and lipid levels can change acutely with liver damage, we measured plasma cholesterol and triglyceride levels.

Table 1 shows a substantial decline in cholesterol level, which is significant for both doses at all three time points. Triglycerides showed similar large and statistically significant decreases (data not shown). In Fig. 2 the ratio of α-tocopherol to cholesterol is shown to normalize α-tocopherol levels relative to changes in blood lipids. Increases in this ratio were seen at the high dose; the difference was significant at 16 h. The ratio of α-tocopherol to triglycerides also showed increases, with significant differences for the high dose at 2 h and for both doses at 7 h (data not shown).

Plasma level of total CoQ

As compared to the controls, both CoQ_9 and CoQ_{10} concentrations in the plasma of CCl_4-treated rats were generally lower after the administration of CCl_4 for both doses at the three time points studied. The differences were significant for the high CCl_4 doses at 2 and 7 h time points and for the low doses at 7 and 16 h after CCl_4 treatment. Total CoQ_{10} was significantly lower 16 h after treatment with a high dose of CCl_4 (Table 1). In contrast to α-tocopherol, the ratio of total CoQ_9 to cholesterol (Fig. 3) was significantly changed with treatment for high doses at all three time points and for the low dose at the 7 h time point. The total CoQ_{10}/cholesterol ratio was significantly decreased only for the high dose at the early time point (data not shown).

Table 1. Effect of CCl₄ on Plasma Biomarkers of Antioxidants

	2 h			7 h			16 h		
	Control	120 mg/kg	1200 mg/kg	Control	120 mg/kg	1200 mg/kg	Control	120 mg/kg	1200 mg/kg
Alanine aminotransferase (IU/L)	41 ± 3	50 ± 4	202 ± 29*	31 ± 1	311 ± 18*	2164 ± 612*	46 ± 4	407 ± 273*	10 676 ± 326*
α-Tocopherol (µM)	7.1 ± 0.5	5.7 ± 0.4	6.4 ± 0.5	6.3 ± 0.9	6.0 ± 0.3	3.6 ± 0.5*	5.4 ± 0.5	3.6 ± 0.6*	3.9 ± 0.4*
Coenzyme Q₉ (nM)	161 ± 12	138 ± 11	77 ± 8*	192 ± 10	98 ± 15*	64 ± 12*	197 ± 10	89 ± 6*	179 ± 31
Coenzyme Q₁₀ (nM)	61 ± 6	48 ± 6	31 ± 3*	62 ± 5	46 ± 9*	24 ± 5*	81 ± 2	39 ± 3*	52 ± 8*
Cholesterol (mM)	1.47 ± 0.05	1.31 ± 0.04*	1.09 ± 0.03*	1.41 ± 0.08	1.12 ± 0.02*	0.66 ± 0.07*	1.32 ± 0.05	0.90 ± 0.07*	0.70 ± 0.06*
Ascorbic acid (µM)	6.9 ± 0.6	6.4 ± 1.4	11.4 ± 1.3*	5.5 ± 0.7	8.7 ± 1.3	8.5 ± 1.9	5.4 ± 0.2	5.6 ± 1.8	6.9 ± 1.2
Uric acid (µM)	65 ± 5	50 ± 2	72 ± 8	70 ± 6	62 ± 3	61 ± 6	44 ± 4	83 ± 16	62 ± 23
GSH (µM)	13.7 ± 1.3	11.4 ± 0.5	15.0 ± 1.5	17.4 ± 2.0	19.4 ± 1.4	19.1 ± 1.0	17.5 ± 1.5	18.1 ± 0.9	73.3 ± 7.2*
GSSG (µM)	1.7 ± 0.3	2.1 ± 0.3	3.8 ± 1.3	2.4 ± 0.1	3.9 ± 0.7	3.3 ± 0.1	2.4 ± 0.2	3.5 ± 0.4	26.1 ± 3.7*
Total antioxidant capacity (mM)	3.08 ± 0.06	3.19 ± 0.02	3.01 ± 0.07	3.29 ± 0.06	3.28 ± 0.11	3.50 ± 0.09	3.14 ± 0.08	3.37 ± 0.08	6.44 ± 1.20*

Biomarkers were measured 2, 7, and 16 h after i.p. administration of 120 or 1200 mg/kg CCl₄ in canola oil to male Fisher rats. Control rats received an equal volume of canola oil. Values are means ± SE for five rats.

* Statistically significant differences from respective controls.

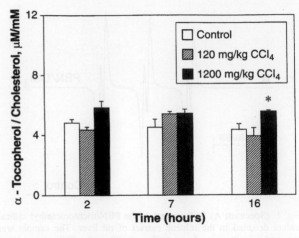

Fig. 2. Ratio of plasma α-tocopherol to cholesterol. The ratios of α-tocopherol to cholesterol were calculated from data in Table 1. Values are means ± SE. Asterisks indicate statistically significant difference from respective controls.

Plasma level of ascorbic acid and uric acid

As shown in Table 1, plasma levels of ascorbic acid tended to be higher in the treatment group at early times, with the difference being significant at 2 h (2-fold) for the high dose. No significant effect on uric acid was found for either dose at any time point studied (Table 1).

Plasma levels of GSH and GSSG

GSH was unaffected by the low dose, but a significant increase (more than 4-fold) was seen at the high dose at 16 h (Table 1). GSSG showed a similar pattern, with the

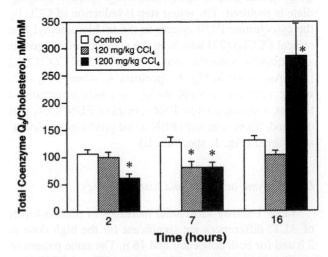

Fig. 3. Ratio of plasma coenzyme Q₉ to cholesterol. The ratios of coenzyme Q₉ to cholesterol were calculated from data in Table 1 and separate determinations of cholesterol (data not shown). Values are means ± SE. Asterisks indicate statistically significant ($p < .05$) differences from respective controls.

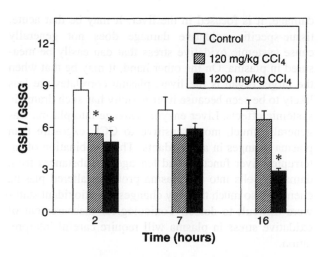

Fig. 4. Ratio of plasma GSH to GSSG. The ratios of GSH to GSSG were calculated from data in Table 1. Values are means ± SE. Asterisks indicate statistically significant ($p < .05$) differences from respective controls.

increase for the high dose at 16 h being more than 10-fold (Table 1).

The ratio of GSH to GSSG decreased with treatment, with statistically significant decreases for both doses at 2 h and for the high dose at 16 h (Fig. 4).

Total antioxidant capacity of plasma

The total antioxidant capacity of plasma was unaffected by the low dose, but a significant increase was seen at the high dose at 16 h (Table 1).

DISCUSSION

In this study, major antioxidant substances in blood plasma were measured as part of a larger study to evaluate the efficacy of a panel of noninvasive indicators of oxidative stress. The CCl_4-derived reactive free radical metabolites can covalently bind to macromolecules and also initiate lipid peroxidation by hydrogen abstraction [10–14]. The involvement of lipid peroxidation [13–17] and Kupffer cells [29] in carbon tetrachloride hepatotoxicity has been demonstrated. Although ESR analysis provides the most definitive method of monitoring hepatic free radical adduct formation in the living animal, the requirement for the administration of a spin trapping agent makes it inapplicable for human studies. We anticipated that carbon tetrachloride dosing, with its generation of the trichloromethyl radical and the resultant lipid peroxidation, would be accompanied by a decrease of antioxidants in the plasma.

In this study, liver damage could be seen in histopathology specimens of liver tissue as early as 2 h after

CCl_4 treatment, even in the low-dose group. Confirmation of functional liver damage was seen by significant increases in serum liver enzymes by 7 h in the low-dose group and at all time points in the high-dose group. Other studies also found this rapid rise in liver enzymes after single-dose CCl_4 administration; the rise continues for at least 24 h and requires several days to resolve [30]. The liver damage was severe by 16 h in our high-dose group as indicated by the very high serum liver enzyme levels. In a previous study in the same age and strain of rats [31], liver content of ALT was found to be 60 IU/g wet weight in livers weighing an average of 9 g/rat. Calculating plasma concentrations resulting from equilibration of all liver enzymes with plasma and the ruptured liver, ALT would be about 32,000 IU/L of plasma. The most severely affected rats treated with high-dose CCl_4 in this study had ALT plasma levels of ~10,000 IU/L, indicating severe damage to a large portion of the liver.

Vitamin E is a fat-soluble antioxidant available only from dietary sources. Prior treatment with α-tocopherol hemisuccinate can increase liver α-tocopherol and protect against CCl_4-mediated liver damage [32]. Thus, we hypothesized that a decrease in plasma α-tocopherol would occur with treatment. As hypothesized, plasma levels in our experiment were lower in the treated group, and the effect was stronger for high-dose treatment and for longer times after dosing. However, when we adjusted for cholesterol or triglyceride, major components of the matrix for α-tocopherol [33], the concentration of the antioxidant was not significantly lower in CCl_4-treated rats. It appears that the direct toxicity of CCl_4 to the lipid-packaging mechanisms of the liver and the accompanying decreases in plasma triglycerides and cholesterol [34] would explain the observed decrease in plasma α-tocopherol. Reports of increased hepatic α-tocopherol after CCl_4 treatment [35] may also be a result of not accounting for the known rapid increase of liver lipid vacuoles in treated animals and the accompanying decrease in all plasma lipids. Our findings are consistent with the lipoprotein-based transport of α-tocopherol by acute mobilization from stores [36, 37].

CoQ_n, along with α-tocopherol, is a major lipid-soluble antioxidant synthesized in the liver. Studies have indicated that the antioxidative role of CoQ (CoQ_9 is predominant in rodents) involves an interaction with α-tocopherol [23,38]. Loss of total CoQ_9 from the plasma among the experimental groups may suggest that the oxidative stress caused by CCl_4 is associated with the decrease in the total CoQ_9 concentration. This interpretation is supported by the significant decrease in the total Q_9/cholesterol ratio. It may be relevant to point out that the similarity in the patterns of changes in the total CoQ and α-tocopherol in the experimental groups may be a reflection, in part, of the interaction between them demonstrated previously [38] as well as more general decreases in plasma lipids. However, our findings suggest that the increased consumption of total CoQ is more sensitive to CCl$_4$ exposure than that of α-tocopherol.

Ascorbic acid is a water-soluble antioxidant produced endogenously in rat liver tissue [39]. Its antioxidant properties result in part from the fact that its oxidation product, semidehydroascorbate radical, is unreactive and, therefore, not damaging [40–45]. Enzymatic systems exist in vivo to recycle the semidehydroascorbate to ascorbate at the expense of reduced nicotinamide adenine dinucleotide/reduced nicotinamide adenine dinucleotide phosphate or GSH [40]. A single large dose of ascorbic acid can be protective in CCl$_4$ liver toxicity [46]. Thus, we expected to see a decline in plasma ascorbic acid concentrations in the treated groups. Instead, levels increased at the 2 h high dose. We have no explanation for this effect.

Glutathione acts as an antioxidant both intracellularly and extracellularly in conjunction with various enzymatic processes that reduce hydrogen peroxide and hydroperoxides as GSH is oxidized to GSSG and other mixed disulfides. GSH is produced in the liver and maintained in high concentration in most tissues. In this study neither GSH nor GSSG changed significantly except for the high-dose group at 16 h when the liver damage was extreme. Intracellular GSH and GSSG, normally found at much higher concentrations compared with plasma, are most likely being released from damaged hepatic cells at this time.

The ratio of GSH to GSSG would be expected to be a more sensitive marker of oxidative stress, because small increases in GSSG and decreases in GSH can appear more amplified by examining the ratio than by measuring either one separately. The significant decrease in this ratio at 2 h in both dose groups suggests early oxidative stress.

There was no evidence of change in total antioxidant capacity in the treated groups except for the 2-fold increase observed in the high-dose group at 16 h when liver damage was severe. This correlates with the large increase in plasma GSH at this dose and time, but GSH is not responsible for the increase because the increase in GSH was only in the micromolar range, whereas the increase in total antioxidant capacity was in the millimolar range. What substances from the severely damaged liver tissue might account for this antioxidant activity are unknown.

These results indicate that measures of oxidative stress in the circulation must be interpreted with great caution. Acute local toxicity (whether or not it is mediated by free radical damage) may have dramatic effects on systemic measures of antioxidant status. Even though the primary mechanism of CCl$_4$ toxicity is oxidative

damage, it is focused in the liver. It may be that acute, tissue-specific oxidative damage does not generally cause systemic oxidative stress that can easily be measured in plasma. On the other hand, it may be that when the target organ is the liver, plasma correlates are less likely to be seen because liver toxicity has such dramatic systemic effects. Liver enzyme leakage into plasma was generally much more sensitive to CCl$_4$ exposure than plasma changes in antioxidants. The combination of interrupted liver function and leakage of substances from damaged cells into the plasma probably altered plasma chemistry so much that any changes in antioxidant status were difficult to detect. Toxicological measurement of oxidative stress in plasma will require careful interpretation.

Acknowledgements — The authors thank Jean Corbett, Yang Fann, Ralph Wilson, John Seely, Ralph Slade, Robert McConnaughey, Kay Crissman, Judy Richards, Linda K. Kwong, and Qu Feng for excellent technical support and acknowledge the contributions of all members of the committee on Biomarkers of Oxidative Stress Study (BOSS) at NIEHS for their helpful suggestions, comments, and discussion. The authors also thank Ms. Mary J. Mason for editorial assistance.

REFERENCES

[1] Rice-Evans, C. A.; Burdon, R. H., eds. *Free radical damage and its control.* Amsterdam-London-New York-Tokyo: Elsevier; 1994.
[2] Halliwell, B. Oxidants and human disease: some new concepts. *FASEB J.* **1:**358–364; 1987.
[3] Freeman, B. A.; Crapo, J. D. Biology of disease: free radicals and tissue injury. *Lab. Invest.* **47:**412–426; 1982.
[4] Marx, J. L. Oxygen free radicals linked to many diseases. *Science* **235:**529–531; 1987.
[5] Pryor, W. A. Free radical biology: xenobiotics, cancer and aging. *Ann. N.Y. Acad. Sci.* **393:**1–22; 1982.
[6] Sohal, R. S., ed. *Age pigments.* Amsterdam: Elsevier; 1981.
[7] Chow, C. K. Nutritional influence on cellular antioxidant defense systems. *Am. J. Clin. Nutr.* **32:**1066–1081; 1979.
[8] de Zwart, L. L.; Meerman, J. H. N.; Commandeur, J. N. M.; Vermeulen, N. P. E. Biomarkers of radical damage: applications in experimental animals and in humans. *Free Radic. Biol. Med.* **26:** 202–226; 1999.
[9] de Zwart, L. L.; Hermanns, R. C. A.; Meerman, J. H. N.; Commandeur, J. N. M.; Salemink, P. J. M.; Vermeulen, N. P. E. Evaluation of urinary biomarkers for free radical–induced liver damage in rats treated with carbon tetrachloride. *Toxicol. Appl. Pharmacol.* **148:**71–82; 1998.
[10] Recknagel, R. O. A new direction in the study of carbon tetrachloride hepatotoxicity. *Life Sci.* **33:**401–408; 1983.
[11] Recknagel, R. O.; Glende, E. A.; Dolak, J. A.; Waller, R. L. Mechanisms of carbon tetrachloride toxicity. *Pharmacol. Ther.* **43:**139–154; 1989.
[12] Williams, A. T.; Burk, R. F. Carbon tetrachloride hepatotoxicity: an example of free radical mediated injury. *Semin. Liver Dis.* **10:**279–284; 1990.
[13] Link, B.; Durk, H.; Thiel, D.; Frank, H. Binding of trichloromethyl radicals to lipids of the hepatic endoplasmic reticulum during tetrachloromethane metabolism. *Biochem. J.* **223:**577–586; 1984.
[14] Biasi, F.; Albano, E.; Chiarpotto, E.; Corongiu, F. P.; Pronzato, M. A.; Marinari, U. M.; Parola, M.; Dianzani, M. U.; Poli, G. *In vivo* and *in vitro* evidence concerning the role of lipid peroxidation in the mechanism of hepatocyte death due to carbon tetrachloride. *Cell Biochem. Funct.* **9:**111–118; 1991.

[15] Albano, E.; Lott, K. A. K.; Slater, T. F.; Stier, A.; Symons, M. C. R.; Tomasi, A. Spin trapping studies on the free radical products formed by metabolic activation of carbon tetrachloride in rat liver microsomal fractions, isolated hepatocytes and in vivo in the rat. *Biochem. J.* **204**:593–603; 1982.

[16] Reinke, L. A.; Towner, R. A.; Janzen, E. G. Spin trapping of free radical metabolites of carbon tetrachloride in vitro and in vivo: effect of acute ethanol administration. *Toxicol. Appl. Pharmacol.* **112**:17–23; 1992.

[17] Cagen, S. Z.; Klaassen, C. D. Hepatotoxicity of carbon tetrachloride in developing rats. *Toxicol. Appl. Pharmacol.* **50**:347–354; 1979.

[18] Wacker, W. E. C.; Ulmer, D. D.; Vallee, B. L. Metalloenzymes and myocardial infarction. II. Malic and lactic dehydrogenase activities and zinc concentrations in serum. *N. Engl. J. Med.* **255**:449–456; 1956.

[19] Wroblewski, F.; La Due, J. S. Serum glutamic pyruvic transaminase in cardiac and hepatic disease. *Proc. Soc. Exp. Biol. Med.* **91**:565–571; 1956.

[20] Asada, M.; Galambos, J. T. Sorbitol dehydrogenase and hepatocellular injury: an experimental and clinical study. *Gastroenterology* **44**:578–587; 1963.

[21] Mashige, F.; Tanaka, N.; Maki, A.; Kamei, S.; Yamanaka, M. Direct spectrophotometry of total bile acids in serum. *Clin. Chem.* **27**:1352–1356; 1981.

[22] Vandewoude, M.; Claeys, M.; De Leeuw, I. Determination of α-tocopherol in human plasma by high-performance liquid chromatography with electrochemical detection. *J. Chromatogr.* **311**:176–182; 1984.

[23] Lass, A.; Forster, M. J.; Sohal, R. S. Effects of coenzyme Q$_{10}$ and α-tocopherol administration on their tissue levels in the mouse: elevation of mitochondrial α-tocopherol by coenzyme Q$_{10}$. *Free Radic. Biol. Med.* **26**:1375–1382; 1999.

[24] Kutnink, M. A.; Hawkes, W. C.; Schaus, E. E.; Omaye, S. T. An internal standard method for the unattended high-performance liquid chromatographic analysis of ascorbic acid in blood components. *Anal. Biochem.* **166**:424–430; 1987.

[25] Jones, D. P.; Carlson, J. L.; Samiec, P. S.; Sternberg, Jr., P.; Mody, Jr., V. C.; Reed, R. L.; Brown, L. A. S. Glutathione measurement in human plasma: evaluation of sample collection, storage and derivatization conditions for analysis of dansyl derivatives by HPLC. *Clin. Chim. Acta* **275**:175–184; 1998.

[26] Miller, N. J.; Rice-Evans, C.; Davies, M. J.; Gopinathan, V.; Milner, A. A novel method for measuring antioxidant capacity and its application to monitoring the antioxidant status in premature neonates. *Clin. Sci.* **84**:407–412; 1993.

[27] Poyer, J. L.; McCay, P. B.; Lai, E. K.; Janzen, E. G.; Davis, E. R. Confirmation of assignment of the trichloromethyl radical spin adduct detected by spin trapping during ^{13}C-carbon tetrachloride metabolism *in vitro* and *in vivo*. *Biochem. Biophys. Res. Commun.* **94**:1154–1160; 1980.

[28] Packer, J. E.; Slater, T. F.; Willson, R. L. Reactions of the carbon tetrachloride–related peroxy free radical with (CCl$_3$O$_2$·) amino acids: pulse radiolysis evidence. *Life Sci.* **23**:2617–2620; 1978.

[29] Edwards, M. J.; Keller, B. J.; Kauffman, F. C.; Thurman, R. G. The involvement of Kupffer cells in carbon tetrachloride toxicity. *Toxicol. Appl. Pharmacol.* **119**:275–279; 1993.

[30] Ohta, Y.; Nishida, K.; Sasaki, E.; Kongo, M.; Ishiguro, I. Attenuation of disrupted hepatic active oxygen metabolism with the recovery of acute liver injury in rats intoxicated with carbon tetrachloride. *Res. Commun. Mol. Pathol. Pharmacol.* **95**:191–207; 1997.

[31] Baker, H. J.; Lindsey, J. R.; Weisbroth, S. H., eds. *The laboratory rat.* New York: Academic Press; 1979.

[32] Tirmenstein, M. A.; Leraas, T. L.; Fariss, M. W. α-Tocopheryl hemisuccinate administration increases rat liver subcellular α-tocopherol levels and protects against carbon tetrachloride-induced hepatotoxicity. *Toxicol. Lett.* **92**:67–77; 1997.

[33] Bjorneboe, A.; Bjorneboe, G.-E. A.; Bodd, E.; Hagen, B. F.; Kveseth, N.; Drevon, C. A. Transport and distribution of α-tocopherol in lymph, serum, and liver cells in rats. *Biochim. Biophys. Acta* **889**:310–315; 1986.

[34] Lipid peroxidation: Part II. Pathological implications. *Chem. Phys. Lipids.* **45**:364–368; 1987.

[35] Miyazawa, T.; Suzuki, T.; Fujimoto, K.; Kaneda, T. Phospholipid hydroperoxide accumulation in liver of rats intoxicated with carbon tetrachloride and its inhibition by dietary α-tocopherol. *J. Biochem.* **107**:689–693; 1990.

[36] Vitamin E deficiency without fat malabsorption. *Nutr. Rev.* **46**:189–194; 1988.

[37] Traber, M. G.; Sies, H. Vitamin E in humans: demand and delivery. *Annu. Rev. Nutr.* **16**:321–347; 1996.

[38] Lass, A.; Sohal, R. S. Electron transport-linked ubiquinone-dependent recycling of α-tocopherol inhibits autooxidation of mitochondrial membranes. *Arch. Biochem. Biophys.* **352**:229–236, 1998.

[39] Banhegyi, G.; Braun, L.; Csala, M.; Puskas, F.; Mandl, J. Ascorbate metabolism and its regulation in animals. *Free Radic. Biol. Med.* **23**:793–803; 1997.

[40] Halliwell, B. How to characterize a biological antioxidant. *Free Radic. Res. Commun.* **9**:1–32; 1990.

[41] Bendich, A.; Machlin, L. J.; Scandurra, O.; Burton, G. W.; Wayner, D. D. M. The antioxidant role of vitamin C. *Adv. Free Radic. Biol. Med.* **2**:419–444; 1986.

[42] Chow, C. K. Vitamins and related dietary antioxidants. In: Dreosti, I. E. ed. *Trace elements, micronutrients, and free radicals.* Clifton, NJ: The Humana Press; 1991:129–147.

[43] Chou, P.-T.; Khan, A. U. L-ascorbic acid quenching of singlet delta molecular oxygen in aqueous media: generalized antioxidant property of vitamin C. *Biochem. Biophys. Res. Commun.* **115**:932–937; 1983.

[44] Cabelli, D. E.; Bielski, B. H. J. Kinetics and mechanism for the oxidation of ascorbic acid/ascorbate by HO$_2$/O$_2^-$ radicals: a pulse radiolysis and stopped-flow photolysis study. *J. Phys. Chem.* **87**:1809–1812; 1983.

[45] Bielski, B. H. J.; Richter, H. W.; Chan, P. C. Some properties of the ascorbate free radical. *Ann. N. Y. Acad. Sci.* **258**:231–237; 1975.

[46] Ademuyiwa, O.; Adesanya, O.; Ajuwon, O. R. Vitamin C in CCl$_4$ hepatotoxicity: a preliminary report. *Human Exp. Toxicol.* **13**:107–109; 1994.

ABBREVIATIONS

LDH—lactic dehydrogenase

ALT—alanine aminotransferase

ALP—alkaline phosphatase

AST—aspartate aminotransferase

SDH—sorbitol dehydrogenase

5′-NT—5′-nucleotidase

TBA—total bile acids

CoQ—Coenzyme Q

ANALYSIS OF HEPATIC OXIDATIVE STRESS STATUS BY ELECTRON SPIN RESONANCE SPECTROSCOPY AND IMAGING

HITOSHI TOGASHI,* HARUHIDE SHINZAWA,* TAKU MATSUO,* YOSHIO TAKEDA,* TSUNEO TAKAHASHI,*
MASAAKI AOYAMA,† KAZUO OIKAWA,‡ and HITOSHI KAMADA†

*The Second Department of Internal Medicine, Yamagata University School of Medicine, Iida-Nishi, Yamagata; †Institute for Life Support Technology, Yamagata Technopolis Foundation, Matsuei, Yamagata; and ‡Yamagata Research Institute of Technology, Matsuei, Yamagata, Japan

(Received 15 September 1999; Revised 16 November 1999; Accepted 14 December 1999)

Abstract—Real-time detection of free radicals generated within the body may contribute to clarify the pathophysiological role of free radicals in disease processes. Of the techniques available for studying the generation of free radicals in biological systems, electron spin resonance (ESR) has emerged as a powerful tool for detection and identification. This article begins with a review of spin trapping detection of oxygen-centered radicals using X-band ESR spectroscopy and then describes the detection of superoxide and hydroxyl radicals by the spin trap 5,5-dimethyl-1-pyrroline-N-oxide and ESR spectroscopy in the perfusate from isolated perfused rat livers subjected to ischemia/reperfusion. This article also reviews the current status of ESR for the in vivo detection of free radicals and in vivo imaging of exogenously administered free radicals. Moreover, we show that in vivo ESR–computed tomography with 3-carbamoyl-2,2,5,5-tetramethylpyrrolidine-1-oxyl may be useful for noninvasive anatomical imaging and also for imaging of hepatic oxidative stress in vivo. © 2000 Elsevier Science Inc.

Keywords— Carbamoyl-PROXYL, ESR-CT, ESR imaging, ESR measurement, Free radical, Oxidative stress

INTRODUCTION

Reactive oxygen species (ROS) such as superoxide and hydroxyl radicals and hydrogen peroxide play an important role in killing both bacteria and tumor cells as well as signal transduction molecules [1–3]. Overproduction of ROS in the body inversely causes tissue damage, however [4]. Various pathological conditions are associated with ROS-mediated events, including ischemia/reperfusion (I/R) injury, endotoxin-mediated injury, and disseminated intravascular coagulation [5–7]. Some biochemical effects of ROS resulting in progressive cell damage include lipid peroxidation, oxidative modification of proteins, and DNA alterations.

Electron spin resonance (ESR) spectroscopy can be used to detect and identify superoxide and hydroxyl radicals and other radical species. Owing to limitations of ESR sensitivity, only those free radical species with relatively long half-lives are measurable. With the development of spin trapping methods, it has become possible to detect short-lived free radical species using conventional X-band ESR instrumentation. Recently, ESR instruments operating at low frequencies (L-band ESR at

Hitoshi Togashi is an Associate Professor. He graduated from Yamagata University School of Medicine in 1980. He is a gastroenterologist, and his research interest is electron spin resonance imaging.

Haruhide Shinzawa is an Associate Professor. He graduated from Tohoku University School of Medicine in 1974. He is interested in the epidemiological study and genetic analysis of hepatitis viruses.

Taku Matsuo is an Instructor. He graduated from Yamagata University School of Medicine in 1992.

Yoshio Takeda is an Instructor. He graduated from Yamagata University School of Medicine in 1984.

Tsuneo Takahashi is a Professor. He graduated from Tohoku University School of Medicine in 1960 and received his Ph.D. degree from the Graduate School of Tohoku University in 1965.

Masaaki Aoyama is Research Director, Division of Organic Chemistry, Institute for Life Support Technology, Yamagata Technopolis Foundation.

Kazuo Oikawa is a Researcher at Yamagata Research Institute of Technology.

Hitoshi Kamada is Director General, Institute for Life Support Technology, Yamagata Technopolis Foundation.

Address correspondence to: Hitoshi Togashi, M.D., Ph.D., The Second Department of Internal Medicine, Yamagata University School of Medicine, 2-2-2 Iida-Nishi, Yamagata 990-9585, Japan; Tel: +81 (23) 628-5309; Fax: +81 (23) 628-5311; E-Mail: htogashi@med.id.yamagata-u.ac.jp.

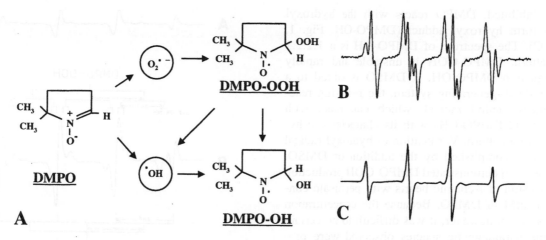

Fig. 1. (A) Diagram of DMPO spin trapped adduct formation after the interaction with superoxide and hydroxyl radicals. Note that DMPO-OH may form as a degradation product of DMPO-OOH via two separate mechanisms. (Spectrum B) ESR spectrum of DMPO-OOH adduct. (Spectrum C) ESR spectrum of DMPO-OH adduct.

<1 GHz) have made it possible to measure radical species in the whole animal [8]. Therefore, it is expected that L-band ESR can be adapted to measure oxidative stress in vivo. This article summarizes recent developments in the ESR measurement of oxidative stress using mostly our own data.

Spin-trapping technique of ROS

Endogenous free radicals produced in vivo have extremely short half-lives and are in low concentrations, making direct detection difficult as a result of the limitations of ESR instrumentation. To overcome these difficulties, spin trapping has been used to measure unstable free radicals. Spin trapping is a technique in which a nitrone or nitroso compound is allowed to react with a free radical to produce a nitroxide whose stability is considerably greater than that of the parent free radical [9]. The resulting ESR spectrum exhibits a hyperfine splitting pattern that is characteristic of spin trapped free radical. Consequently, spin trapping permits the detection and identification of a number of different free radicals generated during the same process. Nevertheless, numerous factors need to be considered before undertaking spin trapping studies. These include: (i) the rate of primary as well as secondary free radical production, (ii) the rate of spin trapping the various free radicals generated, and (iii) the rate of spin trapped adduct decomposition. Of the major ROS thought to be generated in biological systems, only superoxide and hydroxyl radicals contain unpaired electrons. Among the current methods for free radical detection, only spin trapping offers investigators the opportunity to measure and distinguish these ROS simultaneously.

Among the several nitrones used as spin traps, α-phe-

nyl-N-*tert*-butylnitrone (PBN) and 5,5-dimethyl-1-pyrroline N-oxide (DMPO) are commonly employed [10,11]. PBN is a stable compound and forms relatively long-lived spin adducts with various types of radicals. Garlick et al. [12] studied rat hearts perfused in the Langendorff mode with a buffer containing PBN and noted that reperfusion after global ischemia resulted in a burst of PBN radical adduct formation. Bolli et al. [13] demonstrated PBN radical adduct generation in the stunned myocardium of intact dogs. The ESR spectra were consistent with the trapping by PBN of secondary oxygen- and carbon-centered radicals such as alkoxy and alkyl radicals, which could be formed by reactions of primary oxygen radicals with membrane lipids. Kadiiska et al. [14] demonstrated in vivo hydroxyl radical generation in the bile of rats 10 weeks after the rats were fed an iron-loading diet and 40 min after the rats were injected via the intraperitoneal route with PBN dissolved in dimethyl sulfoxide (DMSO). Hydroxyl radicals generated in vivo are converted to methyl radicals via reaction with DMSO [14]. The methyl radical is then detected as its adduct with PBN by ESR spectroscopy. As mentioned in these reports, PBN is more suitable for the detection of carbon-centered radicals than for the detection of superoxide and hydroxyl radicals.

Conversely, DMPO is water-soluble, and its reaction rate constant for superoxide and hydroxyl radicals is high. Moreover, spin trapping analysis using DMPO makes it possible to distinguish simultaneously among a variety of important biologically generated free radicals [9]. The chemical structure of DMPO is shown in Fig. 1A. Although superoxide is chemically unstable, it reacts with DMPO to form superoxide spin trapped adduct (DMPO-OOH; Fig. 1, spectrum B). In the presence of superoxide dismutase (SOD), DMPO-OOH adduct formation is inhibited. DMPO reacts with the hydroxyl radical to form hydroxyl adduct (DMPO-OH, Fig. 1, spectrum C). The spectrum of DMPO-OH is a 1:2:2:1 quartet pattern. DMPO-OOH is unstable and rapidly decomposes into DMPO-OH. If DMSO is added to a hydroxyl radical–generating system, the resulting reaction produces methyl radical, which can react with DMPO to yield DMPO-CH₃ with its characteristic hyperfine splitting pattern. Verification of hydroxyl radical formation is accomplished by the addition of DMSO. Arroyo et al. [15] demonstrated DMPO-OOH production after reperfusion of ischemic hearts with perfusate containing 100 mM of DMPO. Because the concentration used was so high, however, it was difficult to be certain whether the paramagnetic species observed were produced by I/R or by nonspecific toxic effects of DMPO on the myocardium. In the same experimental model as that used by Arroyo and his colleagues, Nakazawa et al. [16] demonstrated the DMPO-OOH adduct on reperfusion of ischemic hearts with DMPO at a low concentration (10 mM). DMPO can be easily applied for the measurement of oxygen-centered free radicals in biological systems at room temperature; however, DMPO-OOH and DMPO-OH adducts are unstable, and their half-lives are short. Recently, a new spin trap, 5-diethoxyphosphoryl-5-methyl-1-pyrroline *N*-oxide (DEMPO) was reported to have improved properties for superoxide detection. DEMPO has an advantage that the half-life of its superoxide adduct is longer than that of DMPO [17,18].

Application of spin trapping technology to the study of ROS production during hepatic I/R

It has been demonstrated that reperfusion of ischemic tissue aggravates ischemic damage in spite of the restoration of blood flow in a variety of tissues [19]. ROS, including superoxide and hydroxyl radicals, have been strongly implicated in the onset of reperfusion injuries [20,21]. After a rupture of esophageal varices or hepatocellular carcinoma in cirrhotic patients, liver damage progresses in association with hemodynamic restoration. The gravest consequences of hepatic I/R injury are liver and multiple organ failures, which are associated with a high mortality [22]. From a therapeutic point of view, therefore, it is essential to elucidate the source of ROS and the molecular mechanisms by which ROS mediate liver injury.

Using X-band ESR spectroscopy, we examined whether ROS were produced after reperfusion of isolated and perfused rat livers subjected to global ischemia [23]. Details of the experiment are described in the legend of Fig. 2 and in a report by Togashi et al. [23]. Spectrum B of Fig. 2 shows a radical trapped by DMPO after 6 min of reperfusion. Two characteristic

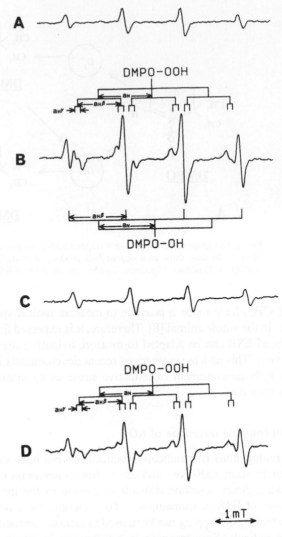

Fig. 2. ESR spectra of the effluent from a reperfused liver that has been subjected to warm ischemia for 90 min. A few drops of hepatic vein effluent were collected in glass test tubes at the times indicated after reperfusion. Two hundred microliters of the effluent was mixed with 30 μl of DMPO (Dojindo Laboratories, Kumamoto, Japan) and 50 μl of 5 mM diethylenetriaminepentaacetic acid. The mixture was aspirated into a flat quartz cell (60 × 10 × 0.25 mm), and ESR spectra were recorded at room temperature with a JEOL RE-3X spectroscope (JEOL, Ltd., Tokyo, Japan) operating at the X-band. Spectra A through D were obtained from the same reperfused liver. (Spectrum A) Background spectrum before subjection to ischemia. (Spectrum B) ESR spectrum of the effluent collected 6 min after reperfusion. DMPO-OOH and DMPO-OH were identified. (Spectrum C) ESR spectrum of allopurinol containing effluent collected 12 min after reperfusion. The spectrum returned to basal level. (Spectrum D) ESR spectrum of effluent collected 20 min after reperfusion. When the effect of allopurinol was eliminated, DMPO-OOH and DMPO-OH reappeared. ESR spectroscopy settings were as follows: receiver gain = 1 × 1000, modulation width = 0.1 mT, time constant = 0.3 s, microwave power = 8 mW, sweep time = 2.5 mT/min, magnetic field = 335 ± 5 mT. Reproduced with kind permission from PJD Publications Limited, Westbury, NY, USA from [47]. Copyright by PJD Publications Ltd.

spectra of spin trapped free radicals were formed. One of the spin adducts seemed to be a DMPO-OOH adduct. The *g* value obtained was 2.006. The hyperfine

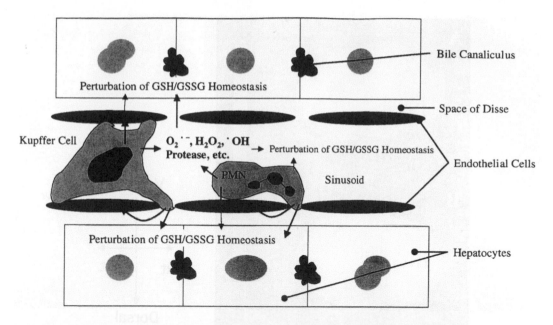

Fig. 3. Schematic diagram of the hepatic sinusoid depicting the potential avenues of phagocyte-released products. Kupffer cells proliferate, and monocytes are recruited in the immediate vicinity of sinusoidal endothelial cells by priming with heat-killed *Corynebacterium parvum*. Three hours after injection of a low dose of LPS, hepatic oxidative stress occurs before the appearance of hepatic injury. The hepatic oxidative stress reduces GSH, increases glutathione disulfide (GSSG), and decreases the GSH/GSSG ratio in the plasma and hepatocytes. The carbamoyl-PROXYL reduction system, including GSH, is impaired by the stress. Thus, the delayed carbamoyl-PROXYL signal inactivation reflects the in vivo hepatic oxidative stress.

splitting constant of the spin trapped adducts was $a_N = 1.43$ mT, $a_H^\beta = 1.13$ mT, and $a_H^\gamma = 0.14$ mT. Addition of SOD to the effluent collected after reperfusion removed the DMPO-OOH spectrum. The other spin trapped radical seemed to be a DMPO-OH adduct. As demonstrated in spectrum B of Fig. 2, a 1:2:2:1 quartet pattern was simultaneously observed. The *g* value obtained was 2.006, and the hyperfine splitting constant was $a_N = a_H^\beta = 1.49$ mT. The DMPO-OH spectrum was attenuated in the presence of DMSO, and DMPO-CH3 was formed by the reaction of the hydroxyl radical with DMSO.

In reperfused organs, ROS can be generated by several mechanisms, including xanthine/xanthine oxidase (XOD), the electron transport system of mitochondria and microsomes, and cyclooxygenase [20,24–26]. We investigated the implications of XOD, microsomal mixed-function oxidase, the electron transport system of mitochondria, and cyclooxygenase as the possible origins of superoxide in our I/R model. We examined the effects of various enzyme inhibitors on superoxide and hydroxyl radical production. The inhibitors of microsomal mixed-function oxidase (50 μM of SKF-525A), the electron transport system of mitochondria (5 mM of amobarbital), and cyclooxygenase (50 μM of indomethacin) did not affect the production of DMPO-OOH or DMPO-OH. Nevertheless, allopurinol, an inhibitor of XOD, completely suppressed the production of superox-

ide and the superoxide-derived hydroxyl radical (Fig. 2, spectrum C). DMPO-OOH and DMPO-OH were detected once again after replacement of the perfusate with one lacking allopurinol (Fig. 2, spectrum D). We showed clearly that superoxide and the superoxide-derived hydroxyl radical originated from a xanthine/XOD system in the ex vivo hepatic I/R model. Further study is required to clarify the pathophysiological role of the generated ROS after reperfusion in subsequent liver injury occurring in vivo.

In vivo detection of oxidative stress by ESR

If the spatial distribution of free radicals generated in organs could be visualized in living animals, it would provide information on the pathophysiological role of free radicals in the body. Because only a limited number of free radicals are generated and they are short lived, it is difficult to obtain their direct images in the body. Halpern et al [27] first measured the production of free radicals in the tissue of a living animal using a low-frequency 260 MHz spectrometer and α-(4-pyridyl-1-oxide)-N-*tert*-butylnitrone (4-POBN) plus ethanol. They detected an ESR spectrum characteristic of 4-POBN-CH(OH)-CH3 originating from hydroxyl radicals produced during radiation in a leg tumor (12 mm diameter) in living mice.

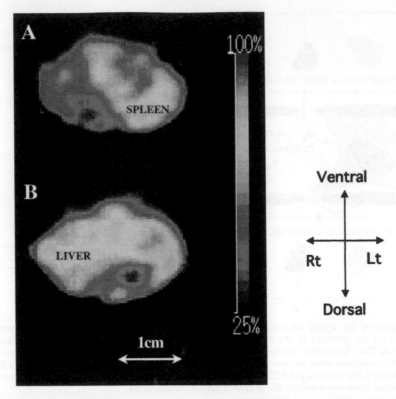

Fig. 4. ESR-CT images of mouse abdomen before and after LPS injection. The ESR-CT system makes it possible to obtain CT images at a measurement time of 40 s. Thus, we can conduct a spatiotemporal analysis of carbamoyl-PROXYL in the liver at the same slice level. The ESR-CT images were obtained 280 s after carbamoyl-PROXYL injection. (A) Before LPS injection, the carbamoyl-PROXYL signal in the liver has nearly disappeared. (B) Three hours after LPS injection, the reduction of carbamoyl-PROXYL in the liver is delayed.

ESR imaging of exogenous free radicals

Conventional X-band ESR spectroscopy is sensitive to free radicals only to within the first millimeter of the surface of living tissue. To achieve tissue penetration and free radical detection in organs, L-band ESR, which can minimize the dielectric loss, is used for imaging [8,28–39]. Most ESR images have been obtained by administration of exogenous spin probes to experimental animals. At an early stage in the development of the technology, approximately 40 min was necessary to obtain an image of the rat head using an ESR–computed tomography (CT) system [8]. The ESR-CT system was later improved, making it possible to shorten the measurement time to 40 s. Nitroxides such as carbamoyl-PROXYL are reduced to the corresponding hydroxylamine in vivo, resulting in the disappearance of their ESR signals [40,41]. During the process, GSH (the reduced form of glutathione) and ascorbic acid are speculated to work cooperatively in the reduction of nitroxides [42]. Moreover, tissue pO_2 affects the in vivo stability of nitroxides [43]. Kuppsusamy et al. [36] demonstrated that carbamoyl-PROXYL reduction was significantly higher in tumor tissue than in normal tissue by two-

dimensional ESR projection imaging. ESR-CT imaging may be useful for both anatomical and functional imaging in the living body.

In vivo imaging of increased oxidative stress in the liver by ESR-CT

Acute hepatic injury occurs by a priming of heat-killed *Corynebacterium parvum* followed by a low-dose injection of lipopolysaccharide (LPS) to specific pathogen-free mice [44]. By priming the specific pathogen-free animals, proliferation of Kupffer cells and recruitment of monocytes into the liver occur. After injection of a small amount of LPS into mice primed with *C. parvum*, activated phagocytes produced and released large quantities of superoxide, hydrogen peroxide, nitric oxide, and their derivatives, which target the surrounding tissue (Fig. 3). Three hours after LPS injection, liver injury was not obvious by biochemical and histological examination; however, significant hepatic oxidative stress had already occurred. This was confirmed by an increase in the GSSG level, in the production of ascorbic acid radicals in the plasma, and in the GSSG level and heme oxygenase 1 expression in the liver 3 h after the LPS injection (data not presented). Six hours after the LPS injection, acute liver injury was obvious both biochemically and histologically. We showed that the oxidative stress occurred before the onset of the obvious hepatic necrosis.

Pharmacokinetic studies of administered nitroxides have been performed in rat models with silicosis and streptozotocin-induced diabetes [45,46]. According to these reports, kinetic clearance of nitroxides was significantly affected by oxidative stress occurring in vivo. The carbamoyl-PROXYL reduction system, including GSH, is impaired by the stress. The kinetic clearance was significantly slower in the oxidative stress group than in the control group. Moreover, the kinetic clearance and concentration of GSH in plasma showed a significantly positive relation. If ESR-CT can show the difference in kinetic clearance of nitroxides between the time before oxidative stress and during it, we can determine increased oxidative stress in the liver by comparing ESR-CT images. We explored the possibility that oxidative stress in the liver could be shown by ESR-CT in living mice before the onset of obvious hepatic injury. Carbamoyl-PROXYL solution (200 mM; 2 ml/kg of body weight) was injected rapidly through the tail vein, and a series of ESR-CT images of the abdomen was taken (after 120, 200, 280, and 360 s) before the LPS injection. Immediately after the carbamoyl-PROXYL injection, a high-intensity signal was visible mainly in the liver and the spleen. The signal in the liver disappeared rapidly because the carbamoyl-PROXYL was rapidly reduced to its ESR silent hydroxylamine. Figure 4A shows ESR-CT images obtained 280 s after carbamoyl-PROXYL injection. The spleen was still strongly imaged, but the signal in the liver had nearly disappeared. A low concentration of LPS was subsequently injected, and hepatic oxidative stress was thus induced. Three hours after the LPS injection, the same volume of carbamoyl-PROXYL solution was injected into the tail vein again, and another series of ESR-CT images of the abdomen was taken. The image in Fig. 4B was obtained from the same mouse as that used in Fig. 4A. The high-intensity signal at 280 s was still clearly visible in the liver and spleen, indicating the delayed inactivation of the carbamoyl-PROXYL signal (Fig. 4B). Significant oxidative stress may impair the reduction of carbamoyl-PROXYL, including GSH. In vivo ESR-CT with carbamoyl-PROXYL clearly showed increased hepatic oxidative stress before the appearance of hepatic injury.

CONCLUSIONS AND FUTURE PERSPECTIVES

In the last few years, ESR spectroscopy and imaging have evolved considerably so that they now provide useful information for analyzing the role of free radicals in the living body. ESR-CT images provide information that has never before been available; however, a number of problems must be overcome before in vivo ESR measurement can be used in human beings. One member of our group (M.A.) developed 3-(D-glucopyranosyl-oxymethyl-) 2,2,5,5-tetramethylpyrroline-1-oxyl (G-PROXYL), which is incorporated into cells more easily than carbamoyl-PROXYL. After incorporation into cells, G-PROXYL is reduced to form the ESR silent hydroxylamine. Under oxidative stress, the hydroxylamine is reoxidized by ROS, resulting in reactivation of the ESR signal. Therefore, ESR measurement with G-PROXYL may be useful for the in vivo imaging of oxidative stress status in the tissues of living animals. The development and application of spin labels should allow the pathophysiological analysis of oxidative stress in vivo.

Acknowledgement — The authors wish to thank PJD Publications for giving us permission to reproduce Fig. 2 from the *Research Communications in Biochemistry and Cell & Molecular Biology*.

REFERENCES

[1] Rose, H.; Klebanoff, S. J. Bactericidal activity of a superoxide anion-generating system. A model for the polymorphonuclear leukocyte. *J. Exp. Med.* **149:**27–39; 1979.

[2] Handa, K.; Sato, S. Stimulation of microsomal NADPH oxidation by quinone group-containing anticancer chemicals. *Gann* **67:** 523–528; 1976.

[3] Satriano, J. A.; Shuldiner, M.; Hora, K.; Xing, Y.; Shan, Z.; Schlondorff, D. Oxygen radicals as second messengers for expression of the monocyte chemoattractant protein, JE/MCP-1, and the monocyte colony-stimulating factor, CSF-1, in response to tumor necrosis factor-alpha and immunoglobulin G. Evidence for involvement of reduced nicotinamide adenine dinucleotide phosphate (NADPH)-dependent oxidase. *J. Clin. Invest.* **92:**1564–1571; 1993.

[4] Fantone, J. C.; Ward, P. A. Role of oxygen-derived free radicals and metabolites in leukocyte-dependent inflammatory reactions. *Am. J. Pathol.* **107:**395–418; 1982.

[5] Atalla, S. L.; Toledo-Pereyra, L. H.; MacKenzie, G. H.; Cederna, J. P. Influence of oxygen-derived free radical scavengers on ischemic livers. *Transplantation* **40:**5847–590; 1985.

[6] Jaeschke, H.; Farhood, A.; Smith, C. W. Contribution of complement-stimulated hepatic macrophages and neutrophils to endotoxin-induced liver injury in rats. *Hepatology* **19:**973–979; 1994.

[7] Yoshikawa, T.; Murakami, M.; Yoshida, N.; Seto, O.; Kondo, M. Effects of superoxide dismutase and catalase on disseminated intravascular coagulation in rats. *Thromb. Haemost.* **50:**869–872; 1983.

[8] Ishida, S.; Matsumoto, S.; Yokoyama, H.; Mori, N.; Kumashiro, H.; Tsuchihashi, N.; Ogata, T.; Yamada, M.; Ono, M.; Kitajima, T.; Kamada, H.; Ogata, T.; Yamada, M.; Ono, M.; Kitajima, T.; Yoshida, E. An ESR-CT imaging of the head of a living rat receiving an administration of a nitroxide radical. *Magn. Reson. Imaging* **10:**109–114;1992.

[9] Britigan, B. E.; Cohen, M. S.; Rosen, G. M. Detection of the production of oxygen-centered free radicals by human neutrophils using spin trapping techniques: a critical perspective. *J. Leukoc. Biol.* **41:**349–362; 1987.

[10] Finkelstein, E.; Rosen, G. M.; Rauckman, E. J. Spin trapping of superoxide and hydroxyl radical: practical aspects. *Arch. Biochem. Biophys.* **200:**1–16; 1980.

[11] Mason, R. P.; Hanna, P. M.; Burkitt, M. J.; Kadiiska, M. B.

Detection of oxygen-derived radicals in biological systems using electron spin resonance. *Environ. Health Perspect.* **102**(Suppl.): 33–36; 1994.

[12] Garlick, P. B.; Davies, M. J.; Hearse, D. J.; Slater, T. F. Direct detection of free radicals in the reperfused rat heart using electron spin resonance spectroscopy. *Circ. Res.* **61**:757–760; 1987.

[13] Bolli, R.; Patel, B. S.; Jeroudi, M. O.; Lai, E. K.; McCay, P. B. Demonstration of free radical generation in "stunned" myocardium of intact dogs with the use of the spin trap alpha-phenyl N-*tert*-butyl nitrone. *J. Clin. Invest.* **82**:476–485; 1988.

[14] Kadiiska, M. B.; Burkitt, M. J.; Xiang, Q. H.; Mason, R. P. Iron supplementation generates hydroxyl radical in vivo. An ESR spin-trapping investigation. *J. Clin. Invest.* **96**:1653–1657; 1995.

[15] Arroyo, C. M.; Kramer, J. H.; Dickens, B. F.; Weglicki, W. B. Identification of free radicals in myocardial ischemia/reperfusion by spin trapping with nitrone DMPO. *FEBS Lett.* **221**:101–104; 1987.

[16] Nakazawa, H.; Arroyo, C. M.; Ichimori, K.; Saigusa, Y.; Minezaki, K. K.; Pronai, L. The demonstration of DMPO superoxide adduct upon reperfusion using a low non-toxic concentration. *Free Radic. Res. Commun.* **14**:297–302; 1991.

[17] Roubaud, V.; Sankarapandi, S.; Kuppusamy, P.; Tordo, P.; Zweier, J. L. Quantitative measurement of superoxide generation using the spin trap 5-(diethoxyphosphoryl)-5-methyl-1-pyrroline-N-oxide. *Anal. Biochem.* **247**:404–411; 1997.

[18] Liu, K. J.; Miyake, M.; Panz, T.; Swartz, H. Evaluation of DEPMPO as a spin trapping agent in biological systems. *Free Radic. Biol. Med.* **26**:714–721; 1999.

[19] McCord, J. M. Oxygen-derived free radicals in postischemic tissue injury. *N. Engl. J. Med.* **312**:159–163; 1985.

[20] Granger, D. N.; McCord, J. M.; Parks, D. A.; Hollwarth, M. E. Xanthine oxidase inhibitors attenuate ischemia-induced vascular permeability changes in the cat intestine. *Gastroenterology* **90**: 80–84; 1986.

[21] Adkison, D.; Hollwarth, M. E.; Benoit, J. N.; Parks, D. A.; McCord, J. M.; Granger, D. N. Role of free radicals in ischemia-reperfusion injury to the liver. *Acta Physiol. Scand. Suppl.* **548**: 101–107; 1986.

[22] Keller, G. A.; West, M. A.; Cerra, F. B.; Simmons, R. L. Macrophage-mediated modulation of hepatic function in multiple-system failure. *J. Surg. Res.* **39**:555–563; 1985.

[23] Togashi, H.; Shinzawa, H.; Yong, H.; Takahashi, T.; Noda, H.; Oikawa, K.; Kamada, H. Ascorbic acid radical, superoxide, and hydroxyl radical are detected in reperfusion injury of rat liver using electron spin resonance spectroscopy. *Arch. Biochem. Biophys.* **308**:1–7; 1994.

[24] Ambrosio, G.; Zweier, J. L.; Duilio, C.; Kuppusamy, P.; Santoro, G.; Elia, P. P; Tritto, I.; Cirillo, P.; Condorelli, M.; Chiariello, M. Evidence that mitochondrial respiration is a source of potentially toxic oxygen free radicals in intact rabbit hearts subjected to ischemia and reflow. *J. Biol. Chem.* **268**:18532–18541; 1993.

[25] Paller, M. S.; Jacob, H. S. Cytochrome P-450 mediates tissue-damaging hydroxyl radical formation during reoxygenation of the kidney. *Proc. Natl. Acad. Sci. USA* **91**:7002–7006, 1994.

[26] Kontos, H. A.; Wei, E. P.; Ellis, E. F.; Jenkins, L. W.; Povlishock, J. T.; Rowe, G. T.; Hess, M. L. Appearance of superoxide anion radical in cerebral extracellular space during increased prostaglandin synthesis in cats. *Circ. Res.* **57**:142–151, 1985.

[27] Halpern, H. J.; Yu, C.; Barth, E.; Peric, M.; Rosen, G. M. In situ detection, by spin trapping, of hydroxyl radical markers produced from ionizing radiation in the tumor of a living mouse. *Proc. Natl. Acad. Sci. USA* **92**:796–800; 1995.

[28] Takeshita, K.; Utsumi, H.; Hamada, A. ESR measurement of radical clearance in lung of whole mouse. *Biochem. Biophys. Res. Commun.* **177**:874–880; 1991.

[29] Quaresima, V.; Alecci, M.; Ferrari, M.; Sotgiu, A. Whole rat electron paramagnetic resonance imaging of a nitroxide free radical by a radio frequency (280 MHz) spectrometer. *Biochem. Biophys. Res. Commun.* **183**:829–835; 1992.

[30] Kuppusamy, P.; Chzhan, M.; Vij, K.; Shteynbuk, M.; Lefer, D. J.; Giannella, E.; Zweier, J. L. Three-dimensional spectral-spatial

EPR imaging of free radicals in the heart: a technique for imaging tissue metabolism and oxygenation. *Proc. Natl. Acad. Sci. USA* **91**:3388–3392; 1994.

[31] Hiramatsu, M.; Oikawa, K.; Noda, H.; Mori, A.; Ogata, T.; Kamada, H. Free radical imaging by electron spin resonance computed tomography in rat brain. *Brain Res.* **697**:44–47; 1995.

[32] Halpern, H. J.; Peric, M.; Yu, C.; Barth, E. D.; Chandramouli, G. V.; Makinen, M. W.; Rosen, G. M. In vivo spin-label murine pharmacodynamics using low-frequency electron paramagnetic resonance imaging. *Biophys. J.* **71**:403–409; 1996.

[33] Yoshimura, T.; Yokoyama, H.; Fujii, S.; Takayama, F.; Oikawa, K.; Kamada, H. In vivo EPR detection and imaging of endogenous nitric oxide lipopolysaccharide-treated mice. *Nat. Biotechnol.* **14**:992–994; 1996.

[34] Kazama, S.; Takashige, G.; Yoshioka, H.; Tanizawa, H.; Ogata, T.; Koscielniak, J.; Berliner L. J. Dynamic electron spin resonance (ESR) imaging of the distribution of spin labeled dextran in a mouse. *Magn. Reson. Med.* **36**:547–550; 1996.

[35] Yokoyama, H.; Fujii, S.; Yoshimura, T.; Ohya-Nishiguchi, H.; Kamada, H. In vivo ESR-CT imaging of the liver in mice receiving subcutaneous injection of nitric oxide-bound iron complex. *Magn. Reson. Imaging* **15**:249–253; 1997.

[36] Kuppusamy, P.; Afeworki, M.; Shankar, R. A.; Coffin, D.; Krishna, M. C.; Hahn, S. M.; Mitchell, J. B.; Zweier, J. L. In vivo electron paramagnetic resonance imaging of tumor heterogeneity and oxygenation in a murine model. *Cancer Res.* **58**:1562–1568; 1998.

[37] Togashi, H.; Shinzawa, H.; Ogata, T.; Matsuo, T.; Ohno, S.; Saito, K.; Yamada, N.; Yokoyama, H.; Noda, H.; Oikawa, K.; Kamada, H.; Takahashi, T. Spatiotemporal measurement of free radical elimination in the abdomen using an in vivo ESR-CT imaging system. *Free Radic. Biol. Med.* **25**:1–8; 1998.

[38] Masumizu, T.; Fujii, K.; Kohno, M.; Nagai, S.; Odagaki, Y.; Imanari, M.; Mori, A.; Packer, L. Three dimensional electron spin resonance (ESR) imaging of internal organs in living mice. *Biochem. Mol. Biol. Int.* **46**:707–717; 1998.

[39] He, G.; Shankar, R. A.; Chzhan, M.; Samouilov, A.; Kuppusamy, P.; Zweier, J. L. Noninvasive measurement of anatomic structure and intraluminal oxygenation in the gastrointestinal tract of living mice with spatial and spectral EPR imaging. *Proc. Natl. Acad. Sci. USA* **96**:4586–4591; 1999.

[40] Couet, W. R.; Brasch, R. C.; Sosnovsky, G.; Tozer, T. N. Factors affecting nitroxide reduction in ascorbate solution and tissue homogenates. *Magn. Reson. Imaging* **3**:83–88; 1985.

[41] Iannone, A.; Hu, H. P.; Tomasi, A.; Vannini, V.; Swartz, H. M. Metabolism of aqueous soluble nitroxides in hepatocytes: effects of cell integrity, oxygen, and structure of nitroxides. *Biochim. Biophys. Acta* **991**:90–96; 1989.

[42] Fuchs, J.; Groth, N.; Herrling, T.; Zimmer, G. Electron paramagnetic resonance studies on nitroxide radical 2,2,5,5-tetramethyl-4-piperidin-1-oxyl (TEMPO) redox reactions in human skin. *Free Radic. Biol. Med.* **22**:967–976; 1997.

[43] Chen, K.; Glockner, J. F.; Morse, P. D., II; Swartz, H. M. Effects of oxygen on the metabolism of nitroxide spin labels in cells. *Biochemistry* **28**:2496–2501; 1989.

[44] Sato, T.; Shinzawa, H.; Abe, Y.; Takahashi, T.; Arai, S.; Sendo, F. Inhibition of Corynebacterium parvum-primed and lipopolylysaccharide-induced hepatic necrosis in rats by selective depletion of neutrophils using a monoclonal antibody. *J. Leukoc. Biol.* **53**:144–150; 1993.

[45] Sano, T.; Umeda, F.; Hashimoto, T.; Nawata, H.; Utsumi, H. Oxidative stress measurement by in vivo electron spin resonance spectroscopy in rats with streptozotocin-induced diabetes. *Diabetologia* **41**:1355–1360; 1998.

[46] Vallyathan, V.; Leonard, S.; Kuppusamy, P.; Pack, D.; Chzhan, M.; Sanders, S. P.; Zweir, J. L. Oxidative stress in silicosis: evidence for the enhanced clearance of free radicals from whole lungs. *Mol. Cell. Biochem.* **168**:125–132; 1997.

[47] Ohno, S.; Togashi, H.; Shinzawa, H.; Matsuo, T.; Sugahara, K.; Shao, L.; Kamada, H.; Takahashi, T. Reactive oxygen species originate from xanthine oxidase after reperfusion of the ischemic

rat liver and activate nuclear factor kappa B in the liver. *Res. Commun. Biochem. Cell & Mol. Biol.* **3**:69–94; 1999.

ABBREVIATIONS

carbamoyl-PROXYL—3-carbamoyl-2,2,5,5-tetramethyl-pyrrolidine-1-oxyl

DEMPO—5-diethoxyphosphoryl-5-methyl-1-pyrroline *n*-oxide

DMPO—5,5-dimethyl-1-pyrroline *n*-oxide

DMSO—dimethyl sulfoxide

ESR—electron spin resonance

G-PROXYL—3-(D-glucopyranosyloxymethyl-) 2,2,5,5-tetramethylpyrroline-1-oxyl

GSH—reduced glutathione

GSSG—glutathione disulfide

L-band—low-frequency band

LPS—lipopolysaccharide

PBN—α-phenyl-N-*tert*-butylnitrone

4-POBN—α-(4-pyridyl-1-oxide)-N-*tert*-butylnitrone

SOD—superoxide dismutase

XOD—xanthine oxidase

NONINVASIVE STUDY OF RADIATION-INDUCED OXIDATIVE DAMAGE USING IN VIVO ELECTRON SPIN RESONANCE

YURI MIURA* and TOSHIHIKO OZAWA[†]

*Department of Biochemistry and Isotopes, Tokyo Metropolitan Institute of Gerontology, Tokyo, Japan and [†]Department of Bioregulation Research, National Institute of Radiological Sciences, Chiba, Japan

(Received 30 September 1999; Revised 29 December 1999; Accepted 10 January 2000)

Abstract—Nitroxyl radicals injected into a whole body indicate the disappearance of signal intensity of in vivo electron spin resonance (ESR). The signal decay rates of nitroxyl have reported to be influenced by various types of oxidative stress. We examined the effect of X-irradiation on the signal decay rate of nitroxyl in the upper abdomen of mice using in vivo ESR. The signal decay rates increased 1 h after 15 Gy irradiation, and the enhancement was suppressed by preadministration of cysteamine, a radioprotector. These results suggest that the signal decay of nitroxyl in whole mice is enhanced by radiation-induced oxidative damage. The in vivo ESR system probing the signal decay of nitroxyl could provide a noninvasive technique for the study of oxidative stress caused by radiation in a living body.   © 2000 Elsevier Science Inc.

Keywords—In vivo ESR, Oxidative stress, Nitroxyl radical, ROS, Spin probe, Signal decay, Redox, Free radicals

INTRODUCTION

As it has become apparent that oxygen free radicals are involved in numerous pathophysiological conditions, there is a growing interest in the in vivo detection of free radicals by electron spin resonance (ESR). ESR was developed as an in vivo detector of bioradicals in the late 1980s. Because bioradicals such as nitric oxide, superoxide radical, and hydroxyl radical are present in ex-tremely low concentrations in tissue and organs and because their lifetimes are extremely short, indirect detection employing spin trapping agents or spin probes is more useful.

The most widely used spin probe is a nitroxyl radical, which is relatively stable at room temperature and has low toxicity to organisms (Fig. 1). Nitroxyl radical is not stable in biological systems, however. In hepatic microsomes and various cultured cells, the reduction and reoxidation of nitroxyl compounds have been reported since the 1970s, suggesting that various biomolecules and enzymes such as cytochrome P-450, cytochrome P-450 reductase, mitochondrial electron transport systems, and ascorbic acid, are involved in the redox reaction of nitroxyl. Furthermore, it has become apparent that oxygen concentration, antioxidant content, and oxygen free radicals influence the rate of the redox reaction of nitroxyl in biological systems. Therefore, in in vitro or cell-level experiments, a nitroxyl compound may be used as a sensitive redox indicator by monitoring its redox reaction. In whole-body experiments, there have been several recent reports noting the redox change in organisms in pathophysiological conditions such as oxidative stress and examining the effects on the redox reaction of nitroxyl in vivo. Here, we review the recent in vivo ESR studies on the redox reaction of nitroxyl and discuss the

Yuri Miura graduated with a degree in pharmaceutical sciences from the University of Tokyo. She was a Research Assistant at the School of Pharmaceutical Sciences, Showa University, and received her Ph.D. degree from the University of Tokyo in 1994. She completed a postdoctoral fellowship with Dr. Toshihiko Ozawa at the National Institute of Radiological Sciences. She is currently a Research Scientist at Tokyo Metropolitan Institute of Gerontology. Her research interests include radiation biology of neuronal and glial cells. Toshihiko Ozawa graduated with a degree in pharmaceutical sciences from the University of Tokyo. In 1974, he received his Ph.D. degree from the same university and performed his postdoctoral work with Dr. James P. Collman at Stanford University before becoming a Senior Researcher at the National Institute of Radiological Sciences. He was then a Section Head, and in 1993, he assumed the position of Director of the Department of Bioregulation Research. He is also a Visiting Professor at the Graduate School of Natural Sciences at Chiba University. His research interests include in vivo detection and electron spin resonance imaging of active oxygens and free radicals.

Address correspondence to: Toshihiko Ozawa, Department of Bioregulation Research, National Institute of Radiological Sciences, Chiba 263-8555, Japan; Tel: +81 (43) 206 3120; Fax: +81 (43) 255 6819; E-Mail: ozawa@nirs.go.jp

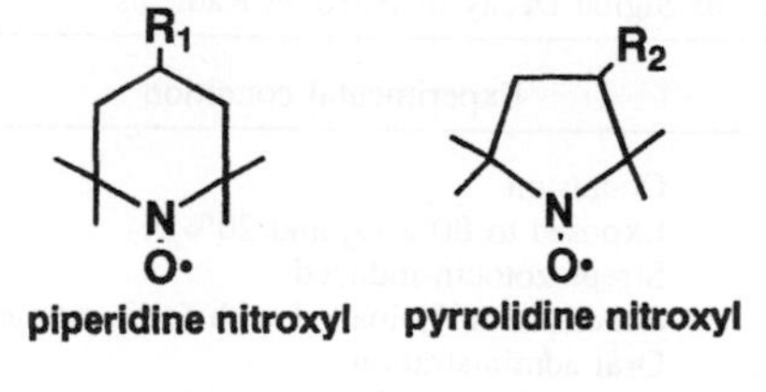

	Functional Group	Acronym
R_1	–OH	hydroxy-TEMPO
	$-N^+(CH_3)_4I^-$	Cat1
	$-NH_2$	amino-TEMPO
R_2	$-CONH_2$	carbamoyl-PROXYL
	–COOH	carboxy-PROXYL
	$-COOCH_3$	MCPROXYL

Fig. 1. Chemical structures of nitroxyl compounds.

possibility of applying in vivo ESR to the in vivo non-invasive study of oxidative damage.

PHARMACOKINETICS OF NITROXYL RADICALS INJECTED TO WHOLE BODY

In in vivo ESR measurement, the signal intensity of a nitroxyl radical decreases with time after injection. The pharmacokinetics of a nitroxyl compound (carbamoyl-PROXYL) using L-band (approximately 1 GHz microwave) ESR was first reported by Berliner and Wan [1]. They detected the nitroxyl signal of carbamoyl-PROXYL and measured the signal disappearance in rat tails. Since then, a several groups have reported spectra and the signal decay of nitroxyl radical in the abdomen or head of living animals by L-band ESR [2–7].

The signal decay of nitroxyl in the whole body is ascribed to two factors: one is the metabolism of nitroxyl to a diamagnetic molecule and the other is the diffusion and excretion from the measured region to other organs.

In vitro, there are many reports of the reduction of nitroxyl radicals to the corresponding hydroxylamine in microsomes, mitochondria, and whole cells. In vivo, nitroxyl compounds also seem to be reduced to hydroxylamine, based on the fact that the signal intensity of the collected blood was recovered by the addition of potassium ferricyanide, which oxidized hydroxylamine to nitroxyl radical [8]. Takeshita et al. [9] extracted the metabolite of hydroxy-TEMPO from mouse lung and identified it as its hydroxylamine by means of thin-layer chromatography. The reduction rates of nitroxyl depend on the chemical structure of nitroxyl compounds. Komarov and Lai [10] have measured the in vivo reduction kinetics of about 20 different nitroxyl compounds by S-band (3.5 GHz microwave) ESR. They examined the effect of the chemical structures of nitroxyl compounds on the half-life of in vivo reduction, showing that pyrrolidine nitroxyl was more resistant to cellular metabolism in vivo than piperidine nitroxyl. In vivo, the reducing enzymes and ascorbate should be involved in the reduction of a nitroxyl radical similar to what occurs in vitro. Vianello et al. [11] calculated the ascorbate contribution to various nitroxyl removals on the basis of the ascorbate concentration in organs and the second-order kinetic constants of nitroxyl reduction measured in phosphate-buffered saline. They speculated that the disappearance of piperidine nitroxyl in the brain was controlled mainly by ascorbate, although that of pyrrolidine nitroxyl was not.

On the other hand, Bacic et al. [3] have reported the detection of an increase in nitroxyl signal in the bladder of a mouse injected with Cat1 using L-band ESR. Takechi et al. [12] examined the urine of rats that were administered 3-carboxy-PROXYL, and their results suggested that parent nitroxyl and its hydroxylamine were excreted into the urine 2 to 10 h after administration. These data suggest that water-soluble nitroxyl compounds were excreted into the urine.

EFFECTS OF VARIOUS TYPES OF OXIDATIVE STRESS ON SIGNAL DECAY RATES OF NITROXYL RADICALS

Because various types of oxidative stress are known to change the redox state and metabolic capacity of organisms, it is expected that they would affect the pharmacokinetics of nitroxyl radical in vivo. Thus, the effects of oxidative stress or the administration of various drugs were studied by examining the signal decay rates of various nitroxyls [8,13–21] (Table 1). The results are divided into two categories depending on whether the treatment caused an increase or decrease in the signal decay rate of nitroxyl radicals. In cases in which signal decay was inhibited, the results were attributed to a decrease in the reducing capacity of the organism [8,20, 21]. In contrast, treatment with idebenone or chronic administration of vitamin C was concluded to enhance the radical reducing ability in the living body, resulting in an increase in the decay rates of nitroxyl [18,19]. In cases of oxidative stress such as ischemia-reperfusion, hyperoxia, diabetes, iron overload, and CCl_4 administration [13–17], it was speculated that a free radical reaction occurring during the oxidative damage may have been involved in the enhancement. This was because the enhancement was suppressed by in vivo antioxidants and because in vitro nitroxyl radical reacts with various ROS such as superoxide, hydroxyl radicals, and peroxyl rad-

Table 1. Effects of Various Treatments in Animals on the Rates of Signal Decay of Nitroxyl Radicals

Effect	Treatment	Measured region	Spin probe	Experimental condition	Reference
Increase	Ischemia-reperfusion	Femoral	Amino-TEMPO	Occlusion	[13]
Increase	Hyperoxia	Abdomen	Carbamoyl-PROXYL	Exposed to 80% O_2 and 20% N_2	[14]
Increase	Diabetes	Abdomen	Carbamoyl-PROXYL	Streptozotocin-induced	[15]
Increase	Iron overload	Abdomen	Carbamoyl-PROXYL	Subcutaneously loaded with ferric-citrate	[16]
Increase	CCl_4	Abdomen	Carbamoyl-PROXYL	Oral administration	[17]
Increase	Idebenone	Head	Carbamoyl-PROXYL	Intracerebroventricular injection	[18]
Increase	Vitamin C	Head	Hydroxy-TEMPO	Vitamin C–containing food	[19]
Decrease	Aging	Head	Carbamoyl-PROXYL	6, 30, and 39 month old mice	[8]
Decrease	Seizure	Head	Carbamoyl-PROXYL	Kainic acid–induced	[20]
Decrease	CCl_4	Hepatic and pelvic domain	Hydroxy-TEMPO	Intraperitoneal administration	[21]

icals, leading to the signal disappearance. The results of CCl_4 administration were different between the report of Inaba et al. [21] and that of Utsumi et al. [17]. Inaba et al. [21] injected CCl_4 intraperitoneally and measured in vivo ESR 48 h after injection, and Utsumi et al. [17] used oral administration and measured in vivo ESR 30 min after injection. CCl_4 injected into an organism is metabolized by cytochrome P-450 in liver microsomes, which accompanies the generation of ROS, causing hepatic damage and a decrease in metabolic capacity. It seems that there is a change in the degree of the damage between 30 min and 48 h after administration. Therefore, these results suggest that different stages of oxidative stress should yield different effects on the signal decay rate of nitroxyl radical.

EFFECTS OF RADIATION ON SIGNAL DECAY RATES OF CARBAMOYL-PROXYL IN THE ABDOMENS OF MICE

Radiation produces various ROS such as hydroxyl radicals, superoxide, and hydrogen peroxide in the whole body not only directly but also indirectly through a subsequent free radical reaction and inflammation, resulting in a change in the redox status of the organism. We have examined the effects of radiation on the decay rate of a nitroxyl radical (carbamoyl-PROXYL) using L-band ESR [22].

In experiments in noncysteamine-treated mice, the mice were separated into six groups. Groups 1 and 2 were treated by sham irradiation as a control. Groups 3 and 4 were treated by X-irradiation at a dose of 7.5 Gy, which is approximately the $LD_{50/30}$ of mice, and groups 5 and 6 were treated by 15 Gy irradiation. In vivo ESR measurement was performed 1 h after irradiation in groups 1, 3, and 5, and 4 or 5 d after irradiation in groups 2, 4, and 6. In cysteamine-treated mice, cysteamine was injected into the mice intraperitoneally 20 min before irradiation (2.0 mM/kg). Whole-body irradiation was performed at a dose rate of approximately 0.6 Gy/min. Sham irradiation of the controls included comparable immobilization in the same irradiation chamber. Anesthetic was not administered to either irradiated or sham-irradiated mice. The in vivo ESR spectra of the nitroxyl radical were measured as follows. A mouse was anesthetized using pentobarbital and placed in a loop-gap resonator. The solution of carbamoyl-PROXYL was injected into the tail vein, and ESR spectra were measured in the upper abdomen of the mouse repeatedly beginning immediately after injection. The rate constants of the signal decay of nitroxyl were calculated from the signal decay curves, which were determined from semilogarithmic plots of the peak heights of the ESR signal at the lower magnetic field.

Table 2 summarizes the kinetic constants of signal decay in the abdomens of the mice. One hour after

Table 2. Radiation Effects on the Signal Decay Rates (Gy/min) of Carbamoyl-PROXYL in the Abdomens of Mice

	1 h After irradiation		4 or 5 d After *irradiation	
	Without cysteamine	With cysteamine	Without cysteamine	With cysteamine
0 Gy	0.109 ± 0.015	0.102 ± 0.010	0.125 ± 0.031	—
7.5 Gy	0.130 ± 0.029	—	0.165 ± 0.029	0.138 ± 0.004
15.0 Gy	$0.145 \pm 0.021*$	0.100 ± 0.005	$0.075 \pm 0.008^\dagger$	—

* $p < .001$, different from that 1 h after 0 Gy (without cysteamine).
† $p < .05$, different from that 4 or 5 d after 0 Gy (without cysteamine).

irradiation, the signal decay rates increased by 15 Gy irradiation; 5 d after irradiation, these rates significantly decreased by 15 Gy irradiation. These data suggest that there are at least two factors that affect the signal decay rate of nitroxyl: one causing enhancement and another causing inhibition. Five days after irradiation, the mice exposed to 15 Gy irradiation were severely damaged; about half of the mice died within 5 d. The factor causing inhibition was believed to be the degeneration of the systemic condition, including metabolic and excretive capacities, as the result of high-dose irradiation.

To study the factor causing enhancement, we examined the effect of cysteamine, which is a radical scavenger and radioprotector, on the enhancement of the signal decay rate in nitroxyl (Table 1). It seemed that preadministration of cysteamine significantly suppressed the enhancement of the signal decay rate of nitroxyl as a result of X-irradiation. Other radioprotectors such as 5-HT, WR2721, hydroxy-TEMPO, IL-1β, and SCF also suppressed the enhancement of the signal decay rate of nitroxyl at the appropriate doses.

BIOLOGICAL MECHANISM OF THE ENHANCEMENT OF NITROXYL DECAY IN THE WHOLE BODY

What is the factor causing enhancement in this case? There are two possibilities: the induction of reducing capacity by oxidative stress and the participation of free radical reaction as a result of X-irradiation. We examined the activity of reducing enzymes (cytochrome P-450 and cytochrome P-450 reductase) and antioxidative enzymes (superoxide dismutase, catalase, and glutathione peroxidase) as well as the contents of endogenous antioxidants (vitamins E and C) under the present conditions. The results indicated that neither the activities of the reducing and antioxidative enzymes nor the contents of endogenous antioxidants increased 1 h after irradiation, suggesting that the reducing capacities in the mice were not induced 1 h after irradiation. Accordingly, free radical reactions in tissue caused by X-irradiation might be involved in the enhancement of the signal decay rate of nitroxyl. Although the lifetimes of primarily formed ROS are quite short as a result of X-irradiation, they should subsequently cause biological and chemical chain reactions, which, in turn, would produce fresh ROS. The fact that the radical scavenger cysteamine suppressed the enhancement seems to support the hypothesis that free radical reaction induced by X-irradiation participates in the enhancement of nitroxyl decay similar to what occurs in other types of oxidative stress [13–17]. Because in vivo ESR study is a whole-body experiment, however, physiological factors that affect the rate of tissue distribution or excretion of nitroxyl (e.g., blood pressure, blood flow rate, body temperature) cannot be ruled out.

EFFECT OF RADIATION ON SIGNAL DECAY RATES OF MCPROXYL IN THE HEADS OF MICE

The spin probe injected into peripheral blood cannot be distributed to brain tissue because of the blood brain-barrier (BBB). Thus, we synthesized BBB-permeable spin probe, and radiation damage to the brain was examined using in vivo ESR [23].

MCPROXYL was more lipophilic than carbamoyl-PROXYL and well distributed in brain tissue after intravenous injection. The signal decay rate of nitroxyl radical in the head region decreased 1 h after irradiation unlike that of carbamoyl-PROXYL in the upper abdomen. We examined the effect of radiation on the reducing activity of nitroxyl in the brain homogenate and the content of ascorbic acid in the brain, and the results showed that the reducing capacity was not decreased 1 h after irradiation. Although the biological mechanism of the radiation effect on MCPROXYL disappearance remains unclear, it is possible that the BBB-permeable spin probe might provide some information on radiation damage in the brain.

TOPICS

The analysis of in vivo ESR data is quite difficult because of the many factors affecting the signal decay rate of nitroxyl radicals in the whole body. Nevertheless, we can say for certain that the signal decay rate of nitroxyl radicals in vivo reflects the pathophysiological and/or physiological state of a living body. Therefore, we believe that in vivo ESR can be used for clinical applications so as to probe the change of the signal decay rate in nitroxyl.

Recently, there have been many in vivo ESR studies aiming to improve in vivo imaging. Nicholson et al. [24] have reported the in vivo imaging of nitroxyl clearance using a LODESR imaging apparatus to demonstrate that carboxy-PROXYL injected into rats shifted from the liver to the kidneys with time after injection. Yokoyama et al. [25] have reported ESR imaging for the signal decay of nitroxyl in the brains of rats. They used MCPROXYL, a BBB-permeable spin probe, and analyzed the effect of kainic acid–induced seizures on the signal decay rates in the hippocampus and cerebral cortex, showing that the half-life of nitroxyl radicals was significantly prolonged in the hippocampus but not in the cerebral cortex by kainic acid–induced seizure. In vivo imaging of the signal decay rate of nitroxyl should provide more information on the pathophysiology of the living body.

Conversely, it is also necessary to develop the spin probe, whose spectrum yields pathophysiological or physiological information on organisms. Here, we summarize the recent reports on such spin probes. Gallez et al. [26] have developed a pH-sensitive nitroxyl compound, which is manifested in the ESR spectrum as a decrease in hyperfine coupling constant with pH-induced change, and Sotgiu et al. [27] have performed pH-sensitive imaging using this nitroxyl. Dragutan et al. [28] also synthesized pH-sensitive spin probes. Using these spin probes, in vivo ESR can provide a noninvasive technique for monitoring pH in tissue or organs. Furthermore, Yamaguchi et al. [29] synthesized spin-labeled triglyceride (SL-TG) and analyzed in vivo ESR spectra of lipid emulsion containing SL-TG in the chests of mice. Immediately after administration, ESR signal derived from lipid particles was observed; after that, ESR signal derived from free and immobilized fatty acids to which lipoprotein lipase in the blood hydrolyzed lipid particles was superimposed on the spectra. These investigators demonstrated that in vivo ESR can determine the pharmacokinetics of lipid emulsion in a noninvasive fashion, suggesting that oxidative damage of lipoproteins or blood cells in the living body may be analyzed by this method.

CONCLUSIONS

Nitroxyl radicals injected into an organism indicate the disappearance of signal intensity of in vivo ESR. It is shown that the signal decay rates of nitroxyl depend on various types of oxidative stress, including X-irradiation. Thus, in vivo ESR could provide a noninvasive technique for the study of oxidative stress in a living body.

Acknowledgements — This study was supported in part by a Grant-in-Aid for Scientific Research (No. 10357021) from the Ministry of Education, Science, Sports, and Culture of Japan, and by a Grant from the Cosmetology Research Foundation.

REFERENCES

[1] Berliner, L. J.; Wan, X. In vivo pharmacokinetics by electron magnetic resonance spectroscopy. *Magn. Reson. Med.* 9:430–434; 1989.

[2] Ishida, S.; Kumashiro, H.; Tsuchihashi, N.; Ogata, T.; Ono, M.; Kamada, H.; Yoshida, E. In vivo analysis of nitroxide radicals injected into small animals by L-band ESR technique. *Phys. Med. Biol.* 34:1317–1323; 1989.

[3] Bacic, G.; Nilges, M. J.; Magin, R. L.; Walczak, T.; Swartz, H. M. In vivo localized ESR spectroscopy reflecting metabolism. *Magn. Reson. Med.* 10:266–272; 1989.

[4] Ferrari, M.; Colacicchi, S.; Gualtieri, G.; Santini, M. T.; Sotgiu, A. Whole mouse nitroxide free radical pharmacokinetics by low frequency electron paramagnetic resonance. *Biochem. Biophys. Res. Commun.* 166:168–173; 1990.

[5] Utsumi, H.; Muto, E.; Masuda, S.; Hamada, A. In vivo ESR measurement of free radicals in whole mice. *Biochem. Biophys. Res. Commun.* 172:1342–1348; 1990.

[6] Takeshita, K.; Utsumi, H.; Hamada, A. ESR measurement of radical clearance in lung of whole mouse. *Biochem. Biophys. Res. Commun.* 177:874–880; 1991.

[7] Miura, Y.; Utsumi, H.; Hamada, A. Effects of inspired oxygen concentration on in vivo redox reaction of nitroxide radicals in whole mice. *Biochem. Biophys. Res. Commun.* 182:1108–1114; 1992.

[8] Gomi, F.; Utsumi, H.; Hamada, A.; Matsuo, M. Aging retards spin clearance from mouse brain and food restriction prevents its age-dependent retardation. *Life Sci.* 52:2027–2033; 1993.

[9] Takeshita, K.; Utsumi, H.; Hamada, A. Whole mouse measurement of paramagnetism—loss of nitroxide free radical in lung with a L-band ESR spectrometer. *Biochem. Mol. Biol. Int.* 29:17–24; 1993.

[10] Komarov, A. M.; Lai, C. S. In vivo pharmacokinetics of nitroxides in mice. *Biochem. Biophys. Res. Commun.* 201:1035–1042; 1994.

[11] Vianello, F.; Momo, F.; Scarpa, M.; Rigo, A. Kinetics of nitroxide spin label removal in biological systems: an in vitro and in vivo ESR study. *Magn. Reson. Imaging* 13:219–226; 1995.

[12] Takechi, K.; Tamura, H.; Yamaoka, K.; Sakurai, H. Pharmacokinetic analysis of free radicals by in vivo BCM (blood circulation monitoring)-ESR method. *Free Radic. Res.* 26:483–496; 1997.

[13] Masuda, S.; Utsumi, H.; Hamada, A. In vivo ESR studies on radical reduction in femoral ischemia-reperfusion of whole mice. In: Yagi, K.; Kondo, M.; Niki, E.; Yoshikawa, T., eds. *Oxygen radicals.* Amsterdam: Elsevier, 1992:175–178.

[14] Miura, Y.; Hamada, A.; Utsumi, H. In vivo ESR studies of antioxidant activity on free radical reaction in living mice under oxidative stress. *Free Radic. Res.* 22:209–214; 1995.

[15] Sano, T.; Umeda, F.; Hashimoto, T.; Nawata, H.; Utsumi, H. Oxidative stress measurement by in vivo electron spin resonance spectroscopy in rats with streptozotocin-induced diabetes. *Diabetologia* 41:1355–1360; 1998.

[16] Phumala, N.; Ide, T.; Utsumi, H. Noninvasive evaluation of in vivo free radical reactions catalyzed by iron using in vivo ESR spectroscopy. *Free Radic. Biol. Med.* 26:1209–1217; 1999.

[17] Utsumi, H.; Ichikawa, K.; Takeshita, K. In vivo ESR measurements of free radical reactions in living mice. *Toxicol. Lett.* 82/83:561–565; 1995.

[18] Yokoyama, H.; Tsuchihashi, N.; Ogata, T.; Hiramatsu, M.; Mori, N. An analysis of the intracerebral ability to eliminate a nitroxide radical in the rat after administration of idebenone by an in vivo rapid scan electron spin resonance spectrometer. *MAGMA* 4:247–250; 1996.

[19] Matsumoto, S.; Mori, N.; Tsuchihashi, N.; Ogata, T.; Lin, Y.; Yokoyama, H.; Ishida, S. Enhancement of nitroxide-reducing activity in rats after chronic administration of vitamin E, vitamin C, and idebenone examined by an in vivo electron spin resonance technique. *Magn. Reson. Med.* 40:330–333; 1998.

[20] Ueda, Y.; Yokoyama, H.; Ohya-Nishiguchi, H.; Kamada, H. ESR spectroscopy for analysis of hippocampal elimination of a nitroxide radical during kainic acid-induced seizure in rats. *Magn. Reson. Med.* 40:491–493; 1998.

[21] Inaba, K.; Nakashima, T.; Shima, T.; Mitsuyoshi, H.; Sakamoto, Y.; Okanoue, T.; Kashima, K.; Hashiba, M.; Nishikawa, H.; Watari, H. Hepatic damage influences the decay of nitroxide radicals in mice—an in vivo ESR study. *Free Radic. Res.* 27:37–43; 1997.

[22] Miura, Y.; Anzai, K.; Urano, S.; Ozawa, T. In vivo electron paramagnetic resonance studies on oxidative stress caused by X-irradiation in whole mice. *Free Radic. Biol. Med.* 23:533–540; 1997.

[23] Miura, Y.; Anzai, K.; Takahashi, S.; Ozawa, T. A novel lipophilic spin probe for the measurement of radiation damage in mouse brain using in vivo electron spin resonance (ESR). *FEBS Lett.* 419:99–102; 1997.

[24] Nicholson, I.; Foster, M. A.; Robb, F. J. L.; Hutchison, J. M. S.; Lurie, D. J. In vivo imaging of nitroxide-free-radical clearance in the rat, using radiofrequency longitudinally detected ESR imaging. *J. Magn. Reson. (Series B.)* 113:256–261; 1996.

[25] Yokoyama, H.; Lin, Y.; Itoh, O.; Ueda, Y.; Nakajima, A.; Ogata, T.; Sato, T.; Ohya-Nishiguchi, H.; Kamada, H. EPR imaging for in vivo analysis of the half-life of a nitroxide radical in the hippocampus and cerebral cortex of rats after epileptic seizures. *Free Radic. Biol. Med.* 27:442–448; 1999.

[26] Gallez, B.; Mader, K.; Swartz, H. M. Noninvasive measurement

of the pH inside the gut by using pH-sensitive nitroxides. An in vivo EPR study. *Magn. Reson. Med.* **36**:694–697; 1996.

[27] Sotgiu, A.; Mader, K.; Placidi, G.; Colacicchi, S.; Ursini, C. L.; Alecci, M. pH-sensitive imaging by low-frequency EPR: a model study for biological applications. *Phys. Med. Biol.* **43**:1921–1930; 1998.

[28] Dragutan, H.; Caragheorgheopol, A.; Chiralen, F.; Mehlhorn, R. J. New amino-nitroxide spin labels. *Bioorg. Med. Chem.* **4**:1577–1583; 1996.

[29] Yamaguchi, T.; Itai, S.; Hayashi, H.; Soda, S.; Hamada, A.; Utsumi, H. In vivo ESR studies on pharmacokinetics and metabolism of parenteral lipid emulsion in living mice. *Pharm. Res.* **13**:729–733; 1996.

ABBREVIATIONS

ESR—electron spin resonance

ROS—reactive oxygen species

carbamoyl-PROXYL—3-carbamoyl-2,2,5,5-tetramethyl pyrrolidine-1-oxyl

hydroxy-TEMPO—4-hydroxy-2,2,6,6-tetramethyl piperidine-1-oxyl

Cat1—4-trimethylamino-2,2,6,6-tetramethyl piperidine-1-oxyl

carboxy-PROXYL—3-carboxy-2,2,5,5-tetramethyl pyrrolidine-1-oxyl

MCPROXYL—3-methoxymethyl-2,2,5,5-tetramethyl pyrrolidine-1-oxyl

CCl$_4$—carbon tetrachloride

LD$_{50/30}$—50% lethal dose for 30 d

5-HT—5-hydroxytryptamine

WR2721—S-2-(3-aminopropylamino)ethylphosphorothioic acid

IL-1β—interleukin 1β

SCF—stem cell factors

LODESR—longitudinally detected ESR

Bioassays for Oxidative Stress Status (BOSS). Edited by W.A. Pryor
© 2001 Elsevier Science B.V. All rights reserved.

THE USE OF CYCLIC VOLTAMMETRY FOR THE EVALUATION OF ANTIOXIDANT CAPACITY

SHLOMIT CHEVION,* MATTHEW A. ROBERTS,[†] and MORDECHAI CHEVION*[†]

*The Hebrew University of Jerusalem, Jerusalem, Israel and [†]Nestle Research Center, Lausanne, Switzerland

(Received 3 September 1999; Revised 11 November 1999; Accepted 11 January 2000)

Abstract—Low–molecular weight antioxidants (LMWAs) play a major role in protecting biological systems against reactive oxygen–derived species and reflect the antioxidant capacity of the system. Cyclic voltammetry (CV), shown to be convenient methodology, has been validated for quantitation of the LMWA capacity of blood plasma, tissue homogenates, and plant extracts. Analysis of the CV tracing yields the values of (i) the biological oxidation potential, E and $E_{1/2}$, which relate to the nature of the specific molecule(s); (ii) the intensity (Ia) of the anodic current; and (iii) the area of the anodic wave (S). Both Ia and S relate to the concentration of the molecule(s). LMWA components of human plasma and animal tissues were identified and further validated by reconstruction of the CV tracing and by high-performance liquid chromatography–electrochemical detection. To reflect the oxidative stress status, the use of an additional parameter, R, has been proposed. R represents the level (%) of oxidized ascorbate (compared with total ascorbate) and is measured by high-performance liquid chromatography–electrochemical detection. All these parameters were monitored in healthy human subjects as well as in chronic (diabetes mellitus) and acute care patients (subjected to total body irradiation before bone marrow transplantation). The electroanalytical methodologies presented here could be widely employed for rapid evaluation of the status of subjects (in health and disease) for monitoring of their response to treatment and/or nutritional supplementation as well as for screening of specific populations.   © 2000 Elsevier Science Inc.

Keywords—Free radical, Low–molecular weight antioxidant, Cyclic voltammetry, Ascorbate, Dehydroascorbate, Plant antioxidants, Diabetes, Total body irradiation, Total antioxidant capacity

INTRODUCTION

Living cells, including those of man, animals, and plants, are continuously exposed to a variety of challenges that exert oxidative stress. These could stem from endogenous sources through normal physiological processes such as mitochondrial respiration. Alternatively, they could result from exogenous sources such as exposure to pollutants and ionizing irradiation.

Oxidative stress arises in a biological system after an increased exposure to oxidants, a decrease in the antioxidant capacity of the system, or both. It is often associated with or leads to the generation of reactive oxygen species (ROS), including free radicals [1], which are strongly implicated in the pathophysiology of disease [2,3].

Cells are equipped with several defense systems,

Shlomit Chevion received her M.A. degree from Villa Nova University (Mathematics) and her Ph.D. degree from The Hebrew University of Jerusalem (Medical Sciences), studying the antioxidant status of human subjects in health and disease states. She was a postdoctoral fellow at the Beltsville Human Nutrition Research Center, United States Department of Agriculture, Beltsville, MD.

Matthew A. Roberts obtained a B.S. degree from Purdue University (1991) and then a Ph.D. degree from Cornell University (1996), specializing in analytical toxicology. He received postdoctoral training at the Laboratory for Electrochemistry, Ecole Polytechnique Federal de Lausanne, and he now works at the Nestle Research Center for Food and Life Science in Lausanne, Switzerland.

Mordechai Chevion got his formal education at The Hebrew University of Jerusalem. He is the incumbent of the "Dr. W. Ganz Chair of Heart Studies" at The Hebrew University of Jerusalem. Recently, he served as a visiting scientist at the National Heart, Lung, and Blood Institute at the National Institutes of Health in Bethesda, MD, and he is now on sabbatical leave at Nestle Research Center.

Address correspondence to Professor Mordechai Chevion, The Dr. William Ganz Chair of Heart Studeies, The Hebrew University of Jersualem, The Hebrew University–Hadassah Schools of Medicine and Dental Medicine, Ein Kerem Campus, P.O. Box 12272, Jerusalem IL-91120, Israel; Tel: +972 (2) 675-8158 or 8160; Fax: +972 (2) 641-5848, E-Mail:chevion@cc.huji.ac.il.

which act through a variety of mechanisms [1–4]. These can be categorized into protection via enzymatic activities and protection through low–molecular weight antioxidants (LMWAs) [5]. The LMWA family consists of many compounds, each of which acts as a direct chemical scavenger neutralizing ROS component(s) [4] or indirectly through transition metal chelation [6]. LMWAs are small molecules that frequently infiltrate cells, accumulate (at high concentrations) in specific compartments associated with oxidative damage, and then are regenerated by the cell [1]. In human tissues, cellular LMWAs are obtained from various sources. Glutathione (GSH), nicotinamide adenine dinucleotide (reduced form), and carnosine [7] are synthesized by the cells; uric acid (UA) [8] and bilirubin [9] are waste products of cellular metabolism; and ascorbic acid (AA) [10] tocopherols and polyphenols are antioxidants obtained from the diet.

Edible plants constitute a major nutritional source for LMWAs in mammals, including man. Consumption of fresh fruits and vegetables has been shown to be associated with lower incidences and lower mortality rates of cancer, heart diseases and atherosclerosis, brain degenerative diseases, aging, and other pathologies [11–18].

Plasma is often used for the evaluation of (free radical) induced damage. Plasma contains critical targets for oxidative damage such as lipoproteins, and low-density lipoproteins in particular. It also contains important antioxidants such as AA and UA and represents a biologically relevant milieu. It reflects the integrated antioxidant status, including the recent "history" of oxidative stress (see below), which arises from both body tissues (synthesis and reservoirs) and nutrition. Hence, to evaluate the interrelations between disease, diet, free radicals, and vitamin supplementation, methods based on the characterization of plasma parameters are considered to be bona fide representatives of the antioxidant status of the whole organism.

Several methods for measuring the total antioxidant capacity of a biological sample, including blood plasma, body fluids, and plant extracts, have been proposed and recently reviewed [19–22]. These are related to the capacity of a sample to compete for and scavenge a specific ROS. They provide useful information, which is not sufficient for the evaluation of the overall antioxidant profile of the biological fluid, tissue homogenate, or plant extract.

We have previously suggested the use of cyclic voltammetry (CV) as an instrumental tool for the evaluation of the total antioxidant capacity of the LMWAs of human and animal plasma, other body fluids, and animal tissue homogenates [23–27]. Likewise, we have used CV for the evaluation of the total antioxidant capacity of edible plants [28]. A cyclic voltammogram (CV tracing) provides information describing the integrated antioxidant capacity, which arises in large part from the LMWAs, without the specific determination of the contribution of each individual component. It is based on the analysis of the anodic current (AC) waveform (Fig. 1), which is a function of the reductive potential of a given compound in the sample and/or a mixture of compounds. The CV tracing indicates the ability of the compound to donate electron(s) around the potential of the anodic wave. Most of the LMWAs are reducing agents, which quench ROS through electron donation. Therefore, evaluation of the overall reducing power of a biological sample by CV would reflect its LMWA activity.

The total antioxidant capacity of the sample is a function combining two sets of parameters. The first is the biological oxidation potentials (OPs), characterized by the $E_{1/2}$ value (Figs. 1B and 1C), which reflect the specific reducing power of a component (or components with similar potential). The second is the intensity of the AC current (Ia), reflecting the concentration of the component(s). Integration yields a value equivalent to the total antioxidant capacity of the sample, without the necessity of determining the antioxidant activity of each of its components.

Recently, we have proposed using the area under the AC wave (S; related to the charge) rather than the Ia as a better parameter reflecting the total antioxidant capacity of a sample [28]. This provides a marked advantage in some cases, particularly when an AC wave represents more than a single component. Changes in S better reflect a change in a single component within such an anodic wave than the corresponding change in Ia.

The CV of plasma could provide information about the in vivo exposure of the subject to oxidative stress (before blood collection) and about the antioxidant pool of the subject. It can also be used for monitoring the antioxidant status of patients and the success (or failure) of the treatment they receive. Likewise, CV allows the determination of the total antioxidant capacity of edible plants before ingestion and could be used for food quality monitoring over a product shelf life.

CV METHODOLOGY

Principle

CV methodology is used extensively in electroanalytical chemistry to determine redox properties of molecules in solution. Experimentally, the potential of a working electrode is linearly scanned (vs. a reference electrode, typically Ag/AgCl) from an initial value to a final value and back while recording the AC (Ia). Figure 1A shows CV of a ferrocyanide standard. The current is obtained when the potential excitation signal (in biolog-

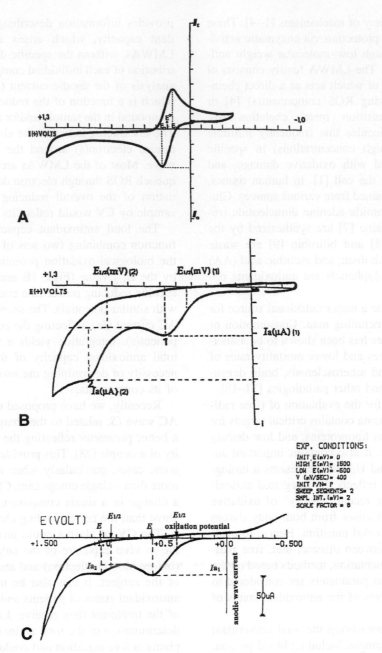

Fig. 1. Analysis of CV tracing. (A) Potassium ferrocyanide in the range of −0.3 to 1.3 V at a rate of 100 mV/s versus Ag/AgCl reference electrode. (B) Healthy human plasma in the range of −0.3 to 1.3 V at a rate of 100 mV/s versus Ag/AgCl. (C) Edible plant (cauliflower) in the range of −0.5 to 1.5 V at a scan rate of 400 mV/s versus Ag/AgCl. Although Ea is the OP at the peak of Ia, the standard and compound parameter $E_{1/2}$ has been used to identify a component and to determine its ability to donate reducing equivalents. $E_{1/2}$ represents the OP at Ia / 2, at half the height of the peak of the AC wave.

ical systems, typically from −0.5–+1.5 V) is applied to the working electrode. As the potential is scanned positively (lower part of the trace), the electrode becomes sufficiently positive to oxidize Fe^{+2} to Fe^{+3} and gives rise to the recorded anodic current (Ia). The Ia increases rapidly until the reductive potential is reached and then falls off as Fe^{+2} is depleted at the anodic surface. At 1.3 V, the scan direction is switched to negative (upper part of scan), giving rise to the analogous cathodic current (Ic), completing the (first) cycle. Thus, CV is a method capable of rapidly generating a new species during the forward scan and then monitoring its fate on the reverse scan.

The important parameters obtained from a cyclic voltammogram are the Ia, Ic, anodic OP (Ea), and cathodic OP (Ec). All these values can be readily obtained from the voltammogram. The peak current, I, for an electrochemically reversible system (rapid transport of electrons on the surface within the framework of the experiment) is described by the Randles-Sevcik equation: $I = (2.69 \times 10^5)n^{3/2}$ A $D^{1/2}$ $V^{1/2}$ C, where I is the peak current in amperes, n is the electron stoichiometry in equivalents per mole, A is the electrode area in square centimeters; D is the diffusion coefficient in square centimeters per second, C is the concentration in moles per cubic centimeter, and V is the potential scan rate in volts per second.

In an electrochemically reversible couple, both species are stable and rapidly exchange electrons with the working electrode. The formal redox potential ($E^{\circ\prime}$) for such a reversible couple is centered between Ea and Ec of the peak (Fig. 1A):

$$E^{\circ\prime} = (Ea + Ec)/2$$

The number of electrons transferred (n) during the electrode reaction for a reversible couple can be determined from the separation between the anodic and cathodic peak potentials by:

$$Ea - Ec = 0.059/n$$

Instrument

Analyses of samples by CV could be carried out on a variety of electrochemical analyzers. When an analog instrument is used, special attention must be given to determination of the scan range as well as the start point. This prevents a horizontal shift of the biological OPs as was mistakenly reported [29–31].

Typically, a modified low-volume cell is used, and it contains a ~0.5 to 2.0 ml sample. A three-electrode system is employed: (i) the reference electrode (Ag/AgCl), (ii) the working electrode (a glassy carbon disc [BAS] 3.2 mm in diameter), and (iii) a counterelectrode (platinum wire). CV tracings are recorded from −0.5 to +1.5 V versus the reference electrode at a scan rate of 100 or 400 mV/s.

Dependency of Ia and S on the concentration

The CV tracings in Fig. 2 show the dependency of the peak current of the anodic wave (Ia) as well as the area under the anodic peak (S) as a function of the concentration of commonly used standards AA and trolox. Both parameters have a linear relation with the concentration.

For pure AA and UA in phosphate buffer, $E_{1/2}$ was 380 and 420 mV, respectively [23,24,26]. This order is in accord with their antioxidant capacity. AA, which is a more potent antioxidant, donates its electrons at a lower OP. Assuming that AA and UA are its sole major components, the maximal change in the OP of the first anodic

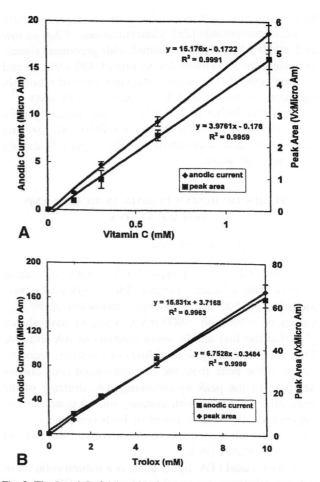

Fig. 2. The Ia and S of AA and trolox versus their concentrations. (A) AC amplitude (Ia) and peak area S for increasing concentrations of ascorbate (0–1.25 mM) in water/acetic acid/acetonitrile mixture (40/30/30), pH 1.9. (B) Ia and S for increasing concentrations (0–10 mM) of trolox in 0.2 M of phosphate buffer, pH 7.4.

wave could be 40 mV, which is the difference between the OP of AA and UA.

As an AC wave of a biological sample often represents more than a single component, each of which could donate electron(s) around the same potential, we have proposed [28] that S could prove more useful than Ia for monitoring changes in these components. An example is the AC wave of plasma, which has an $E_{1/2}$ of ~400 mV. We have shown [23,24,26] that this anodic peak is comprised of AA and UA. The changes in the concentration of one of these components, particularly AA, which is a minor component (by concentration), could be better monitored through the changes in S rather than by Ia. As the typical potentials, $E_{1/2}$, for AA and UA are similar but different, the change in the concentration of AA significantly affects the width of the anodic peak and S, although only marginally affecting Ia.

Higher CV scan rates give rise to higher Ia; Ia is proportional to the scan rate$^{1/2}$ according to the Randles-

Sevcik equation and as has been shown experimentally for these compounds [28]. Concentrations of AA as low as 5 to 10 μM can be measured with precision (signal-to-noise ratio > 5) using a scan rate of 400 mV/s. *Ia* and *S* values can also be used for quantitation of relatively high concentrations of AA (verified up to 10 mM).

Phosphate buffer (0.2 M, pH 7.4) and solvent systems used, including water/acetic acid (70/30, pH 1.9) and water/acetic acid/ acetonitrile (40/30/30, pH 1.9), did not show any AC wave.

STUDIES OF HUMAN PLASMA IN HEALTH AND DISEASE STATES

CV tracing of human plasma

Figure 1B shows a typical CV tracing of human plasma from a healthy subject. Two major peaks were identified at 500 and 1000 mV (characterized by $E_{1/2}$ values of ~400 and ~900 mV). Also, we have shown [24] that the first anodic wave consists of AA and UA. This was verified by several lines of evidence: reconstitution of the peak from the two individual components, decrease of the peak by oxidizing AA, decrease of the peak after treatment with uricase, and its complete disappearance after the removal of both components. The tracing of such a peak can be deconvoluted into two (or more) AC waves [26].

Thioctic acid (TA, lipoic acid) is a natural cofactor of α-keto dehydrogenases whose biological OP is at the same range as the second anodic wave [25]. The AC *Ia* was shown to depend linearly on the concentration of added TA in both buffer and human plasma. Also, diabetic patients, who ingest TA daily, displayed a higher AC in the second wave than normal controls or patients who did not receive the supplement. The detailed composition of this AC wave still awaits further investigation. GSH, a major intracellular antioxidant, was considered to be a contributor to the second peak. It was found that GSH, cysteine (also as a representative of other sulfhydryl-containing components), and oxidized glutathione demonstrated a low CV response and could not be studied by this methodology [26]. Using ultramicroelectrodes (see below) (with increased electrode sensitivity) or other specially designed electrodes (with high selectivity) may allow the monitoring of GSH in tissue homogenates. Leakage of reduced nicotinamide adenine dinucleotide phosphate (NADPH) from red blood cells has been considered and ruled out as a contributor to the second anodic wave. NADPH also demonstrated a relatively low CV response, and this second wave is present in all samples, including those that were carefully handled and did not show any other sign of leakage or hemolysis (i.e., hemoglobin levels below 3 μM).

Often, changes in the pH of a sample cause alterations in its redox properties as reflected by a horizontal potential (voltage) shift [24,28]. Thus, it is important to characterize the appropriate OP according to the pH of the sample. This should be of concern when deproteination of a biological sample (by acid or organic solvent) is employed, particularly for plant samples, which are often acidic.

An important practical consideration is that the working electrode has to be intensively polished before each measurement so as to remove biological residues from its surface and to maintain its sensitivity.

Special attention should be given during the collection of plasma samples. Ethylenediaminetetraacetic acid, which is often used as an anticoagulant in blood samples, exhibits its specific AC wave at around 900 mV and thus interferes with the second AC wave of the plasma. Conversely, heparin is CV-silent [23,24].

Changes in the CV tracing of plasma subjected to oxidative stress

Oxidative stress was exerted on plasma in vitro [23, 24]. It was expected that challenges by ROS would lead to a decrease in the antioxidant capacity of the plasma and that this, in turn, would be reflected by a parallel decrease in *Ia*. Plasma samples were exposed to

1) Peroxyl radicals: Incubation of plasma samples at 37°C with AAPH [2,2'-azobis(2-amidinopropane) dihydrochloride] led to reciprocal time dependencies of *Ia* on the concentration (0–200 mM) of AAPH and on the duration of incubation (0–3 h). In general, a linear dependency was found between log *Ia* and the extent of the various types of oxidative stress [25].
2) Copper ions: Incubation of plasma samples with copper(II) sulfate for 1 h showed reciprocal time dependencies between *Ia* and [Cu(II)] (0–400 μM) and *Ia* and the duration of incubation (0–3 h).
3) Ionizing irradiation: Plasma samples were subjected to ionizing irradiation (0, 1200, 100,000, and 1,000,000 cGy).

Table 1 shows the marked decrease in the two AC waves as determined by CV. It also shows the selective decreases in the levels of AA and UA. It is noteworthy that the changes observed in the two AC waves were specific to the type of oxidative stress. No significant changes in $E_{1/2}$ of the CV tracing were found.

We have systematically characterized the use of the AC parameters *Ia* and *S* rather than $E_{1/2}$ alone as the major parameters determining the antioxidant capacity of plasma LMWAs as reflected by CV [23,24,28].

After a broad variety of sample treatments, the

Table 1. Changes in Levels of LMWA After Oxidative Stress as Evaluated by Ia of CV and by HPLC-ECD

Inducer of oxidative stress	Concentration	Percentage of change by CV		Percentage of change by HPLC	
		at 420 mV	at 900 mV	AA	UA
AAPH	200 mM	-92 ± 2	-48 ± 3	-96 ± 3	-83 ± 2
CuSO$_4$	400 μM	-31 ± 2	-31 ± 2	-94 ± 2	No change
γ-irradiation	1,000,000 rad	-79 ± 3	-27 ± 2	-32 ± 3	-92 ± 4

Stress with AAPH, CuSO$_4$, H$_2$O$_2$, and γ-irradiation was exerted on human plasma samples as described. The CV tracings were recorded and analyzed, and the changes were evaluated from the AC heights at around 400 and 900 mV. AA and UA were analyzed by HPLC-ECD using an amperometric detector (with an electrode potential of +0.5 V vs. an Ag/AgCl reference electrode).

changes observed were in the values of Ia and reflect analogous changes in the concentrations of specific LMWAs. Although changes in $E_{1/2}$ are not expected to occur with age [26] or as a result of treatment [24–27], this claim has been the main focus of a series of CV studies [29–31]. This claim was probably based on measurements using an analog instrument, which could have led to inaccurate determinations of $E_{1/2}$ as can be identified in the published CV tracings (Figs. 1–3 in a report by Kohen et al. [29]). "Oxidation of saliva leads to a higher potential recorded by CV; the older the donor, the higher the potential recorded, indicating a lower ability of the saliva to donate electron" [29]. Similar results were reported for rats undergoing total body irradiation (TBI): "the anodic peaks shown in the CV were shifted to higher potentials" as presented in Fig. 2 of another report by Kohen et al. [30]. Likewise, "the induction of oxidative stress results in higher OP values recorded by the CV in three biological systems tested (*Escherichia coli* cells, rat jejunal mucosa, lactate dehydrogenase)" [31]. It is unfortunate that in a recent review [32], these authors added confusion by referring to their old work without clarifying the change in perception.

Antioxidant status of healthy subjects

To determine the effects of vitamin C on cardiovascular risk factors, the effect of dietary enrichment with vitamin C was studied in 40 young healthy Yeshiva students consuming a high-saturated fatty acid diet [18, 26]. After a 1 month run-in period during which they consumed 50 mg/d of AA (Lo-C), half of the subjects were randomized to receive 500 mg/d of AA (Hi-C) for a further 2 months, although the other half remained on the original diet (Lo-C). Plasma AA increased from 13 to 52 μmol/L ($p < .01$) on the Hi-C diet. At the same time, we monitored an additional plasma parameter of these students, R, which we consider to be a marker of the oxidant stress:

$$R = [DHAA] / \{[AA] + [DHAA]\}$$

The ratio between the oxidized form of AA, dehydroascorbate (DHAA), and total AA significantly decreased in this group, indicating a better antioxidant status. In the Lo-C diet group, the level of AA, which decreased during the run-in period from 21 to 13 μM/L, remained at this level, although the R ratio remained high. An increase in the lag period of the in vitro oxidation of low-density lipoproteins was observed and correlated well with the plasma AA concentrations. The changes in the antioxidant capacity of the plasma by CV showed a high positive correlation between AA levels and the Ia of the first AC wave. Because the level of UA in plasma is up to 10 times higher than the level of AA, and as they are both components of the first AC wave, we aimed to eliminate the contribution of UA by treating the plasma with uricase and then recognizing the difference. Addition of excess cysteine to the plasma after removal of UA allows the determination of total AA and DHAA. Subsequently, R can be determined by CV and yields results similar to the values obtained by high-performance liquid chromatography–electrochemical detection (HPLC-ECD).

Antioxidant status of patients

Two groups of patients were studied. The first consisted of patients subjected to TBI before bone marrow transplantation (BMT) [26,27], and the other consisted of patients suffering from diabetes mellitus (DM) [25,26].

TBI. TBI is a routine preconditioning procedure for the treatment of leukemia and aplastic anemia before BMT. Ionizing irradiation generates ROS that react with cellular components, inactivating bone marrow cells. Plasma of 14 patients undergoing TBI before BMT was examined and evaluated by CV for the antioxidant status of these patients (Fig. 3). In addition, the levels of AA and UA were determined by HPLC-ECD, and R was calculated. The antioxidant capacity (as reflected by Ia of the CV tracings) was lower by 36% ($p < .02$) than that of

Table 2. Total Amounts of LMWA in the Edible Plants Tested, Expressed as Equivalents of AA per 100 g of Plant Sample[a]

Plant	n	Peak 1		Peak 2		Peak 3		Vitamin C (mg) equivalent units	Vitamin C (mg) USDA[b]
		E (mV)	$\Sigma_i (S_i \times V_i)$	E (mV)	$\Sigma_i (S_i \times V_i)$	E (mV)	$\Sigma_i (S_i \times V_i)$		
Broccoli	5	654 ± 12	954.0	1261 ± 13	385	—	—	117.5	93.2
Cauliflower	5	568 ± 31	840.0	1255 ± 15	534	—	—	120.5	46.4
Strawberry	2	749 ± 27	1934.0	—	—	—	—	169.6	56.7
Tomato	4	588 ± 63	186.0	1017 ± 30	74	—	—	22.8	19.1
Potato	4	612 ± 16	620.0	1225 ±16	777	—	—	122.5	19.7[c]
Corn	4	615 ± 11	7.4	1279 ± 12	135	962 ± 14	3.3	13.0	6.8

[a] The sum of the products of S multiplied by the volume of the extract of the edible plants tested.

[b] *USDA Nutrient Database for Standard Reference*, edn. 12 [33]. The actual vitamin C content in each of the edible plants.

[c] USDA value represents the content of vitamin C in the flesh and skin of raw potato, although we have used only the flesh of raw potato.

The CV instrument automatically calculated the values of E and S for each AC wave (peak).

i = 1 − n, the number of extraction steps; E, = OP; S = area of AC wave; V, = volume of sample.

V was normalized to 100 g of plant equivalent. The product was calculated from the area, S, of the CV tracing and the volume of the extract for each step.

Vitamin C equivalent units (mg of vitamin C per 100 g of sample) represent the actual vitamin C in addition to the other LMWAs in the plant extract.

normal matched controls; 4 months after the TBI treatment, *Ia* recovered to a level 22% higher than before the treatment ($p < .05$). Both AA and UA decreased after irradiation by 84 ($p < .02$) and 24% ($p < .05$), respectively, but returned to a level of 21 and 320% after 4 months compared to the baseline values. The changes in [UA] were probably affected by allopurinol (a xanthine oxidase inhibitor) given as routine pretransplant therapy until day −1. The R value was 45%, and the range for normal controls was 13.2 ± 1.5%. The R value increased by 69% after TBI.

It is interesting to note that to induce a decrease in antioxidant capacity in vitro in plasma (as indicated by *Ia*), which is comparable with the in vivo decrease (similar to that demonstrated for the patients), a 1000-fold higher dose of irradiation was required. TBI alters antioxidant homeostasis and dramatically enhances the damage inflicted on the cells. Employing CV measurements has led to a better understanding of the balance between oxidative stress and antioxidant use as well as to a reconsideration of the routine use of allopurinol as a pretreatment for TBI patients and to the prescription of antioxidant support before and/or after TBI treatment.

DM. Ninety-two DM patients were included in a study. Their blood samples were analyzed by HPLC for the levels of AA, DHAA, and UA and by CV for their antioxidant capacity (by *Ia*). The patients were divided into groups according to the severity of their nephropathy or according to the severity of their total complication count. Severe DM patients who were under treatment with TA showed a higher level of the second AC wave than those not taking the drug. The level of AA was found to be lower with increasing albuminuria and with a complication score over 60%. Patients with severe DM

without TA had significantly higher R values ($p < .02$), although for the group taking TA, R values were not different from those of nondiabetic controls. Generally, when comparing the two groups of patients with the same number of complications, those who did not receive TA had significantly lower AA levels combined with higher R values than those taking TA. Thus, based on the use of these parameters, it could be demonstrated that the antioxidant status of these patients is significantly improved by treatment with TA. This positive correlation between TA and AA, and the negative correlation between AA and R values, strongly indicates that TA or other antioxidant(s) should be considered as drugs for improving the antioxidant status of DM patients.

STUDIES OF EDIBLE PLANTS BY CV

The evaluation of the total antioxidant capacity of edible plants has been studied [28]. Vegetables and fruits that are often consumed in the United States have been examined. The LMWAs were extracted from these edible plants. It was found that complete extraction of LMWAs could be obtained typically by three to five steps using water/acetic acid/acetonitrile (40/30/30) ("extraction solution"). The recovery rate of ascorbate was 93%.

Samples weighing 50 to 100 g were used, and the total content of LMWAs was evaluated from S values of the CV tracing and then translated into equivalents of ascorbate in 100 g of edible plant (Table 2). The CV revealed up to three AC waves. Figure 1C shows a typical CV tracing of a plant sample (cauliflower). Table 2 also shows the plant antioxidant equivalents calculated from the area of the peaks and expressed as vitamin C equivalent units (milligrams of vitamin). This represents the actual vitamin C as well as the other LMWAs in the plant extract. For comparison, the last column of Table 2 depicts the actual levels of vitamin C in these edible plants as published by the United States Department of Agriculture [33]. The total antioxidant capacity of each plant (vitamin C equivalent units) was higher than the vitamin C levels. The difference between the last two columns of Table 2 could reflect the antioxidant capacity of the other LMWAs (except ascorbate) in each plant.

STUDIES OF ANIMAL TISSUES BY CV

Recent reports [30,34,35] have documented the CV tracings of animal tissues. In the rat, two anodic waves were identified in aqueous homogenates of brain (324 and 867 mV), heart (350 and 834 mV), liver (366 and 904 mV), lung (419 and 897 mV), kidney (490 and 921 mV), and intestine (486 and 941 mV) [34]. In rat skin homogenate, a third anodic wave was observed (426, 853, and 1078 mV). AA and UA were identified as the major constituents of the first anodic wave in brain homogenate, using an approach analogous to the methodology that we have shown [24]. Also, spiking brain samples with relatively high concentrations (0.5 mM) of other antioxidants (carnosine, tryptophan, and melatonin) yielded an increase in the amplitude of the second anodic wave [34]. These findings suggest but do not provide "conclusive evidence" for possible contributors to the second anodic wave. Lipoic acid (TA) has also been shown to increase the second anodic wave [34] as suggested in a previous report by Chevion et al. [25]. Lipophilic extracts of brain tissue showed two or three anodic waves (at 200–250, 850–900, and approximately 1100 mV) [34].

The antioxidant status of various organs after closed-head injury was studied by CV. Changes in the total reducing capacities of water-soluble (phosphate-buffered saline homogenates) and lipid extracts (acetonitrile/methanol with *tetra*-butyl ammonium perchlorate) were detected at 5 min [35] and at 1 and 24 h [34] after the injury. These reflected analogous changes in the antioxidant level.

FUTURE DEVELOPMENTS FOR CV OF ANTIOXIDANT STATUS

Use of ultramicroelectrodes

It has been shown here that CV is able to determine general antioxidant status as well as that of specific biological compounds such as AA and UA. All the experiments so far described have used standard elec-trode geometry or microelectrodes[1] (i.e., electrodes with dimensions in the millimetric scale). These types of electrodes are easily fabricated and/or can be obtained through commercial sources, which makes them a practical solution for many laboratories. It has been known for some time that ultramicroelectrodes, electrodes with dimensions in the micrometric or submicrometric scale, have a number of inherent measurement advantages, however [36]. As the general importance and public awareness of antioxidant status continues to grow, the need for more sophisticated, sensitive, and rapid instrumentation can easily be imagined. This need will provide an opportunity to apply new types of electrode configurations for enhanced analytical performance.

Measurement advantages are realized with electrodes of extremely small dimensions because of the theoretical relations between electrode geometry and ionic diffusion. The micron-sized dimensions of ultramicroelectrodes allow "three-dimensional" or hemispherical diffusion as opposed to planar diffusion, which is characteristic of much larger electrodes. Therefore, scaling down the size of an individual electrode has the advantage of increasing the rate of mass transport, increasing the signal-to-noise (Faradaic/charging current) ratio, and decreasing ohmic signal loss [37]. Furthermore, it is possible to connect large numbers of ultramicroelectrodes "fingers" into single but separate working electrodes and counterelectrodes [38]. If these finger sets are then interdigitated as shown in Fig. 4, the resulting pattern is termed an *interdigitated array* (IDA) [39]. IDA configurations have been shown to provide a number of advantages in electrochemical detection, including rapid response times, high signal-to-noise ratio, and the ability to use extremely small sample volumes [40]. Of particular importance to IDA electrode designs is the ability to induce redox cycling of reversible test species. Because of the close proximity of cathode and anode finger sets, molecules may be alternatively oxidized and reduced, thereby generating more signal electrons than n per molecule (an approximately 10–200-fold increase in sensitivity can be achieved depending on geometry, electrochemical technique, and analyte chemistry) [41–43]. Furthermore, the sophisticated electronics usually needed to detect the extremely small currents associated with individual ultramicroelectrode filaments are not necessary because of the summation of current from the large array of electrodes. It has also been reported that IDAs can be used for fast-scan voltammetry (several volts per second) [38,39]. Normal scan rates (10–100 mV/s) do not require extensive analysis time for single

---

[1] It is of historical significance that the term *microelectrode* does not refer to a scientific dimension but simply means that this class of electrodes are smaller than the large instrumental setups used in the early part of this century.

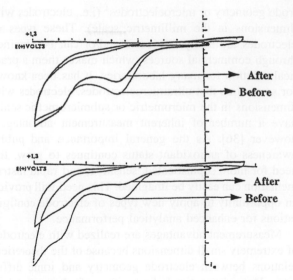

Fig. 3. CV of plasma of patients before and after receiving TBI before BMT. The range is from -0.3 to 1.3 V at a rate of 100 mV/s versus the Ag/AgCl reference electrode. The plasma was collected in heparin. Significant changes can be seen only in the Ia and not in the OP $E_{1/2}$ [27].

point measurements. In automated flow-based system such as HPLC or flow-injection analysis systems, however, many measurements need to be taken over time and integrated after analysis; therefore, fast scanning may be yet another advantage of the IDA configuration.

Such enhanced electrode design would offer marked antioxidant measurement advantages and would directly build on the current work in electrochemical characterization of antioxidant status in biological specimens.

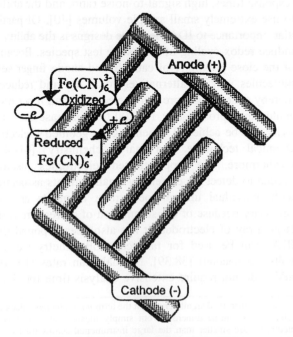

Fig. 4. Redox cycling at IDA electrodes.

Integrated value of antioxidant capacity

Representing the total antioxidant capacity of a system or organism with a single figure or set of limited number of figures rather than a complete tracing can prove advantageous, particularly in clinical settings. Although both Ia and S represent the integrated value of the reductive capacities of the components, neither one is a true representative of the antioxidant ability stemming from the OP. For instance, anodic waves at 400 and 900 mV both assign identical weight when the antioxidant capacity is evaluated. It can be visualized that components with either very high or very low OPs could yield a pro-oxidant effect. This could take place via the reduction of oxygen and the production of superoxide radical anion and hydrogen peroxide. Alternatively, direct oxidation of cellular components could occur, yielding only limited beneficial effects. Thus, a correction function should be applied to the current intensity values for each OP so as to normalize the contribution of Ia values according to potentials of biological relevance. A Gaussian function symmetrically centered from 300 to 600 mV, which rapidly decays, could be considered and experimentally tested. Likewise, a function of the general form $Ia / [(OP)**n]$ ($n > 1$) could also be examined.

CONCLUSIONS

We have shown several lines of evidence that the total reducing power of human plasma of patients as measured by CV correlates well with the severity of the disease. CV parameters also correlate well with the severity of an acute treatment that the subject receives. This is based on the fact that ex vivo plasma represents the in vivo antioxidant status (capacity) before the collection of blood.

Enrichment of the diet of healthy subjects with an antioxidant such as ascorbate or TA led to concomitant changes in Ia and R, an increase in the antioxidant capacity of the plasma (expressed as an increase in Ia), and a decrease in the oxidative stress as reflected by a decrease in R.

It has been repeatedly confirmed that a human diet rich in fruits and vegetables provides a marked health benefit through lowering the incidence of cancer, heart disease, and other pathologies. It has further been proposed that this effect stems from the increased nutritional supply of vitamins (and minerals). These vitamins include ascorbate (vitamin C) as well as vitamin E (the tocopherols), carotenoids, polyphenols, and others. Among these vitamins, the water-soluble ones are represented in the family of LMWAs as reported directly by the CV tracing. The lipid-soluble antioxidants can also be visualized by the CV methodology. CV studies of animal tissue homogenates and extracts also correlate with CHI [34].

In summary:

—CV is an efficient instrumental tool for evaluating the total (integrated) antioxidant capacity of LMWAs in human plasma and animal tissues [23–27] as well as in edible plants [28].

—The total reducing power of the sample is provided by the CV tracing, without the necessity of measuring the specific antioxidant capacity of each component alone.

—CV does not require the labor-intensive characterization of the activity of each component against a specific ROS, which, in turn, serves as the basis for other methodologies measuring the total antioxidant capacity.

—Preparation of a sample for CV measurement is simple and rapid and does not require advanced procedures.

—CV methodology allows rapid screening of many samples and preparations and is especially suitable for screening studies.

—The sensitivity of CV is sufficient for determining the physiological concentrations of antioxidants.

—The area under the AC wave (S) rather than the AC itself (Ia) should be used for evaluation of the total antioxidant capacity of a system when the AC wave comprises several components with similar but different $E_{1/2}$ values. The CV of such a combination is represented by a broader anodic wave rather than by an increase in Ia.

—S rather than Ia is associated with monitoring of the changes in the concentration of a specific LMWA component within a wave as a result of stress. The changes for individual components within a single AC wave could be different. The changes in S would better represent the residual antioxidant capacity and allow better quantitation of the loss in the specific component.

CV tracings can be obtained in aqueous medium as well as in organic solvents like acetonitrile, water/acetonitrile, and acetonitrile/methanol mixtures provided that there are redox-active components and enough electrolytes in the solution to support redox reactions on the electrode surface. This modification allows the quantitation of lipid-soluble components as well.

The combination of acetic acid, acetonitrile, and water is optimal for the quantitative extraction of antioxidants (both water- and lipid-soluble) and for quantitative evaluation of the antioxidant capacity.

In most cases, both blood plasma and edible plant extracts show AC waves at similar biological OPs. This is in accord with the fact that plants serve as sources for some antioxidant components of the human body and

that there is some similarity between the redox biochemical pathways of plants and mammalian tissues.

Individual components of the first AC wave in human plasma have been identified. The complete identification and assignment of the individual component(s) of the second wave in human plasma as well as in other tissues are of marked importance and await further investigation.

Acknowledgements — The financial support of the United States–Israel Binational Science Foundation (Grant 95-234 to M.C.) is acknowledged. The authors thank Dr. E. Berenshtein (The Hebrew University of Jerusalem) for his assistance and Dr. J. Rossier (Ecole Polytechnique Federal de Lausanne) for critical reading of this paper.

REFERENCES

[1] Halliwell, B; Gutteridge, J. M. C. Protection against oxidants in biological systems: the superoxide theory of oxygen toxicity. *Free radicals in biology and medicine.* Oxford: Clarendon Press; 1989: 86–179.

[2] Gutteridge, J. M. C. Free radicals in disease processes: a complication of cause and consequence. *Free Radic. Res. Commun.* **19:**141–158; 1993.

[3] Ames, B. M.; Shigenaga, M. K.; Hagen, T. M. Oxidants, antioxidants and the degenerative diseases of aging. *Proc. Natl. Acad. Sci. USA* **90:**7915–7922; 1993.

[4] Halliwell, B. Free radicals and antioxidants: a personal view. *Nutr. Rev.* **52:**253–263; 1994.

[5] Halliwell, B. How to characterize a biological antioxidant. *Free Radic. Res. Commun.* **9:**1–32; 1990.

[6] Chevion, M. A site-specific mechanism for free radical induced biological damage: the essential role of redox active transition metals. *Free Radic. Biol. Med.* **5:**27–37; 1988.

[7] Chance, P. A.; Sies, H.; Boveris, A. A hydroperoxide metabolism in mammalian organs. *Physiol. Rev.* **59:**527–605; 1979.

[8] Ames, B. M.; Cathcart, R.; Schwiers, E.; Hochstein, P. Uric acid produces an antioxidant defense in humans against oxidant and radical-caused aging and cancer: a hypothesis. *Proc. Natl. Acad. Sci. USA* **73:**6858–6862; 1981.

[9] Stocker, R.; Yamamoto, Y.; McDonagh, A.; Glazer, A. N.; Ames, B. N. Bilirubin is an antioxidant of possible physiological importance. *Science* **235:**1043–1045; 1987.

[10] Frei, B.; England, L.; Ames, B. N. Ascorbate is an outstanding antioxidant in human blood plasma. *Proc. Natl. Acad. Sci. USA* **86:**6377–6381; 1989.

[11] Doll, R. An overview of the epidemiological evidence linking diet and cancer. *Proc. Nutr. Soc.* **49:**119–131; 1990.

[12] Dragsted, L. O.; Strube, M.; Larsen, J. C. Cancer protective factors in fruits and vegetables: biochemical and biological background. *Pharmacol. Toxicol.* **72** (Suppl. 1):116–135; 1993.

[13] Willett, C. W. Diet and health: what should we eat? *Science* **264:**532–537; 1994.

[14] Willett, C. W. Micronutrients and cancer risk. *Am. J. Clin. Nutr.* **59**(Suppl.):162S–165S; 1994.

[15] La Vecchia, C.; Decarli, A.; Pagano, R. Vegetable consumption and risk of chronic disease. *Epidemiology* **9:**208–210; 1998.

[16] Gey, K. F. The antioxidant hypothesis of cardiovascular disease: epidemiology and mechanisms. *Biochem. Soc. Trans.* **18:**1041–1045; 1990.

[17] Posner, B. M.; Franz, M.; Quatromoni, P. The Interhealth Steering Committee. Nutrition and the global risk for chronic diseases: the interhealth nutrition initiative. *Nutr. Rev.* **52:**201–207; 1994.

[18] Harats, D.; Chevion, S.; Nahir, M.; Norman, Y.; Sagee, O.; Berry, E. M. Citrus fruit supplementation reduces lipoprotein oxidation in young men ingesting a diet high in saturated fat: presumptive evidence for an interaction between vitamins C and E in vivo. *Am. J. Clin. Nutr.* **67:**240–245; 1998.

[19] Rice-Evans, C.; Miller, N. J. Total antioxidant status in plasma and body fluids. *Methods Enzymol.* **234:**279–293; 1994.

[20] Wang, H.; Cao, G.; Prior, R. L. Total antioxidant capacity of fruits. *J. Agric. Food Chem.* **44:**701–705; 1996.

[21] Cao, G.; Sofic, E.; Prior, R. L. Antioxidant capacity of tea and common vegetables. *J. Agric. Food Chem.* **44:**3426–3431; 1996.

[22] Cao, G.; Prior, R. L. Comparison of different analytical methods for assessing total antioxidant capacity of human serum. *Clin. Chem.* **44:**1309–1315; 1998.

[23] Chevion, S.; Berry, E. M.; Kitrossky, N.; Kohen, R. Evaluation of plasma low molecular weight antioxidant capacity by cyclic voltammetry (abstract). In: Proceedings of the International Meeting on Free Radicals in Health and Disease, Istanbul, 1995, abstract no. 300.

[24] Chevion, S.; Berry, E. M.; Kitrossky, N.; Kohen, R. Evaluation of plasma low molecular weight antioxidant capacity by cyclic voltammetry. *Free Radic. Biol. Med.* **22:**411–421; 1997.

[25] Chevion, S.; Hofmann, M.; Ziegler, R.; Chevion, M.; Nawroth, P. P. The antioxidant properties of thioctic acid: characterization by cyclic voltammetry. *Biochem. Mol. Biol. Int.* **41:**317–327; 1997.

[26] Chevion, S. The antioxidant status of plasma in health and disease (Ph.D thesis). Jerusalem: The Hebrew University of Jerusalem; 1998.

[27] Chevion, S.; Or, R.; Berry, E. M. The antioxidant status of patients subjected to total body irradiation. *Biochem. Mol. Biol. Int.* **47:**1019–1027; 1999.

[28] Chevion, S.; Chevion, M.; Chock, P. B.; Beecher, G. R. The antioxidant capacity of edible plants: extraction protocol and direct evaluation by cyclic voltammetry. *J. Med. Food* **2:**1–11; 1999.

[29] Kohen, R.; Tirosh, O.; Kopolovich, K. The reductive capacity index of saliva obtained from donors of various ages. *Exp. Gerontol.* **27:**161–168; 1992.

[30] Kohen, R.; Tirosh, O.; Gorodetsky, R. The biological reductive capacity of tissues is decreased following exposure to oxidative stress: a cyclic voltammetry study of irradiated rats. *Free Radic. Res. Commun.* **17:**239–248; 1992.

[31] Kohen, R. The use of cyclic voltammetry for the evaluation of oxidative damage in biological samples. *J. Pharmacol. Toxicol. Methods* **29:**185–193; 1993.

[32] Kohen, R.; Beit Yannai, E.; Berry E. M., Tirosh O. Overall low molecular weight antioxidant activity of biological fluids and tissues by cyclic voltammetry. *Methods Enzymol.* **300:**285–296; 1999.

[33] USDA. *USDA nutrient database for standard reference* (12th ed.). http://www.nal.usda.gov/fnic/foodcomp.

[34] Shohami, E.; Gati, I.; Beit Yannai, E.; Trembovler, V.; Kohen, R. Closed head injury in the rat induces whole body oxidative stress: overall reducing antioxidant profile. *J. Neurotrauma* **16:**365–376; 1999.

[35] Beit Yannai, E.; Kohen, R.; Horowitz, M.; Trembovler, V.; Shohami, E. Changes of biological reducing activity in rat brain following closed head injury: a cyclic voltammetry study in normal and heat-acclimated rats. *J Cereb. Blood Flow Metab.* **17:**273–279; 1997.

[36] Brett, C.; Brett, A. *Electrochemistry: principles, methods, and applications.* New York: Oxford University Press; 1993:174–198.

[37] Fleischmann, M.; Pons, S.; Rolison, D. R.; Schmidt, P. P. *Ultramicroelectrodes.* Morganton, NC: Datatech Systems; 1987, p 363.

[38] Thormann, W.; van den Bosch, P.; Bond, A. M. Voltammetry at linear gold and platinum microelectrode arrays produced by lithographic techniques. *Anal. Chem.* **57:**2764–2770; 1985.

[39] Chidsey, C.; Feldman, B. J.; Lundgren, C.; Murray, R. W. Micrometer-spaced platinum interdigitated array electrode: fabrication, theory, and initial use. *Anal. Chem.* **58:**601–607; 1986.

[40] Niwa, O.; Morita, M.; Tabei, H. Electrochemical behavior of reversible redox species at interdigitated array electrodes with different geometries: consideration of redox cycling and collection efficiency. *Anal. Chem.* **62:**447–452; 1990.

[41] Niwa, O.; Morita, M.; Tabei, H. Highly sensitive and selective voltammetric detection of dopamine with vertically separated interdigitated array electrodes. *Electroanalysis* **3:**163–168; 1991.

[42] Niwa, O.; Xu, Y.; Halsall, B. H.; Heineman, W. R. Small-volume detection of 4-aminophenol with interdigitated array electrodes and its application to electrochemical enzyme immunoassay. *Anal. Chem.* **65:**1559–1563; 1993.

[43] Niwa, O.; Tabei, H. Voltammetric measurements of reversible and quasi-reversible redox species using carbon film based interdigitated array microelectrodes. *Anal. Chem.* **66:**285–289; 1994.

ABBREVIATIONS

AA—ascorbic acid

AAPH-2,2′—azobis(2-amidinopropane) dihydrochloride

AC—anodic current

BMT—bone marrow transplantation

CV—cyclic voltammetry

DHAA—dehydroascorbate

DM—diabetes mellitus

E, $E_{1/2}$, Ea, Ec—OP maximal, OP at half of the peak height, OP anodic, OP cathodic

HPLC-ECD—high-performance liquid chromatography–electrochemical detection

Ia, Ic—anodic current, cathodic current

IDA—interdigitated array

LMWA—low molecular weight antioxidant/s

OP—oxidation potential

R—[DHAA]/{[AA] + [DHAA]},

ROS—reactive oxygen–derived species

S—the area of the anodic current wave

TA—thioctic acid, lipoic acid

TBI—total body irradiation

UA—uric acid

Bioassays for Oxidative Stress Status (BOSS). Edited by W.A. Pryor

QUANTIFICATION OF THE OVERALL REACTIVE OXYGEN SPECIES SCAVENGING CAPACITY OF BIOLOGICAL FLUIDS AND TISSUES

RON KOHEN,* ELANGOVAN VELLAICHAMY,[†] JAN HRBAC,* IRITH GATI,* and OREN TIROSH[‡]

*Department of Pharmaceutics, School of Pharmacy, The Hebrew University of Jerusalem, Jerusalem, Israel; [†]Department of Molecular Cardiology, Lerner Research Institute, The Cleveland Clinic Foundation, Cleveland, OH, USA; and [‡]Department of Molecular and Cell Biology, University of California at Berkeley, Berkeley, CA, USA

(Received 3 September 1999; Revised 27 January 2000; Accepted 3 February 2000)

Abstract—A method has been developed for measuring and evaluating the overall antioxidant activity derived from the low–molecular weight antioxidants (scavengers). The principle governing this method is based on a common chemical characteristic of the scavengers, their reducing properties. It was hypothesized and then demonstrated that an evaluation of the overall reducing power of a biological sample correlates with the overall scavenging activity of the sample. In order to quantify the total reducing power, the cyclic voltammetry methodology was applied. The resulting measurements correlated with the antioxidant activity of both hydrophilic and lipophilic scavengers. The method is suitable for use in biological fluids and in tissue homogenates, and can supply information concerning the type of antioxidants and their total concentration without having to determine specific compounds. A noninvasive procedure for determining skin overall scavenging activity is also described. This method is based on a well containing an extraction solution that is attached to the skin's surface. Following incubation time the extraction solution is analyzed using the cyclic voltammeter instrument and other methods. We have found these methods suitable for evaluating the reducing capacity status in various clinical conditions such as diabetes, ionizing and nonionizing irradiation, brain degenerative diseases, head trauma, and inflammatory bowel diseases. This method is also an efficient tool for evaluating the overall antioxidant capacity of mixtures of antioxidant preparations in vitro. The measurements themselves are simple and rapid. Furthermore, they do not require manipulation of the samples. © 2000 Elsevier Science Inc.

Keywords—Invasive and noninvasive techniques, Cyclic voltammetry, Total antioxidant capacity, Free radicals, Antioxidant profile, Oxidative stress, Low–molecular weight antioxidants

BACKGROUND

The group of low–molecular weight antioxidants (LMWA) is much larger than the group of antioxidant enzymes and possesses several attributes that contribute to their biological functions [1,2]. These attributes include the ability of these compounds to penetrate and reach specific locations in the cell where oxidative stress may occur. Therefore, these compounds can give site-specific protection against the deleterious reactive oxygen species (ROS) in places where the antioxidant enzymes do not have access. These compounds possess a wide spectrum of activity towards a variety of ROS and can act in many different mechanisms. A distinction

Address correspondence to: Dr. Ron Kohen, Hebrew University of Jerusalem, School of Pharmacy, Department of Pharmaceutics, P.O. Box 12065, 91120 Jerusalem, Israel; Tel: +972 (2) 675-8659; Fax: +972 (2) 675-7246; E-Mail: ronko@cc.huji.ac.il.

Ron Kohen received his B. Pharm. degree in pharmacy and Ph.D. in biochemistry from the Hebrew University of Jerusalem. He did a postdoctoral fellowship with Dr. Bruce Ames at the University of California, Berkeley. In 1988 he joined the Department of Pharmaceutics at the Faculty of Medicine of The Hebrew University of Jerusalem. His research interests include aging, biological redox regulation and quantification, the role of antioxidants and oxidative stress in pathological disorders, and development of new methodologies and biosensors.

Elangovan Vellaichamy received his Ph.D. in biochemistry from the University of Madras, India. From 1997 to 1999 he was a postdoctoral fellow with Drs. Kohen and Shohami at the Department of Pharmaceutics of The Hebrew University of Jerusalem. Currently he holds a postdoctoral position at the Cleveland Clinic Foundation, Department of Molecular Cardiology, Lerner Research Institute, USA.

Jan Hrbac is a Ph.D. student at Palacky University, Czech Republic. He is currently a visiting research fellow at the Hebrew University of Jerusalem.

Irith Gati received her B.Sc. degree in biology from The Hebrew University of Jerusalem. She is a senior laboratory technician and is involved in research in various fields of pharmaceutics.

Oren Tirosh received his Ph.D. degree in pharmaceutics from the Hebrew University of Jerusalem. In 1997, he joined the laboratory of Dr. Laster Packer at the University of California, Berkeley, as a postdoctoral fellow.

Reprinted from: *Free Radical Biology & Medicine*, Vol. 28, No. 6, pp. 871–879, 2000

must be made between the LMWA, which act directly with the ROS (chemical scavengers), and the LMWA, which act indirectly (e.g., chelating agents). The method described in this paper is related to the chemical scavengers. There are numerous methods for evaluating LMWA [3–5]. However, it is not technically practical to measure all the LMWA at the same time at a specific tissue. To overcome this difficulty, several methods have been suggested for evaluating the total LMWA [5–13]. In addition to the existing methods we have suggested a new approach for the quantification of the overall LMWA, which act directly with the ROS of both hydrophilic and lipophilic nature in biological fluids and tissues [14,15].

Principle of quantification of LMWA that act directly with ROS (scavengers)

An examination of the chemical structure and properties of the scavengers reveals that these molecules are reducing agents [1,14–16]. These compounds donate electrons to the reactive species, thus acting as a reducing agent and oxidizing the free radicals. The scavengers themselves become radicals, however, they are stable radicals. Therefore, it was assumed that evaluation of the overall reducing power (the sum of the contribution of the various reducing antioxidants [scavengers] present in a specific environment) of a biological fluid or tissue homogenate would reflect its overall antioxidant activity [2,14–16].

Principle and procedure of voltammetric measurements

Cyclic voltammetry measurements have been conducted for many years to evaluate electron transfer between molecules [17–19]. These measurements are usually carried out for isolated compounds under specific conditions. In these techniques, an electrical potential gradient is applied (relative to a reference electrode) across an electrode-solution interface (working electrode) to oxidize or reduce species present in solution (in case of cyclic voltammetry a linear potential gradient is applied). The resulting current vs. potential is recorded (cyclic voltammogram). The voltammograms can supply information concerning thermodynamic, kinetic, and analytical features of the electrochemical couple under investigation (for review, see [20]). Although many factors determine the current shape and value (form of applied potential, electrode size and geometry, size of energy barrier for electron transfer, interaction between the electrode surface, and the electroactive molecules), voltammetric waves are usually obtained in a peak shaped or sigmoidal mode. Thus, the position of the current wave on the voltage axis (x-axis of the voltammogram) can be

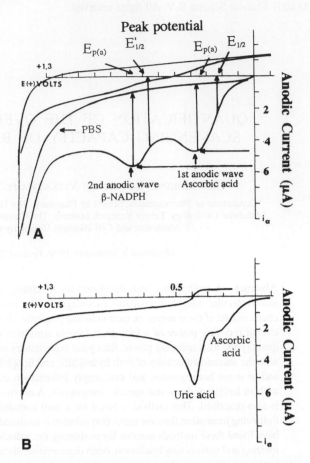

Fig. 1. Cyclic voltammograms of LMWA in vitro. The measurement conditions are described in the text. The peak potential is calculated from the x-axis (type of LMWA) and anodic current from the y-axis (total concentration of the LMWA). (A) A typical cyclic voltammogram of ascorbic acid (0.5 mM) and β-NADPH (0.5 mM). (B) A typical cyclic voltammogram of a combination of uric acid (0.5 mM) and ascorbic acid (0.5 mM).

determined and is referred to as the potential where the peak current (peak potential $E_{p(a)}$) or inflection point (half-wave potential, $E_{1/2}$) occurs (Fig. 1A). The oxidation potential of a compound may be defined phenomenologically as the potential where $E_{p(a)}$ is observed for a given set of conditions. The size of the current (anodic current) (AC) (Ia) (calculated from the y-axis of the voltammogram) depends on a number of parameters (area of the electrode, the rate in which the potential is applied) and is proportional, at any potential, to the concentration of the substrate in the bulk solution [18–20]. It is used analytically to monitor concentration [14,15,17–25]. These two parameters (peak potential and anodic current) can be characterized as the reducing power parameter and can supply information concerning the type of reducing antioxidants (their ability to donate electrons) and their total concentration. Although these measurements are taken in biological systems, which are composed of many electrochemical couples and are not reversible, we found the results to be so reliable and highly reproducible that they qualified for comparison studies.

CYCLIC VOLTAMMETRY MEASUREMENT OF BIOLOGICAL SAMPLES

Instrumental setup

The biological samples (biological fluids, tissue homogenates, and organic extraction of these two) are placed in a cyclic voltammeter cell (200 μl–3 ml) [2,14–16] into which three electrodes have been introduced: the working electrode, 3.3 mm in diameter (e.g., glassy carbon or mercury film electrodes), the reference electrode (e.g., Ag/AgCl), and the auxiliary electrode (platinum wire). The potential is applied linearly to the working electrode at a constant rate (e.g., 100mV/s) either towards the positive potential (evaluation of reducing equivalents) or towards the negative potential (evaluation of oxidizing species). We have used a cyclic voltammeter (CV) CV-1B and electrochemical working station CV-50W (Bioanalytical Systems; West Lafayette, IN, USA). During operation of the CV, a potential current curve is recorded (cyclic voltammogram). As described above, the reducing power's parameter as determined from the cyclic voltammogram is composed of two parameters: (i) the peak potential [$E_{p(a)}$] and (ii) the anodic current (AC).

Figure 1 [14] shows a typical cyclic voltammogram of combinations of several antioxidants measured in vitro. Figure 1A shows a cyclic voltammogram of a combination of ascorbic acid (0.5 mM) and β-reduced nicotinamide adenine dinucleotide phosphate (NADPH) (0.5 mM). The phosphate-buffered saline (PBS), which is used as the solvent in the hydrophilic cyclic voltammetry measurements, does not possess any reducing power and serves as the base line for the calculation of the anodic current (Fig. 1A). Each scavenger has its characteristic peak potential (Fig. 1A). If two scavengers possess close or similar peak potentials, the two anodic waves (AW) may overlay each other. Figure 1B shows a cyclic voltammogram of a combination of a mixture of ascorbic acid and uric acid (0.5 mM). If the concentration of one of the components composing the same AW increases, the resulting cyclic voltammogram demonstrates one AW containing the two compounds. This is usually the case when biological samples are measured. Each AW detected is composed of many scavengers, sharing a close peak potential [14]. Although the evaluation of the $E_{p(a)}$ of the detected AWs may indicate the type of different reducing antioxidants (scavengers) present in the sample tested (their peak potentials), the overall concentration of the reducing antioxidants can also be calculated from the cyclic voltammogram. This can be accomplished by measuring the AC from the y-axis of each wave, as described above and shown in Fig. 1. Changes in the concentration of one or more LMWA in each group of molecules result in a change in the anodic current of the AW of this group [14–16,21–25].

Since biological samples contain various compounds capable of absorbing to the electrode surface after applying potential (e.g., proteins), it is essential to polish the electrode surface after each measurement as described by the manufacturer (e.g., polishing of glassy carbon by the use of polishing alumina). It is essential that a measurement of a standard compound, which possesses known reversible redox properties, is conducted routinely. A standard of potassium ferrocyanide is often used [18].

Sensitivity and the spectrum of antioxidants that can be detected by the cyclic voltammeter

The cyclic voltammetry methodology allows for the determination of both hydrophilic and lipophilic LMWA. Figure 2 shows a typical cyclic voltammogram of two lipid-soluble antioxidants: (i) α-tocopherol (1 mM) and (ii) an oxidized form of lipoic acid (0.9 mM) in vitro.

Theoretically, every reducing agent can be detected by the cyclic voltameter, depending on the conditions used. Throughout our measurements we used a glassy carbon electrode, which is capable of reacting with many scavengers possessing oxidation potential in the positive zone vs. Ag/AgCl. Therefore, glutathione, one of the most important biological antioxidants, which possess its oxidation potential in the negative zone vs. Ag/AgCl, can not be detected using this electrode. Mercury film electrodes, which allow the application of a negative potential, can be used to evaluate glutathione and other thiol containing compounds [14].

The limit of detection for isolated compounds is in the range of 1–10 μM. This range of sensitivity is sufficient for determining physiological concentrations of biological relevant scavengers. In fact the limited sensitivity can be a positive aspect of this method because low concentrations of redox-active compounds will not interfere with the measurements. For example, brain neurotransmitters, which can also be extracted from brain and possess redox-active properties, will not be detected due to their low concentration (dilution by the overall brain homogenate) compared to the high concentration of the LMWA present in the brain [20].

Preparation of biological samples for CV measurements [14]

The cyclic voltammetry analysis is suitable for both biological fluids and tissues. The minimum volume re-

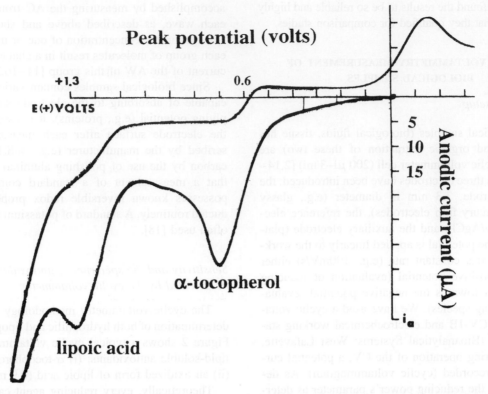

Fig. 2. Cyclic voltammogram of lipid-soluble antioxidants: α-tocopherol (1 mM) and lipoic acid (oxidized form) (0.9 mM).

quired is about 200–300 ml of biological fluid (e.g., saliva, plasma, cerebrospinal fluid, semen fluid). Once collected the biological fluid is diluted with an equal volume (1:1) of a suitable buffer (for example, PBS pH 7.2). The buffer is essential in order to keep the pH constant and to supply a sufficient concentration of electrolytes needed for performing the measurement. Different types of buffers can be used according to the specific needs, as long as there is sufficient buffer to keep the pH constant and ensure sufficient ionic strength. The samples can be kept under nitrogen for up to 6 months in −70°C without any significant change in their CV tracing.

Biological tissues (30–500 mg) can be removed and immediately deep frozen in liquid air. This biological tissue can be kept in −70°C for storage or can be homogenized and immediately analyzed. The homogenization should be carried out in 1 ml PBS on ice using a mechanical homogenizer. Following homogenization the samples should be centrifuged at 1000 × g for 10 min at 4°C to remove large insoluble particles. The pH of all homogenates is then measured and adjusted to 7.2.

Extraction and measurement of lipophilic LMWA (scavengers)

After the measurements of the water-soluble LMWA are completed, the biological fluids and tissue homogenates can undergo further lipophilic extraction to determine the lipophilic LMWA (it is important to keep an aliquot of 30 μl from the PBS homogenate for protein determination). Three to four milliliters of the PBS homogenate or PBS-biological fluid mixture is combined 1:4 (V:V) with a mixture of ethanol:hexan (1:5) [5]. After rough shaking (5 min on vortex) the mixture is then centrifuged at 1000 × g for 15 min at 4°C. The upper layer is separated and the lower layer is extracted again. The two upper layers are combined and the organic solvents removed by evaporation. The residue obtained is dissolved in an acetonitrile solution containing 1% *tert*-butylamonium perchlorate as a supporting electrolyte. Other extraction solvents are also possible depending on the nature of the lipophilic scavenger to be extracted. It is also possible to perform the extraction on the tissue itself (and not on the PBS homogenate of the tissue). The procedure is basically the same except for the first stage in which the tissue is homogenized on ice using a mechanical homogenizer in a solution of ethanol:hexan (1:5) in a ratio of 1:4 (w/v). It is important for the samples be protected from light and kept on ice and under nitrogen.

Evaluation of water- and lipid-soluble LMWA (scavengers) (LMWA profile) in biological fluids

Figure 3 shows a few examples of water and lipophilic LMWA present in biological fluids. Plasma, for

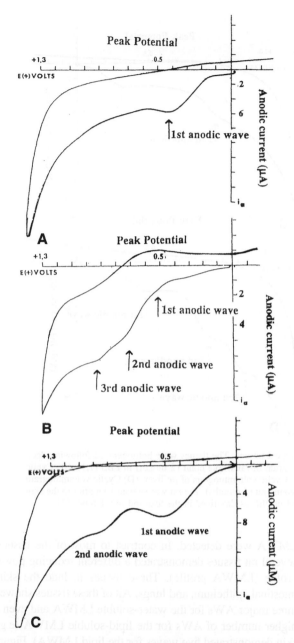

Fig. 3. Hydrophilic and lipophilic cyclic voltammograms of some biological fluids. The measurements were carried out as described in the text. The organic extraction was performed on 4 ml of plasma. (A) Cyclic voltammogram of human plasma from an individual who is 25 years old. (B) Cyclic voltammogram of the lipophilic extraction of the same plasma shown in panel A. (C) Cyclic voltammogram of human semen fluid.

example, demonstrated one major anodic wave at a peak potential of 400 ± 50 mV [14,23] (Fig. 3A). It was found that this wave is composed mostly from ascorbic acid and uric acid (see below). Appearance of a second wave may occur as a contamination from hemolysis, which occurred during the separation of the plasma from the blood. The major component, contributing to the second wave is NADPH, which originated from the red blood

cells (see Fig. 1). The presence of other LMWA in a low concentration (< 1 μM) can not be detected by the CV. In order to extract the plasma lipophilic LMWA at least 4 ml is needed. The extraction is carried out as described above for the lipophilic extraction. Figure 3B shows a typical cyclic voltammogram of plasma lipophilic extraction. Three anodic waves were detected at peak potentials of 490, 690, and 880 mV, indicating three lipophilic LMWA or three groups of LMWA. The first anodic wave probably relates to the compounds, coenzyme Q_{10} and carotene, the second wave relates to tocopherol compounds, and the third group is still unidentified and may be related to the oxidized form of lipoic acid. Human semen fluid (Fig. 3C) demonstrates two major anodic waves at peak potentials of 520 and 920 mV for the water-soluble LMWA.

Although plasma and its overall LMWA activity is often used as the most convenient biological fluid to assess the effect of oxidative stress and to evaluate the overall antioxidant status of the individual, we found that this is not always appropriate. We believe that plasma does not necessarily reflect the antioxidant profile of tissues [26], and therefore is not suitable for such evaluation. We showed that plasma samples collected from healthy volunteers are significantly different in the AC values and may reflect only the dietary habits of the individual. Evaluation of the reducing power of other biological fluids such as saliva may be more reliable because it is not subject to dietary influence and interference.

In addition, some changes in the peak potential (not just in the anodic current) of the same biological fluid or tissue between various samples might occur. For example, the peak potential of plasma obtained from healthy donors may shifted within the range of 150 mV [23]. In other tissues it may be even higher. This is due to changes in the relative concentration of the compounds composing the wave. A shift in the potential may occur depending on the peak potential of the compound present in high concentrations and possessing high electroactive responses.

Evaluation of water- and lipid-soluble LMWA (scavengers) in biological tissues

The overall scavenging ability of biological tissues can be quantified by the cyclic voltammeter after their homogenization as described above. It was found that biological tissues behave differently from each other and that each tissue possesses its own characteristic reducing power profile (LMWA profile). Figure 4 shows a few cyclic voltammograms of water- and the lipid-soluble extraction of LMWA of some tissues [14,21,24,25,27–29]. Most of the tissues tested demonstrated two anodic

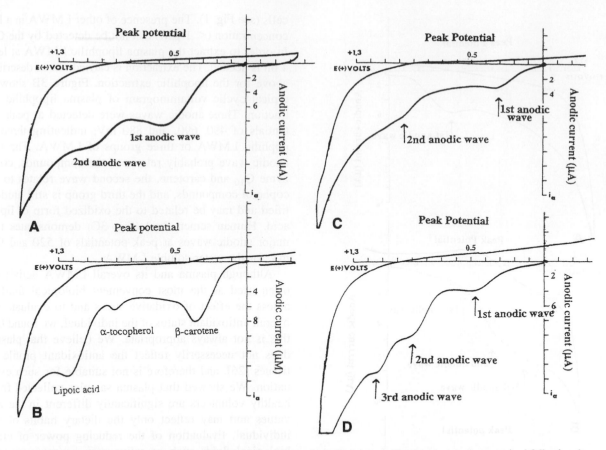

Fig. 4. Hydrophilic and lipophilic cyclic voltammograms of some biological tissues. The tissues were homogenized following deep freeze in liquid air on ice as described in the text. (A) Cyclic voltammogram of rat whole brain (2 month) in PBS (1:1). (B) Cyclic voltammogram of lipophilic extraction of rat brain PBS homogenate. (C) Cyclic voltammogram of rat liver. (D) Cyclic voltammogram of hydrophilic homogenization of rat skin. The skin was removed from young rat (2 months). The rat was shaved 10 h prior to the skin removal. Following deep freeze in liquid air, the skin was homogenized in PBS as described in the text and in reference [2].

waves, indicating two major groups of LMWA. Figure 4A shows a cyclic voltammogram of rat (2 month) whole brain that was homogenized in PBS buffer. This voltammogram profile is typical for the brain where two AWs were detected in peak potentials ($E_{1/2}$) of 420 mV and 980 mV. Figure 4B shows the cyclic voltammogram of lipophilic extraction of the rat whole brain. It can be seen that the profile of the lipid-soluble LMWA of rat brain demonstrated 3 AWs, indicating three groups of lipophilic LMWA at peak potentials of 250, 980, and 1110 mV. This is different from the profile of water-soluble LMWA. Using high-performance liquid chromatography (HPLC) and other analytical methods [14], it was found that the first AW is composed of β-carotene and coenzyme Q_{10}, the second AW is composed of tocopherols and melatonin, and the third group may be composed of oxidized lipoic acid. From the HPLC measurements it is clear that there are other compounds responsible for the different waves, but their structure is still unknown.

Figure 4C shows a typical cyclic voltammogram of rat liver. Again, two anodic waves for the water-soluble

LMWA were detected. In contrast to most of the tissues, several rat tissues demonstrated a different reducing power profile (LMWA profile). These tissues include the skin, intestinal epithelium, and lungs. All of these tissues showed three major AWs for the water-soluble LMWA and even a higher number of AWs for the lipid-soluble LMWA (e.g., skin demonstrated five waves for the lipid LMWA). Figure 4D shows an example of water-soluble LMWA of skin [2,25]. Three AWs can be determined at peak potentials ($E_{1/2}$) of 390, 850, and 1020 mV. The different cyclic voltammogarams obtained for the different tissues may serve as a characteristic feature of the different tissues. It may be speculated that the high number of LMWA groups detected for these three tissues (skin, intestinal epithelium, and lungs) is derived from the fact that these tissues are exposed to oxidative stress both from endogenous and exogenous sources. Therefore, it was hypothesized that this method may be used to assess the susceptibility of a biological tissue to oxidative stress.

In order to quantify the reducing LMWA composing the different anodic waves, several methods were applied

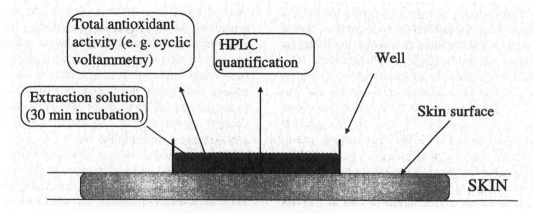

Fig. 5. Schematic representation of the well containing the extraction solution placed on the surface of the skin. The well was designed to measure the antioxidant surface activity of the skin. Following an incubation time of 30 min, the extraction fluid (1ml) was taken for cyclic voltammetry measurement and other analysis.

(data not shown) [14]. These included the use of HPLC instrument equipped with an electrochemical detector [4], the use of specific enzymes such as ascorbate oxidase (EC 1.10.3.3) and uricase (EC 1.7.3.3) [14], and the spiking of various LMWA into the homogenate to determine their peak potential in the biological environment of the tested sample [14].

Noninvasive measurement of skin-scavenging capacity [2,27,28]

In order to apply the cyclic voltammetry methodology for noninvasive evaluation of the total antioxidant capacity we chose the skin as the suitable target. The outer layer of skin is most susceptible to oxidative stress from exogenous and endogenous sources. We have previously [2,27,28] shown that there is an accumulation of organic peroxide on the outer surface (stratum corneum) during the aging process. While performing this procedure it was found that skin releases scavengers from its outer layer into an outside solution [2,27,28]. Therefore, we designed a well (1 cm in diameter) that was placed on the surface of the skin (Fig. 5). In humans the well is placed on the inner side of the wrist and attached to the arm using an adhesive pad (Parafilm, American National Cam, Chicago, IL, USA). In rat, the hair from the back of the animal is shaved 10 h before the measurement and the well is attached to the back of the animal using silicone rubber. An extraction solution of PBS at pH 7.2 (1–1.5 ml) (or another buffer at a different pH) is introduced into the well for a 30 min incubation period. After incubation the extraction solution is analyzed by the cyclic voltammetry method to evaluate the overall profile (type and concentration) of the LMWA secreted from the upper layer of the skin or specifically by HPLC equipped with electrochemical detectors. It is recommended that

oxygen be removed from the extraction buffers prior to its placement on the surface of the skin. This is done by bubbling nitrogen (99.999%) for 20 min in the buffer stock solution and flushing above the extraction solution in the well during the experiment.

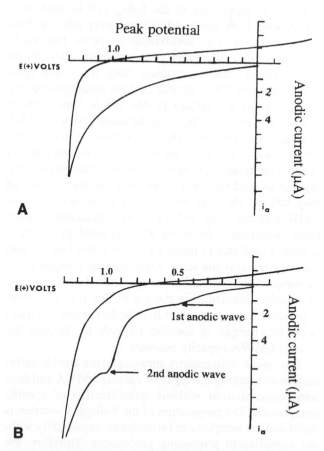

Fig. 6. Cyclic voltammograms of the extraction solution, which was placed on the back of rat for a 30 min incubation period. (A) Phosphate-buffered saline (PBS) (prior to its placement on the skin surface). (B) Phosphate-buffered saline following 30 min of incubation.

Figure 6 shows cyclic voltammograms of the extraction solution prior to and following the incubation period on skin surface. It can be seen that while there was no reducing power for the extraction solution itself, two AWs were detected following an incubation of 30 min on skin surface. The peak potentials detected for the two waves were at 430 ± 50 and 940 ± 40. These two waves suggest that two groups of water-soluble reducing LMWA can be extracted from the skin surface. It was found [27,28] that 30 min are required for full extraction and 24 h are needed for the skin to recover its LMWA at the site of the extraction.

Humans demonstrated a different profile of LMWA on the skin surface. It was found that humans lack the first anodic wave (e.g., ascorbic acid). Using HPLC it was found that uric acid, also a component of the first anodic wave, can be extracted from the skin; however, its concentration is low, below the detection limit of the cyclic voltammeter.

SUMMARY

The group of LMWA that act directly, the scavengers, is of great importance to the living cell because these compounds have many advantages over other antioxidant molecules (e.g., antioxidant enzymes) (see Background). Cyclic voltammetry measurements allow for the evaluation of these compounds (both water- and lipid-soluble nature). The outcome of these measurements can be used as an antioxidant profile and is convenient for comparison purposes between different systems and following exposure to oxidative stress. Although this method does not allow for the direct determination of specific compounds, it may supply information concerning the overall behavior of the scavenging system and may help to elucidate the regulation mechanism of the LMWA in the body and the interrelationships between these compounds. By using this methodology, we have already found that in many cases the overall antioxidant capacity remains the same while specific changes occur in some of the antioxidant compounds. In other words, the results obtained from the use of this approach make it possible for us to suggest that the living cell can adapt itself to changes in specific LMWA by keeping the overall LMWA capacity constant.

The cyclic voltammetry measurements provide information concerning the types of various LMWA and their total concentration without quantification of specific compounds. The preparation of the biological samples is rapid and the samples can be analyzed immediately without complicated processing procedures. Therefore, the method is suitable for screening a large number of samples and may help to identify unusual pathological profiles. The sensitivity of this procedure is relatively low

and was found to be in the range of 1 μM; however, such sensitivity is sufficient for determining physiological concentrations of biologically relevant scavengers without interference of other redox-active compounds present in low concentration. It is important to use similar conditions throughout the measurements. Changes in the peak potential of up to ± 50 mV are acceptable due to changes in the relative concentrations of the reducing antioxidants composing the wave.

These methods (invasive and noninvasive) have been utilized already in several systems including diabetes [29,30], ulcerative colitis [31], irradiation therapy [32], brain degenerative diseases, and head trauma [24,25,33], skin status and pathologies [2,27,28], and also in the study of the aging process [2,14,15,27,28]. This methodology may also be used as an indicator for oxidative damage [34]. In conclusion, the LMWA profile as provided by the cyclic voltammetry measurements is an efficient tool to compare different populations, as well as to estimate the ability of a biological sample to cope with ROS and its susceptibility to oxidative stress.

Acknowledgements — We would like to thank Ms. Madelyn Segev for her assistance in editing and writing this manuscript. R. Kohen is affiliated with the David R. Bloom Centre of Pharmacy. This research was supported by a grant from the Israeli Ministry of Health.

REFERENCES

[1] Halliwell, B.; Gutteridge, J. M. C., eds. *Free radical in biology and medicine.* Oxford: Clarendon Press, Oxford University Press; 1989.
[2] Kohen, R. Skin antioxidants: their role in aging and in oxidative stress. New approaches for their evaluation. *Biomed. Pharmacother.* **53:**181–192; 1999.
[3] Halliwell, B.; Aeschbach, R.; Loliger, J.; Aruoma, O. I. The characterization of antioxidants. *Food Chem. Toxicol.* **33:**601–617; 1995.
[4] Motchnik, P. A.; Frei, B.; Ames, B. N. Measurement of antioxidants in human blood plasma. *Meth. Enzymol.* **234:**269–279; 1994.
[5] Rice-Evans, C.; Miller, N. J. Total antioxidant status in plasma and body fluids. *Meth. Enzymol.* **234:**279–293; 1994.
[6] Whitehead, T. P.; Thorpe, G. H. G.; Maxwell, S. R. J. Enhanced chemiluminescent assay for antioxidant capacity in biological fluids. *Anal. Biochem. Acta* **266:**257–264; 1992.
[7] Cooper, M. J.; Engel, R. R. Carbon monoxide production from L-3, 4-dihydroxyphenylalanine: a method for assessing the oxidant/antioxidant properties of drugs. *Clin. Chim. Acta* **202:**105–109; 1991.
[8] Lissi, E.; Salim-Hanna, M.; Pascual, C.; del Castillo, M. D. Evaluation of total antioxidant potential and total antioxidant reactivity from luminol-enhanced chemiluminescence measurements. *Free Radic. Biol. Med.* **18:**153–158; 1995.
[9] Cao, G.; Alessio, H. M.; Cutler, R. G. Oxygen-radical absorbance capacity assay for antioxidants. *Free Radic. Biol. Med.* **14:**303–311; 1993.
[10] Miller, N. J.; Rice-Evans, C.; Davies, M. J.; Gopinathan V.; Milner, A. D. A new automated method for estimating plasma antioxidant activity and its application to the investigation of antioxidant status in premature neonates. *Clin. Sci.* **84:**407–412; 1993.
[11] Ghiselli, A.; Serafini, M.; Giuseppe, M.; Azzini E.; Ferro-Luzzi,

A. A fluorescence-based method for measuring total plasma antioxidant capability. *Free Radic. Biol. Med.* **18**:29–36; 1995.

[12] Benzie, I. F. F.; Strain, J. J. Ferric reducing/antioxidant power: direct measure of the total antioxidant activity of biological fluids and modified version for simultaneous measurement of total antioxidant power and ascorbic acid concentration. *Meth. Enzymol.* **299**:15–27; 1999.

[13] Cao, G.; Prior, R. L. Measurement of antioxidants in human blood plasma. *Meth. Enzymol.* **299**:50–62; 1999.

[14] Kohen, R.; Beit-Yannai, E.; Berry, E. M.; Tirosh, O. Evaluation of the overall low molecular weight antioxidant activity of biological fluids and tissues *Meth. Enzymol.* **300**:285–290; 1999.

[15] Kohen, R. Evaluation of oxidant/antioxidant capacity of biological tissues and fluids by cyclic voltammetry: a new approach. *International Symposium on Antioxidant and Disease Prevention.* Stockholm, Sweden: Int. Life Sci. Inst. (ILSI); 1993:abstr. 100.

[16] Kohen, R.; Tirosh, O.; Gorodetzky, R. The biological reductive capacity of tissues is decreased following exposure to oxidative stress: a cyclic voltammetry study of irradiated rats. *Free Radic. Res. Commun.* **17**:239–248; 1992.

[17] Kissinger, P. T.; Jonathan, B. H.; Adams, R. N. Voltammetry in brain tissue: a new neurophysiological measurement. *Brain Res.* **55**:209–213; 1973.

[18] Heineman, W. R.; Kissinger, P. T. *Laboratory techniques in electroanalytical chemistry.* New York: Marcel Dekker; 1984.

[19] Bard, A.; Faulkner, L. R. *Electrochemical methods.* New York: John Wiley & Sons; 1980.

[20] O'Neill, R. D.; Lowry, J. P.; Mas, M. Monitoring brain chemistry in vivo: voltammetric techniques, sensors, and behavioral applications. *Crit. Rev. Neurobiol.* **12**:69–127; 1998.

[21] Blau, S.; Rubinstein, A.; Bass, P.; Singaram. C.; Kohen, R. Differences in the reducing power along the rat GI tract: lower antioxidant capacity of the colon. *Mol. Cell. Biochem.* **194**:185–191; 1999.

[22] Chevion, S.; Berry, E. M.; Kitrossky, N.; Kohen, R. Evaluation of plasma low molecular weight antioxidant capacity by cyclic voltammetry. *Free Radic. Biol. Med.* **22**:411–421; 1997.

[23] Kohen, R.; Chevion, S.; Schwartz, R.; Berry, E. M. Evaluation of the total low molecular weight antioxidant activity of plasma in health and diseases: a new approach. *Cell Pharmacol.* **3**:355–359; 1996.

[24] Shohami, E.; Beit-Yannai, E.; Horowitz, M.; Kohen, R. Oxidative stress in closed head injury: brain antioxidant capacity as an indicator of functional outcome. *J. Cereb. Blood Flow Metab.* **17**:1007–1019; 1997.

[25] Shohami, E.; Beit-Yannai, E.; Gati, I.; Trembovler, V.; Kohen K. Closed head injury in the rat induces whole body oxidative stress: overall reducing antioxidant profile. *J. Neurotrauma* **16**:365–376; 1999.

[26] Kohen, R.; Berry, E. What does the total antioxidant status of plasma reflect? *Redox Report* **3–4**:253–254; 1997.

[27] Kohen, R.; Fanberstein, D.; Zelkowicz, A.; Tirosh, O.; Farfouri, S. Noninvasive in vivo evaluation of skin antioxidant activity and oxidation status. *Meth. Enzymol.* **300**:428–437; 1999.

[28] Kohen, R.; Fanberstein, D.; Tirosh, O. Reducing equivalents in the aging process. *Arch. Gerontol. Geriat.* **24**:103–123; 1997.

[29] Elangovan, V.; Shohami, E.; Gati, I.; Kohen, R. Increased hepatic lipid-soluble antioxidant capacity as compared to other organs of streptozotocine-induced diabetic rats: a cyclic voltammetry study. *Free Radic. Res.* **32**:125–134; 2000.

[30] Ornoy, A; Zaken, V.; Kohen, R. Role of reactive oxygen species in the diabetic induced anomalies in rat embryos in vitro: a reduction in antioxidant enzymes and low molecular weight antioxidants (LMWA) may be the causative factor for increased anomalies. *Teratology* **60**:376–386; 1999.

[31] Blau, S.; Rubinstein, A.; Bass, P.; Kohen, R. The relation between colonic inflammation severity and total low molecular weight antioxidant profiles in experimental colitis. *Digest. Dis. Sci.* in press; 2000.

[32] Chevion, S.; Or, R.; Berry E. M. The antioxidant status of patients subjected to total body irradiation. *Biochem. Mol. Biol. Int.* **47**:1019–1027; 1999.

[33] Beit-Yannai, E.; Kohen, R.; Horowitz, M.; Trembovler, V.; Shohami, E. Changes of biological reducing activity in rat brain following closed head injury: a cyclic voltammetry study in normal and heat-acclimated rats. *J. Cereb. Blood Flow Metab.* **17**:273–279; 1997.

[34] Kohen, R. The use of cyclic voltammetry for the evaluation of oxidative damage in biological samples. *J. Pharm. Meth.* **29**:185–193; 1993.

Bioassays for Oxidative Stress Status (BOSS). Edited by W.A. Pryor

BREATH ALKANES AS A MARKER OF OXIDATIVE STRESS IN DIFFERENT CLINICAL CONDITIONS

ELAHEH AGHDASSI and JOHANE P. ALLARD

Department of Medicine, The Toronto Hospital, General Division, Toronto, Canada

(Received 3 September 1999; Revised 14 December 1999; Accepted 25 January 2000)

Abstract—We assessed oxidative stress in three different clinical conditions: smoking, human immunodeficiency virus (HIV) infection, and inflammatory bowel disease, using breath alkane output and other lipid peroxidation parameters such as plasma lipid peroxides (LPO) and malondialdehyde (MDA). Antioxidant micronutrients such as selenium, vitamin E, C, β-carotene and carotenoids were also measured. Lipid peroxidation was significantly higher and antioxidant vitamins significantly lower in smokers compared to nonsmokers. Beta-carotene or vitamin E supplementation significantly reduced lipid peroxidation in that population. However, vitamin C supplementation had no effect. In HIV-infected subjects, lipid peroxidation parameters were also elevated and antioxidant vitamins reduced compared to seronegative controls. Vitamin E and C supplementation resulted in a significant decrease in lipid peroxidation with a trend toward a reduction in viral load. In patients with inflammatory bowel disease, breath alkane output was also significantly elevated when compared to healthy controls. A trial with vitamin E and C is underway. In conclusion, breath alkane output, plasma LPO and MDA are elevated in certain clinical conditions such as smoking, HIV infection, and inflammatory bowel disease. This is associated with lower levels of antioxidant micronutrients. Supplementation with antioxidant vitamins significantly reduced these lipid peroxidation parameters. The results suggest that these measures are good markers for lipid peroxidation.   © 2000 Elsevier Science Inc.

Keywords—Lipid peroxidation, Breath alkanes, Oxidative stress, Pentane, Ethane, Free radicals

INTRODUCTION

Lipid peroxidation in biological tissues has been investigated by measurement of different peroxidation products such as lipid hydroperoxides, conjugated dienes, and malondialdehyde. Lipid peroxidation also produces hydrocarbon gases, ethane and pentane, which are respectively produced from the peroxidation of omega-3 or omega-6 polyunsaturated fatty acids (PUFA). Measurement in the exhaled breath of these volatile alkanes is another way of assessing lipid peroxidation and the effect of diseases and antioxidant supplementation on oxidative stress. The goals of this article are (i) to review the advantages and limitations of the breath alkane output and describe the method and (ii) to report the results of our studies in certain clinical conditions such as smoking, HIV infection, and inflammatory bowel disease.

Breath alkane output as an index of lipid peroxidation

Evidence that volatile hydrocarbons arise after peroxidation of unsaturated fats was provided by in vitro studies where it was shown that hydrocarbon gases evolved during the decomposition of fatty acid hydroperoxides by β-scission [1,2]. Volatile hydrocarbons were also shown to arise during lipid peroxidation from in vivo studies. In these studies, dietary antioxidants and lipid sources significantly influenced pentane and ethane output, supporting the concept that breath alkane output is a good index of lipid peroxidation [3–6]. This method was also used to study lipid peroxidation induced by halogenated hydrocarbons [7–9], ozone [6], and ethanol [10–12].

In humans, breath alkane output was initially difficult to measure and only a few reports were published [13, 14]. Sensitivity and accuracy were insufficient mainly because subjects were breathing room air with variable ambient concentrations of alkanes. This noninvasive method of measuring breath alkane was finally standard-

---

Address correspondence to: Dr. Johane P. Allard, Toronto General Hospital, University Health Network, 200 Elizabeth Street, Eaton Wing, 9th Floor, 217a, Toronto, Ontario M5G-2C4 Canada; Tel: (416) 340-4413; Fax: (416) 348-0065; E Mail: johane.allard@utoronto.ca.

Reprinted from: *Free Radical Biology & Medicine, Vol. 28, No. 6, pp. 880-886, 2000*

ized and validated as a measure of lipid peroxidation in humans [15–17]. Some of these initial studies reported only on BPO [15,18,19] but later BEO was also measured [20,21–25]. A significant negative correlation between BPO and plasma vitamin E was demonstrated [19]. The influence of feeding on BPO was also assessed after an overnight fast and then after an oral intake of either 75 g glucose or 50 g of fat. No significant change was observed between fasting and fed states [26]. BPO was also measured before and after 10 d supplementation of vitamin E (1000 IU/d) [19] and there was a significant decrease in BPO. BPO was used to determine the recommended dietary allowances for vitamin E and elucidate the optimal relationship between polyunsaturated fatty acid PUFA and vitamin E intake [19,27]. In addition to the effect of antioxidant and lipid sources on BPO, xenobiotic exposure such as smoking was studied [20,28]. Cigarette smoke contains large amount of free radicals that are capable of initiating lipid peroxidation. It also contains high concentrations of hydrocarbons [28]. Results showed that smoking increased both BPO and BEO [20]. Chronic smokers still had increased BPO and BEO 5 d after the last cigarette, suggesting that the increase in breath alkane output was due to increased lipid peroxidation rather than high hydrocarbon content in cigarette smoke [28].

Strengths and weaknesses of breath alkane output

The method of measuring alkane output is not without potential sources of error. These should be taken into account when designing a study. One of the problems is the accumulation of alkanes in the lungs due to exposure to hydrocarbons in the environment. The lungs must be adequately washed-out with hydrocarbon-free air (HCFA) and breath samples should not be contaminated with ambient air. It was found that 4 min of wash-out was sufficient [15]. The site of hydrocarbon production could also alter the proportion of gases exhaled. Peroxidation in the lungs could potentially give higher values as compared with peroxidation elsewhere because hydrocarbons from the lungs would be exhaled without body distribution and metabolism. Another potential source of breath hydrocarbon is the gastrointestinal tract, from the action of intestinal bacteria on linoleate hydroperoxide [29]. Human colonic flora can also produce pentane but only if it is supplied with an appropriate substrate like corn oil [30]. Finally, changes in hydrocarbon metabolism and distribution can both potentially alter the relationship between breath alkane output and lipid peroxidation [31]. Ethane metabolism in humans is slower than pentane which is more influenced by liver function. Therefore liver disease and alcohol consumption can also affect BPO and BEO [10,11,28,32].

We designed all our studies with these potential confounding variables in mind. Study subjects followed a controlled diet that provides a PUFA to saturated fatty acid ratio of 0.3:1. They are also screened for smoking, liver dysfunction, ethanol intake and antioxidant supplementation. In addition, in smokers, BPO and BEO are measured 12 h after the last cigarette so that there is no influence from the hydrocarbon content of the cigarette smoke. A 12 h fast is also maintained to allow for other measurements.

Breath alkane output methodology

Breath sample collection. Subjects breathed for 4.0 min through a cardboard mouthpiece connected to a Rudolph valve from a Tedlar bag (Aerovironment, Monrovia, CA, USA) containing hydrocarbon-free air (HCFA) (0.1 ppm total hydrocarbons expressed as methane, Ultra-Zero Air, Prax Air, Canada). The exhaled air was discarded. Thus the initial atmospheric air with its hydrocarbon contents was flushed from the lungs. During the succeeding 2.0 minutes, while the HCFA was inspired, breath was collected into a second Tedlar bag for alkane measurements. The volume of air collected was recorded by a Medishield spirometer attached to the output portion of the Rudolph valve. Throughout the 6 min the nose was closed by a clip.

Analysis of breath for alkanes. The analysis was performed according to the method described by Lemoyne et al. [15] on a Shimadzu GC 14A gas-solid chromatograph (Mandel Scientific Company Limited, Guelph, Ontario) set at range 4 and sensitivity 10^4 and equipped with a flame ionization detector (FID). Hydrocarbon-free nitrogen gas was used as a carrier at a flow of 20 ml/min. Pentane and ethane gases were analyzed in a column filled with Porasil-D (Chromatographic Specialties Inc, Brockville, ON, Canada) in a stainless steel coil held isothermally at 60°C with a calibration curve derived from known concentrations of pentane and ethane standard (1 ppm ethane; 1 ppm pentane, Prax Air). One flame of double FID was used, at 200°C, with hydrogen at a flow rate of 30 ml/min (0.5 kg/cmxcm) and medical compressed air at a flow rate of 0.8 L/min for maximum sensitivity. Finally, the output of the FID was led to a recorder (Shimadzu CR 501 Chromatopac, Mandel Scientific Company, Ltd.). The peak areas under the curve for ethane and pentane were measured using 50 ml of the breath sample. The concentrations of ethane and pentane production were then calculated and expressed in pmol/kg/min.

Other measurements of oxidative stress used in our laboratory

Plasma malondialdehyde (MDA). Plasma MDA was assessed by the thiobarbituric acid (TBA) method as described by Draper et al. [33]. The procedures involves denaturing the plasma protein with 10% trichloroacetic acid solution and adding 0.05% BHT solution before digestion on a heating block at 95°C for 30 min. The mixture is then centrifuged and the supernatant is combined with the TBA solution in a 1:1 v/v ratio and heated at 95°C for 30 min. After cooling, the reaction mixture is extracted with 1 ml of n-butanol using a vortex mixture. A 0.5 ml aliquot of the extract is mixed with 0.25 ml methanol and 0.25 ml of mobile phase (15% acetonitrile and 0.6% tetrahydrofuran in 5 mM phosphate buffer) and 20 μl is injected onto a reverse phase C18 uBondpak high-performance liquid chromatography (HPLC) column (Chromatographic Specialties Inc., Brockville, ON, Canada).

Plasma lipid peroxide levels. Plasma lipid peroxides were assessed using a commercially available kit (Kamiya lipid peroxide kit, Biomedical Co., Thousand Oaks, CA, USA).

Plasma α- and γ-tocopherol. These were analyzed using an isocratic reverse phase HPLC and fluorescence spectrophotometry at 294 nm according to the method of Nata et al. [34].

Plasma ascorbic acid. Samples were analyzed for ascorbic acid by spectrophotometry [35]. In this method, total biologically active ascorbic acid concentrations were determined by spectrophotometry at 521 nm using 2,4-dinitrophenyl hydrazine as a chromogen.

Plasma α- and β-carotene. Plasma levels of carotenoids were determined by the method of Sapuntzakis et al. [36]. In this method, a reversed phase C18 column was used with an isocratic solvent system (methanol/acetonitrile/tetrahydrofuran, 50:45:5, v/v/v) after a hexane extraction using 200 μl of plasma sample.

OUR PREVIOUS WORK

Smokers

Cigarette smoke is a major environmental source of free radicals and other oxidant species [37] that can recruit and activate phagocytes to produce more free radicals endogenously [38]. This results in oxidative cellular damage [39–41] that is believed to contribute to the pathogenesis of chronic lung disease and perhaps

lung cancer. Smokers have been reported to have increased oxidative stress as measured by lipid peroxidation products in plasma [18]. In addition they have lower plasma concentrations of antioxidant vitamins such as vitamin C, β-carotene, and vitamin E [42].

The ability of β-carotene (BC) to reduce lipid peroxidation was investigated in our laboratory [21]. In this randomized double-blind controlled trial, 42 nonsmokers and 28 smokers received either 20 mg BC or placebo (p) daily for 4 weeks. At baseline lipid peroxidation measured by BPO was significantly higher in the two smoking groups (BC: 8.8 ± 1.1, P: 9.4 ± 1.4 pmol/kg/min) than in the two nonsmoking groups (BC: 5.7 ± 0.5, P: 5.9 ± 0.6 pmol/kg/min) ($p < .005$). BPO decreased significantly only in smokers receiving BC (6.5 ± 0.7 pmol/kg/min; $p < .04$). Changes in BEO were not significant. This may be attributed to the greater variability of ethane measurements and a too small sample size to detect a significant change in BEO since sample size was calculated for BPO.

In another study [18], vitamin E supplementation, 800 IU/d for 2 weeks, decreased significantly BPO in 13 smokers. Smokers were also shown to have higher BPO when compared to 19 healthy nonsmokers (16.3 ± 1.9 vs. 5.8 ± 0.5 pmol/kg/min; $p < .001$). However, both groups had comparable plasma vitamin E and selenium concentrations.

Finally, the effect of vitamin C supplementation on lipid peroxidation and pulmonary function tests was investigated [22]. In this randomized double blind placebo controlled trial, 56 smokers received either 500 mg of vitamin C or placebo daily for 4 weeks. At baseline, BPO (C: 7.5 ± 1.4 vs. P: 7.0 ± 1.3 pmol/kg/min) and plasma MDA were similar between groups and did not change significantly after 4 weeks of vitamin C supplementation (C: 5.3 ± 0.9 vs. P: 5.5 ± 0.9 pmol/kg/min). No changes were detected in the pulmonary function tests. The population, in this trial, smoked a lower number of cigarettes per day. This may explain why BPO values were lower than those reported in the vitamin E and BC studies [18,21]. This may also explain the lack of effect from vitamin C supplementation, because the response to antioxidant supplementation is usually greater in those with higher lipid peroxidation indices at baseline [18,21].

In summary, smokers have elevated levels of breath alkane output. This is often associated with lower plasma antioxidant vitamins. Antioxidant supplementation with either vitamin E or BC reduces lipid peroxidation.

HIV infection

HIV infection induces a wide array of immunologic alterations resulting in the progressive development of opportunistic infections and malignancy, which results in acquired immunodeficiency syndrome (AIDS). Of the mechanisms contributing to this progression, oxidative stress induced by the production of reactive oxygen species (ROS) may play a critical role in the stimulation of HIV replication and the development of immunodeficiency [43,44]. In vitro experiments [44] have demonstrated that ROS can specifically activate the transcription factor nuclear factor (NF)-κB to induce the expression and replication of HIV. Addition of antioxidant vitamins to the system blocked this activation and inhibited HIV replication [45–47]. Several studies have now documented an excessive production of ROS in the HIV-infected population, regardless of the extent of their immunosuppression, based on measurements of lipid peroxidation indices in plasma and expired breath [23, 48–53]. This increase in lipid peroxidation may be moderated by a normal antioxidant defense system like antioxidant vitamins and enzymes that scavenge ROS. Previous studies have shown that patients with HIV infection may have deficiencies in many of these components, including selenium [23,54], vitamin A [55], E [56], C [57], and carotenoids [58].

In the studies we conducted, we compared HIV-infected subjects to healthy controls [23] for evidence of oxidative stress. Forty-nine HIV positive patients (25 asymptomatic and 24 with AIDS) with a mean age of 39 years (range: 25–64 years) were recruited from the Immunodeficiency Clinic. Exclusion criteria were acute opportunistic infection, smoking, initiation of antioxidant vitamin therapy before the study, hyperlipidemia, diabetes, kidney or liver dysfunction, intractable diarrhea (more than six liquid stools per day), vomiting or evidence of gastrointestinal bleeding.

At the time of the study, most patients were on antiretroviral treatment (zidovudine [Glaxo Welcome, Miss, Ont, Canada], lamivudine [Glaxo-Biochem, Miss, Ont, Canada], and sequinavir [Hoffman-Laroche, Miss, Ont, Canada]). The control group was composed of 15 (10 men and 5 women) healthy, seronegative nonsmokers with a mean age of 35 years (range: 21–60 years) recruited from the local population. Control subjects had no acute or chronic illnesses and were not taking any medication or nutritional supplements.

Lipid peroxidation determined from breath-alkane output and plasma lipid peroxides (LPO) concentration were significantly higher in the HIV-positive group when compared to control subjects (pentane: 9.05 $\pm$ 1.23 vs. 6.06 $\pm$ 0.56 pmol/kg/min, $p < .05$; ethane:28.1 $\pm$ 3.41 vs 11.42 $\pm$ 0.55 pmol/kg/min, $p < .005$; LPO: 50.7 $\pm$ 8.2 vs 4.5 $\pm$ 0.8 umol/L, $p < .005$). There were no significant differences in lipid peroxidation indices or plasma antioxidant vitamins between HIV-positive asymptomatic subjects and those with AIDS. The increase in lipid peroxidation in the HIV-positive subjects

was also associated with a lower plasma concentration of antioxidant micronutrients such as vitamin C, α-tocopherol, β-carotene, and selenium. These findings were consistent with other studies [48–51].

The mechanism underlying the increased oxidative stress in the HIV population remains unclear. In addition to an excessive production of ROS, which may be explained by polymorphonuclear leukocyte activation during infection episodes or by a pro-oxidant effect of tumor necrosis factor-α produced by activated macrophages [58], a weakened antioxidant defense system may play a role. Based on this, we decided to investigate the effect of antioxidant vitamin supplementation on oxidative stress. This was specifically relevant because, every year, large sums of money are spent by the HIV-infected population on various supplements.

We conducted a randomized double-blind, placebo-controlled trial to investigate the effects of 800 IU/d of dl-α tocopherol acetate and 1000 mg/d of ascorbic acid, or matched placebo, on oxidative stress and viral load in patients with HIV infection [24]. The duration of the supplementation was 3 months.

The same 49 subjects were randomly assigned to either the supplemented or the placebo groups. Compliance was verified by counting leftover pills and by measuring plasma vitamin concentrations. During the study, there was no significant change in body mass index or dietary intake record in either group. The results showed a significant increase in plasma vitamin E and C concentrations in the supplemented group when compared to the placebo group. Plasma retinol, carotenoids, zinc and selenium remained unchanged during the study, suggesting that other supplements were not taken. BPO, BEO, plasma MDA and lipid peroxides were not significantly different between the two groups at baseline. However, there was a significant decrease in these measurements after the 3 month vitamin supplementation when compared to the placebo group. Plasma HIV viral load was similar at baseline between the two groups. The change after 3 months of supplementation was observed to be more favorable in the vitamin-supplemented group with a mean log decrease of 0.45 $\pm$ 0.39 copies/ml compared with a mean log increase of 0.5 $\pm$ 0.4 copies/ml for the placebo group ($p = .1$; 95% confidence interval: -0.21 to -2.14). Linear regression analysis used to evaluate the association between viral load at baseline and the change after 3 months was significant ($p = .026$) indicating that the change after 3 months depended on the initial score.

Therefore, lipid peroxidation, as measured by breath alkane output, plasma MDA and LPO, is increased in the HIV-infected population. Vitamin E and C supplementation significantly reduced this oxidative stress, which was associated with a trend toward a reduction in viral load.

Inflammatory bowel disease

During acute intestinal inflammation, either in ulcerative colitis (UC) or Crohn's disease (CD), there is an intense flux of neutrophils out of circulation, into the inflamed mucosa and the intestinal lumen [59]. Large amounts of toxic ROS are then released from the phagocytic leukocytes and a pro-oxidative imbalance is created by this overproduction of ROS [60,61]. Increased mucosal production of ROS related to disease activity has been shown in colorectal biopsy specimens [62,63] and in stimulated mucosal phagocytes [64] from patients with inflammatory bowel disease compared with controls. In addition, plasma concentrations of vitamin A, C, E, and β-carotene are decreased in this population with 40–50% of patients at risk of developing hypovitaminosis [65–67].

We measured breath pentane output in a group of nonsmoking patients with Crohn's disease (CDAI 145 ± 28, range 28–296) [68] and found a higher level of breath pentane output when compared to healthy controls (11.6 ± 1.7 vs. 5.8 ± 0.5 pmol/kg/min, $p < .05$). These levels were in the same range as those observed in oxidatively stressed smokers (9.1 ± 1.4 pmol/kg/min). Plasma vitamin C was also lower. These data along with previous studies [69] provide additional support for the suggestion that this patient population is oxidatively stressed. Assessments of other plasma antioxidant vitamins and the effect of vitamin E and C supplementation on oxidative stress are currently being studied.

SUMMARY

We used breath alkane output, along with plasma MDA and LPO, as markers of lipid peroxidation in three different clinical settings: smoking, HIV infection, and inflammatory bowel disease. In these human populations, we documented an increase in lipid peroxidation when diseased patients were compared to healthy controls. This increase in lipid peroxidation was associated with lower plasma antioxidant micronutrients. Breath alkane output and the other lipid peroxidation parameters, were significantly reduced with antioxidant vitamin supplementation. These results suggest that breath alkane output, along with plasma MDA and LPO, are good indices of lipid peroxidation. Whether a reduction in lipid peroxidation with antioxidant supplementation will improve clinical outcome remains to be studied.

REFERENCES

[1] Mounts, T. L.; McWeeny, D. J.; Evans, C. D.; Dutton, H. J. Decomposition of linoleate hydroperoxides: precursors of oxidative dimers. *Chem. Phys. Lipids* **4**:197–202; 1970.

[2] Dumelin, E. E.; Tappel, A. L. Hydrocarbon gases produced during in vitro peroxidation of polyunsaturated fatty acids and de-composition of performed hydroperoxides. *Lipids* **12**:894–900; 1977.

[3] Dillard, C. J.; Dumelin, E. E.; Tappel, A. L. Effect of dietary vitamin E on expiration of pentane and ethane by the rat. *Lipids* **12**:109–114; 1977.

[4] Dillard, C. J.; Litov, R. E.; Tappel, A. L. Effect of dietary vitamin E, selenium, and polyunsaturated fats on in vivo lipid peroxidation in the rat as measured by pentane production. *Lipids* **13**:396–402; 1978.

[5] Hafeman, D. G.; Hoekstra, W. G. Lipid peroxidation in vivo during vitamin E and selenium deficiency in the rat as monitored by ethane evolution. *J. Nutr.* **107**:666–672; 1977.

[6] Dumelin, E. E.; Dillard, C. J.; Tappel, A. L. Effect of vitamin E and ozone on pentane and ethane expired by rats. *Arch. Environ. Health* **33**:129–135; 1978.

[7] Hafeman, D. G.; Hoekstra, W. G. Protection against carbon tetrachloride-induced lipid peroxidation in the rat by dietary vitamin E, selenium and methionine as measured by ethane evolution. *J. Nutr.* **107**:656–665; 1977.

[8] Gavino, V. C.; Dillard, C. J.; Tappel, A. L. Release of ethane and pentane from tissue slices: effect of vitamin E, halogenated hydrocarbons and iron overload. *Arch. Biochem. Biophys.* **233**:741–747; 1984.

[9] Sagai, M.; Tappel, A. L. Lipid peroxidation induced by some halomethanes as measured by in vivo pentane production in the rat. *Toxicol. Appl . Pharmacol.* **49**:283–291; 1979.

[10] Litov, R. E.; Gee, D. L.; Downey, J. E.; Tappel, A. L. The role of peroxidation during chronic and acute exposure to ethanol as deter-mined by pentane expiration in the rat. *Lipids* **167**:52–63; 1981.

[11] Litov, R. E.; Irving, D. H.; Downey, J. E.; Tappel, A. L. Lipid peroxidation: a mechanism involved in acute ethanol toxicity as demonstrated by in vivo pentane production in the rat. *Lipids* **13**:305–307; 1978.

[12] Herschberger, L. A.; Tappel, A. L. Effect of vitamin E on pentane exhaled by rats treated with methyl ethyl ketone peroxides. *Lipids* **17**:686–691; 1982.

[13] Dillard, C. J.; Litov, R. E.; Savin, W. M.; Dumelin, E. E.; Tappel, A. L. Effects of exercise, vitamin E, and ozone on pulmonary function and lipid peroxidation. *J. Appl. Physiol.* **45**:927–932; 1978.

[14] Hempel, V.; May, R.; Frank, H.; Remmer, H.; Koster, U. Isobutene formation during halothane anaesthesis in man. *Br. J. Anaesth.* **52**:989–992; 1980.

[15] Lemoyne, M.; Van Gossum, A.; Kurian, R.; Ostro, M.; Axler, J.; Jeejeebhoy, KN. Breath pentane analysis as an index of lipid peroxidation: a functional test of vitamin E status. *Am. J. Clin. Nutr.* **46**:267–272; 1987.

[16] Dumelin, E. E.; Tappell, A. L. Hydrocarbon gases produced during in vitro peroxidation of polyunsaturated fatty acids and decomposition of preformed hydroperoxides. *Lipids* **12**:894–900; 1977.

[17] Pryor, W. A.; Godber, S. S. Noninvasive measures of oxidative stress status in humans. *Free Radic. Biol. Med.* **10**:177–84; 1991.

[18] Hoshino, E.; Shariff, R.; Van Gossum, A.; Allard, J. P.; Pichard, C.; Kurian, R.; and Jeejeebhoy, K. N. Vitamin E suppresses increased lipid peroxide in cigarette smokers. *J. Parent. Ent. Nutr.* **14**:300–305; 1990.

[19] Van Gossum, A.; Shariff, R.; Lemoyne, M.; Kurian, R.; Jeejeebhoy, K. N. Increased lipid peroxidation after lipid infusion as measured by breath pentane output. *Am. J. Clin. Nutr.* **48**:1394–1399; 1988.

[20] Sakamoto, M. Ethane expiration among smokers and non smokers. *Nippon Eiseigaku Zasshi* **40**:835–840; 1985.

[21] Allard, J. P.; Royall, D.; Kurian, R.; Muggli, R.; Jeejeebhoy, K. N. Effects of B-carotene supplementation on lipid peroxide in humans. *Am. J. Clin. Nutr.* **59**:884–90; 1994.

[22] Aghdassi, E.; Royall, D.; Allard, J. P. Oxidative stress in smokers supplemented with vitamin C. *Int. J. Vitam. Nutr. Res.* **69(1)**:45–51; 1999.

[23] Allard, J. P.; Aghdassi, E.; Chau, J.; Salit, I.; walmsley, S. Oxidative stress and plasma antioxidant micronutrients in humans with HIV infection. *Am. J. Clin. Nutr.* **67**:143–147; 1998.

[24] Allard, J. P.; Aghdassi, E.; Chau, J.; Tam, C.; Kovacs, C.; Salit, I. E.; Walmsley, S. Effects of vitamin E and C supplementation on oxidative stress and viral load in HIV-infected subjects. *AIDS* **12**:1653–1659; 1998.

[25] Allard, J. P.; Kurian, R.; Aghdassi, E.; Muggli, R.; Royall, D. Lipid peroxidation during n-3 fatty acid and vitamin E supplementation in humans. *Lipids* **32**:535–541; 1997.

[26] McGee, C. D.; Mascarenhas, M. G.; Ostro, M. J.; Tsallas, G.; Jeejeebhoy, K. N. Selenium and vitamin E stability in parenteral solution. *JPEN* **9(5)**:568–570; 1985.

[27] Horwitt, M. K. Supplementaion with vitamin E. *Am. J. Clin. Nutr.* **47**:1088–1089; 1988.

[28] Wade, C. R.; Van Rij, A. M. In vivo lipid peroxidation in man as measured by the exhalation of volatile hydrocarbons: the effect of cigarette smoke inhalation. *Proc. Univ. Otago. Med. Sch.* **64**:75–76; 1986.

[29] Gelmont, D.; Stein, R. A.; Mead, J. F. The bacterial origin of rat breath pentane. *Biochem. Biophys. Res. Commun.* **102**:932–936; 1981.

[30] Hiele, M.; Ghoos, Y.; Rutgeerts, P.; Vantrappen, G.; Schoorens, D. Influence of nutritional substrates on the formation of volatiles by the fecal flora. *Gastroenterology* **100**:1597–1602; 1991.

[31] Van Rij, A.; Wade, C. R. The metabolism of low molecular weight hydrocarbon gases in man. *Free Rad. Res. Comms.* **4**:99–103; 1987.

[32] Hartmut, F.; Hintze, T.; Bimboes, S.; Remmer, H. Monitoring lipid peroxidation by breath analysis: endogenous hydrocarbons and their metabolic elimination. *Toxicol. Appl. Pharmacol.* **56**:337–44; 1980.

[33] Draper, H. H.; Squires, E. J.; Mahmoodi, H.; Wu, J.; Agarwal, S.; Hadley, M. A comparative evaluation of thiobarbituric acid methods for the determination of malondialdehyde in biological materials. *Free Radic. Biol. Med.* **15**:353–63; 1993.

[34] Nata, C.; Sapuntzakis, M. S.; Bhagavan, H.; Bowen, P. Low serum levels of carotenoids in sickle cell anaemia. *Eur. J. Haematol.* **41**:131–135; 1988.

[35] Health and nutrition survey, laboratory procedures used by the clinical chemistry division, centres for disease control for the 2nd health and nutrition examination survey (NHANES II), 1976–1980, USDHHS, public health services. IV-analytical methods, vitamin C, pp. 17–19, Atlanta.

[36] Sapuntzakis, M. S.; Bowen, P. E.; Kikendall, J. W.; Burgess, M. Simultaneous determination of serum retinol and various carotenoids: their distribution in middle-aged men and women. *J. Micronutr. Anal.* **3**:27–45; 1987.

[37] Pryor, W. A.; and Stone, K. Oxidants in cigarette smoke. *Ann. NY Acad. Sci.* **686**:12–27; 1993.

[38] Van antwerpen, L.; Theron, A. J.; and Myer, M. S.; Richards, G. A.; Wolmarans, L.; Booysen, U.; Vandermewe, C. A.; Sluis-Cremer, G. K.; Anderson, R. Cigarette smoke-mediated oxidant stress, phagocytes, vitamin C, vitamin E and tissue injury. *Ann. NY Acad. Sci.* **686**:53–65; 1993.

[39] Pryor, W. A. Free radical biology: xenobiotics, cancer and aging. *Ann. NY Acad. Sci.* **339**:1–22; 1982.

[40] Freeman, B. A.; and Crapo, J. D. Biology of disease. Free radical and tissue injury. *Lab. Invest.* **47**:412–426; 1982.

[41] Halliwell, B. Oxidants and human disease: some new concepts. *FASEB J.* **1**:358–364; 1987.

[42] Bolton-Smith, C. Antioxidant vitamin intakes in Scottish smokers and non-smokers. *Ann. NY Acad. Sci.* **686**:347–358; 1993.

[43] Wong, G. H. W.; McHugh, T.; Weber, R.; Goeddel, D. V. Tumour necrosis factor alpha selectively sensitizes human immunodeficiency virus infected cells to heat and radiation. *Proc. Natl. Acad. Sci. USA* **88**:4372–4376; 1991.

[44] Schreck, R.; Rieber, P.; Baeuerle, P. A. Reactive oxygen intermediates as apparently widely used messengers in the activation of the NF-kβ transcription factor and HIV-1. *EMBO J.* **10**:2247–2258; 1991.

[45] Harakeh, S.; Jariwalla, R. J.; Pauling, L. Suppression of human immunodeficiency virus replication by ascorbate in chronically

and acutely infected cells. *Proc. Natl. Acad. Sci. USA* **87**:7245–7249; 1990.

[46] Harakeh, S.; Jariwalla, R. J. Comparative study of the anti-HIV activities of ascorbate and thiol-containing reducing agents in chronically HIV-infected cells. *Am. J. Clin. Nutr.* **54(6)**:1231S–1235S; 1991.

[47] Cagu,S. R.; Beckman, B. S.; Rangan, S. R. S.; Agrawal, K. C. Increased therapeutic efficacy of Zidovudine in combination with vitamin E. *Biochem. Biophys. Res. Commun.* **165**:401–407; 1989.

[48] Halliwell, B.; Cross, C. E. Reactive oxygen species, antioxidants, and acquired immunodeficiency syndrome. *Arch. Intern. Med.* **151**:29–31; 1991.

[49] Rivillard, J. P.; Vincent, C. M. A. Lipid peroxide in human immunodeficiency virus infection. *JAIDS* **5**:637–638; 1992.

[50] Postaire, E.; Massias, L.; Lopez, O.; Mollereau, M.; Hazebroucq, G. Alkane measurements in human deficiency virus infection. In: Pasquier Birkhauser, P., ed. *Oxidative stress, cell activation and viral infection.* Basel: Verlag; 1994:333–340.

[51] Sonnerborg, A.; Carlin, G.; Akerlund, B.; Jarstrand, C. Increased production of malondialdehyde in patients with HIV infection. *Scand. J. Infect. Dis.* **20**:287–290; 1988.

[52] Malvy, D. J. M.; Richard, M. J.; Arnaud, J.; Favier, A.; Amedee-Manesme, O. Relationship of plasma malondialdehyde, vitamin E and antioxidant micronutrients to human immunodeficiency virus-1 seropositivity. *Clin. Chim. Acta* **224**:89–94; 1994.

[53] Coutellier, A.; Bonnefont-Rousselot, D.; Delattre, J.; Lopez, O.; Jaudon, M. C.; Herson, S.; Emerit, J. Stress oxidatif chez 29 sujets seropositifs. *Presse Med.* **21**:1809–1812; 1992.

[54] Cirelli, A.; Ciardi, M.; de simone, C.; Sorice, F.; Giordano, P.; Ciaralli, L.; Constantini, S. Serum selenium concentration and disease progress in patients with HIV infection. *Clin. Biochem.* **24**:211–214; 1991.

[55] Simba, R. D.; Graham, N. M. H.; Caiaffa, W. T. Increased mortality associated with vitamin A deficiency during human immunodeficiency virus type 1 infection. *Arch. Intern. Med.* **153**:2149–2154; 1993.

[56] Javier, J. J.; Fodyce-Baum, M. k.; Beach, R. S. Antioxidant micronutrients and immune function in HIV-1 infection. *FASEB. Proc.* **4(4)**:A940; 1990.

[57] Bogden, J. D.; Baker, H.; Frank, O.; Porez, G.; Kemp, F.; Bruening, K.; Louria, D. Micronutrient status and human immunodeficiency virus (HIV) infection. *Ann. NY Acad. Sci.* **587**:189–195; 1990.

[58] Das, U. N.; Podma, M.; Sogar, P. S.; Ramesh, G.; Koratkar, R. Stimulation of free radical generation in human leukocytes by various agents including tumour necrosis factor is a calmodulin-dependent process. *Biochem. Biophys. Res. Commun.* **167**:1030–6; 1990.

[59] Saverymuttu, S. H.; Peters, A. M.; Lavender, J. P.; Chadwick, V. S.; Hodgson, H. J. In vivo assessment of granulocyte migration in diseased bowel in Crohn's disease. *Gut* **26(4)**:378–83; 1985.

[60] Harris, M. L.; Schiller, H. J.; Reilly, P. M.; Donowitz, M.; Grisham, M. B.; Bulkley, G. B. Free radicals and other reactive oxygen metabolites in inflammatory bowel disease. Cause, consequence, or epiphenomena? *Pharmacol. Ther.* **53**:375–408; 1992.

[61] Klebanoff, S. J. Oxygen metabolites from phagocytes. In: Gallin, J. I.; Goldstein, I. M.; Snyderman, R., eds. *Inflammation: basic principles and clinical correlates.* New York: Raven Press; 1992: 541–588.

[62] Simmond, N. J.; Allen, R. E.; Stevens, T. R. J.; Van Someren, R. N. M.; Blake, D. R.; Rampton, D. S. Chemiluminescence assay of reactive oxygen metabolites in inflammatory bowel disease. *Gastroenterology* **103**:186–196; 1992.

[63] Keshavarzian, A.; Sedghi, S.; Kanofsky, J.; List, T.; Robinson, C.; Ibrahim, C.; Winship, D. Excessive production of reactive oxygen metabolites by inflamed colon: analysis by chemiluminescence probe. *Gastroenterology* **103**:177–185; 1992.

[64] Williams JG. Phagocytes, toxic oxygen metabolites and inflammatory bowel disease: implications for treatment. *Ann. R. Coll. Surg. Eng.* **72**:253–262; 1990.

[65] Abad-Lacruz, A.; Fernandez-Banares, F.; Cabre, E.; Gil, A.; Esteve, M.; Gonzales-Huix, F.; Xiol, X.; Gassule, M. A. The effect of total enteral tube feeding on the vitamin status of malnourished patients with inflammatory bowel disease. *Int. J. Vitam. Nutr. Res.* **58:**428–435; 1988.

[66] Fernandez-Banares, F.; Abad-Lacruz, A.; Xiol, X.; Gine, J. J.; Dolz, C.; Cobre, E.; Esteve, M.; Gonzales-Huix, F.; Gassull, M. A. Vitamin status in patients with inflammatory bowel disease. *Am. J. Gastroenterol.* **84:**744–788; 1989.

[67] D'Odorico, A.; Pozzato, F.; Minotto, M.; Bonvicini, P.; Venturic, C.; Nianqing, L.; D'inca, R.; Di leo, V.; Sturniolo, C. G. Plasma carotenoids and other antioxidant levels in Crohn's disease. *Gastroenterology* 110:A897; 1996.

[68] Lau, H.; Ball, A.; Aghdassi, E.; Wellend, B.; Habal, F.; Wolman, S.; Allard, J. P. Evidence for increased lipid peroxide in Crohn's disease. CDDW Abstr. (1997).

[69] Kokoszka, J.; Nelson, R. L. Determination of inflammatory bowel disease activity by breath pentane analysis. *Dis. Colon Rectum* **36:**597–601; 1993.

ABBREVIATIONS

BPO—breath pentane output
BEO—breath ethane output
IBD—inflammatory bowel disease
ROS—reactive oxygen species
BC—β-carotene
P—placebo
MDA—malondialdehyde
NF-κB—nuclear factor kappa-β
HIV—human immunodeficiency virus
LPO—lipid peroxides
UC—ulcerative colitis
CD—Crohn's disease

Bioassays for Oxidative Stress Status (BOSS). Edited by W.A. Pryor

FORUM ON OXIDATIVE STRESS STATUS (OSS) AND ITS MEASUREMENT

WILLIAM A. PRYOR

Biodynamics Institute, Louisiana State University, Baton Rouge, LA, USA

(Received 22 June 2000; Accepted 22 June 2000)

This is the fourth installment in our Forum on OSS. The first three groups of papers on oxidative stress can be found in *FRBM* as follows:

- First group: Volume 27, pp. 1135–1196, December 1999
- Second group: Volume 28, pp. 503–536, February 15, 2000
- Third group: Volume 28, pp. 837–886, March 15, 2000.

The Forum on OSS is the largest we have published and it has attracted a great deal of interest. It is clear that measurement of the OSS of individuals has great promise and gobs of scientists are working on methods of development and verification. If and when OSS measurement becomes both reliable and routine, it will be a very important tool in the management of healthy living and a validation of the importance of oxidants in general, and radicals in particular, in stress.

This fourth group of papers in this forum opens with an interesting contribution from Barbacanne and her colleagues on ways to detect superoxide release from

Address correspondence to: Dr. William Pryor, Biodynamics Institute, 711 Choppin Hall, Louisiana State University, Baton Rouge, LA 70803, USA.

endothelial cells. Superoxide release from endothelial cells may be involved in the early stages of atherosclerosis and plaque development and represents a critical analytical problem.

The second paper, by Pastorino et al., deals with the lingering problem of measuring lipid hydroperoxides in plasma. We have come a long way from the days when the TBARS test was taken as equivalent to lipid hydroperoxides, but the analysis for these critical oxidative intermediates remains fraught with difficulty and controversy.

The third paper, by Winterbourn and Kettle, presents a method for quantitation of bioactivity of hypochlorous acid from myeloperoxidase. It is becoming increasingly clear that HOCl is an important and recurring theme in the opera called Reactive Oxygen Species (ROS).

The last contribution by Mueller presents a new method for the quantitation of hydrogen peroxide. Since hydrogen peroxide measurement occurs in many biooxidative processes, this new method could be a very helpful new tool.

There will be at least one more Forum devoted to OSS methods. Persons who would like to submit a paper to this series should send a letter of inquiry to me (at the Biodynamics Institute, Fax: 225-388-4936).

Reprinted from: *Free Radical Biology & Medicine, Vol. 29, No. 5, pp. 387, 2000*

Bioassays for Oxidative Stress Status (BOSS). Edited by W.A. Pryor
© 2001 Elsevier Science B.V. All rights reserved.

DETECTION OF SUPEROXIDE ANION RELEASED EXTRACELLULARLY BY ENDOTHELIAL CELLS USING CYTOCHROME C REDUCTION, ESR, FLUORESCENCE AND LUCIGENIN-ENHANCED CHEMILUMINESCENCE TECHNIQUES

Marie-Aline Barbacanne,*,† Jean-Pierre Souchard,† Benoit Darblade,*
Jean-Pierre Iliou,‡ Françoise Nepveu,† Bernard Pipy,* Francis Bayard,* and Jean-François Arnal*

*INSERM U397 et Laboratoire de Physiologie, et CJF-9107, Toulouse Cedex, France; †Laboratoire Pharmacophores Redox, Phytochimie et Radiobiologie, Université Paul Sabatier; Toulouse Cedex, France; and ‡Institut de Recherches Servier, Suresnes, France

(Received 24 August 1999; Revised 25 April 2000; Accepted 18 May 2000)

Abstract—Endothelium produces oxygen-derived free radicals (nitric oxide, $NO^{\bullet}$; superoxide anion, $O_2^{\bullet-}$) which play a major role in physiology and pathology of the vessel wall. However, little is known about endothelium-derived $O_2^{\bullet-}$ production, particularly due to the difficulty in assessing $O_2^{\bullet-}$ when its production is low and to controversies recently raised about the use of lucigenin-enhanced chemiluminescence. We compared four techniques of $O_2^{\bullet-}$ assessment when its production is low. In the present study, we have compared ferricytochrome c reduction, electron spin resonance (ESR) spectroscopy using DMPO as spin trap, hydroethidine fluorescence, and lucigenin-enhanced chemiluminescence to assess $O_2^{\bullet-}$ production in cultured bovine aortic endothelial cells (BAEC). We focused our study on extracellular $O_2^{\bullet-}$ production because the specificity of the signal is provided by the use of superoxide dismutase, and this control cannot be obtained intracellularly. We found that the calcium ionophore A23187 dose-dependently stimulated $O_2^{\bullet-}$ production, with a good correlation between all four techniques. The signals evoked by postconfluent BAEC were increased 2- to 7-fold in comparison to just-confluent BAEC, according to the technique used. Ferricytochrome c 20 μm rather than at 100 μm appears more suitable to detect $O_2^{\bullet-}$. However, in the presence of electron donors such as NADH or NADPH, lucigenin-enhanced chemiluminescence generated high amounts of $O_2^{\bullet-}$. Thus, ferricytochrome c reduction, electron spin resonance (ESR), and hydroethidine fluorescence appear as adequate tools for the detection of extracellular endothelium-derived $O_2^{\bullet-}$ production, whereas lucigenin may be artifactual, even when a low concentration of lucigenin is employed.   © 2000 Elsevier Science Inc.

Keywords—Superoxide anion, Lucigenin, Chemiluminescence, Cytochrome c, Electron spin resonance, Endothelial cell, Free radicals

INTRODUCTION

In the late sixties, McCord and Fridovich showed that superoxide free radical anion ($O_2^{\bullet-}$) could be produced enzymatically in mammalian tissues, and demonstrated that superoxide dismutase (SOD) catalyzed the dismutation of $O_2^{\bullet-}$ [1]. The role of $O_2^{\bullet-}$ in nonspecific host defence has been recognized for a long time, and more

recently in the signal transduction of physiological communications as well as in the pathophysiological mechanisms of various processes [2]. Often, the implication of reactive oxygen species (ROS) in these processes is indirectly demonstrated through the use of antioxidant molecules. Indeed, the half-life of ROS as $O_2^{\bullet-}$ is very short and the assessment of its production is not an easy task, particularly in nonphagocytic cells such as endothelium. However, endothelium-derived ROS appears to play a crucial role in the pathophysiology of many processes, such as aging or atherosclerosis [3].

Address correspondence to: Jean-François Arnal, INSERM U397, CHU Rangueil, 31403 Toulouse Cedex, France; Tel: +33 (5) 61 32 21 47; Fax: +33 (5) 61 32 21 41; E-Mail: arnal@rangueil.inserm.fr.

Reprinted from: *Free Radical Biology & Medicine, Vol. 29, No. 5, pp. 388–396, 2000*

Four techniques were independently used to assess the relatively low production of $O_2^{\bullet-}$ by endothelium. The simplest and easiest technique consisted of following spectrophotometrically the reduction of ferricytochrome c by $O_2^{\bullet-}$. This technique was used to evaluate basal and stimulated $O_2^{\bullet-}$ generation from cultured endothelial cells [2,4–7] (to quote only the initial reports). In contrast, only very few studies employed electron spin resonance (ESR) and the spin trap 5,5-dimethyl-1 pyrroline-N-oxide (DMPO) to detect the endothelial production of $O_2^{\bullet-}$. Uncoupling of cellular reductase by menadione [8] and anoxia followed by reoxygenation [9] were the two conditions allowing detection of $O_2^{\bullet-}$ from cultured endothelial cells. Using ESR, we recently were able to detect $O_2^{\bullet-}$ from cultured bovine aortic endothelial cells (BAEC) using the calcium ionophore A23187 as a stimulus [10]. Finally, the most sensitive techniques for detecting the low $O_2^{\bullet-}$ production from endothelial cells seem to be fluorescence and lucigenin-enhanced chemiluminescence. Hydroethidine fluorescence has been applied as a detector of intracellular $O_2^{\bullet-}$ in endothelial cells [11], but is less frequently used than the lucigenin chemiluminescence technique, which has wide application in the assessment of extra- and intracellular $O_2^{\bullet-}$ production [12–15]. However, this technique was recently questioned because reduced lucigenin can itself generate $O_2^{\bullet-}$ [16–19] although others have rehabilitated lucigenin luminescence [20–22].

Thus, it would seem urgent to compare these techniques for assessing $O_2^{\bullet-}$ when its production is low, as in endothelial cells. In the present study, we applied ferricytochrome c reduction, ESR spectroscopy using DMPO as spin trap, hydroethidine fluorescence, and lucigenin-enhanced chemiluminescence to assess $O_2^{\bullet-}$ production in cultured BAEC. We focused our study on extracellular $O_2^{\bullet-}$ production because the specificity of the signal is provided from the use of SOD, and this control cannot be obtained intracellularly. The calcium ionophore A23187 was used to stimulate $O_2^{\bullet-}$ production and the effect of cell confluency on $O_2^{\bullet-}$ production was determined. As an NAD(P)H oxidase is suspected to be one of the major sources of extracellular $O_2^{\bullet-}$ production, and as these electron donors are often used to enhance $O_2^{\bullet-}$ production [23,24], we investigated the effect of NADH and NADPH on the signals.

METHODS

Cell culture and materials

BAEC were obtained as described previously [25] and cultured in Dulbecco's modified Eagle's (DME) medium supplemented with 10% heat-inactivated calf serum (CS) at 37°C and 1 ng/ml basic fibroblast growth factor

(bFGF) under a 10% CO_2 humidified atmosphere. The cells used in this study were between the fifth and fifteenth passage. To avoid confounding effects produced by differences in cell density upon initial seeding, all passages were made using a splitting ratio of 1:6. This ensured that the number of mitoses necessary to bring the cells to confluence was similar between experiments. Cell passages were always performed 3–6 d after confluency. This allowed for maturation of the cell phenotype before passage. When different sized culture dishes (20 cm^2), flasks (25 cm^2), or wells (9.6 cm^2) were employed, the amount of cells added per cm^2 of the culture surface was identical (about 17,000/cm^2) under each culture condition. Confluency was determined by visual inspection of the cells, and deemed present when > 95% of the cells were in contact with adjacent cells, as previously described [26]. Under these culture conditions, the cells invariably reached confluence 3 d after passage. Using this approach, just-confluent cells 3 d after passage (D0) and postconfluent cells (D0+4 or D0+6) were obtained by simply allowing 9 d in culture after passage. To allow comparison between just-confluent (D0) and postconfluent (D0+4 or D0+6) cells, BAEC were counted and the results indexed by the number of cells per cm^2: 0.6×10^5 at D0 and 10^5 at D0+6.

Except when specified, all reagents were purchased from Sigma (St. Louis, MO, USA): superoxide dismutase (SOD, ref. S-2515), catalase (ref. C-40), cytochrome c (ref. C 2506), bis-N-methylacridinium (lucigenin, ref. M 8010), β-nicotinamide adenine dinucleotide (NADH, ref. 340-105), b-nicotinamide adenine dinucleotide phosphate (NADPH, ref. 201-205), 5,5-dimethyl-1 pyrroline-N-oxide (DMPO), diethylenetriaminepenta-acetic acid (DTPA, ref. D-751), and activated charcoal. DMPO was purified before use by treatment with activated charcoal for 15 min as reported in the literature [27], passed through a membrane filter (0.2 μm, Sartorius), aliquoted, protected from light, and stored frozen at $-20°C$. Hydroethidine (HE) was purchased from Acros Organics (Geel West Zone 2, Belgium). Stock solutions of hydroethidine (10^{-2} m) were prepared by dissolving HE in dimethyl sulfoxide and stored under nitrogen at $-20°C$ until use.

Cytochrome c reduction

Cytochrome c reduction was then used to assess the $O_2^{\bullet-}$ production in BAEC (400,000 cells/well). BAEC were washed once with sodium phosphate buffer, and incubated with 1 ml of a medium containing 1 g/l glucose, 0.2 g/l $CaCl_2$ 0.0059 g/l DTPA, 4.54 g/l NaCl, 0.37 g/l KCl in sodium phosphate buffer (2.35 g/l NaH_2PO_4/ 7.61 g/l Na_2HPO_4, pH 7.4) with 20 μm cytochrome c at 37°C and with or without SOD (100 UI/ml). Cytochrome c reduction was determined at zero time to obtain basal values for 15 min. The absorbance of the medium was read spectrophotometrically at 550 nm against a distilled water blank. Reduction of cytochrome c in the presence of SOD was subtracted from the values without SOD. The absorbance differences between comparable wells with or without SOD were converted to equivalent $O_2^{\bullet-}$ release by using the molecular extinction coefficient for cytochrome c of 21 mM^{-1} cm^{-1}[28]. The $O_2^{\bullet-}$ production was expressed in pmol per min per 10^6 cells.

ESR measurements

BAEC cultured in a 25 cm^2 flask were washed with PBS, and then incubated with a mix containing 150 mM DMPO, 1 g/l glucose, 0.2 g/l CaCl$_2$, 0.0059 g/l DTPA, 0.15 g/l NaCl, 0.37 g/l KCl in sodium phosphate buffer (2.35 g/l NaH$_2$PO$_4$/7.61 g/l Na$_2$HPO$_4$, pH 7.4), A23187 for 15 min. The supernatant was then transferred to a flat quartz cell that was inserted in a TM 110 Bruker cavity. ESR spectra were recorded at room temperature with an ER 200 D Bruker spectrometer by starting a 3 min scan 5 min after the end of the incubation with the cells. The ESR spectrometer operated at 9.66 GHz with high frequency at 100 kHz, modulation amplitude 1 Gauss, time constant 0.5 s, microwave power 10 mW, field : mid range at 3500 Gauss, and scan range 200 Gauss. The intensity of the ESR signal was calculated by adding the heights of the four peaks, and expressed in arbitrary units [10,27].

Hydroethidine fluorescence

BAEC cultured in a 25 cm^2 flask were washed with PBS, and then incubated with a mix containing 10 μm HE, 1 g/l glucose, 0.2 g/l CaCl$_2$, 0.0059 g/l DTPA, 0.15 g/l NaCl, 0.37 g/l KCl in sodium phosphate buffer (2.35 g/l NaH$_2$PO$_4$/7.61 g/l Na$_2$HPO$_4$, pH 7.4), A23187 with and without SOD (100 U/ml) in 2 ml final volume. The supernatant was then transferred to a quartz cell and the oxidation of HE to E$^+$ was monitored fluorimetrically by exciting at 488 nm with a 5 nm slit and following emission at 610 nm with a 20 nm slit on a Jobin Yvon Fluoromax spectrofluorometer. The integration time was 1 ms. The blank corresponding to HE alone in the buffer was subtracted from the value.

Lucigenin-enhanced chemiluminescence

$O_2^{\bullet-}$ production was measured by chemiluminescence in the presence of the chemiluminogenic probe lucigenin. The medium was removed from the 20 cm^2 Petri dishes and the cells washed once with sodium phosphate buffer. The cells were then scraped off using a rubber policeman into 1 ml of a mix containing 1 g/l glucose, 0.2 g/l CaCl$_2$, 0.0059 g/l DTPA, 4.54 g/l NaCl, 0.37 g/l KCl in sodium phosphate buffer (2.35 g/l NaH$_2$PO$_4$/7.61 g/l Na$_2$HPO$_4$, pH 7.4) at 37°C. The big cellular aggregates were then dissociated by soft mild pipetting. Glass scintillation vials containing 1 ml of cells were placed in a scintillation counter (Packard Tricarb 4640, Packard, Meriden, CT, USA) switched to the out-of-coincidence mode because the photomultiplier tube of this apparatus provided the most sensitive tool to detect the low level of luminescence generated by the cells using a low lucigenin concentration (5 μm). Lucigenin (5 or 250 μm final) was then added. Counts were obtained at 1 min intervals at room temperature for 15 min. The background determined from vials containing all the components except lucigenin was subtracted from the values. Chemiluminescence was triggered with the calcium ionophore A23187 (3 and 10 μm). SOD (100U/ml) was added in the vial to assess extracellular $O_2^{\bullet-}$ production.

In preliminary experiments, cells had been detached by trypsinization (and two-step centrifugation in order to remove trypsin) instead of scraping, but as the reproducibility of the chemiluminescent signal provided by trypsinization was not as good as that given by scraped cells, the latter technique was used.

The lucigenin-enhanced chemiluminescence was also studied using a thermostatically (37°C) controlled 1251 LKB luminometer. The BAEC were scraped off using a rubber policeman in 400 μl HBSS and transferred into a luminometer cuvette, which was then placed in the luminometer. Chemiluminescence was triggered with A23187 (600 μl of final volume in the cuvette, 3 or 10 μm final concentration) and continuously monitored for 10 min.

Isolated vascular rings experiments

Studies were performed with isolated aortas from New Zealand White male rabbits. The animals were killed by an overdose of sodium pentobarbital. The thoracic aorta was isolated and kept in cold Krebs-Hepes buffer with the following millimolar composition : NaCl 118.3, KCl 4.69, CaCl$_2$ 1.25, MgSO$_4$ 1.17, K$_2$HPO$_4$ 1.18, NaHCO$_3$ 25.0, and glucose 11.1, pH 7.40. The aorta was cleaned of adherent adventitial tissue and cut into ring segments of 3 mm length. Aortic rings were suspended within less than 1 h after sacrifice in individual organ chambers filled with Krebs buffer (20 ml) continuously aerated by 95% O$_2$-5% CO$_2$ and maintained at 37°C. Care was taken not to injure the endothelium during ring preparation. Tension was recorded with a linear force transducer. The resting tension was increased to 1.5 g over a period of 1 h, and the ring segments were exposed to 80 mM KCl until the optimal tension for generating force during isometric contraction was reached. The vessels were preconstricted with phenylephrine (PE) to achieve 70% of maximal PE contraction (10^{-6} M). When a plateau was reached, the rings were exposed cumulatively to the endothelium-dependent vasodilator acetylcholine (ACh, 1 nM-3 μm) [29]. After 30 min incubation with lucigenin (5 or 250 μm) $\pm$ NAD(P)H 100 μm, responses to ACh were again assessed.

Statistical analysis

The data are expressed as mean $\pm$ standard error. Comparisons of data between two groups were made using Student t-test and between more than two groups by ANOVA and a Scheffe's post-hoc test when appropriate. P values $< .05$ were considered significant.

RESULTS AND DISCUSSION

Characterization of the reduction of cytochrome c signals

As various concentrations of cytochrome c have been reported in the literature (70, 20, 140, 80, and 100 μm) [4–7,30], respectively, and as different preparations of cytochrome c are commercialy available (purified with trichloroacetic acid (TCA) or not, acetylated or not), we first compared three preparations of cytochrome c at two different concentrations (20 and 100 μm) to detect $O_2^{\bullet-}$ generated by postconfluent BAEC (D0+6). We found that, at the concentration of 20 μm, C2506 (prepared with TCA), C7752 (prepared with acetic acid instead of TCA), and C4186 (acetylated on the lysine residue) provided similar SOD-inhibitable values (ΔDO) (Fig. 1). However, it was not the case at a 5-fold higher concentration (100 μm): whereas C4186 (acetylated) provided similar results as at 20 μm, the C2506 (TCA prepared) could no longer detect $O_2^{\bullet-}$ production at 100 μm, the C7752 (TCA-free) providing intermediate results (Fig. 1). These variations could be explained by the contamination by antioxidant present at various levels, which interference is obvious at high concentration (100 μm) but does not apparently disturb the measurement at a lower concentration (20 μm).

Thus, we applied the 20 μm concentration to detect $O_2^{\bullet-}$ generated by BAEC at various stages of confluency. Cytochrome c 20 μm had an absorbance of 0.152, and on incubation for 15 min with unstimulated postconfluent BAEC (D0+6), the absorbance increased to 0.165 in the absence of SOD and to 0.162 in the presence of SOD, the SOD-inhibitable signal being 0.003 $\pm$ 0.001

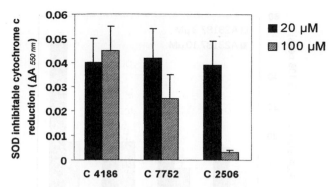

Fig. 1. Detection of $O_2^{\bullet-}$ generated by BAEC using three different preparations of cytochrome c at concentrations (20 and 100 μm).

(Fig. 2). Thus, the basal $O_2^{\bullet-}$ production could be detected from 10^6 BAEC, and is represented in Fig. 2. Cells exposed to 10 μm A23187 for 15 min induced a signal of 0.202 $\pm$ 0.02 in the absence of SOD and of 0.159 $\pm$ 0.003 in the presence of SOD (Fig. 2). Cytochrome c reduction from 10 μm A23187-stimulated growth arrested cells (D0+6) BAEC were 6.7-fold greater than those in just-confluent cells (D0) (Fig. 2). Analogous results were found using A23187 3 μm. The estimated $O_2^{\bullet-}$ production by A23187 (10μm) stimulated postconfluent cells was 136 pmoles/min/10^6 cells.

In 1996, we reported that proliferating endothelial cells released more $O_2^{\bullet-}$ than postconfluent cells [30]. However, the concentration of cytochrome c (C2506) was 100 μm. We know now that the lack of sensitivity is important in this case. The estimated production of $O_2^{\bullet-}$ appears to have been influenced more by the denominator (number of cells, lower in proliferating cells) than by the numerator (total amount of $O_2^{\bullet-}$, which did not differ between proliferating and postconfluent cells). Other re-

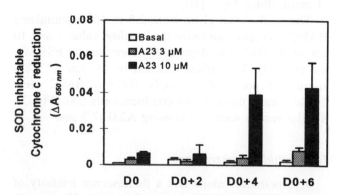

Fig. 2. SOD-inhibitable cytochrome c reduction by BAEC (absorbance/10^6 cells for 15 min). Effect of 0, 3, and 10 μm A23187 on the signals obtained from BAEC at various stages of confluency [from just-confluent (DO) to postconfluent (D0+6) BAEC]. The final concentration of cytochrome c was 20 or 100 μm. Data are the average of triplicate incubations, and representative of two different experiments. D0 and D0+2 are $p < .05$ vs. respective D0+6.

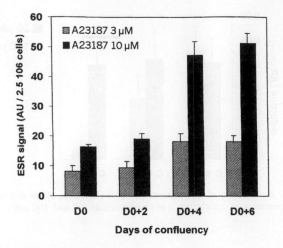

Fig. 3. ESR signal (Arbitrary Units/2.5 × 10⁶ cells for 15 min) from BAEC stimulated either with 3 and 10 μm A23187. Data are the average of triplicate incubations, and representative of two different experiments. D0 and D0+2 are $p < .05$ vs. respective D+6.

ported data where a postconfluent effect was not observed [7] could be explained by the high concentration of cytochrome c (80 μm).

ESR signals detected using DMPO as spin trap

We have previously shown that incubation of DMPO for 15 min with postconfluent BAEC (D0+6) stimulated by the calcium ionophore A23187 10 μm elicited a typical ESR signal resulting from the DMPO-OH adduct [10]. When SOD 30 U/ml was added, the ESR signal was completely suppressed, whereas coincubation of DMPO, catalase (2000 U/ml), and A23187 10 μm did not alter the ESR signal given by BAEC. Altogether, these results demonstrated that the ESR adduct DMPO-OH detected in the supernatant of BAEC originated from the trapping of extracellular $O_2^{\cdot-}$ [10].

The ESR signal given from 2.5 × 10⁶ unstimulated BAEC was less than twice the baseline value. 3 and 10 μm of A23187 dose-dependently increased the ESR signal (Fig. 3). The signals elicited by 10 μm A23187-stimulated postconfluent cells (D0+6) BAEC were 3-fold greater those in just-confluent cells (D0) (Fig. 3). Similar results were found using A23187 3 μm.

Hydroethidine fluorescence

Hydroethidine alone had a fluorescence intensity of about 500000 ± 25000, and incubation for 15 min with unstimulated postconfluent BAEC (D0+6) increased the fluorescence intensity to 714300 in the absence of SOD and to 599058 in the presence of SOD, the SOD-inhibitable signal being 115242 ± 30000 AU (Fig. 4). Thus, the basal $O_2^{\cdot-}$ production could be detected from 10⁶

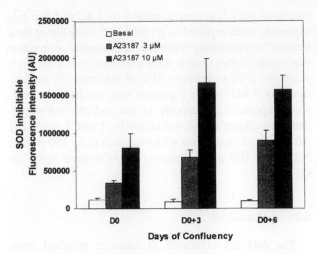

Fig. 4. SOD-inhibitable hydroethidine fluorescence from BAEC (Arbitrary Units/10⁶ cells for 15 min). Effect of 0, 3, and 10 μm A23187 on the signals obtained from BAEC at various stages of confluency [from just-confluent (D0) to postconfluent (D0+6) BAEC]. Data are the average of triplicate incubations, and representative of two different experiments. D0 are $p < .05$ vs. respective D0+2 and D0+6.

BAEC. The signal increased 16.4-fold when BAEC were stimulated with 10 μm A23187 (Fig. 4). The signals elicited by 10 μm A23187-stimulated postconfluent cells BAEC were 2-fold greater than those elicited by just-confluent cells (Fig. 4). Analogous results were found using A23187 3 μm.

Lucigenin-enhanced chemiluminescent signals

We first evaluated the effect of a low concentration (5 μm) of lucigenin on the $O_2^{\cdot-}$ production by endothelial cells using a scintillation counter. The basal chemiluminescent signal from postconfluent cells (D0+4) was inhibited (21%) by SOD (100 U/ml), which corresponded to extracellular $O_2^{\cdot-}$ production, SOD not being taken up by endothelial cells, at least after a short period of incubation [31]. The counterpart of the signal probably corresponds to intracellular $O_2^{\cdot-}$ production. Thus, lucigenin allowed the detection of basal $O_2^{\cdot-}$ production from unstimulated BAEC (Fig. 5A). A23187 dose-dependently increased the chemiluminescent signal in postconfluent cells. This signal was 42% inhibited by SOD. The SOD-inhibitable chemiluminescent signals are shown in Fig. 5A. The signals from 10 μm A23187-stimulated postconfluent cells BAEC were 7-fold greater than those from just-confluent cells (D0) (Fig. 5A).

We then evaluated the effect of a high concentration (250 μm) of lucigenin on $O_2^{\cdot-}$ production from BAEC. This 50-fold higher concentration of lucigenin elicited SOD-inhibitable chemiluminescent signals that were on average 46-fold higher than the low concentration (compare Figs. 5A and B). The SOD-inhibitable chemilumi-

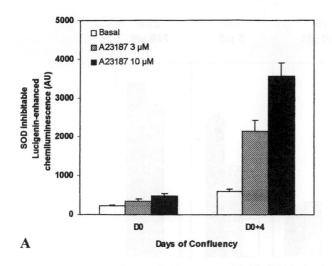

A

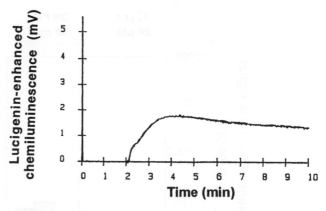

Fig. 6. Representative tracings from lucigenin (250 μm) -enhanced chemiluminescence (using a thermostatically controlled 1251 LKB luminometer) from 2×10^6 postconfluent BAEC stimulated with 10 μm A23187. The basal chemiluminescence (obtained before stimulating the cells) was automatically subtracted, and 30% of the signal was SOD-inhibitable.

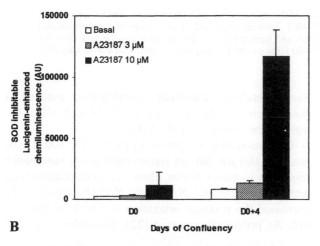

B

Fig. 5. SOD-inhibitable lucigenin-enhanced chemiluminescence from BAEC (Arbitrary Units/10^6 cells for 15 min) using a scintillation counter switched to the out-of-coincidence mode. Effect of 0, 3, and 10 μm A23187 on the signals obtained from just-confluent (DO) and postconfluent (D0+4) BAEC with (A) 5 μm lucigenin and (B) 250 μm lucigenin. Data are the average of triplicate incubations, and representative of two different experiments. D0 are $p < .05$ vs. respective and D0+4.

nescent signals in postconfluent cells (D0+4) represented about 30% of the overall chemiluminescent signal. Again, the signals from 10 μm A23187-stimulated postconfluent BAEC were 10-fold greater than those from just-confluent cells (Fig. 5B).

We also compared the signals given by a scintillation counter switched to the out-of-coincidence mode with those provided by a classical luminometer (1251 LKB). The stimulation of postconfluent BAEC (2.5×10^6) with 10 μm A23187 elicited an immediate chemiluminescent signal in the presence of 250 μm lucigenin. A representative tracing is shown in Fig. 6. The SOD-inhibitable chemiluminescent signals represented 30% of the overall

chemiluminescent signal, and the signals from 10 μm A23187-stimulated postconfluent BAEC were 9.5-fold greater than those from just-confluent cells (not shown). Thus, the data obtained with the two chemiluminescent techniques were very similar. However, the sensitivity of the luminometer 1251 LKB did not allow the detection of a chemiluminescent signal when only 5 μm lucigenin was used.

Comparison of the signals as a function of confluency obtained with the four techniques

To allow comparison between just-confluent (D0) and postconfluent (D0+6) cells, BAEC were counted and the results indexed by the number of cells per cm^2: 0.6×10^5 at D0 and 10^5 at D0+6. The signals given by cytochrome c, ESR, fluorescence, and lucigenin-enhanced chemiluminescence from 10 μm A23187-stimulated growth arrested BAEC were, respectively, 6.7-, 3-, 2-, and 7-fold greater than those from just-confluent cells (compare Figs. 2–5). Similar results were found using A23187 3 μm. Thus, under basal and A23187 stimulation, and although the magnitude of the signals varied, all four techniques provided SOD-inhibitable signals suggesting that they detected roughly similar variations of $O_2^{\cdot-}$ production. Use of hydroethidine could, however, have led to a possible underestimation of extracellular $O_2^{\cdot-}$ production. This possible underestimation has recently been reported and the mechanism(s) analyzed by Benov et al. [32].

Effect of NADH and NADPH on the signals

NADH 100 μm increased the cytochrome c, ESR, and fluorescent signals 2.0-, 2.1-, and 1.8-fold, respectively,

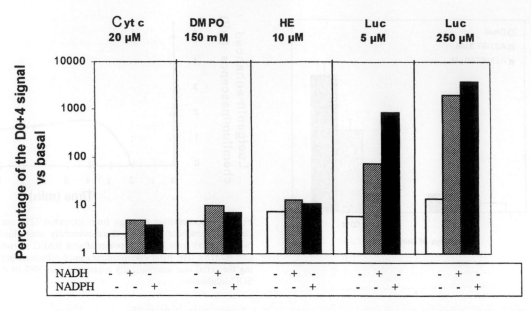

Fig. 7. Effect of NADH (100 μm) and NADPH (100 μm) on the intensity of the cytochrome c, ESR, fluorescence, and lucigenin signal from A23187 (3 μm) -stimulated postconfluent BAEC (2.5 × 10^6 cells). Data are expressed as percentage of the unstimulated postconfluent BAEC signal and are the average of triplicate incubations, and representative of two different experiments. Note the logarithmic scale. All these signals were 100% inhibitable with SOD 100 U/ml except the lucigenin signals, which were only 30% inhibitable.

of A23187-stimulated postconfluent BAEC (Fig. 7). NAPDH 100 μm also significantly increased the A23187-stimulated postconfluent BAEC, but to a lesser extent (1.6-, 1.5-, and 1.5-fold, respectively) (Fig. 7). These signals were completely inhibited by SOD (not shown).

In contrast, NADH 100 μm and NADPH 100 μm dramatically increased the 5 μm lucigenin-enhanced chemiluminescence (74- and 863-fold) compared to basal signals in postconfluent BAEC (Fig. 7). Similarly, NADH 100 μm and NADPH 100 μm dramatically increased the 250 μm lucigenin-enhanced chemiluminescence about >500- and 1000-fold compared to basal signals in post-confluent BAEC (Fig. 7). Only 30% of these signals were inhibited by SOD (not shown). Electron donors such as NADH or NADPH enhance endothelium-derived $O_2^{\bullet -}$ production in the presence of lucigenin [14,23,33]. However, the present data demonstrate that, under these conditions, lucigenin is mainly responsible for this $O_2^{\bullet -}$ production.

The biological importance of lucigenin-derived $O_2^{\bullet -}$ generation was assessed using a bioassay system. Measurement of the isometric tension of vascular rings provides a useful tool for assessment of the bioavailability of $NO^{\bullet}$. As $NO^{\bullet}$ and $O_2^{\bullet -}$ react together, the inactivation of $NO^{\bullet}$ by $O_2^{\bullet -}$ was evaluated by measuring the alteration of endothelium-dependent relaxation of rabbit thoracic aortic ring in the presence or absence of lucigenin. When the rabbit aortic rings were precontracted to 70% of the phenylephrine maximal contraction, the endothe-

lium-dependent vasodilator acetylcholine induced a dose-dependent relaxation maximally by 84± 8% at the highest concentration applied (Fig. 8).

Incubation with lucigenin 5 μm, NADH 100 μm or NADPH 100 μm did not significantly affect vasodilator responses to acetylcholine. However, coincubation with lucigenin 5 μm and NADH or NADPH significantly decreased the maximal relaxation to ~61% of the control. As previously reported [22], incubation with 250

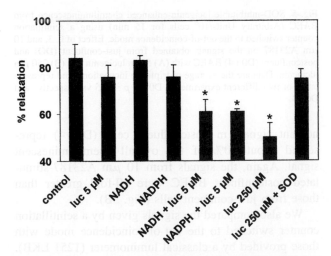

Fig. 8. Effect of lucigenin (luc) and NADH (100 μm) or NADPH (100 μm) in the presence or absence of superoxide dismutase (SOD, 200 U/ml) on endothelium-dependent relaxation of rabbit thoracic rings. Results are expressed in percentage of maximal relaxation in response to acetylcholine. Data are the average of triplicate experiments. *$p <$ .05 vs. control.

μm lucigenin for 30 min decreased the endothelium-dependent relaxation to $50 \pm 6\%$, this alteration being restored by SOD 200 U/ml.

Lucigenin-enhanced chemiluminescence has been extensively used as a tool for assessing $O_2^{\bullet-}$ production but recently became a matter of controversy [16–18,21,22, 34]. From the present data and those available in the literature, we propose that (i) under some very restricted conditions (SOD-inhibitable superoxide production, i.e., extracellular production, and in the absence of extracellular electron donors such as NADH or NADPH), lucigenin-enhanced chemiluminescence could have provided an adequate estimate of superoxide production; however, (ii) in most of the studies, most of the chemiluminescence arose from the intracellular compartment, where electron donors such as NADH or NADPH are very abundant. In addition, controls using SOD-inhibition of luminescence cannot be performed. Experiments using NADH or NADPH lucigenin-enhanced chemiluminescence are definitively artifactual and should be repeated using cytochrome c reduction or ESR technique.

In conclusion, ferricytochrome c reduction, ESR spectroscopy using DMPO as spin trap, and hydroethidine fluorescence appear suitable techniques to assess extracellular $O_2^{\bullet-}$ production in cultured endothelial cells, without major differences in sensitivity between these three techniques. Although more sensitive, lucigenin-enhanced chemiluminescence, in the presence of electron donors such as NADH or NADPH, generated high amounts of $O_2^{\bullet-}$ and this latter technique could not be recommended. The present study could help to interpret published data and to choose among the available techniques those suitable for the detection and quantitation of $O_2^{\bullet-}$ production in cultured endothelial cells or any cell types with a low level of superoxide production.

Acknowledgements — This work was supported by INSERM, Fondation de France, and ARC (national grant 5358).

REFERENCES

[1] McCord, J. M.; Fridovich, I. Superoxide dismutase. An enzymatic function for erythrocuprein (hemocuprein). *J. Biol. Chem.* **244:** 6049–6055; 1969.

[2] Suzuki, Y. J.; Forman, H. J.; Sevanian, A. Oxidants as stimulators of signal transduction. *Free Radic. Biol. Med.* **22:**269–285; 1997.

[3] Diaz, M. N.; Frei, B.; Vita, J. A.; Keaney, J. F. Antioxidants and atherosclerotic heart disease. *N. Engl. J. Med.* **337:**408–416; 1997.

[4] Matsubara, T.; Ziff, M. Superoxide anion release by human endothelial cells: synergism between a phorbol ester and a calcium ionophore. *J. Cell Physiol.* **127:**207–210; 1986.

[5] Steinbrecher, U. P. Role of superoxide in endothelial-cell modification of low-density lipoproteins. *Biochim. Biophys. Acta* **959:** 20–30; 1988.

[6] Holland, J. A.; Pritchard, K. A.; Pappolla, M. A.; Wolin, M. S.; Rogers, N. J.; Stemerman, M. B. Bradykinin induces superoxide anion release from human endothelial cells. *J. Cell Physiol.* **143:** 21–25; 1990.

[7] Murphy, H.; Shayman, J.; Till, G.; Mahroughi, M.; Owens, C.; Ryan, U.; Ward, P. Superoxide responses of endothelial cells to C5a and TNF-α: divergent signal transduction pathways. *Am. J. Physiol.* **263:**L51–59; 1992.

[8] Rosen, G. M.; Freeman, B. A. Detection of superoxide generated by endothelial cells. *Proc. Natl. Acad. Sci. USA* **81:**7269–7273; 1984.

[9] Zweier, J. L.; Kuppusamy, P.; Lutty, G. A. Measurement of endothelial cell free radical generation: evidence for a central mechanism of free radical injury in post ischemic tissues. *Proc. Natl. Acad. Sci. USA* **85:**4046–4050; 1988.

[10] Souchard, J.; Barbacanne, M.; Margeat, E.; Maret, A.; Nepveu, F.; Arnal, J. EPR detection of extracellular superoxide anion released by cultured endothelial cells. *Free Radic. Res.* **29:**441–449; 1998.

[11] Carter, W. O.; Kumar, P.; Narayaman, P.; Robinson, J. Intracellular hydrogen peroxide and superoxide anion detection in endothelial cells. *J. Leuk. Biol.* **55:**253–258; 1994.

[12] Pagano, P. J.; Tornheim, K.; Cohen, R. A. Superoxide anion production by rabbit thoracic aorta: effect of endothelium-derived nitric oxide. *Am. J. Physiol.* **265:**H707–H712; 1993.

[13] Kooy, N. W.; Royall, J. A. Agonist-induced peroxynitrite production from endothelial cells. *Arch. Biochem. Biophys.* **310:**352–359; 1994.

[14] Mohazzab-H, K.; Kaminski, P.; Wolin, M. NADH oxidoreductase is a major source of superoxide anion in bovine coronary artery endothelium. *Am. J. Physiol.* **266:**H2568–H2572; 1994.

[15] Münzel, T.; Sayegh, H.; Freeman, B. A.; Tarpey, M. M.; Harrison, D. G. Evidence for enhanced vascular superoxide anion production in nitrate tolerance: a novel mechanism underlying tolerance and cross-tolerance. *J. Clin. Invest.* **95:**187–194; 1995.

[16] Liochev, S. I.; Fridovich, I. Lucigenin as mediator of superoxide production: revisited. *Free Radic. Biol. Med.* **25:**926–928; 1998.

[17] Liochev, S. I.; Fridovich, I. Lucigenin (Bis-N-methylacridinium) as a mediator of superoxide anion production. *Arch. Biochem. Biophys.* **337:**115–120; 1997.

[18] Liochev, S. I.; Fridovich, I. Lucigenin luminescence as a measure of intracellular superoxide dismutase activity in *Escherichia coli*. *Proc. Natl. Acad. Sci. USA* **94:**2891–2896; 1997.

[19] Vasquez-Vivar, J.; Hogg, N.; Pritchard, K. A. Jr.; Martasek, P.; Kalyanaraman, B. Superoxide anion formation from lucigenin: an electron spin resonance spin-trapping study. *FEBS Lett.* **403:**127–130; 1997.

[20] Vargas, R.; Wroblewska, B.; Rego, A.; Hatch, J.; Ramwell, P. W. Estradiol inhibits smooth muscle cell proliferation of pig coronary artery. *Br. J. Pharmacol.* **109:**612–617; 1993.

[21] Li, Y.; Zhu, H.; Kuppusamy, P.; Roubaud, V.; Zweier, J.; Trush, M. Validation of lucigenin (Bis-N-methylacridinium) as a chemilumigenic probe for detecting superoxide anion radical production by enzymatic and cellular systems. *J. Biol. Chem.* **273:**2015–2023; 1998.

[22] Skatchkov, M.; Sperling, D.; Hink, U.; Mulsch, A.; Harrison, D.; Sindermann, I.; Meinertz, T.; Münzel, T. Validation of lucigenin as a chemiluminescent probe to monitor superoxide as well as basal vascular nitric oxide production. *Biochem. Biophys. Res. Commun.* **254:**319–324; 1999.

[23] Griendling, K. K.; Mineri, C. A.; Ollerenshaw, J. D.; Alexander, R. W. Angiotensin II stimulates NADH and NADPH activity in cultured vascular smooth muscle cells. *Circ. Res.* **74:**1141–1146; 1994.

[24] Griendling, K.; Alexander, R. W. Endothelium control of the cardiovascular system: recent advances. *FASEB J.* **10:**283–292; 1996.

[25] Gospodarowicz, D.; Moran, J.; Braun, D.; Birdwell, C. R. Clonal growth of bovine endothelial cells in tissue culture: fibroblast growth factor as a survival agent. *Proc. Natl. Acad. Sci. USA* **73:**4120–4124; 1976.

[26] Arnal, J.-F.; Yamin, J.; Dockery, S.; Harrison, D. G. Regulation of endothelial nitric oxide synthase mRNA, protein and activity during cell growth. *Am. J. Physiol.* **267:**C1381–C1388; 1994.

[27] Buettner, G. R.; Oberley, L. W. Consideration in the spin trapping of superoxide and hydroxyl radicals in aqueous systems using 5,5-dimethyl-1-pyrroline-1-oxide. *Biochem. Biophys. Res. Commun.* **83:**69–74; 1978.

[28] Van Gelder, B. F.; Slater, E. C. The extinction coefficient of cytochrome c. *Biochem. Biophys.Acta* **58:**593–595; 1962.

[29] Furchgott, R. F.; Vanhoutte, P. M. Endothelium-derived relaxing and contracting factors. *FASEB J.***3:**2007–2018; 1989.

[30] Arnal, J. F.; Tack, I.; Besombes, J. P.; Pipy, B.; Nègre-Salvayre, A. Nitric oxide and superoxide production during endothelial cell proliferation. *Am. J. Physiol.* **271:**C1521–C1526; 1996.

[31] Markey, B.; Phan, S.; Varani, J.; Ryan, U.; Ward, P. Inhibition of cytotoxicity by intacellular superoxide dismutase supplementation. *Free Radic. Biol. Med.* **9:**307–314; 1990.

[32] Bonov, L.; Sztejnberg, L.; Fridovich, I. Critical evaluation of the use of hydroethidine as a measure of superoxide anion radical. *Free Radic. Biol. Med.* **25:**826–831; 1998.

[33] Pagano, P. J.; Yasushi, I.; Tornheim, K.; Gallop, P. M.; Tauber, A. I.; Cohen, R. A. An NADPH oxidase superoxide-generating system in the rabbit aorta. *Am. J. Physiol.* **268:**H2274–H2280; 1995.

[34] Fridovich, I. Superoxide anion radical, superoxide dismutases, and related matters. *J. Biol. Chem.* **272:**18515–18517; 1997.

Bioassays for Oxidative Stress Status (BOSS). Edited by W.A. Pryor

KINETIC ANALYSIS OF LIPID-HYDROPEROXIDES IN PLASMA

ANTONIO M. PASTORINO, MATILDE MAIORINO, and FULVIO URSINI

Department of Biological Chemistry, University of Padova, Padova, Italy

(Received 24 January 2000; Revised 18 April 2000; Accepted 20 April 2000)

Abstract—We increased the precision of chemiluminescent procedure for measuring lipid hydroperoxides in plasma or lipoproteins by (i) escaping from extraction and chromatography of lipids, (ii) using detergent dispersed lipids, and (iii) calculating the results by fitting the photon emission rate with the integrated equation, which describes the model of the series of reactions. The use of kinetics instead of the crude integration of cps increases precision because at each measurement the correct reaction pathway is tested. This was relevant for the optimization of the analytical procedure, contributing to the elimination of possible side reactions. The relationship between lipid hydroperoxide content in the sample and cps is not linear; thus, the calculation of results through internal calibration is carried out using an exponential equation. This is in agreement with the reaction mechanism and raises the point of the linear calibration previously reported in other chemiluminescent procedures. Although sensitive and precise, this procedure suffers for being time consuming, requiring approximately 30 min per sample. Moreover, since no chromatography is used, information about the hydroperoxides in different lipid classes is missing. Obviously this will be solved when a validated procedure for quantitatively extracting lipid hydroperoxides is available.   © 2000 Elsevier Science Inc.

Keywords—Free radical, Lipid hydroperoxide, Luminol, Chemiluminescence

LIPID HYDROPEROXIDES

Lipid hydroperoxides (LOOH) are the main product of autoxidation of fatty acids and are present in virtually all foods containing fats, of which, above a given threshold, they account for spoilage and degradation [1]. Lipid hydroperoxides are also present, obviously at a much lower concentration, in living organisms where they have been suggested to play a role in the molecular mechanism of several pathological processes [2,3]. In fact, the concentration of LOOH and their decomposition products increases in diseases where oxidant free radicals are produced and, further, the level of lipid peroxidation has been adopted as a marker to support the claim about the free radical nature of a given pathophysiological condition [2–7].

The presence of lipid hydroperoxides in plasma lipoproteins is required for further oxidation of lipoprotein particles in the presence of both transition metals and lipoxygenase. Oxidatively modified lipoproteins seem to be involved in the activation of the series of cellular events ultimately resulting in atherosclerosis [8]. This brought to the focus the issue of understanding the source of lipid hydroperoxides in plasma lipoproteins, while the biological effect of oxidatively modified lipoproteins on vascular cells became a major issue in cardiovascular research.

Hydroperoxidation of unsaturated fats is a thermodynamically favorable event, which is kinetically controlled in the foodstuff, as well as in vivo, by several elements which, because of this, are classified as antioxidants. Lipid hydroperoxides have a 2-fold relation-

Antonio M. Pastorino has degrees in Chemistry (1965) and Pharmacy (1984) from the University of Padova, Italy. He served for 30 years as a research scientist at the Glaxo-Wellcome Laboratory for Biomedical Research in Verona. In the framework of studies on the anti-atherosclerotic activity of drugs, he had a long-term collaboration with the Department of Biological Chemistry of Padova, where he now serves as guest scientist.

Matilde Maiorino is Associate Professor of Biochemistry at the School of Medicine of the University of Padova. Her scientific experience ranges from mechanisms of lipid peroxidation and antioxidant defense to enzymology and regulation of gene expression.

Fulvio Ursini is Full Professor of Biochemistry at the School of Medicine of the University of Padova. He has experience on enzymology, lipid peroxidation, and antioxidant mechanism. His major fields are now the function of selenoenzymes and role of nutrition on oxidative resistance of plasma lipoproteins.

Address correspondence to: Fulvio Ursini, M.D., Department of Biological Chemistry, University of Padova, viale G. Colombo 3, I-35121, Padova, Italy; Fax: +39 (049) 8073310; E-Mail: ursini@civ.bio.unipd.it.

ship with antioxidants. Chain-breaking free radical scavengers limit the peroxidative chain reaction by reducing by one electron a lipid peroxy radical to a hydroperoxide, while peroxidolytic compounds, or enzymes, by reducing by two electrons hydroperoxides, prevent their decomposition leading to new initiations [9,10]. Thus, lipid hydroperoxides are the central molecules on which the major antioxidant systems operate and integrate.

Following the discovery of lipoxygenases, a physiological role has been identified for lipid hydroperoxides and the concept of "peroxide tone" has been introduced, which implies that a steady state concentration of hydroperoxides is functionally relevant in maintaining the optimal cellular function, as far as activation of cycloxygenases and lipoxygenases is concerned [11,12].

Finally, the rapidly expanding field of redox regulation of elements involved in the control of gene expression and cellular signaling [13–16] highlighted the concept that, in some disease conditions, a poorly controlled balance between production and reduction of hydroperoxides could generate distorted cellular responses.

REACTIONS OF LIPID HYDROPEROXIDES

The oxidizing potential of lipid hydroperoxides, through one-electron or two-electron redox transitions, drives reactions involved in (i) the damaging effect, (ii) the possible physiological effect, (iii) the antioxidant defense, and (iv) the analytic procedures.

The easy molecule-assisted homolysis related to the low dissociation energy of the O-O bond (44 kcal/mol^{-1}) or, more likely, the one-electron transfer in the presence of metal ions, produces radicals that cause further decomposition of fatty acid hydroperoxide, finally leading to a series of aldehydes [1]. These contribute to cellular damage and are measured in the popular thiobarbituric acid (TBA) test for lipid peroxidation. The oxidation of transition metal complexes in the presence of hydroperoxides has been used in some analytical procedures, where the chromogen is provided by the change of the redox status of the metal. The heterolytic displacement in the presence of a nucleophile—a two-electron redox transition—accounts for biological damage when thiols or methionine residues in proteins are oxidized. The same reaction provides antioxidant protection when the nucleophile belongs to the active center of an enzyme, which, in order to be fully active, has to react much faster than the molecule to be protected and the native form of which is rapidly regenerated by a suitable reductant [17]. It appears reasonable that also the possible physiological effect of hydroperoxides has to deal with oxidation of some targets via nucleophilic displacement, without free radical intermediates, possibly through an enzyme providing specificity to the reaction. A relevant example of

this mechanism has been recently described to take place in the late phase of spermatogenesis, when the mitochondrial capsule of spermatozoa is built up by oxidation of protein thiols catalyzed by the selenoenzyme phospholipid hydroperoxide glutathione peroxidase (PHGPx), which finally becomes a cross-linked component of the structure [18]. The nucleophilic displacement reaction of lipid hydroperoxides is the reaction involved in analytical procedures such as the enzymatic determination with selenium dependent peroxidases and the iodometric titration.

ANALYSIS OF LIPID HYDROPEROXIDES

The analytical procedures based on one-electron redox transition have been widely used in both food chemistry and biomedical research. The oxidation of different ferrous iron complex [4,19,20] by lipid hydroperoxide generates an easily measurable colored complex, but the procedure suffers the limited specificity of the reaction. In fact, the oxidation potential of the ferrous iron complexes allows the electron transfer to other compounds, different from lipid hydroperoxides, possibly present in the reaction mixture, including oxygen. The kinetics of the reaction actually favors the reaction with LOOH, thus allowing their measurement, although in the presence of oxygen. Nevertheless, in our experience the approach provides limited precision when samples, such as plasma, containing a very low level of LOOH, are analyzed. This approach is further biased by the fact that, in the presence of ferrous iron complexes, free radicals are produced from lipid hydroperoxides, which could promote oxidation of lipids, thus leading to an overestimation of the peroxide title. Granted these radicals could be reduced by the excess of ferrous iron and that the reaction with hydroperoxides is rather fast, thus limiting the occurrence of side reaction, the intrinsic bias of the procedure is, in our opinion, not fully overcome. This is in agreement with the fact that plasma values obtained by these procedures are higher than those obtained by our chemiluminescent procedure (see below).

The iodometric titration [21], taking place by a two-electron nucleophilic displacement, is not biased by the production of new radicals. However, the problem of the easy oxidability of the donor substrate is even more relevant. Moreover, the iodine generated in the reaction could be variably adsorbed by proteins present in the sample, thus introducing a bias difficult to overcome also adopting an internal calibration.

The enzymatic titration of lipid hydroperoxides with a selenium-dependent peroxidase active on lipids provides an excellent specificity, but the sensitivity is limited to that of detection of oxidized glutathione [22]. This, together with the absolute requirement of a solution where both the enzymes are active and the lipid substrate is fully soluble or homogeneously dispersed, limits the applicability of the procedure. Practically, using plasma or isolated lipoproteins dispersed with a detergent, the detection limit is, by our calculation, approximately 10 nmol per ml of plasma or per mg of LDL cholesterol, respectively.

DETECTION OF LIPID HYDROPEROXIDES BY CHEMILUMINESCENCE

For all the above reasons, in the framework of our study on plasma lipid hydroperoxides we set up a sensitive procedure that is specific and precise as well [23–25]. The analysis of chemiluminescence emission of luminol oxidized in the presence of lipid hydroperoxides and hemin provides the required sensitivity. The specificity is substantially increased by the use of a ferric iron complex, which, practically, reacts only with hydroperoxides. This reaction generates both a hemin hydroxyl radical and a lipid alkoxy radical [1]. The latter rearranges to an epoxyallylic radical that, on oxygen addition, generates a peroxy radical [1,25]. This is the final oxidant most likely involved in the formation of luminol radical. The limited reactivity of hemin is the first element of specificity of this test, in comparison with procedures involving ferrous iron. Moreover, in setting up and validating the analytical procedure we used PHGPx [24], a selenium peroxidase active on all lipid hydroperoxides [17]. The enzymatic reduction of LOOH completely prevented the photon emission, confirming the specificity for LOOH of the chemiluminescent reaction. We also have evidence that the chemiluminescent reaction takes places with the same kinetics for hydroperoxide derivatives of phospholipids, triglycerides, and cholesterol esters. It appears reasonable that endoperoxides must react as well, although we did not address this issue specifically. Hydrogen peroxide is at least two order of magnitude less efficient than LOOH in producing photon emission. A plausible reason for this is that the homolytic breakdown of the peroxide gives rise to an alkoxy radical and then a peroxy radical in the case of a lipid hydroperoxide, but to a hydroxyl radical on the case of hydrogen peroxide. The very high reactivity of the latter with several molecules could actually limit the oxidation of luminol. However, a different reaction pathway cannot be excluded.

The oxidant generated by the interaction of hemin with the lipid hydroperoxide oxidized the luminol monoanion to the corresponding radical (Fig. 1). This can both reduce oxygen to superoxide and add superoxide in a concerted mechanism where two luminol radicals are required to generate the endoperoxide derivative of luminol [26]. The latter decomposes immediately giving rise to nitrogen and the excited form of aminopthalate. This excited species decays to the ground state by emitting a photon.

The precision of the measurement was optimized by

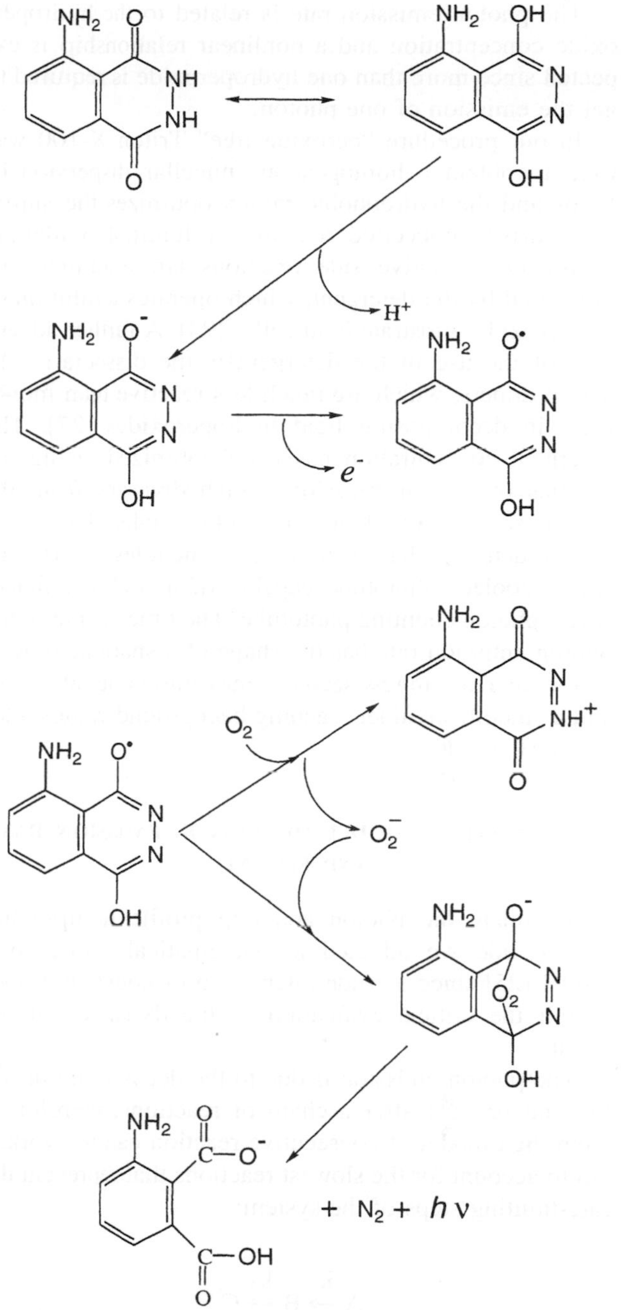

Fig. 1. Reactions of luminol oxidation: The dissociated enolic form of luminol is oxidized to the corresponding radical. This can both reduce oxygen to superoxide and add superoxide. The product of the concerted mechanism is an endoperoxide that decomposes, giving rise to nitrogen and the excited form of aminopthalate. The latter decays to ground state emitting a photon.

adopting a kinetic approach. In fact, the product of the reaction is not a given measurable compound but an electronically excited intermediate, the decay of which generates photons detected in the analytic procedure. Thus, an optimized analysis of LOOH has to take in account the kinetics of the reaction.

The photon emission rate is related to the hydroperoxide concentration and a nonlinear relationship is expected since more than one hydroperoxide is required to get the emission of one photon.

In our procedure "peroxide free" Triton X-100 was used to obtain a homogeneous micellar dispersion of lipids and the hydrophobic milieu optimizes the superoxide-driven concerted reaction of luminol oxidation. Moreover, oxidative side reactions are minimized or prevented by the detergent, which operates a dilution of the peroxidic substrate in micelles [24]. A further advantage of the use of the detergent is the dissociation of hemin dimers, which are much less reactive than monomers in decomposing lipid hydroperoxides [27]. The detergent concentration has been optimized, using the kinetics of photon emission, which diverges from the model (see below) when side reactions take place.

The detection hardware adopted includes a very sensitive, cooled, phototube capable of a highly efficient single photon counting phototube. The time course of the photon emission rate has the shape of a sharp peak with a rapid increase in few seconds, maximum after about 6 s and a smooth shoulder reaching background values after about 50 s.

MATHEMATICAL MODEL OF PHOTON EMISSION RATE AND FITTING

To relate the photon counting profile to lipid hydroperoxide we adopted a mathematical fitting to a model [25] since a crude integration of counts not does permit the critical evaluation of the dynamics of the event.

The photon emission is due to the decay of an unstable end product after a chain of reactions; therefore a simplified model of consecutive reaction can be worked out to account for the slowest reactions that represent the rate-limiting steps of the system:

$$A \xrightarrow{k_1} B \xrightarrow{k_2} C$$

where A = hydroperoxides, B = luminol-superoxide intermediate, and C = aminopthalate excited state.

The integration of the system of differential equations for two consecutive first-order reactions gives the time course of the transient intermediate B as

$$B = A * (k_1/(k_2 - k_1)) * (e^{-k_1 t} - e^{-k_2 t}) \quad (1)$$

where k_1 and k_2 are complex constants actually accounting for more that one reaction.

On the other hand, C decays in a quasi-instantaneous

rate to the fundamental state ($t_{1/2} > 5$ ns) so that the photon emission rate matches directly the rate of formation of compound C.

The photon count is a differential measure and so directly related to the rate of variation of C but not to its accumulation, like in the usual concentration measurement in kinetics experiments:

$$cps = dC/dt = k_2 B \quad (2)$$

Integration of data is therefore not necessary for obtaining B from C, because c.p.s. is directly proportional to the concentration of B. So the value of B of Eqn. 2 can be substituted in Eqn. 1 so obtaining a direct relationship between the photon count and concentration of A:

$$cps = A * k_2 * (k_1/(k_2 - k_1)) * (e^{-k_1 t} - e^{-k_2 t}) \quad (3)$$

This equation describes the relationship between photon emission and concentration of lipid hydroperoxides.

Since different free radical scavengers, possibly present in the sample, could affect the reaction rate and thus the complex constants k_1 and k_2, an internal standardization was necessary. The concentration of lipid hydroperoxide in the sample, corrected for the effect of inhibitors of the chemiluminescent reaction, was extrapolated at zero internal standard concentration. Notably this correction is independent from the nature of inhibitor, and does not require a precise knowledge of the mechanism of inhibition.

The photon emission rates are fitted starting from an initial time identified as the last point giving background reading before the sudden increase of cps. Data were processed using Eqn. 3, to which an offset value can be eventually added, by nonlinear regression analysis using a program operating by successive iteration. The program used starts from initial provisional estimates of k_1 and k_2 that were determined in a series of separate experiments as $k_1 = 0.71 \pm 0.28$ and $k_2 = 9.47 \times 10^{-2} \pm 7.40 \times 10^{-3}$ ($s^{-1} \pm$ SD), using purified phospholipid hydroperoxides. From these, for every iteration the program finds the best parameters that satisfy Eqn. 3, then computes the chi square (χ^2). When the χ^2 difference between two successive iterations is less then a critical value (usually <1%) the iteration stops and the parameters of the last iteration are considered those that best satisfy Eqn. 3.

As expected, the calculated values of A as cps are not a linear function of the actual hydroperoxide content, while data are best fit by an exponential equation, consistent with the concerted mechanism described above:

$$A = qx^n \quad (4)$$

where x is the hydroperoxide concentration and q a constant. The exponential constant n was determined in a series of measurements on pure lipid hydroperoxides preparation as 2.03 ± 0.27.

The value of both q and n in Eqn. 4 is influenced by radical quenchers and decreases as the quencher concentration increases. The mathematical simulation of the equation showed that q and n influence the shape of the calibration curve in a similar way so that it cannot be discriminated which parameter is modified by any given quencher. In the presence of quenchers and inhibitors of the chemiluminescent reaction, the calibration curve approximates to a straight line and the average slope decreases. Because the hydroperoxides of the samples and that of the internal standard are subjected to the same quenching effect, the actual value of n and q is obtained for each sample.

Introducing an internal standardization, Eqn. 4 becomes

$$A = q(x + z)^n \qquad (5)$$

where z is the hydroperoxide content of the sample and x the internal standard.

Due to the low level of lipid hydroperoxides in lipoproteins and the presence in plasma of efficient antioxidants such as ascorbate and urate, samples are prepared by gel filtration of whole plasma with low molecular weight cut-off "desalting" columns, or affinity chromatography with heparin-Sepharose to isolate apoB-containing lipoproteins [24]. Suitable storage of the samples requires immediate freezing in liquid nitrogen and storage at $-80°C$. This "mild" sample preparation procedure guarantees that no measurable loss or generation of lipid hydroperoxides takes place within 2 weeks.

CONCLUSIONS

The described kinetic procedure for measuring low amounts of lipid hydroperoxides is suitable for analysis of human plasma.

The biases related to unspecific or side reactions have been eliminated by optimizing the reaction mixture composition, and this is tested at each measurement by the quality of the fitting of the model reaction. The major disadvantage of the procedure is that LOOH of different lipid classes are measured together. However, the procedure can be applied to single lipid classes or species when a procedure for isolating lipid classes without any artificial loss or generation of hydroperoxides is developed and validated.

In our opinion, the HPLC procedures for measuring lipid hydroperoxides suffer the poorly reproducible extraction of lipid hydroperoxides. Moreover, in HPLC-CL procedures the detection and integration take place in a flow cell, and chemiluminescence data are processed as absorbance or fluorescence of a stable product. This is not fully correct, in our opinion, since the photon emission is a dynamic event, which is completely meaningful only when analyzed kinetically. The effect of the "window of reading," generated by the use of a flow cell could produce a pseudolinear calibration curve of photon emission vs. lipid hydroperoxides, which is in disagreement with the basic chemistry of the reaction. In fact, kinetic analysis suggests a concerted mechanism and a secondorder kinetics of photon emission rate as a function of hydroperoxide concentration.

We recently applied this procedure to demonstrate unequivocally that a "regular" breakfast increases the plasma level of lipid hydroperoxides and that antioxidants such as those of wine, taken with food, dampens the postprandial increase of plasma lipid hydroperoxides [28]. This evidence could be relevant for studying the definition of the impact of different foods containing lipid hydroperoxides and antioxidants on oxidation and oxidability of plasma lipoproteins.

REFERENCES

[1] Terao, J. Reactions of lipid hydroperoxides. In: Vigo-Pelfrey, C., ed. *Membrane lipid oxidation* (Vol. I). Boca Raton: CRC Press; 1990:219–238.

[2] Clark, I. A.; Cowden, W. B.; Hunt, N. H. Free radical-induced pathology. *Med. Res. Rev.* **5**:297–332; 1985.

[3] Halliwell, B.; Gutteridge, J. M. C. Role of free radical and catalytic metal ions in human disease. *Methods Enzymol.* **186**:1–85; 1990.

[4] Barthel, G.; Grosh, W. Peroxide value determination—comparison of some methods. *J. Am. Oil. Chem. Soc.* **51**:540–544; 1974.

[5] Janero, D. R. Malondialdehyde and thiobarbituric acid-reactivity as a diagnostic indices of lipid peroxidation and peroxidative tissue injury. *Free Radic. Biol. Med.* **9**:515–541; 1990.

[6] Pryor, W. A.; Castle, L. Chemical methods for the detection of lipid hydroperoxides. *Methods Enzymol.* **105**:293–299; 1984.

[7] Slater, T. F. Overview of methods used for detecting lipid peroxidation. *Methods Enzymol.* **105**:283–293; 1984.

[8] Berliner, J. A.; Heinecke, J. W. The role of oxidized lipoproteins in atherogenesis. *Free Radic. Biol. Med.* **20**:707–727; 1996.

[9] Halliwell, B. Antioxidant characterization. Methodology and mechanism. *Biochem. Pharmacol.* **49**:1341–1348; 1995.

[10] Ursini, F.; Maiorino, M.; Sevanian, A. Membrane hydroperoxides. In: Sies, H., ed. *Oxidative stress: oxidants and antioxidants*. London: Academic Press; 1994:319–336.

[11] Kulmacz, R. J.; Lands, W. E. M. Requirement for hydroperoxide by the cyclooxigenase and peroxidase activities of prostaglandin H synthase. *Prostaglandins* **25**:531–540; 1983.

[12] Schnurr, K.; Belkner, J.; Ursini, F.; Schewe, T.; Kühn, H. The selenoenzyme Phospholipid Hydroperoxide Glutathione Peroxidase controls the activity of the 15-Lipoxygenase with complex substrates and mantains the specificity of the oxygenation products. *J. Biol. Chem.* **271**:4653–4658; 1996.

[13] Abate, C.; Patel, L.; Rauscer, F. J. III; Curran, T. Redox regulation of fos and jun DNA-binding activity in vitro. *Science* **249**:1157–1161; 1990.

[14] Pahl, H. L.; Baeuerle, P. A. Oxygen and the control of gene expression. *BioEssays* **16**:497–502; 1994.

[15] Schulze-Osthoff, K.; Los, M.; Bauerle, P. A. Redox signalling by

transcription factors NF-κ B and AP-1 in lymphocytes. *Biochem. Pharmacol.* **50**:735–741; 1995.

[16] Suzuki, Y. J.; Forman, H. J.; Sevanian, A. Oxidants as stimulators of signal transduction. *Free Radic. Biol. Med.* **22**:269–285; 1997.

[17] Ursini, F.; Maiorino, M.; Brigelius-Flohé, R.; Aumann, K. D.; Roveri, A.; Schomburg, D.; Flohé, L. The diversity of glutathione peroxidases. *Methods Enzymol.* **252**:38–53; 1995.

[18] Ursini, F.; Heim, S.; Kiess, M.; Maiorino, M.; Roveri, A.; Wissing, J.; Flohé, L. Dual function of the selenoprotein PHGPx during sperm maturation. *Science* **285**:1393–1396; 1999.

[19] Stine, C. M.; Harland, H. A.; Coulter, S. T.; Jenness, R. A. Modified peroxide test for detection of lipid oxidation in dairy products. *J. Dairy Sci.* **37**:202–208; 1954.

[20] Nourooz-Zadeh, J.; Tajaddini-Sarmadi, J.; Wolff, S. P. Measurement of plasma hydroperoxide concentrations by the ferrous oxidation-xylenol orange assay in conjunction with triphenilphosphine. *Anal. Biochem.* **220**:403–409; 1994.

[21] Cramer, G. L.; Miller, J. F.; Pendleton, R. B.; Lands, E. M. Iodometric measurement of lipid hydroperoxides in human plasma. *Anal. Biochem.* **193**:204–211; 1991.

[22] Maiorino, M.; Roveri, A.; Ursini, F.; Gregolin, C. Enzymatic determination of membrane lipid peroxidation. *Free Radic. Biol. Med.* **1**:203–209; 1985.

[23] Zamburlini, A.; Maiorino, M.; Barbera, P.; Pastorino, A. M.; Roveri, A.; Cominacini, L.; Ursini, F. Measurement of lipid hydroperoxides in plasma lipoproteins by a new highly-sensitive "single photon counting" luminometer. *Biochim. Biophys. Acta* **1256**:233–240; 1995.

[24] Zamburlini, A.; Maiorino, M.; Barbera, P.; Roveri, A.; Ursini, F.

[25] Pastorino, A. M.; Zamburlini, A.; Zennaro, L.; Maiorino, M.; Ursini, F. Measurement of lipid hydroperoxides in human plasma and lipoproteins by kinetic analysis of photon emission. *Methods Enzymol.* **300**:33–43; 1998.

[26] Faulkner, K.; Fridovich, I. Luminol and Lucigenin as detectors for superoxide. *Free Radic. Biol. Med.* **15**:447–451; 1993.

[27] Brown, S. B.; Dean, T. C.; Jones, P. Catalytic activity of iron(III)-centered catalysts. Role of dimerization in the catalytic action of ferrihemes. *Biochem. J.* **117**:741–744; 1970.

[28] Ursini, F.; Zamburlini, A.; Cazzolato, G.; Maiorino, M.; Bittolo-Bon, G.; Sevanian, A. Post-prandial lipid hydroperoxides: a possible link between diet and atherosclerosis. *Free Radic. Biol. Med.* **25**:250–252; 1998.

Direct measurement by single photon counting of lipid hydroperoxides in human plasma and lipoproteins. *Anal. Biochem.* **232**:107–113; 1995.

ABBREVIATIONS

LOOH—lipid hydroperoxide

LDL—low-density lipoprotein

PHGPx—phospholipid hydroperoxide glutathione peroxidase

HPLC-CL—high-performance liquid chromatography; chemiluminescence detection

Bioassays for Oxidative Stress Status (BOSS). Edited by W.A. Pryor

BIOMARKERS OF MYELOPEROXIDASE-DERIVED HYPOCHLOROUS ACID

CHRISTINE C. WINTERBOURN and ANTHONY J. KETTLE

Free Radical Research Group, Department of Pathology, Christchurch School of Medicine, Christchurch, New Zealand

(Received 23 November 1999; Revised 7 February 2000; Accepted 24 February 2000)

Abstract—Hypochlorous acid is the major strong oxidant generated by neutrophils. The heme enzyme myeloperoxidase catalyzes the production of hypochlorous acid from hydrogen peroxide and chloride. Although myeloperoxidase has been implicated in the tissue damage that occurs in numerous diseases that involve inflammatory cells, it has proven difficult to categorically demonstrate that it plays a crucial role in any pathology. This situation should soon be rectified with the advent of sensitive biomarkers for hypochlorous acid. In this review, we outline the advantages and limitations of chlorinated tyrosines, chlorohydrins, 5-chlorocytosine, protein carbonyls, antibodies that recognize HOCl-treated proteins, and glutathione sulfonamide as potential biomarkers of hypochlorous acid. Levels of 3-chlorotyrosine and 3,5-dichlorotyrosine are increased in proteins after exposure to low concentrations of hypochlorous acid and we conclude that their analysis by gas chromatography and mass spectrometry is currently the best method available for probing the involvement of oxidation by myeloperoxidase in the pathology of particular diseases. The appropriate use of other biomarkers should provide complementary information. © 2000 Elsevier Science Inc.

Keywords—Free radicals, Myeloperoxidase, Neutrophil oxidant, Hypochlorous acid, Chlorotyrosine, Chlorohydrin, Oxidant biomarker

INTRODUCTION

The characterization of the specific oxidants responsible for modification of biomolecules in disease processes has been challenging. Biomarkers of reactive oxygen species have the potential not only to determine the extent of oxidative injury, but also to identify the source of the

Christine Winterbourn has an M.Sc. in chemistry from the University of Auckland and a Ph.D. in biochemistry from Massey University. After a postdoc in Canada, she returned to New Zealand in 1971 to work at the Christchurch School of Medicine on the autoxidation of hemoglobin. This introduced her to superoxide and led to a long-term involvement with free radicals.

Tony Kettle, also a New Zealander, completed an M.Sc. at Simon Fraser Unviersity, British Columbia, Canada, then in 1986 undertook his Ph.D. with Dr. Winterbourn.

Dr. Winterbourn's interest in hemogloblin, superoxide and mechanisms of hydroxyl radical production led her to the neutrophil, and for his Ph.D. Tony was introduced to another heme enzyme, myeloperoxidase. Thus began his enthusiasm for the enzyme and a long-term collaboration between the two on its enzymology and the reaction of its major product hypochlorous acid. Dr. Kettle's current interest lies in chlorination reactions and identifying the role of myeloperoxidase in inflammatory diseases.

Address correspondence to: Christine Winterbourn, Free Radical Research Group, Department of Pathology, Christchurch School of Medicine, P.O. Box 4345 Christchurch, New Zealand; Tel: + 64 (3) 364 0564; Fax: + 64 (3) 364 1083; E-Mail: christine.winterbourn@chmeds.ac.nz.

oxidant. Such information is important for predicting the consequences of oxidation as well as for providing a basis for designing appropriate interventions to alleviate injury. Neutrophils are a major source of reactive oxidants and are likely contributors to the oxidative damage associated with a variety of diseases in which inflammatory cells participate [1]. A number of biomarkers for reactive oxidants have been identified in recent years, some of which have the potential to distinguish oxidation by neutrophils in these conditions.

Stimulated neutrophils generate superoxide and its dismutation product, hydrogen peroxide, and release the heme enzyme myeloperoxidase [1,2]. Myeloperoxidase, in addition to oxidizing classical peroxidase substrates to radical intermediates, has the unique property of converting chloride to hypochlorous acid (HOCl) [3].

$$H_2O_2 + Cl^- \xrightarrow{\text{myeloperoxidase}} H_2O + HOCl$$

Under most circumstances, HOCl is likely to be the major strong oxidant produced by neutrophils [4]. Myeloperoxidase is also present in monocytes and there is evidence that it is expressed by macrophages [5] and microglia [6] in vivo. Thus, specific reaction products of

Reprinted from: *Free Radical Biology & Medicine*, Vol. 29, No. 5, pp. 403–409, 2000

HOCl should provide biomarkers of the oxidant activity of neutrophils and other myeloperoxidase-containing cells.

HOCl is a highly reactive species that participates in both oxidation and chlorination reactions. It has many biological targets for oxidation. Thiols and thioethers are particularly reactive and other compounds, including ascorbate, urate, pyridine nucleotides, and tryptophan, are oxidized by HOCl, although not as rapidly [7,8]. The main biological chlorination reactions are with amine groups to give chloramines [9–11]; with tyrosyl residues to give ring chlorinated products [12]; with unsaturated lipids and cholesterol to give chlorohydrins [13,14]; and ring chlorination of cytosine resides in nucleic acids [15].

Important characteristics for a good biomarker are specificity for the reactive species in question, and chemical and biological stability. In addition, the target should have a high reactivity and be present at a high enough concentration for the biomarker to be a significant product. For HOCl, oxidative reactions are fastest [16] and therefore most favored in a biological milieu. However, they also tend to be nonspecific. So although thiol oxidation products and methionine sulfoxide are likely to be major products, they are not useful as specific markers for HOCl. A possible exception is a cyclic sulfonamide formed from the oxidation of reduced glutathione (GSH) [17]. Chlorination reactions tend to be several orders of magnitude slower than oxidations and much less favored. However, by incorporating chlorine into the target molecule, they have a greater potential to be more specific.

The most favored chlorination reaction of HOCl is with amine groups. However, the chloramine products are also reactive oxidants [9]. They are readily reduced back to parent amines by biological reductants such as thiols, ascorbate, and methionine, and should be short-lived in biological fluids. In addition, chloramines of α-amino acids spontaneously break down to aldehydes, with the release of ammonia and carbon dioxide [18]. Typical half-lives are about 10 min at 37°C [19]. Chloramines, therefore, are unlikely to be sufficiently stable to be useful biomarkers. They may perhaps have an application for monitoring oxidation that is occurring at the moment of sample collection, but this possibility has not been explored. Chlorination of tyrosyl residues, unsaturated lipids, and cytosine residues is slower and less favored than chlorination of amines [15,20,21]. However, these reactions have the advantage of producing more stable end products. Provided analytical methods of sufficient sensitivity are available, these products show the most promise as biomarkers of HOCl formation.

Proteins will be significant biological targets for HOCl. The reaction gives a variety of products, most of which are poorly characterized. Tyrosine chlorination occurs [12], as do chloramine [22] and carbonyl forma-

Fig. 1. Chlorinated tyrosines formed from the reaction of HOCl with tyrosyl residues in proteins.

tion [23] and, in some cases, crosslinking [24] or fragmentation [22]. Some of the changes are recognized by an antibody that has been raised against HOCl-treated protein (HOP1) [25]. This antibody enables the analytic techniques to be complemented by immunocytochemical investigation of the localization of the modified protein.

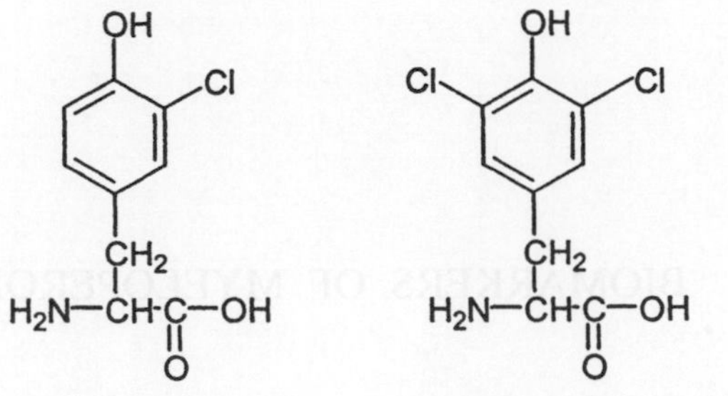

Of the products that have been studied, several have potential to be sensitive and/or selective biomarkers of HOCl activity. Chlorotyrosines, chlorohydrins, protein carbonyls, anti-HOP reactivity, 5-chlorocytosine, and glutathione sulfonamide all have their advantages and disadvantages, as discussed below.

CHLORINATED TYROSINES

HOCl reacts with tyrosyl residues in proteins to form 3-chlorotyrosine [12,20,26] and 3,5-dichlorotyrosine [27]. The chlorinated tyrosines (Fig. 1) are ideal biomarkers for HOCl because they retain chlorine and are stable under the acid conditions required to hydrolyze proteins [28]. They are also produced in proteins exposed to isolated myeloperoxidase and stimulated neutrophils [12]. Myeloperoxidase is the only human enzyme capable of catalyzing the formation of HOCl under physiologic conditions [4]. However, chlorinated tyrosines may not be absolutely specific for myeloperoxidase because nonenzymatic chlorination of tyrosine can occur in the highly acid environment of the stomach [29]. Nothing is currently known about the metabolism of chlorinated tyrosines, but they are potential substrates for dehalogenases or glutathione-S-transferases. Their suitability as biomarkers has been confirmed by the demonstration of markedly elevated levels of 3-chlorotyrosine in low-density lipoproteins (LDL) from human atherosclerotic intima (300 chlorotyrosines/million tyrosyl residues) compared with levels in peripheral blood LDL (10 chlorotyrosines/million tyrosyl residues) [30]. Extensive chlorination of proteins has been observed in bronchoalveolar lavage fluid from patients with acute respiratory distress syndrome (ARDS) (4.8 nmol/mg protein) [31]. However, chlorination in proteins from healthy controls was also very high (0.3 nmol/mg protein). Assuming the protein in lavage fluid was predominantly albumin, these levels correspond to about 2 and 0.15% of tyrosyl residues, respectively. The control values were 150 times higher than those reported by Hazen and Heinecke [30], which suggests that further work is required to confirm the results in ARDS.

Chlorinated tyrosines are minor products of HOCl-treated proteins [12]. Therefore, sensitive analytical procedures are required to detect changes in proteins that have been exposed to low amounts of HOCl. Several methods are available to measure chlorinated tyrosines in proteins, which can also be used to detect the free chlorinated amino acids. These include high-performance liquid chromatography (HPLC) with either fluorescence [12] or electrochemical detection [12,32], and gas chromatography with mass spectrometry (GC/MS) [28,33]. The preferred method uses stable isotope dilution mass spectrometry with selective ion monitoring [28]. GC/MS is the most sensitive and specific method giving characteristic fragmentation patterns of the analytes, thereby eliminating the problems with coeluting peaks that can occur with HPLC. Also, use of stable isotope-labeled analogues as internal standards enables correction for losses that occur during sample preparation and analysis.

For GC/MS analysis, proteins must be isolated from clinical samples and hydrolyzed to liberate amino acids. These amino acids are then isolated by solid phase extraction and derivatized with either heptafluorobutyric acid anhydride or trifluoroacetic acid anhydride [28]. It is important to stress that adequate care must be taken with these procedures to avoid artefactual formation of chlorinated tyrosines. This caution is essential when trying to measure levels of chlorination of less than about 1 chlorinated tyrosine per 1000 tyrosine residues. Chloride and nitrite should be removed from samples prior to hydrolysis because under acid conditions, they can be converted to chlorinating and nitrating species [28]. For the same reason, hydrolysis and derivatization of samples with hydrochloric acid should be avoided [34]. Alternatives to hydrochloric acid include hydrobromic acid or methanesulfonic acid. The latter is recommended when brominated tyrosines, which are biomarkers for eosinophil peroxidase-derived hypobromous acid (HOBr) [35], are also being assayed. Phenol should be added to hydrolysis tubes to scavenge any halogenating or nitrating species. However, in the HPLC assay, phenol interferes with electrochemical detection so that an alternative trap must be used [32].

Most current assays measure only 3-chlorotyrosine. However, this compound is an intermediate in the chlorination of proteins because it is more reactive than

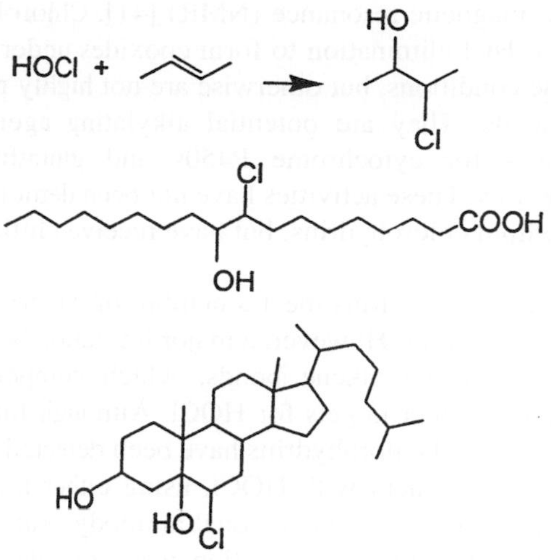

Fig. 2. Route to chlorohydrin formation and isomeric examples of oleic acid and cholesterol chlorohydrin.

tyrosine with HOCl. Therefore, at high concentrations of HOCl, levels of 3-chlorotyrosine plateau as it is converted to 3,5-dichlorotyrosine. Measurement of both of these chlorinated tyrosines will provide a more accurate assessment of extent of protein chlorination. GC/MS methods and HPLC with electrochemical detection have the potential to measure chlorinated tyrosines, 3-nitrotyrosine, o-tyrosine, m-tyrosine, and dityrosine concurrently. This detection makes it possible to readily assess the contributions various reactive oxygen species have made to protein damage. This is especially useful because 3-nitrotyrosine is derived from myeloperoxidase-dependent oxidation of nitrite as well as peroxynitrite [36]. Thus, similar levels of chlorinated and nitrated tyrosines would indicate involvement of myeloperoxidase, whereas a predominance of 3-nitrotyrosine would suggest that oxidative stress was dependent on mainly peroxynitrite. Alternatively, participation of other oxidants, such as hydroxyl radical, can be assessed from the levels of o-tyrosine, m-tyrosine, and dityrosine [37].

LIPID CHLOROHYDRINS

Chlorohydrins are formed by the addition of HOCl to alkene bonds (Fig 2). This reaction has been demonstrated for unsaturated fatty acids and cholesterol [13,14, 38–40] and the derivatives have been characterized using mass spectrometry. Mono- and poly-unsaturated fatty acids and cholesterol all react at comparable rates [38], and bischlorohydrins as well as monochlorohydrins of the polyunsaturated fatty acids can be formed [13,38]. This gives rise to a complex pattern of positional and stereo isomers, which can be characterized from their fragmentation patterns on mass spectrometry and by nuclear magnetic resonance (NMR) [41]. Chlorohydrins undergo HCl elimination to form epoxides under mildly alkaline conditions, but otherwise are not highly reactive compounds. They are potential alkylating agents and substrates for cytochrome P450s and glutathione-S-transferases. These activities have not been demonstrated for the lipid chlorohydrins, but have received little attention.

Lipid chlorohydrins meet a number of the criteria for a good biomarker. However, a major limitation is the low reactivity of the alkene bonds, which compete very poorly with other targets for HOCl. Although fatty acid and cholesterol chlorohydrins have been detected in cells following treatment with HOCl, using either thin layer chromatography or a monoclonal antibody against oleic acid chlorohydrin, this detection was possible only at concentrations well above cytotoxic concentrations [42, 43]. The case was similar for LDL: chlorohydrins were detected only at HOCl concentrations well above those required for protein modification [40].

These observations raise doubts about whether chlorohydrins are detectable at physiologically relevant HOCl doses and whether they have any use as markers. However, this may be a limitation of the sensitivity of the analytical methods that have been used. Fatty acid and cholesterol chlorohydrins have been detected in most studies by GC/MS after derivatization of the carboxyl and hydroxyl groups. The method is sensitive, but recoveries may be low. Loss of parent fatty acids appears to be greater than chlorohydrin recovery, especially with more complex lipid mixtures [38,43], and cholesterol readily breaks down to epoxide and dihydroxy derivatives during the conditions of the assay [38]. Instability has also limited detection of fatty acid bromohydrins by GC/MS [44]. Greater success has been achieved with these compounds using the gentler technique of electrospray mass spectrometry, which does not require derivatization [39, 44]. This approach, combined with online HPLC separation, has the promise of greater recovery and sensitivity and is worthy of further investigation. This potential application is born out by a recent study in which chlorohydrin derivatives of phosphatidyl cholines in LDL that had been treated with HOCl were identified using LC/electrospray MS separation with selective ion monitoring [45]. This suggests that with LC/MS detection, chlorohydrins may have potential as biomarkers of HOCl.

5-CHLOROCYTOSINE

Studies on the reactions of HOCl with nucleotides and nucleic acid bases indicate that the most-favored reactions are with amino groups to give chloramines [10,11]. Compounds such as thymidine and uridine that undergo ring chlorination are approximately two orders of magnitude more reactive than primary amino groups. The heterocyclic chloramines are themselves reactive and able to act as chlorine donors. Therefore, these products are not suitable biomarkers. However, chlorination of C5 of cytosine has recently been described for deoxycytidine and nucleic acids exposed to a myeloperoxidase system [15]. 5-Chlorocytosine does appear to be a stable end product, although a minor one, particularly at near-neutral pH. However, with suitably sensitive detection methods, 5-chlorocytosine may also prove to be a useful biomarker of HOCl.

PROTEIN CARBONYLS

Carbonyls form readily when proteins are treated with HOCl. One established mechanism is via the breakdown of chloramines to aldehydes [18,46], but other routes are also possible. Carbonyls can be measured colorimetrically with dinitrophenylhydrazine [47], or immunologically by enzyme-linked immunosorbent assay (ELISA) [48] or Western blot [47] with an antibody against the dinitrophenylhydrazine-derivatized protein. The immunologic methods offer considerably greater sensitivity. The ELISA assay can detect carbonyls on albumin treated with as little as 20 nmol of HOCl per milligram of protein [27] and has been used to show elevations in bronchoalveolar lavage fluid and plasma from critically ill patients with ARDS [49]. A variety of oxidation mechanisms lead to formation of carbonyls [50], so their presence cannot be taken as specific evidence of HOCl or myeloperoxidase involvement. However, a high degree of correlation was seen between carbonyls and myeloperoxidase in bronchoalveolar lavage specimens in ARDS [49], in tracheal aspirates from premature infants, and in sputum from asthmatic patients (unpublished results). Since the ELISA carbonyl assay is sensitive and relatively easy to perform, it can provide a useful measure of oxidation at inflammatory sites.

ANTI-HOP REACTIVITY

Two monoclonal antibodies have been raised against chlorinated LDL and recognize HOCl-specific epitopes; they are accordingly called anti-HOP antibodies [25]. They do not cross-react with aldehyde-modified or Cu^{2+}-oxidized proteins. One recognizes only HOCl-treated LDL. The other also recognizes albumin that has reacted with HOCl and must bind epitopes that are commonly formed on proteins. These epitopes are unknown and may not be unique to HOCl-treated proteins. Indeed, we have shown that anti-HOP1 also recognizes albumin that has been treated with HOBr [27]. These antibodies

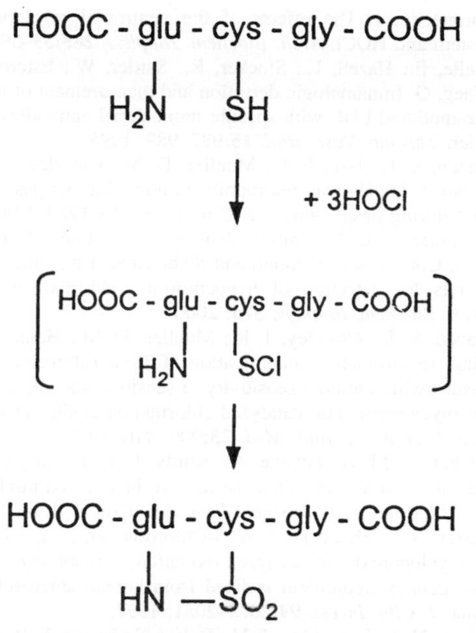

HOOC - glu - cys - gly - COOH
| |
H₂N   SH

↓ + 3HOCl

[HOOC - glu - cys - gly - COOH]
[| |]
[H₂N   SCl]

↓

HOOC - glu - cys - gly - COOH
| |
HN —SO₂

Fig. 3. Proposed route for the conversion of glutathione to glutathione sulfonamide (MW 317) on reaction with HOCl.

have been used to demonstrate the presence of HOCl-modified proteins in human atherosclerotic lesions [51] and to probe the role of myeloperoxidase in kidney disease [52]. The pronounced immunostaining observed in diseased renal tissue provided strong evidence for the involvement of HOCl in glomerular and tubulointerstitial injury. Immunocytochemistry and immunoblotting methods that employ the anti-HOP antibodies should be useful for locating the sites in vivo where HOCl promotes damage, and for identifying which proteins are modified. Concurrent use of the antibodies with GC/MS assays for chlorinated and brominated tyrosines would be optimal to confirm involvement of either HOCl or HOBr.

GLUTATHIONE SULFONAMIDE

Thiols are the most reactive biological substrates for HOCl so far identified [8,16]. GSH, therefore, should be a major low–molecular-weight target and any unique product from this reaction would be most attractive as a biomarker for HOCl production. Although the main product is oxidized glutathione (GSSG), a significant proportion of the GSH is oxidized to a cyclic sulfonamide (Fig. 3) [17]. This structure has been assigned on the basis of its mass and chemical properties, and the reaction leading to its formation is thought to involve cyclization of the sulfenyl or sulfonyl chloride [10,17]. There is strong suggestive evidence that it is released by cells after exposure to HOCl [53]. Whether the sulfonamide is unique to HOCl has yet to be established, and

detection methods that are sufficiently sensitive for probing biological samples are not available. However, the attributes of glutathione sulfonamide as a biomarker of HOCl justify its further investigation.

CONCLUSION

The availability of techniques for measuring specific reaction products of HOCl has opened the way for using these biomarkers to assess the contribution of myeloperoxidase-derived products to oxidative tissue injury. To date, there have been relatively few applications of these methods to the analysis of clinical material. Positive reactivity to anti-HOP antibody has been detected in atherosclerotic lesions [51] and in inflamed kidney [52], and 3-chlorotyrosine has been measured in atherosclerotic lesions [30] and in ARDS samples [31]. However, the study of links between biomarkers and disease severity is still in its infancy and the clinical relevance of these findings has yet to be established. Of the markers of HOCl, measurement of chlorinated tyrosine is the method of choice. GC/MS with selective ion monitoring is the most powerful approach, but HPLC methods can also be applied. The GC/MS assay has the added advantage of being able to measure other modified tyrosine derivatives in the same assay and so distinguish HOCl from other oxidants. Protein reactivity with anti-HOP antibody can be used in conjunction with analytic procedures for examining tissue localization. Protein carbonyl measurements, while not specific, provide a simple way of monitoring the extent of oxidation. Other biomarkers need further exploration. Lipid chlorohydrins have potential, but are not favored products of HOCl and further development of sensitive assays such as LC/MS is required before their value can be assessed. Glutathione sulfonamide has considerable potential as a novel biomarker and also warrants further investigation.

Acknowledgements — This work was supported by grants from the Health Research Council of New Zealand and the New Zealand Lotteries Board.

REFERENCES

[1] Klebanoff, S. J. Oxygen metabolites from phagocytes. In: Gallin, J. I.; Snyderman, R., eds. *Inflammation: basic principles and clinical correlates.* Philadelphia: Lippincott Williams & Wilkins; 1999:721–768.

[2] Hampton, M. B.; Kettle, A. J.; Winterbourn, C. C. Inside the neutrophil phagosome: oxidants, myeloperoxidase and bacterial killing. *Blood* **92:**3007–3017; 1998.

[3] Harrison, J. E.; Shultz, J. Studies on the chlorinating activity of myeloperoxidase. *J. Biol. Chem.* **251:**1371–1374; 1976.

[4] Kettle, A. J.; Winterbourn, C. C. Myeloperoxidase: A key regulator of neutrophil oxidant production. *Redox Rep.* **3:**3–15; 1997.

[5] Daugherty, A.; Dunn, J. L.; Rateri, D. L.; Heinecke, J. W. Myeloperoxidase, a catalyst for lipoprotein oxidation, is expressed in human atherosclerotic lesions. *J. Clin. Invest.* **94:**437–444; 1994.

[6] Nagra, R. M.; Becher, B.; Tourtellotte, W. W.; Antel, J. P.; Gold, D.; Paladino, T.; Smith, R. A.; Nelson, J. R.; Reynolds, W. F. Immunohistochemical and genetic evidence of myeloperoxidase involvement in multiple sclerosis. *J. Neuroimmunol.* **78**:97–107; 1997.

[7] Albrich, J. M.; McCarthy, C. A.; Hurst, J. K. Biological reactivity of hypochlorous acid: implications for microbicidal mechanisms of leukocyte myeloperoxidase. *Proc. Natl. Acad. Sci. USA* **78**: 210–214; 1981.

[8] Winterbourn, C. C. Comparative reactivities of various biological compounds with myeloperoxidase-hydrogen peroxide-chloride, and similarity of the oxidant to hypochlorite. *Biochim. Biophys. Acta* **840**:204–210; 1985.

[9] Thomas, E. L.; Grisham, M. B.; Jefferson, M. M. Preparation and characterization of chloramines. *Methods Enzymol.* **132**:569–585; 1986.

[10] Prutz, W. A. Hypochlorous acid interactions with thiols, nucleotides, DNA, and other biological substrates. *Arch. Biochem. Biophys.* **332**:110–120; 1996.

[11] Prutz, W. A. Interactions of hypochlorous acid with pyrimidine nucleotides, and secondary reactions of chlorinated pyrimidines with GSH, NADH, and other substrates. *Arch. Biochem. Biophys.* **349**:183–191; 1998.

[12] Kettle, A. J. Neutrophils convert tyrosyl residues in albumin to chlorotyrosine. *FEBS Lett.* **379**:103–106; 1996.

[13] Winterbourn, C. C.; van den Berg, J. J. M.; Roitman, E.; Kuypers, F. A. Chlorohydrin formation from unsaturated fatty acids reacted with hypochlorous acid. *Arch. Biochem. Biophys.* **296**:547–555; 1992.

[14] Heinecke, J. W.; Li, W.; Mueller, D. M.; Bohrer, A.; Turk, J. Cholesterol chlorohydrin synthesis by the myeloperoxidase-hydrogen peroxide-chloride system: potential markers for lipoproteins oxidatively damaged by phagocytes. *Biochemistry* **33**: 10127–10136; 1994.

[15] Henderson, J. P.; Byun, J.; Heinecke, J. W. Molecular chlorine generated by the myeloperoxidase-hydrogen peroxide-chloride system of phagocytes produces 5-chlorocytosine in bacterial RNA. *J. Biol. Chem.* **274**:33440–33448; 1999.

[16] Folkes, L. K.; Candeias, L. P.; Wardman, P. Kinetics and mechanisms of hypochlorous acid reactions. *Arch. Biochem. Biophys.* **323**:120–126; 1995.

[17] Winterbourn, C. C.; Brennan, S. O. Characterisation of the oxidation products of the reaction between reduced glutathione and hypochlorous acid. *Biochem. J.* **326**:87–92; 1997.

[18] Zgliczynski, J. M.; Stelmaszynska, T.; Ostrowski, W.; Naskalski, J.; Sznajd, J. Myeloperoxidase of human leukaemic leucocytes: oxidation of amino acids in the presence of hydrogen peroxide. *Eur. J. Biochem.* **4**:540–547; 1968.

[19] Hazen, S. L.; d'Avignon, A.; Anderson, M. M.; Hsu, F. F.; Heinecke, J. W. Human neutrophils employ the myeloperoxidase-hydrogen peroxide-chloride system to oxidize alpha-amino acids to a family of reactive aldehydes. Mechanistic studies identifying labile intermediates along the reaction pathway. *J. Biol. Chem.* **273**:4997–5005; 1998.

[20] Domigan, N. M.; Charlton, T. S.; Duncan, M. W.; Winterbourn, C. C.; Kettle, A. J. Chlorination of tyrosyl residues in peptides by myeloperoxidase and human neutrophils. *J. Biol. Chem.* **270**: 16542–16548; 1995.

[21] Carr, A. C.; van den Berg, J. J. M.; Winterbourn, C. C. Differential reactivities of hypochlorous and hypobromous acids with purified *E. coli* phospholipid: formation of haloamines and halohydrins. *Biochim. Biophys. Acta* **1392**:254–264; 1998.

[22] Hawkins, C. L.; Davies, M. J. Hypochlorite-induced damage to proteins: formation of nitrogen-centered radicals from lysine residues and their role in protein fragmentation. *Biochem. J.* **332**: 617–625; 1998.

[23] Buss, H.; Chan, T. P.; Sluis, K. B.; Domigan, N. M.; Winterbourn, C. C. Protein carbonyl measurement by a sensitive ELISA method. *Free Radic. Biol. Med.* **23**:361–366; 1997.

[24] Vissers, M. C. M.; Winterbourn, C. C. Oxidative damage to fibronectin. I. The effects of the neutrophil myeloperoxidase system and HOCl. *Arch. Biochem. Biophys.* **285**:53–59; 1991.

[25] Malle, E.; Hazell, L.; Stocker, R.; Sattler, W.; Esterbauer, H.; Waeg, G. Immunologic detection and measurement of hypochlorite-modified LDL with specific monoclonal antibodies. *Arterioscler. Thromb. Vasc. Biol.* **15**:982–989; 1995.

[26] Hazen, S. L.; Hsu, F. F.; Mueller, D. M.; Crowley, J. R.; Heinecke, J. W. Human neutrophils employ chlorine gas as an oxidant during phagocytosis. *J. Clin. Invest.* **98**:1283–1289; 1996.

[27] Chapman, A. L. P.; Senthilmohan , R.; Winterbourn, C. C.; Kettle, A. J. Comparison of mono and dichlorinated tyrosines with carbonyls for detection of hypochlorous acid-modified proteins. *Arch. Biochem. Biophys.* **376**; 2000.

[28] Hazen, S. L.; Crowley, J. R.; Mueller, D. M.; Heinecke, J. W. Mass spectrometric quantification of 3-chlorotyrosine in human tissues with attomole sensitivity: a sensitive and specific marker for myeloperoxidase-catalyzed chlorination at sites of inflammation. *Free Radic. Biol. Med.* **23**:909–916; 1997.

[29] Nickelsen, M. G.; Nweke, A.; Scully, F. E. J.; Ringhand, H. P. Reactions of aqueous chlorine in vitro in stomach fluid from the rat: chlorination of tyrosine. *Chem. Res. Toxicol.* **4**:94–101; 1991.

[30] Hazen, S. L.; Heinecke, J. W. 3-Chlorotyrosine, a specific marker of myeloperoxidase-catalyzed oxidation, is markedly elevated in low density lipoprotein isolated from human atherosclerotic intima. *J. Clin. Invest.* **99**:2075–2081; 1997.

[31] Lamb, N. J.; Gutteridge, J. M.; Baker, C.; Evans, T. W.; Quinlan, G. J. Oxidative damage to proteins of bronchoalveolar lavage fluid in patients with acute respiratory distress syndrome: evidence for neutrophil-mediated hydroxylation, nitration, and chlorination. *Crit. Care Med.* **27**:1738–1744; 1999.

[32] Crow, J. P. Measurement and significance of free and protein-bound 3-nitrotyrosine, 3-chlorotyrosine, and free 3-nitro-4-hydroxyphenylacetic acid in biologic samples: a high-performance liquid chromatography method using electrochemical detection. *Methods Enzymol.* **301**:151–160; 1999.

[33] van der Vliet, A.; Jenner, A.; Eiserich, J. P.; Cross, C. E.; Halliwell, B. Analysis of aromatic nitration, chlorination, and hydroxylation by gas chromatography-mass spectrometry. *Methods Enzymol.* **301**:471–483; 1999.

[34] Chowdhury, S. K.; Eshraghi, J.; Wolfe, H.; Forde, D.; Hlavac, A. G.; Johnston, D. Mass spectrometric identification of amino acid transformations during oxidation of peptides and proteins: modifications of methionine and tyrosine. *Anal. Chem.* **67**:390–398; 1995.

[35] Wu, W.; Chen, B. K.; d'Avignon, A.; Hazen, S. L. 3-Bromotyrosine and 3,5-dibromotyrosine are major products of protein oxidation by eosinophil peroxidase: potential markers for eosinophil-dependent tissue injury in vivo. *Biochemistry* **38**:3538–3548; 1999.

[36] Eiserich, J. P.; Hristova, M.; Cross, C. E.; Jones, A. D.; Freeman, B. A.; Halliwell, B.; van der Vliet, A. Formation of nitric oxide-derived inflammatory oxidants by myeloperoxidase in neutrophils. *Nature* **391**:393–397; 1998.

[37] Heinecke, J. W.; Hsu, F. F.; Crowley, J. R.; Hazen, S. L.; Leeuwenburgh, C.; Mueller, D. M.; Rasmussen, J. E.; Turk, J. Detecting oxidative modification of biomolecules with isotope dilution mass spectrometry: sensitive and quantitative assays for oxidized amino acids in proteins and tissues. *Methods Enzymol.* **300**:124–144; 1999.

[38] van den Berg, J. J. M.; Winterbourn, C. C.; Kuypers, F. A. Hypochlorous acid-mediated oxidation of cholesterol and phospholipid: analysis of reaction products by gas chromatography-mass spectrometry. *J. Lipid Res.* **34**:2005–2012; 1993.

[39] Carr, A. C.; van den Berg, J. J. M.; Winterbourn, C. C. Chlorination of cholesterol in cell membranes by hypochlorous acid. *Arch. Biochem. Biophys.* **332**:63–69; 1996.

[40] Hazen, S. L.; Hsu, F. F.; Duffin, K.; Heinecke, J. W. Molecular chlorine generated by the myeloperoxidase-hydrogen peroxide-chloride system of phagocytes converts low density lipoprotein cholesterol into a family of chlorinated sterols. *J. Biol. Chem.* **271**: 23080–23088; 1996.

[41] Carr, A. C.; Winterbourn, C. C.; Blunt, J. W.; Phillips, A. J.; Abell, A. D. Nuclear magnetic resonance characterisation of 6^{α}-chloro-5^{β}-cholestane-3^{β},5-diol formed from the reaction of hypochlorous acid with cholesterol. *Lipids* **32**:363–367; 1997.

[42] Domigan, N. M.; Carr, A. C.; Elder, P. A.; Lewis, J. G.; Winterbourn, C. C. A monoclonal antibody recognising the chlorohydrin derivatives of oleic acid for probing hypochlorous acid involvement in tissue injury. *Redox Rep.* **3**:57–63; 1997.

[43] Carr, A. C.; Domigan, N. M.; Vissers, M. C. M.; Winterbourn, C. C. Modification of red cell membrane lipids by hypochlorous acid and haemolysis by preformed lipid chlorohydrins. *Redox Rep.* **3**:263–271; 1997.

[44] Carr, A. C.; Winterbourn, C. C.; van den Berg, J. J. M. Peroxidase mediated bromination of unsaturated fatty acids to form bromohydrins. *Arch. Biochem. Biophys.* **327**:227–233; 1996.

[45] Jerlich, A.; Pitt, A. R.; Schaur, R. J.; Spickett, C. M. Pathways of phospholipid oxidation by HOCl in human LDL detected by LC-MS. *Free Radic. Biol. Med.* **28**:673–682; 2000.

[46] Zgliczynski, J. M.; Stelmaszynska, T.; Domanski, J.; Ostrowski, W. Chloramines as intermediates of oxidation reaction of amino acids by myeloperoxidase. *Biochim. Biophys. Acta* **235**:419–424; 1971.

[47] Levine, R. L.; Williams, J. A.; Stadtman, E. R.; Shacter, E. Carbonyl assays for determination of oxidatively modified proteins. *Methods Enzymol.* **233**:346–357; 1994.

[48] Winterbourn, C. C.; Buss, H. Protein carbonyl measurement by enzyme-linked immunosorbent assay. *Methods Enzymol.* **300**:106–111; 1999.

[49] Winterbourn, C. C.; Buss, I. H.; Chan, T. P.; Plank, L. D.; Clark, M. A.; Windsor, J. A. Protein carbonyl measurements show evidence of early oxidative stress in critically ill patients. *Crit. Care Med.* **28**:143–149; 2000.

[50] Stadtman, E. R.; Oliver, C. N. Metal-catalyzed oxidation of proteins. Physiological consequences. *J. Biol. Chem.* **266**:2005–2008; 1991.

[51] Hazell, L. J.; Arnold, L.; Flowers, D.; Waeg, G.; Malle, E.; Stocker, R. Presence of hypochlorite-modified proteins in human atherosclerotic lesions. *J. Clin. Invest.* **97**:1535–1544; 1996.

[52] Malle, E.; Woenckhaus, C.; Waeg, G.; Esterbauer, H.; Grone, E. F.; Grone, H.-J. Immunological evidence for hypochlorite-modified proteins in human kidney. *Am. J. Pathol.* **150**:603–615; 1997.

[53] Carr, A. C.; Winterbourn, C. C. Oxidation of neutrophil glutathione and protein thiols by myeloperoxidase-derived hypochlorous acid. *Biochem. J.* **327**:275–281; 1997.

ABBREVIATIONS

ARDS—adult respiratory distress syndrome

GC—gas chromatography

HOP—hypochlorous acid-oxidized protein

LC—liquid chromatography

LDL—low-density lipoprotein

MS—mass spectrometry

SENSITIVE AND NONENZYMATIC MEASUREMENT OF HYDROGEN PEROXIDE IN BIOLOGICAL SYSTEMS

Sebastian Mueller

Department of Internal Medicine IV, University of Heidelberg, Heidelberg, Germany

(*Received* 3 *August* 1999; *Revised* 7 *February* 2000; *Accepted* 15 *March* 2000)

Abstract—The increasing demand in detecting H_2O_2 under various experimental conditions is only partly fulfilled by most conventional peroxidase-based assays. This article describes a sensitive and nonenzymatic H_2O_2 assay that is based on the chemiluminescence reaction of luminol with hypochlorite. It allows the determination of H_2O_2 down to nanomolar concentrations. Actual H_2O_2 concentrations rather than a turnover of H_2O_2 can be determined in monolayer cultures, perfusates, suspensions of intact cells, organelles, and crude homogenates. One of the strengths of this assay is that it may be used to assess fast enzyme kinetics (catalase, glutathione peroxidase, oxidases) at very low H_2O_2 concentrations. Its use together with a glucose oxidase/catalase system appears to be a powerful tool in studying signal functions of H_2O_2 in various biological systems on a quantitative basis. Several applications are discussed in detail to demonstrate the technical requirements and analytical potentials.   © 2000 Elsevier Science Inc.

Keywords—Hydrogen peroxide, Free radicals, Catalase, Oxidase, Glutathione peroxidase, Chemiluminescence, Luminol, Glucose oxidase

INTRODUCTION

H_2O_2 is a central oxygen metabolite, produced in several cellular compartments and the source of other reactive oxygen species, e.g., the highly reactive hydroxyl radical. Besides its role in cellular toxicity, H_2O_2 has recently gained much attention as a possible signaling molecule involved in signal transduction pathways [1–3]. The increasing need to detect H_2O_2 under several experimental conditions is only partly fulfilled by conventional peroxidase-based assays using different probes as electron donors. These assays may remove or generate

reactive oxygen species themselves and, therefore, give rise to severe artifacts [4,5]. This article describes a nonenzymatic H_2O_2 assay that has been recently applied to various biological systems [5]. Advantages of this assay are that it may be used (i) to assess fast enzymatic kinetics entailed in catalase-, glutathione peroxidase–, and oxidase-catalyzed reactions at low H_2O_2 concentrations; (ii) to measure actual H_2O_2 concentrations in monolayer cultures and perfusates; and (iii) to estimate diffusion barriers for H_2O_2 [5–9]. A computer-driven chemiluminometer, an apparatus routinely found in biomedical laboratories, is required for these measurements.

PRINCIPLE

The assay is based on the oxidation of luminol (5-amino-2,3-dihydro-1,4-phthalazinedione) by sodium hypochlorite (NaOCl) [5]. Luminol is oxidized by NaOCl to diazaquinone in a two-electron oxidation, which is further specifically converted by H_2O_2 to an excited aminophthalate via an α-hydroxy-hydroperoxide [5,10, 11]. The short luminescence signal (less than 2 s) of this reaction has a maximum wavelengh at 431 nm; it is linearly dependent on H_2O_2 down to the 10^{-9} M range. One-electron transfers inherent in peroxidase-based as-

Sebastian Mueller: Born 1967 in Dresden (Germany). Medical studies in Leipzig and Strasbourg. MD-PhD 1994 in Leipzig. From 1994–1997 clinical specialization and research at Department of Gastroenterology/University of Heidelberg with W. Stremmel. Visiting scholar 1992 at Chemistry Department/University of Denver with G. Eaton and 1997–1998 at Department of Molecular Pharmacology and Toxicology/ University of Southern California in Los Angeles with E. Cadenas as Feodor-Lynen fellow of the Alexander-von-Humboldt foundation. S.M. is now at the Dept. of Gastroenterology/University of Heidelberg. Research interests: biological functions of hydrogen peroxide, hydrogen peroxide metabolism, iron regulation and oxidative stress, oxidative metabolism of peroxisomes.

Address correspondence to: Dr. Sebastian Mueller, University of Heidelberg, Department of Internal Medicine IV, Bergheimer Strasse 58, 69115 Heidelberg, Germany; Tel: +49 (6221) 56 8612; Fax: +49 (6221) 40 8366; E-Mail: sebastian.mueller@urz.uni-heidelberg.de.

says, which may cause redox-cycling reactions or reduction of oxygen to superoxide [11–13], are avoided by this method. The luminol/hypochlorite-dependent chemiluminescence exceeds by far the unspecific so-called luminol-dependent chemiluminescence (generated by other oxidation pathways) [14]. Additionally, the assay procedure permits a simple subtraction of unspecific signals [5]. When used as a flow system, rapid kinetics of H_2O_2-removing or -generating enzymes (catalase, GPO, oxidases) can be studied at physiologically low H_2O_2 concentrations. Due to the short time of measurement, it actually determines the H_2O_2 (M) concentration rather than an H_2O_2 generation rate (mol per time).

DETERMINATION OF FAST H₂O₂ KINETICS IN ENZYMATIC REACTIONS USING A FLOW SYSTEM

The sample is continuously pumped out from a reaction reservoir and luminol and NaOCl are continuously added to the sample allowing a real-time registration of H_2O_2, e.g., during fast enzyme kinetics. One of the advantages is that the sample in the reservoir is not in contact with any of the reagents. The procedure requires a large sample volume (up to 100 ml). This is usually not a problem since enzyme solutions can be highly diluted due to the sensitivity of the assay.

Equipment

The procedure requires the following equipment:

1) chemiluminometer—any luminometer that allows the installation of a flow cell in front of the photomultiplier (e.g., the AutoLumat LB 953 from Berthold EG&G; Wildbad, Germany) can be used; the luminometer is controlled by a computer equipped with software for further processing of time/luminescence intensity data;
2) perfusion pump for NaOCl and luminol;
3) peristaltic pump for sample aspiration;
4) flow cell—a flow cell that allows separate and continuous addition of luminol and NaOCl buffered in PBS at pH 7.4, and the continuous addition of the sample is required; in the author's laboratory a peristaltic pump is used with a 3 mm polyethylene pipeline to continuously aspirate the sample solution (ca. 4 ml/min); black 50 ml plastic syringes are loaded with luminol and NaOCl work solutions and both reactions are continuously pumped via the same perfusion pump into the polyethylene pipeline (ca. 12 ml/h);
5) graduated cylinder of 100 ml for sample solution;
6) magnetic stirrer to continuously mix sample solution; and
7) temperature control unit (if necessary).

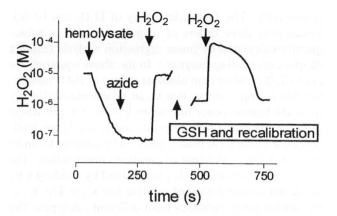

Fig. 1. Parallel determination of catalase and glutathione peroxidase (GPO) activities in a hemolysate at very low H_2O_2 concentrations. Addition of highly diluted hemolysate (1:1000) is followed by the catalase-mediated exponential degradation of H_2O_2. Catalase is subsequently inhibited by sodium azide (1 mM), which can be demonstrated by further incubation with H_2O_2. Finally, glutathione (2 mM) is added. This step requires a recalibration of the system. After a new bolus of H_2O_2 (10^{-4} M) the GPO-mediated decay of H_2O_2 is visible. Both, catalase and GPO activity can be calculated from this experiment.

Reagents

The procedure requires the following reagents:

1) 50 ml 10^{-4} M luminol in 10 mM PBS at pH 7.4 (working solution);
2) 50 ml 10^{-4} M NaOCl in tridistilled water (working solution); and
3) 100 ml 10^{-2} M H_2O_2 in tridistilled water for calibration.

Procedure

The syringes are loaded with the working solutions of NaOCl and luminol and 50–100 ml PBS is added into the graduated cylinder. All pumps are switched on and the system is allowed to equilibrate for about 5 min. The optimal measuring range is found by adjusting the perfusion pump and calibrated by addition of 10^{-5} M H_2O_2 and catalase, respectively.

Example 1: determination of catalase activity at physiological H₂O₂ concentrations [6]

The sample (tissue homogenate, intact cells, purified enzyme) is added into 50 ml 10 mM PBS (graduated cylinder) containing 10^{-5} H_2O_2. A magnetic stirrer is used to ensure a rapid mixing of the enzyme substrate solution. The exponential, catalase-mediated decomposition of H_2O_2 can be followed down to 10^{-9} M H_2O_2 (see Fig. 1). Catalase activity is described by the rate constant $k = \ln (S_1/S_2)/dt$ where dt is the measured time interval, and S_1 and S_2 are H_2O_2 concentrations at time t_1 and t_2, respectively. The first-order decay of H_2O_2 can be followed over three orders of magnitude and k is subsequently calculated by linear regression analysis (using a simple curve fitting program). In the above equation, the ratio (S_1/S_2) rather than absolute values of H_2O_2 concentrations is important so that k can be calculated directly from the luminescence intensities $k = ln\ (I_1/I_2)/dt$ where dt is the measured time interval, and I_1 and I_2 are luminescence integrals at time t_1 and t_2. The constant k can be used as a direct measure of catalase concentration. The specific catalase activity k'_1 is obtained by dividing k by the molar concentration of catalase (e): $k'_1 = k/e$. k'_1 is known for many catalases from different cell types. The value k'_1 for purified catalase from human erythrocytes is $3.4 \times 10^7\ M^{-1}s^{-1}$. This value is used to calculate the absolute concentration of enzyme in blood and tissues [15].

The hypochlorite/luminol technique provides several advantages in comparison to conventional spectrophotometric and titrimetric catalase assays: (i) due to the low H_2O_2 concentrations used, molecular oxygen is completely dissolved and not liberated in gaseous form; (ii) since maximal extracellular H_2O_2 concentrations are known to reach only micromolar levels, determinations of catalase activity at submicromolar concentrations much better reflect physiological conditions; and (iii) repetitive measurements for more than 30 min are possible without loss of enzyme activity and cell viability. The assay has been successfully used to compare catalase activity of intact and homogenated cells/organelles [6]. From these data, the diffusion coefficient for H_2O_2 can be calculated with respect to different membranes [16].

Example 2: parallel determination of catalase and glutathione peroxidase activity in human erythrocytes [8]

This procedure allows the determination of both enzyme activities at low, noninactivating H_2O_2 concentrations and can be applied to all crude cell and tissue homogenates. Highly diluted fresh hemolysate (1:1000) is added into a graduated cylinder containing 50 ml of 10^{-5} M H_2O_2 in PBS (Fig. 1). The solution is permanently stirred. The high dilution insures a decrease in the concentration of glutathione to undetectable concentrations. After addition of the diluted hemolysate, an exponential decay of H_2O_2 is observed corresponding to catalase activity. Catalase activity is then calculated as the rate constant k of the exponential decay of H_2O_2 by linear regression analysis as described above [6,15]. In the next step, catalase is inhibited by addition of 1 mM NaN_3 and 2 mM GSH is added. This amount represents the intracellular GSH concentration in erythrocytes. The

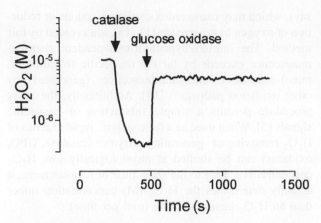

Fig. 2. Generation of H_2O_2 steady state concentrations with catalase and glucose/glucose oxidase (GOX). The steady state concentration of H_2O_2 is determined by the GOX/catalase ratio and can be maintained over hours. This tool appears to be very useful in studying signal functions of H_2O_2 on a quantitative basis in various biological systems.

system is recalibrated by addition of 10^{-4} M H_2O_2 and GPO-mediated H_2O_2 decay is observed. Based on the ping-pong kinetics with infinite limiting maximum velocities and Michaelis-Menton constants established for GPO, the maximum velocity needs to be determined for all individual conditions [17].

Example 3: steady state generation of H_2O_2 with a glucose/glucose oxidase/catalase system to study H_2O_2-dependent signaling pathways [7,9]

The glucose/glucose oxidase system appears to be a powerful tool in studying signal functions of H_2O_2 on a quantitative basis. During the oxidation of glucose by glucose oxidase, H_2O_2 is generated following a zero-order kinetic with $dH_2O_2/dt = k_{GOX}$ (k_{GOX} = rate constant) if dioxygen and glucose are maintained at a constant concentration. Accumulation of H_2O_2 can be controlled by adding appropriated amounts of catalase. H_2O_2 degradation rate by catalase is described by $dH_2O_2/dt = k_{CAT} \times [H_2O_2]$. Thus, steady state levels of H_2O_2 are generated when $k_{GOX} = k_{CAT} \times [H_2O_2]$, and at a constant glucose and dioxygen concentration $[H_2O_2] = k_{GOX}/k_{CAT}$. By varying the enzyme activities, the H_2O_2 concentration can be adjusted and maintained over hours. The luminol/hypochlorite assay assists by measuring this steady state as shown in Fig. 2. Steady state generation in turn allows exact time and dose-dependent studies on redox-sensitive signaling pathways instead of simply adding H_2O_2 as bolus. The GOX/catalase system was successfully used to study the regulation of iron protein 1 (IRP-1) by H_2O_2. It was shown that 10 μM H_2O_2 (steady state) suffice to activate IRP-1 within 20 min by a still unknown signaling cascade [7,9,18].

SINGLE TIME POINT DETERMINATION OF H₂O₂ USING AN INJECTION SYSTEM

In this procedure, luminol is premixed with the sample (e.g., culture medium or perfusate sample). At the appropriate time, NaOCl is added and the luminescence intensity is measured immediately. The measurement is fast and only a small sample volume is needed. An injection device in measuring position is a requisite condition because the luminescence reaction reaches completion within less than 2 s. As an advantage, sample materials are saved and the procedure can be fully automated.

Equipment

The procedure requires the following equipment:

1) chemiluminometer with injection device in measuring position (e.g., AutoLumat LB 953 from Berthold EG&G; Wildbad, Germany); other injection devices are helpful for complete automatization of the experiment (e.g., addition of cell stimulators); the luminometer should be controlled by a computer equipped with software allowing automated performance; and
2) polystyrene tubes for chemiluminometer.

Reagents

The procedure requires the following reagents:

1) stock solution of 10^{-3} M luminol in 10 mM PBS at pH 7. 4 (final concentration of luminol between 10^{-5} and 10^{-4} M);
2) stock solution of 10^{-4} M NaOCl in tridistilled water (final concentration of NaOCl between 10^{-6} and 10^{-5} M); and
3) 10^{-3} M H₂O₂ in tridistilled water for calibration.

Procedure

The injector in measuring position is loaded with NaOCl solution and washed. For optimal measuring range, samples with PBS and luminol are loaded containing catalase (e.g., 10^{-7} M final concentration) and 10^{-5} or 10^{-6} M H₂O₂. If necessary, the NaOCl concentration needs to be adjusted. In a typical experiment, the injection device adds 50 μl of NaOCl (10^{-6}–10^{-5} M final concentration) into 950 μl sample with luminol (5×10^{-5} M final concentration). Usually, samples are measured together with an H₂O₂ calibration solution at the beginning and the end of the batch.

Determination of H₂O₂ release by stimulated neutrophils [5]

In the experiment shown in Fig. 3, 10 μl of luminol

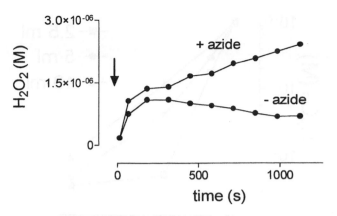

Fig. 3. Increase of H₂O₂ concentration in suspension of neutrophils (oxygen burst) after stimulation with 1 μM FMLP (arrow). Addition of sodium azide inhibits cellular myeloperoxidase and catalase and subsequently leads to H₂O₂ accumulation.

stock solution was added into 20 polystyrene tubes containing 1 ml of 10^5 PMN/ml in HANK's buffer. Every second sample contains sodium azide (1 mM final concentration). In a fully automated experiment, the neutrophils are stimulated by addition of 10 μM FMLP. At different time points, 50 μl NaOCl is added (5×10^{-6} M final concentration). In parallel, the luminescence intensity is recorded for 2 s. The system is calibrated using known amounts of H₂O₂. A rapid increase of H₂O₂ is observed after stimulation of neutrophils. H₂O₂ is later removed by myeloperoxidase and catalase, which can be inhibited by sodium azide. Using this method, changes of H₂O₂ concentration could be detected with less than 3000 neutrophils/ml.

H₂O₂ removal from culture medium of B6 fibroblasts [7,9]

Figure 4 shows the removal of H₂O₂ from culture medium (three different volumes) by fibroblasts growing in monolayer culture. B6 fibroblasts were cultured in RPMI medium in 10 cm culture dishes at 37°C. 10^{-4} M H₂O₂ (final concentration) is added to the medium and 500 μl of the medium is transferred at different time points into polystyrene tubes. 10 μl luminol stock solution is added. After the addition of 50 μl NaOCl (5×10^{-6} M final concentration), luminescence intensity is recorded for 2 s. The culture medium should be used without serum addition. The determination of H₂O₂ removal in cell culture medium is required at all conditions where H₂O₂ is applied to cells to study its signal functions.

GENERAL COMMENTS

The assay is very sensitive and detects H₂O₂ at concentrations as low as 10^{-9} M. It should be noted that

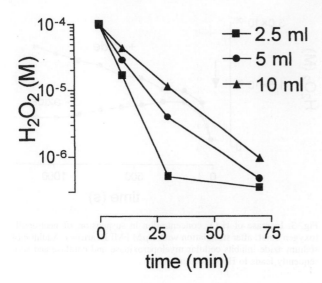

Fig. 4. H_2O_2 removal by cultured B6 fibroblasts (monolayer) in different volumes of culture medium. After bolus addition, H_2O_2 is rapidly decomposed by the cells within minutes. This instability needs to be considered when studying the effects of H_2O_2 on cellular functions.

normal tridistilled water already contains traces of H_2O_2 sometimes as high as up to 10^{-7} M. This is one of the reasons to calibrate the assay with at least 10^{-6} M H_2O_2 and to add catalase. The samples should be kept in the dark since photoreactions generate H_2O_2 (e.g., UV light). Any unspecific chemiluminescence can be detected upon H_2O_2 removal with catalase. The sample volume should be at least 10 times higher than the volume of reagents, because small contamination of reagents with H_2O_2 can significantly decrease the sensitivity of the assay.

NaOCl working solutions should be prepared with tridistilled water to minimize its degradation and they should be freshly prepared and kept in the dark. For each system, the optimal NaOCl concentration should be detected separately. Too high NaOCl concentrations favor unspecific oxidation reactions of luminol. Therefore, NaOCl concentration should be chosen as low as possible (usually 10^{-6}–10^{-5} M). Any compound containing, for example, sulfhydryl- or aminogroups will compete with luminol for NaOCl leading to a decrease in sensitivity [10]. In these cases, samples are diluted (e.g., whole blood 1:100 to 1:10,000) and/or higher concentrations of NaOCl should be used. pH is critical for the reaction and should be kept stable at a value of 7.4 with any appropriate buffer. No luminescence develops at values below 6.5. The chemiluminescence duration increases with pH above 7.4.

Troubleshooting should include: (i) correct concentrations of reactants, (ii) proper installation of the flow cell in front of the photomultiplier, (iii) proper injection of NaOCl in front of the photomultiplier, (iv) low content of NaOCl-reactive compounds (e.g., solutions of sulfhydryl

group–containing proteins should be diluted at least to less than 10^{-6} M), and (v) stability of NaOCl solutions.

CALIBRATION

The luminol/hypochlorite assay is calibrated with known concentrations of H_2O_2 usually between 10^{-3}–10^{-5} M taken from stock solutions. Commercial stock solutions of 30% H_2O_2 are stable for many weeks once kept at 4°C and in the dark. Commercial stock solutions of NaOCl are also stable at 4°C and in the dark. Stock solutions of NaOCl and H_2O_2 can be determined spectrophotometrically at $\epsilon_{290} = 350$ M^{-1}cm^{-1} at pH 12 and $\epsilon_{230} = 74$ M^{-1}cm^{-1}, respectively [19,20]. A routine calibration for established conditions requires at least one sample with a known H_2O_2 concentration and a sample without H_2O_2 upon removal by catalase. These measurements provide the available measuring range. If necessary, the hypochlorite/luminol assay allows the subtraction of the unspecific luminol-dependent chemiluminescence. It simply needs to be measured prior to NaOCl addition and subtracted from the overall luminescence intensity. The same is valid for any non-H_2O_2-related luminescence that is detected after addition of catalase.

SUMMARY

This article describes a sensitive and nonenzymic H_2O_2 assay that is based on the chemiluminescence reaction of luminol with hypochlorite. Actual H_2O_2 concentrations can be measured in monolayer cultures, perfusates, suspensions of intact cells, and crude homogenates. One of the strengths of this assay is that it may be used to assess fast enzyme kinetics at very low H_2O_2 concentrations. The luminol/hypochlorite assay opens new niches in studying the functions of H_2O_2 in biological systems and its metabolism.

Acknowledgements — This work was supported financially by a Feodor-Lynen fellowship of the Alexander-von-Humboldt foundation and by a grant from the Deutsche Forschungsgemeinschaft (SFB601/C2 D.10049500).

REFERENCES

[1] Sen, C. K.; Sies, H.; Baeurle, P. A., eds. *Antioxidant and redox regulation of genes.* San Diego: Academic Press; 2000.
[2] Chance, B.; Sies, H.; Boveris, A. Hydroxyperoxide metabolism in mammalian organs. *Physiol. Rev.* **59:**527–605; 1979.
[3] Khan, A. U.; Wilson, T. Reactive oxygen species as cellular messengers. *Chem. Biol.* **2:**437–445; 1995.
[4] Misra, H. P.; Squatrito, P. M. The role of superoxide anion in peroxidase-catalyzed chemiluminescence luminol. *Arch. Biochem. Biophys.* **215:**59–65; 1982.
[5] Mueller, S.; Arnhold, J. Fast and sensitive chemiluminescence

determination of H₂O₂ concentration in stimulated human neutrophils. *J. Biolumin. Chemilumin.* **10**:229–237; 1995.

[6] Mueller, S.; Riedel, D. H.; Stremmel, W. Determination of catalase activity at physiological H₂O₂ concentrations. *Anal. Biochem.* **245**:55–60; 1997.

[7] Mueller, S.; Pantopoulos, K.; Hentze, M. W.; Stremmel, W. A chemiluminescence approach to study the regulation of iron metabolism by oxidative stress. In: Hastings, J. W.; Kricka, L. J.; Stanley, P. E., eds. *Bioluminescence and chemiluminescence: molecular reporting with photons.* Baffins Lane, Chichester, Sussex: John Wiley & Sons Ltd.; 1997:338–341.

[8] Mueller, S.; Riedel, H. D.; Stremmel, W. Direct evidence for catalase as the predominant H₂O₂-removing enzyme in human erythrocytes. *Blood* **90**:4973–4978; 1997.

[9] Pantopoulos, K.; Mueller, S.; Atzberger, A.; Ansorge, W.; Stremmel, W.; Hentze, M. W. Differences in the regulation of IRP-1 (iron regulatory protein 1) by extra- and intracelullar oxidative stress. *J. Biol. Chem.* **272**:9802–9808; 1997.

[10] Arnhold, J.; Mueller, S.; Arnold, K.; Sonntag, K. Mechanisms of inhibition of chemiluminescence in the oxidation of luminol by sodium hypochlorite. *J. Biolumin. Chemilumin.* **8**:307–313; 1993.

[11] Merényi, G.; Lind, J.; Eriksen, T. E. Luminol chemiluminescence: chemistry, excitation, emitter. *J. Biolumin. Chemilumin.* **5**:53–56; 1990.

[12] Merényi, G.; Lind, J.; Eriksen, T. E. The reactivity of superoxide (O₂⁻) and its ability to induce chemiluminescence and luminol. *Photochem. Photobiol.* **41**:203–208; 1985.

[13] Faulkner, K.; Fridovich, I. Luminol and lucigenin as detectors for O₂⁻. *Free Radic. Biol. Med.* **15**:447–451; 1993.

[14] Allen, R. C.; Loose, L. D. Phagocytic activation of a luminol-dependent chemiluminescence in rabbit alveolar and peritoneal macrophages. *Biochem. Biophys. Res. Commun.* **69**:245–252; 1976.

[15] Aebi, H. Catalase in vitro. *Methods Enzymol.* **105**:121–126; 1984.

[16] Nicholls, P. Activity of catalase in the red cell. *Biochim. Biophys. Acta* **99**:286–297; 1965.

[17] Flohé, L.; Günzler, W. Assays of glutathione peroxidase. *Methods Enzymol.* **105**:114–121; 1984.

[18] Pantopoulos, K.; Hentze, M. W. Activation of iron regulatory protein 1 by oxidative stress in vitro. *Proc. Natl. Acad. Sci. USA* **95**:10559–10563; 1998.

[19] Beers, R. F.; Sizer, I. W. A spectrometric method for measuring the breakdown of hydrogen peroxide by catalase. *J. Biol. Chem.* **195**:133–140; 1952.

[20] Morris, J. C. The acid ionization constant of HOCl from 5 to 35°. *J. Phys. Chem.* **70**:2798–3806; 1966.

ABBREVIATIONS

GSH—glutathione

GPO—glutathione peroxidase

NaOCl—sodium hypochlorite

IRP-1—iron regulatory protein-1

FMLP—N-formyl-methionine-leucine-phenylalanine (chemotactic tripeptide)

OXIDATIVE STRESS STATUS—THE FIFTH SET

WILLIAM A. PRYOR

Biodynamics Institute, Louisiana State University, Baton Rouge, LA, USA

(Received 5 September 2000; Accepted 5 September 2000)

This issue of *FRBM* presents the fifth group of articles in our ongoing Forum on Oxidative Stress Status (OSS). Previous segments of this Forum can be found in earlier issues as follows:

- The first set of seven articles in *FRBM* 27:11/12 (December, 1999).
- The second set of four articles in *FRBM* 28:4 (February, 2000)
- The third set of six articles in *FRBM* 28:6 (March, 2000).
- The fourth set of four articles in *FRBM* 29:5 (September, 2000).

This fifth segment of the OSS Forum has six contributions, beginning with an article by Charles Coudray

and Alain Favier entitled, "Determination of salicylate hydroxylation products as an in vivo oxidative stress marker." This is followed by "Urinary aldehydes as indicators of lipid peroxidation in vivo" by Harold Draper, A. Saari Csallany, and Mary Hadley. Third in this set is a paper by Bernard Gallez and Karsten Mäder, "Accurate and sensitive measurements of pO_2 in vivo using low frequency EPR spectroscopy: how to confer biocompatibility to the oxygen sensors." John S. Althaus, Kari Schmidt, Scott Fountain, Michael Tseng, Richard Carroll, Paul Galatsis, and Edward Hall contributed the fourth article, "LC-MS/MS detection of peroxynitrite-derived 3-nitrotyrosine in rat microvessels." Next, J. Frank, A. Pompella, and H. K. Biesalski present "Histochemical visualization of oxidant stress." The final article in this fifth set of the OSS Forum is by A. Ghiselli, M. Serafini, F. Natella, and C. Scaccini, "Total antioxidant capacity as a tool to assess redox status: critical view and experimental data."

Address correspondence to: Dr. William A. Pryor, Biodynamics Institute, 711 Choppin Hall, Louisiana State University, Baton Rouge, LA 70803, USA.

DETERMINATION OF SALICYLATE HYDROXYLATION PRODUCTS AS AN IN VIVO OXIDATIVE STRESS MARKER

CHARLES COUDRAY* and ALAIN FAVIER†

*Centre de Recherche en Nutrition Humaine d'Auvergne, Laboratoire Maladies Métaboliques et Micronutriments, INRA de Clermont-Ferrand/Theix, Saint Genès Champanelle, France; and †Groupe de Recherche et d'Etude sur les pathologies oxidatives, Laboratoire de Biochimie pharmaceutique, UFR de Pharmacie, Domaine de la Merci, La Tronche, France

(Received 15 November 1999; Revised 23 May 2000; Accepted 1 June 2000)

Abstract—The in vivo measurement of highly reactive free radicals, such as the •OH radical, is very difficult. New specific markers, which are based on the ability of •OH to attack the benzene rings of aromatic molecules, are currently under investigation. The produced hydroxylated compounds can be measured directly. In vivo, radical metabolism of salicylic acid produces two main hydroxylated derivatives (2,3- and 2,5-dihydroxybenzoic acids). The latter acid can be also produced by enzymatic pathways through the cytochrome P-450 system, while the former acid is reported to be solely formed by direct hydroxyl radical attack. Therefore, measurement of 2,3-DHBA, following oral administration of the drug acetyl salicylate, could be proposed for assessment of oxidative stress in vivo. In this paper, a sensitive method for the identification and quantification of hydroxylation products from the reaction of •OH with salicylate in vivo is presented. It employs a high performance liquid chromatography and electrochemical detection system. A detection limit of < 1 pmol for the hydroxylation products has been achieved with linear response over at least five orders of magnitude. Using this technique, we measured plasma levels of 2,3- and 2,5-DHBA dihydroxylated derivatives and salicylic acid and determined the ratios following administration of 1 g acetyl salicylate in 20 healthy subjects. © 2000 Elsevier Science Inc.

Keywords—Hydroxyl radical, Free radicals, Oxidative stress, Salicylic acid, Dihydroxybenzoates, HPLC-electrochemical detection, Humans

INTRODUCTION

Due to its high reactivity, the •OH radical has a very short half-life and is therefore present in extremely low

concentrations. Some direct (electron spin resonance) or indirect (measurement of ethylene liberated from 2-keto 4-methylthiobutyrate, measurement of aromatic hydroxylation) methods have been proposed to probe the formation of •OH radical in vivo. •OH radicals are able to attack the benzene rings of aromatic molecules to produce hydroxylated compounds that can be measured directly. Aromatic hydroxylation has already been used for measuring in vitro •OH production [1–3]. If an aromatic compound can be safely administered to humans in doses that produce concentrations in body fluids sufficient to scavenge •OH, then assaying such products would be reasonable evidence that •OH is being formed in vivo, provided that these products are not formed by enzymatic hydroxylation. A suitable candidate for use in humans may be salicylate or its acetylated form.

O-acetyl salicylic acid (ASA) is a commonly used analgesic and anti-inflammatory agent in man. After ingestion, a substantial amount of ASA is hydrolyzed to

Dr. Charles Coudray, Pharmacist, prepared his Ph.D. thesis in 1989 at the University of Grenoble (France). His research was focused on the assessment of oxidative stress status in vitro and in vivo. He studied the antioxidant role of Zinc and Selenium in the prevention of ischemic lesions in rats at the Hospitalo-university center of Grenoble. Recently, he moved to the national institute for Agronomic Research in Clermont-Ferrand (France) and is working on the intake and bioavailability of minerals in humans.

Professor Alain Favier, Pharmacist, obtained his Ph.D. thesis in 1984 at the University of Grenoble (France). He is in charge of the departments of Pharmaceutical Biochemistry at the faculty of Pharmacy and of Biochemistry at the Hospital of Grenoble. His research is focused on the involvement of free radicals in oxidative diseases as well as on the role of trace elements in biology and medicine.

Address correspondence to: Dr. Charles Coudray, Centre de Recherche in Nutrition Humaine d'Auvergne, Laboratoire Maladies Métaboliques et Micronutriments, INRA de Clermont-Ferrand/Theix, 63122 Saint Genès Champanelle, France; Tel: +33 (76) 76 54 84; Fax: +33 (76) 76 56 64; E-Mail: charles.coudray@clermont.inra.fr.

Reprinted from: *Free Radical Biology & Medicine*, Vol. 29, No. 11, pp. 1064-1070, 2000

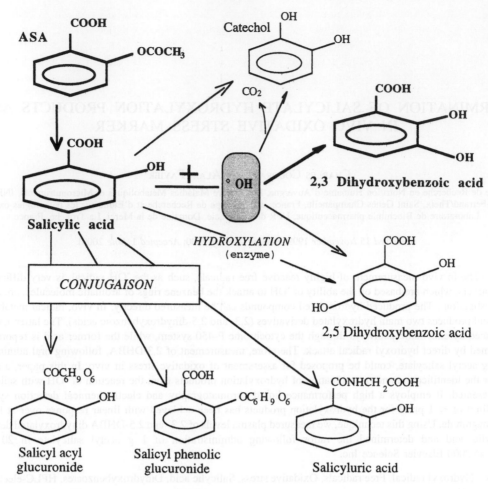

Fig. 1. Major reported enzymatic and oxidative by-products of ASA in human.

salicylic acid (SA) by esterases in the gastrointestinal tract, in the liver, and, to a smaller extent, in the serum [4] (Fig. 1). Salicylate reaches its peak in plasma about 0.5 to 1.5 h after the oral intake of ASA. SA is further metabolized by conjugation to glycine (liver glycine N-acetylase) to form salicyluric acid, by hydroxylation (liver microsomial hydroxylases) to form gentisic acid, and by formation of the phenolic glucuronide and the acyl glucuronide. About 60% of salicylate remains unmodified and can undergo ˙OH attack to produce two derivatives, namely 2,3-dihydroxybenzoate (2,3-DHBA) and, to a much smaller extent, catechol [5], that have not been reported as products of enzymatic metabolism. Thus 2,3-DHBA appears to be a useful marker of in vivo ˙OH production.

MEASUREMENT OF DIHYDROXYBENZOIC ACIDS

Chemicals

Sodium salicylate was obtained from Merck (Darmstadt, Germany); 2,3 dihydroxybenzoic acid, 2,5 dihy-

droxybenzoic acid, 2,6 dihydroxybenzoic acid, and 3,4 dihydroxybenzoic acid were purchased from Sigma (St. Louis, MO, USA). ASA (Aspégic) from Synthelabo, Le Plessis Robinson, France; Acetonitril, methanol, trisodium citrate, sodium acetate, ether, and ethyl acetate were obtained from Prolabo (Lyon, France). All chemicals were analytical reagent grade and used without further purification.

Instrumentation

A Hitachi Liquid Chromatograph pump equipped with an auto-sampler WISP (Waters) was used. Detection was performed with two electrochemical amperometric detectors (Bioanalytical Systems, model LC4B or Shimadzu L-ECD-6A) with plastic cells equipped with glassy carbon electrodes operated at +0.7 V, and an Ag/AgCl reference electrode. The signals from the detector were acquired on Varian DS601 data system and subsequently processed. The analytical stainless-steel column was 150 × 4,6 mm, and packed with octadecyl silane (Spherisorb ODS$_2$, C18) with average particle size of 3 μm (Alltech). A guard column packed with 10 μm spheri-10 PR18 (30 × 4.6 mm I.D., from Alltech) was also used. Peak areas at 0.7 V were recorded with a 1 mV potentiometric screen. All separations were performed at room temperature. For more information, see reference [6].

Optimization procedures

Choice of the oxidation potential. The hydrodynamic voltammograms from 0 to 1400 mV were recorded using 100 mg/l of all studied substances dissolved in the mobile phase (85% Phosphate buffer 100 mM, 0.10 mM SDS, pH 3.3, and 15% methanol, V/V). A CES detector (Erosep Instrument) equipped with 16 electrodes set up at variable potentials was used. We measured the signal/noise ratio of oxidation potential for the concerned substances, especially 2,3-DHBA. The 2,3-, 2,5-, and 3,4-DHBA acids produce a maximum current for a potential of about 0.7 V. However, at this voltage the other possible hydroxybenzoic acids 2,4-and 2,6-DHB acids, catechol, and salicylic acid do not yield a detector response. At a detector potential of 1.1 V, salicylic acid and all the dihydroxybenzoic acid products are recorded. An oxidation potential of 0.7 V vs. the Ag/AgCl reference electrode was chosen to obtain the maximum current response and minimum background noise for the hydroxylated products of salicylate.

Choice of the eluent. An isocratic delivery system was used consisting of a single eluent containing sodium acetate/trisodium citrate (30 mmol/30 mmol) pH 3,90. The flow rate was 0.2 ml/min. We found that addition of 15% of methanol in the mobile phase constitutes a good compromise with a retention time for SA of 7 min and 11 min for the internal standard 2,6 DHBA. The mobile phase was sparged continuously with helium gas during elution. We assayed different potential internal standards previously used by other investigators (3,4 DHBA, 2,4-DHBA, 2,6-DHBA, and catechol). The 2,4- and 2,6-DHBA do not yield a detector response at a detector potential set at 0.7 V, the 3,4-DHBA was thus used for the measurement of DHBAs. For SA assay, the 2,6 DHBA was found to be the most appropriate.

Extraction procedure. Extraction of plasma samples with ether or ethyl acetate were compared. DHBAs were better extracted with ethyl acetate. With double extraction, results were not different, and therefore one simple extraction with ethyl acetate was applied. The extraction with different volumes of ethyl acetate was then examined. We obtained satisfactory results with 3 ml of solvent when 0.4 ml of plasma sample was used.

The optimized procedure was as follows: Aliquots of standard solutions and of plasma samples (400 μl) were mixed with 100 μl of 2.5 μM 3,4-DHBA and acidified by 75 μl of concentrated HCl in 10 × 70 mm glass tubes. The samples were mixed on a vortex-type mixer for 30 s; 3 ml of ethyl acetate were then added and mixed for exactly 2 min. The tubes were finally centrifuged at 1600 × g for 15 min, and two ml of ethyl acetate phase were dried under nitrogen steam. The dry residue was redissolved in 200 μl of mobile phase and 100 μl were injected into the column. Mobile phase consisted of sodium citrate and acetate 30 mmol, pH 3.90. Flow rate was 0.2 ml/min. The concentrations for calibration curves were as follows: 25, 100, 400 nmol 2,3-DHB/l, and 125, 500, 2000 nmol 2,5-DHB/l, using peak-area ratios. Stock solutions of 2,3-DHBA, and 2,5-DHBA were stable for at least 2 months when stored at 4°C in the dark. Saline and plasma blanks were analyzed with each set of standards.

Analytical performance. The detection limit was found to be 0.37 nmol for 2,3-DHBA and 0.62 nmol for 2,5-DHBA. These low limits demonstrate the excellent sensitivity of the proposed method for both DHBAs measurement. The concentrations and peak areas showed linear relationships for 2,3- and 2,5-DHBA in the concentration range investigated (0 to 4000 and 5000 nmol/l, respectively). The absolute recoveries of 2,3-DHBA added at a concentration of 25 nmol/l in six different plasma samples ranged from 86 to 102% with a mean of 94%. Similar studies on 2,5-DHBA (250 nmol/l) yielded recoveries from 94 to 103% with a mean of 98%. The within-run coefficients of variation ranged from 3,3 to 4.1% for 2,3- DHBA and was of 2.1 for 2,5-DHBA. The between-run coefficients of variation ranged between 5.4 and 6.2% for 2.3 DHBA and from 5.7 to 7.4% for 2,5 DHBA. The between-run and recovery assays were conducted during a 15 d period.

Salicylic acid measurement

Salicylic acid assay was performed as previously described [7]. Briefly, aliquots (100 μl) of standard solutions or plasma samples were mixed with 100 μl of 2.5 μM 2,6-DHBA and deproteinized by 200 μl of ethanol in polypropylene conical tubes. The samples were mixed on a vortex-type mixer for exactly 2 min. The tubes were then centrifuged at 1600 × g for 15 min, and 50 μl of the supernatant was diluted with 950 μl of mobile phase. The diluted solution was then filtered on a 0.45 μ filter (Alltech) and 50 μl were injected into the column. The mobile phase consisted of sodium citrate and acetate 30 mmol/l, pH 5.45/methanol (85/15). The flow rate was 1 ml/min and the detector was set up at 295 nm. Standard curves were constructed from measurements of peak-area ratios. The concentrations were as follows: 62.5, 250, 1000 μmol/l of SA.

The human study

To show the applicability of the proposed method, salicylic acid, 2,3-, and 2,5-DHBA concentrations were measured in 20 human subjects after a single oral dose of soluble ASA (Aspégic) (1000 mg were dissolved in 150 ml of water, prepared just prior to administration). Blood samples were drawn before and 2 h following ASA administration. Plasma was separated, quickly frozen and stored at −20°C until required for analysis. Figure 2 shows a chromatogram of a blank plasma sample and a chromatogram of a plasma sample from the subject receiving ASA. Both 2,3- and 2,5-DHBAs were well separated and no interfering peaks were seen in blank plasma samples. Concentrations of 2,3- and 2,5-DHBA as low as 10 nmol/l could be accurately measured. Analysis of samples drawn before ASA administration shows that the plasma fractions were free of salicylic acid or its dihydroxybenzoic acids. The results were obtained from 20 healthy volunteers aged 20 to 40 years with a mean age of 28.2 ± 5.22 years (10 women and 10 men). Two h after ASA administration, the plasma SA level was 487 ± 116; 245–729 μmol/l, that of 2,3-DHBA was 63.2 ± 23.8; 13.6–113 nmol and that of 2,5-DHBA was 832 ± 309; 187–1477 nmol (M ± SD; M ± 2.086 SD).

DISCUSSION

In vivo metabolism of salicylic acid produces two main hydroxylated derivatives (2,3- and 2,5-DHBA) which could be measured. Salicylate reacts with $^{\cdot}OH$ at a rate constant of about $5 \times 10^{+9}$ to 10^{+10} $M^{-1}S^{-1}$ [8]. Low oral doses of ASA may give body-fluid salicylate concentrations, enough to trap $^{\cdot}OH$ radicals in vivo. It should be noted that the principle behind this methodology could be applied to other aromatic compounds such as phenylalanine or benzoic acid [9–11].

Many techniques have already been used to evaluate the hydroxylated compounds in in vitro as well as in in vivo studies. A colorimetric method was initially described by studies [2,12] in which the hydroxylated phenols were treated with sodium tungstate and sodium nitrite in an acidic medium. The pink color was developed in an alkaline medium and measured with a spectrophotometer at 410 nm [13]. HPLC techniques were then developed to separate and quantify the hydroxylated derivatives in animals [1,14–15] or in humans [16–18]. The spectro-photometric and spectro-fluometric methods

were adequate for vitro studies, where high concentrations of $^{\cdot}OH$ are usually formed; however, these detection modes are not sensitive enough to quantify hydroxylated derivatives in in vivo studies. The determination of hydroxyl radicals using salicylate as a trapping agent by gas chromatography-mass spectrometry has also been reported [19]. Consequently, the electrochemical detection mode became popular in most laboratories interested in the quantification of oxidant stress in vivo, especially those concerning human studies.

2,3-DHBA is usually considered a bona fide marker for detecting hydroxyl radicals. However, its level is markedly lower than that of 2,5-DHBA [14]. Our findings confirm a significantly higher level of 2,5-DHBA than 2,3-DHBA (3- to-5-fold). The selective difference between 2,3- and 2,5-DHBA could be associated with differing metabolic rates of production of these adducts, different efficiencies of scavenging $^{\cdot}OH$ radicals at the 3 or the 5 position of the salicylate aromatic ring, and differences in the stability of the adducts against metabolic or biochemical modifications by cellular components [20]. In addition, effect of plasma binding of these adducts on their distribution in blood and into lymph, their transformation and their excretion in the urine has been reported [21]. However, this difference seems to be due to the fact that 2,5-DHBA, but not 2,3-DHBA, is produced by an enzymatic hydroxylation pathway through the cytochrome P-450 microsomal system [22]. The latter seems to be formed exclusively via a radical pathway, which further substantiates the validity of the assumption that 2,3-DHBA authentically reports free radical fluxes in vivo. However, in a cautionary note, Halliwell et al. [23] stressed that 2,3-DHBA derivative may be generated by an unreported minor metabolic pathway, which may be discovered in the future.

The presence of 2,3-DHBA in the plasma of healthy subjects after ASA intake could be due to the baseline rate of intracellular $^{\cdot}OH$ formation from ionizing radiation and the Fenton reactions in vivo [5]. This phenomenon is supported by the detection of baseline values of several markers of $^{\cdot}OH$ production such as allantoin, 8-hydroxy-deoxyguanosine, or thymidine glycols in plasma or urine of healthy subjects.

The 2,3-DHBA derivative appears to be a sensitive and specific marker of in vivo oxidative stress studies. There is now increasing evidence for 2,3-DHBA production after ASA administration in conditions of oxidative stress such as exposure to 100% oxygen, treatment with the redox-cycling drug Adriamycin [14], in vitro neutrophiles activation [24], diabetes [17], alcoholism [25], or rheumatoid arthritis [5].

Because the plasma level of 2,3-DHBA depends on that of salicylate, and because the latter could vary among individuals ingesting the same dose of ASA, the

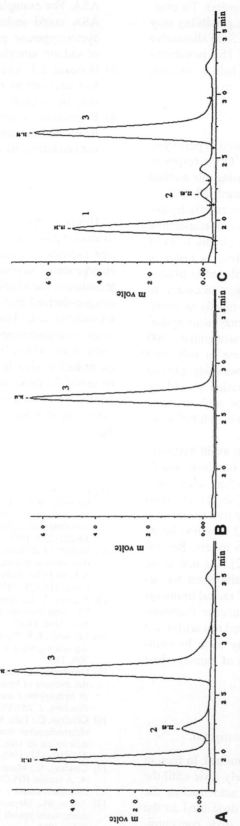

Fig. 2. HPLC-electrochemical detection chromatogram showing separation of isomeric DHBAs. (A) Elution of a standard mixture containing 125 nmol/l 2,3-DHBA, 1000 nmol/l 2,5-DHBA and 500 nmol/l of internal standard 3,4-DHBA. (B) and (C) Plasma samples from a normal subject before and after 1 g per os ASA administration, respectively. The identification of peaks is as follows: (1) 2,5-DHBA; (2) 2,3-DHBA; and (3) 3,4-DHBA.

plasma concentration of 2,3-DHBA should be expressed as 2,3-DHBA/salicylate ratio (mmoles/moles). To overcome such a disadvantage, an alternative possibility may be the intravenous administration of SA. Such alternative has been used in rats and cats [26–27]. This possibility deserves the attention of workers in this field in humans, where nothing has yet been done.

SOME PRECAUTIONS

Salicylate can be hydrolyzed on addition of peroxynitrite, and it is uncertain whether or not 'OH is responsible. Thus, caution should be exercised using this method in systems where $ONOO^-$ is being formed [28–29].

It should be noted that plasma ASA is rapidly hydrolyzed (40% hydrolysis in 120 min at room temperature after sampling) and this could increase plasma level of salicylic acid [30]. Indeed, this hydrolysis is faster in whole blood than in plasma, and is attributed to plasma and red blood cell esterases. It is thus necessary to remove red blood cells from plasma as rapidly as possible. However, it appears that plasma enzymatic hydrolysis is completely inhibited by physostigmine (200 μmol). This should prevent an increase in salicylate plasma level after blood was drawn. In our study, plasma ASA level, 2 h after ASA administration should be negligible. However, at higher doses of ASA or when samples are drawn in the early periods, inhibition of esterases should be considered.

Particular attention should be paid to avoid hydroxylation of aromatic compounds by free transition metals during measurement. Transition metal-free, adequately washed tubes and high pure water should be used throughout the assay. This is especially true for in vitro or ex vivo studies (dialysate, perfusate,) were no chelating agents to neutralize free transition metals. Several teams are working in this field [31–32]. We had some difficulties obtaining reproducible results when we attempted to determine the production of radical hydroxyl in rats after ischemia reperfusion sequence (nonpublished data). Montgomery et al. [33] noted that artifactual production of dihydroxy-benzoates may occur by components of the experimental apparatus of intra-cerebral micro-dialysis system.

FURTHER RESEARCH

Many aspects still remain to be investigated.

1) Larger doses of ASA should be examined. In fact, in humans, ASA is generally not acutely toxic until the plasma concentration of salicylic acid reaches the millimolar range. Thus, the ASA dose used in the present study was well below the toxic concentrations.

2) Salicylic acid has less pharmacological effects than ASA. For example, inhibition of platelet function by ASA could underestimate 'OH production by the cyclooxygenase pathway. The direct administration of sodium salicylate could overcome this concern.

3) Because 2,3- and 2,5-DHBA could be metabolized and excreted in urine, their determination in urine may be of great interest.

4) Evaluation of the relationships between 2,3-DHBA levels and the antioxidant defense systems in different oxidative stress should now be undertaken.

CONCLUSION

Direct evidence for the formation of free radicals in oxidative processes and the causal relationship between free radicals and damage are still lacking. Despite the emergence of electron spin resonance and spin trapping techniques, the identification and the characterization of oxygen-derived free radicals in humans still remains a formidable task. The salicylate hydroxylation products assay as described here provides a simple and convenient method by which 'OH radicals may be detected and quantified in vivo. It is hoped that this methodology will be useful for those attempting to detect and measure 'OH generation in oxidative diseases such as diabetes, myocardial infarction, rheumatoid arthritis, cancer, and aging.

REFERENCES

[1] Floyd, R. A.; Watson, J. J.; Wong, P. K. Sensitive assay of hydroxyl free radical formation utilizing high pressure liquid chromatography with electrochemical detection of phenol and salicylate hydroxylation products. *J. Biochem. Biophys. Methods* **10**:221–235; 1984.

[2] Richmond, R.; Halliwell, B.; Chauhan, J.; Darbre, A. Superoxide-dependent formation of hydroxyl radicals: detection of hydroxyl radicals by the hydroxylation of aromatic compounds. *Anal. Biochem.* **118**:328–335; 1981.

[3] Sagone, A. L.; Husney, R. M.; Davis, B. Biotransformation of para-aminobenzoic acid and salicylic acid by PMN. *Free Radic. Biol. Med.* **14**:27–35; 1993.

[4] Leonards, J. R. Presence of acetylsalicylic acid in plasma following oral ingestion of aspirin. *Proc. Soc. Exp. Biol. Med.* **110**:304–308; 1962.

[5] Grootveld, M.; Halliwell, B. Aromatic hydroxylation as a potential measure of hydroxyl radical formation in vivo. Identification of hydroxylated derivatives of salicylate in human body fluids. *Biochem. J.* **237**:499–504; 1986.

[6] Coudray, C.; Talla, M.; Martin, S.; Fatôme, M.; Favier, A. HPLC-electrochemical determination of salicylate hydroxylation products as an in vivo marker of oxidative stress. *Anal. Biochem.* **227**:101–111; 1995.

[7] Coudray, C.; Mangournet, C.; Bouhadjeb, S.; Faure, H.; Favier, A. A simple HPLC-spectrophotometric assay of salicylic acid in biological fluids. *J. Chromatogr. Sci.* **34**:166–173; 1996.

[8] Anbar, M.; Meyersein, D.; Neta, P. The reactivity of aromatic compounds toward hydroxyl radicals. *J. Phys. Chem.* **70**:2660–2662; 1966.

[9] Kaur, H.; Halliwell, B. Aromatic hydroxylation of phenylalanine

as an assay for hydroxyl radicals. Measurement of hydroxyl radical formation from ozone and in blood from premature babies using improved HPLC methodology. *Anal. Biochem.* **220**:11–15; 1994.

[10] Lamrini, R.; Crouzet, J. M.; Francina, A.; Guilluy, R.; Steghens, J. P.; Brazier, J. L. Evaluation of hydroxyl radicals production using [13]CO2 gas chromatography-isotope ratio mass spectrometry. *Anal. Biochem.* **220**:129–136; 1994.

[11] Sainte-Marie, L.; Boismenu, D.; Vachon, L.; Montgomery, J. Evaluation of sodium 4-hydroxybensoate as an hydroxyl radical trap using gas chromatography-mass spectrometry and high performance liquid chromatography with electrochemical detection. *Anal. Biochem.* **241**:67–74; 1996.

[12] Halliwell, B. Superoxide-dependent formation of hydroxyl radicals in the presence of iron chelates: is it a mechanism for hydroxyl radical production in biochemical systems? *FEBS Lett.* **92**:321–326; 1978.

[13] Singh, S.; Hider, R. C. Colorimetric detection of hydroxyl radical: comparison of the hydroxyl-radical-generating ability of various iron complexes. *Anal. Biochem.* **171**:47–54; 1988.

[14] Floyd, R. A.; Henderson, R.; Watson, J. J.; Wong, P. K. Use of salicylate with high pressure liquid chromatography and electrochemical detection (LEC) as a sensitive measure of hydroxyl free radicals in adriamycin-treated rats. *Free Radic. Biol. Med.* **2**:13–18; 1986.

[15] Powell, S. R.; Hall, D. Use of salicylate as a probe for ˙OH formation in isolated ischemic rat hearts. *Free Radic. Biol. Med.* **9**:133–141; 1990.

[16] Grootveld, M.; Halliwell, B. 2,3 dihydroxybenzoic acid is a product of aspirin metabolism. *Biochem. Pharmacol.* **37**:271–280; 1988.

[17] Ghiselli, A.; Lauranti, O.; De Mattia, G.; Maiani, G.; Ferro-Luzzi, A. Salicylate hydroxylation as an early marker of in vivo oxidative sress in diabetic patients. *Free Radic. Biol. Med.* **13**:621–626; 1992.

[18] Tubaro, M.; Cavallo, G.; Pensa, M. A.; Natale, E.; Ricci, R.; Milazzotto, F.; Tubaro, E. Demonstration of the formation of hydroxyl radicals in acute myocardial infarction in man using salicylate as probe. *Cardiology* **80**:246–251; 1992.

[19] Luo, X.; Lehotay, D. C. Determination of hydroxyl radicals using salicylate as a trapping agent by gas chromatography-mass spectrometry. *Clin. Biochem.* **30**:41–46; 1997.

[20] Powell, S. R. (Review) Salicylate trapping of ˙OH as a tool for studying post-ischemic oxidative injury in the isolated rat heart. *Free Radic. Res.* **21**:355–370; 1994.

[21] Lamka, J.; Laznieck, M.; Gallova, S.; Rudisar, L.; Kvetina, J. Effect of plasma binding of ortho- and para-I-benzoates on their distribution in blood and into lymph, biotransformation and excretion in rat urine. *Eur. J. Metab. Pharmacokinet.* **18**:233–237; 1993.

[22] Ingelman-Sundberg, M.; Kaur, H.; Terelius, Y.; Pearson, J. O.; Halliwell, B. Hydroxylation of salicylate by microsomal fractions and cytochrome P450. Lack of production of 2,3 dihydroxybezoate unless hydroxyl radical formation is permitted. *Biochem. J.* **276**:753–757; 1991.

[23] Halliwell, B.; Kaur, H.; Ingelman-Sunberg, M. Hydroxylation of salicylate as an early assay for hydroxyl radicals: a cautionary note. *Free Radic. Biol. Med.* **10**:439–441; 1991.

[24] Davis, W. B.; Mohammed, B. S.; Mays, D. C.; She, Z. W.; Mohammed, J. R.; Husney, R. M.; Sagone, A. L. Hydroxylation of salicylate by activated neutrophils. *Biochem. Pharmacol.* **38**:4013–4019; 1989.

[25] Thome, J.; Zhang, J.; Davids, E.; Foley, P.; Weijers, H. G.; Wiesberck, G. A.; Boning, J.; Riederer, P.; Gerlach, M. Evidence for increased oxidative stress in alcohol-dependent patients provided by quantification of in vivo salicylate hydroxylation products. *Alcohol Clin. Exp. Res.* **21**:82–85; 1997.

[26] Yang, C. S.; Tsai, P. J.; Chen, W. Y.; Kuo, J. S. Increased formation of interstitial hydroxyl radical following myocardial inschemia: possible relationship to endogenous opioid peptides. *Redox Rep.* **3**:295–301; 1997.

[27] Ophir, A.; Berenshtein, E.; Kitrossky, N.; Averbukh, E. Protection of the transiently ischemic cat retina by zinc-desferrioxamine. *Invest. Ophthalmol. Vis. Sci.* **35**:1212–1222; 1994.

[28] Kaur, H.; Whiteman, M.; Halliwell, B. Peroxynitrite-dependent aromatic hydroxylation and nitration of salicylate and phenylalanine. Is hydroxyl radical involved? *Free Radic. Res.* **26**:71–82; 1997.

[29] Halliwell, B.; Kaur, H. Hydroxylation of salicylate and phenylalanine as assays for hydroxyl radicals: a cautionary note visited for the third time (Review). *Free Radic. Res.* **27**:239–244; 1997.

[30] Cham, B. E.; Ross-Lee, L.; Bochner, F.; Imhoff, D. M. Measurement and pharmacokinetics of acetylsalicylic acid by a novel high performance liquid chromatographic assay. *Ther. Drug Monit.* **2**:365–372; 1980.

[31] Onodera, T.; Ashraf, M. Detection of hydroxyl radicals in the post-ischemic reperfused heart using salicylate as a trapping agent. *J. Mol. Cell. Cardiol.* **23**:365–370; 1991.

[32] Udassin, R.; Ariel, I.; Haskel, Y.; Kitrossky, N.; Chevion, M. Salicylate as an in vivo free radical trap: studies on ischemic insult to the rat intestine. *Free Radic. Biol. Med.* **10**:1–6; 1991.

[33] Montgomery, J.; Sainte-Marie, L.; Boismenu, D.; Vachon, L. Hydroxylation of aromatic compounds as indices of hydroxyl radical production: a cautionary note revisited. *Free Radic. Biol. Med.* **19**:927–933; 1995.

ABBREVIATIONS

ASA—acetyl salicylic acid

2,3-DHBA—2,3-dihydroxybenzoic acid

2,5-DHBA—2,5-dihydroxybenzoic acid

2,6-DHBA—2,6-dihydroxybenzoic acid

3,4-DHBA—3,4-dihydroxybenzoic acid

ECD—electrochemical detection

H_2O_2—hydrogen peroxide

HPLC—high performance liquid chromatography

DHBAS—dihydroxybenzoic acids

˚O2—superoxide anion radical

˙OH—hydroxyl radical

ROS—reactive oxygen species

SA—salicylic acid

Bioassays for Oxidative Stress Status (BOSS). Edited by W.A. Pryor

URINARY ALDEHYDES AS INDICATORS OF LIPID PEROXIDATION IN VIVO

HAROLD H. DRAPER,* A. SAARI CSALLANY,[†] and MARY HADLEY[‡]

*Department of Human Biology and Nutritional Sciences, University of Guelph, Guelph, Ontario, Canada; [†]Department of Food Science and Nutrition, University of Minnesota, St. Paul, MN, USA; and [‡]Department of Food and Nutrition, North Dakota State University, Fargo, ND, USA

(Received 23 November 1999; Revised 30 May 2000; Accepted 15 June 2000)

Abstract—The excretion of malondialdehyde (MDA), lipophilic aldehydes and related carbonyl compounds in rat and human urine was investigated. MDA was found to be excreted mainly in the form of two adducts with lysine, indicating that its predominant reaction in vivo is with the lysine residues of proteins. Adducts with the phospholipid bases serine and ethanolamine and the nucleic acid bases guanine and deoxyguanosine also were found. Except for the adduct with deoxyguanosine (dG-MDA), the excretion of these compounds increased with peroxidative stress imposed in the form of vitamin E deficiency or the administration of iron or carbon tetrachloride. Marked differences in the concentration of dG-MDA in different tissues were correlated with their content of fatty acids having three or more double bonds, the putative source of MDA. Fourteen nonpolar and eleven polar lipophilic aldehydes and other carbonyl compounds were identified as their 2,4-diphenylhydrazine derivatives in rat urine. The excretion of five nonpolar and nine polar compounds was increased under conditions of peroxidative stress. The profile of lipophilic aldehydes obtained for human urine resembled that for rat urine. Except for a reported 4-hydroxynon-2-enal conjugate with mercapturic acid, the conjugated forms of the lipophilic aldehydes excreted in urine remain unidentified. Aldehyde excretion is influenced by numerous factors that affect the formation of lipid peroxides in vivo such as energy status, physical activity and environmental temperature, as well as by wide variations in the intake of peroxides in the diet. Consequently, urinalysis for aldehydic products of lipid peroxidation is an unreliable indicator of the general state of peroxidative stress in vivo.   © 2000 Elsevier Science Inc.

Keywords—Lipid peroxidation, Aldehydes, Ketones, Malondialdehyde, Free radicals

INTRODUCTION

In 1991 [1] it was stated that "one of the greatest needs in the field of free radical biology is the development of

Harold H. Draper received a Ph.D. in nutritional biochemistry from the University of Illinois in 1952 and subsequently pursued research on the function and metabolism of vitamin E at this institution. In 1975, he joined the Department of Nutritional Sciences at the University of Guelph, Ontario, where he conducted research on the metabolism and toxicity of the hydrophilic products of lipid peroxidation with an emphasis on malondialdehyde.

A. Saari Csallany was awarded an Sc.D. in food chemistry by the Technical University of Budapest for work conducted at the University of Illinois on the metabolism of vitamin E. In 1973 she joined the Department of Food Science and Nutrition at the University of Minnesota where she is Professor of Food Chemistry and Nutritional Biochemistry. Her research has been focused on identification of the lipophilic products of lipid peroxidation formed in vivo.

Mary Hadley received a Ph.D. in nutritional biochemistry from the University of Guelph. She is currently Associate Professor in the Department of Food and Nutrition at North Dakota State University.

Address correspondence to: Dr. Harold H. Draper, University of Guelph, Department of Human Biology and Nutritional Sciences, Guelph, Ontario, N1G 2W1 Canada; Tel: (321) 259-5263; Fax: (321) 259-6872; E-Mail: harold_draper@ juno.com.

reliable methods for measuring oxidative stress status (OSS) in humans." A decade later such methods have yet to appear. Also, the clinical value that any such measurement would have in the absence of information on the oxidizing species involved and the metabolic site and nature of its reactions with cell constituents is problematic. Most proposed methods have been limited to the measurement of lipid peroxidative stress in animals using analysis of blood, urine, or solid tissues. Methane and pentane exhalation respond to increased in vivo lipid peroxidation in human subjects fed intravenously [2], but the method of measurement requires specialized equipment and is subject to analytical losses as well as to interference by intestinal bacteria.

Urinalysis for products of in vivo lipid peroxidation is conceptually the most practical means of acquiring information on the reactions of lipid free radicals with cellular compounds in human subjects. Animals fed peroxide-free diets have been shown to excrete increased amounts of malondialdehyde (MDA) [3] and lipophilic

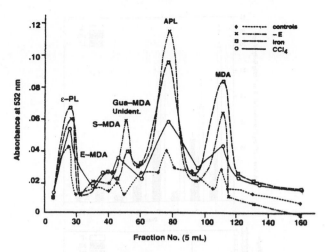

Fig. 1. Effect of vitamin E deficiency and iron or CCl₄ administration on the elution profile of malondialdehyde (MDA) metabolites obtained by anion exchange chromatography of rat urine. In addition to lysine adducts (APL, N-α-acetyl propenal lysine; ε-PL, ε-propenal lysine) and free MDA, the profile reflects minor amounts of adducts with ethanolamine (E-MDA), serine (S-MDA), guanine (Gua-MDA), and unidentified (Unident.) compounds.

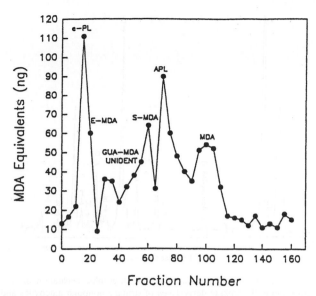

Fig. 2. Elution profile of MDA metabolites obtained by anion exchange chromatography of human urine. See Fig. 1 for abbreviations.

aldehydes [4] after vitamin E depletion or the administration of chemical oxidants but such conditions have little relevance to humans. Urinalysis has yielded valuable information on the products of lipid peroxidative stress in both animals and humans, but it does not include stress exerted by radicals of molecular oxygen except as they may initiate lipid peroxidation. This paper contains an account of the results of this urinalysis.

Excretion of MDA and lipophilic aldehydes

The effect of oxidative stress in the form of iron nitrilotriacetate or carbon tetrachloride administration and of vitamin E deficiency on the excretion of MDA adducts by rats fed an MDA-free diet is shown in Fig. 1 [5]. Total MDA and the proportion excreted in the free form increased with the severity of oxidative stress. MDA is excreted mainly in the form of adducts with lysine and its N-acetylated derivative, indicating that its predominant reaction in vivo is with the ε-amino groups of the lysine residues of proteins. Smaller amounts of adducts with the phospholipid bases serine and ethanolamine and the nucleic acid bases guanine and deoxyguanosine reflect reactions with these compounds. Two additional minor metabolites have been detected and one has been tentatively identified as an adduct with taurine. Analogous MDA adducts were found in the urine of humans consuming a normal diet (Fig. 2) [5].

[14]C-bovine serum albumin that has been exposed to MDA in aqueous solution is subject to an increased rate of proteolysis in vitro by a macroprotease isolated from human erythrocytes by the procedure of Davies and Goldberg [6] (Fig. 3) and a protease isolated from rat liver mitochondria by the procedure of Marcillat et al. [7]. The magnitude of these increases is similar to that observed for proteins exposed to hydroxyl radicals [6,7], indicating that these enzymes hydrolyze proteins damaged not only directly by oxygen radicals but by MDA formed as a result of their reaction with polyunsaturated lipids.

Excretion of the lipophilic products of in vivo lipid peroxidation by fasting rats is illustrated in Figs. 4 and 5 [8]. The sensitivity of the method of measurement was ≤ 0.5 ng. Fourteen nonpolar and 11 polar lipophilic

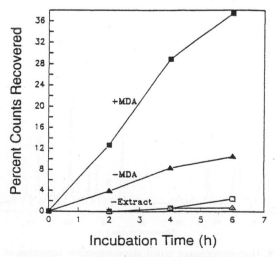

Fig. 3. Effect of exposing ^{14}C-bovine serum albumin to malondialdehyde in solution on its rate of hydrolysis by a red blood cell proteolytic enzyme extract [6].

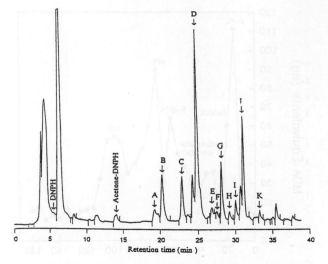

Fig. 4. High-performance liquid chromatographic separation of 2,4-dinitrophenylhydrazine derivatives of urinary nonpolar aldehydes and carbonyl compounds excreted by fasted male rats. A = butanal; B = butan-2-one; D = pentan-2-one; F = hex-2-enal; G = hexanal; H = hepta-2,4-dienal; I = hept-2-enal; K = octanal; C, E, and J unidentified. Absorbance monitored at 378 nm.

aldehydes and related carbonyl compounds were isolated as their 2,4-dinitrophenylhydrazine derivatives by HPLC. Nine polar and three polar compounds were identified. The excretion of polar compounds was much greater than that of nonpolar compounds. Except for a conjugate of 4-hydroxynon-2-enal with mercapturic acid, the conjugated forms of the lipophilic products of in vivo lipid peroxidation excreted in urine have not been identified.

The effect of vitamin E deficiency induced in rats by feeding a diet high in polyunsaturated fatty acids from

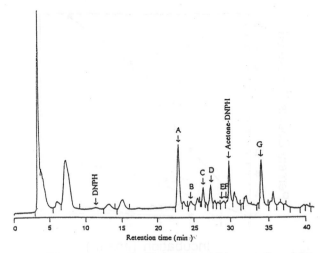

Fig. 5. High-performance liquid chromatographic separation of 2,4-dinitrophenylhydrazine derivatives of urinary polar aldehydes and carbonyl compounds excreted by fasted male rats. E = 4-hydroxyhex-2-enal; A, B, C, D, F, and G unidentified.

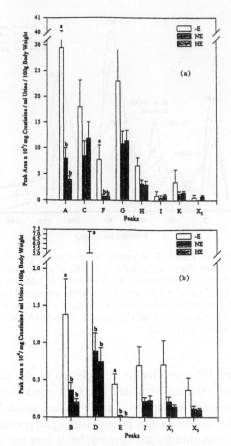

Fig. 6. Comparison of the 2,4-dinitrophenylhydrazine derivatives of urinary nonpolar aldehydes and carbonyl compounds excreted by rats fed diets containing no vitamin E (−E), normal vitamin E (30 mg/kg) (NE), or high vitamin E (300 mg/kg) (HE). (A) A = butanal; C = unidentified; F = hex-2-enal; G = hexanal; H = hepta-2,4-dienal; I = hept-2-enal; K = octanal; X_3 = oct-2-enal. (B) B = butan-2-one; D = pentan-2-one; E, J, X_1, and X_2 unidentified. Values represent the mean ± SEM for 6–9 animals per group. Different superscripts denote significant differences between groups according to the Student-Newman Keuls method ($p < .05$).

distilled corn oil and cod liver oil on the excretion of lipophilic products of in vivo lipid peroxidation is shown in Figs. 6 and 7 [4]. The excretion of five polar and six nonpolar compounds was significantly increased ($p \leq$ .05) in vitamin E deficiency. There was no significant difference in their excretion by animals fed a normal concentration of vitamin E (NE) or ten times this concentration (HE). To the extent that excretion of the lipophilic products of lipid peroxidation in vivo reflects the peroxidative status of rats, it appears that a high intake of vitamin E has no advantage over a normal intake.

The profile of polar and nonpolar aldehydes found in the urine of humans consuming free-choice diets (Figs. 8 and 9) was similar to that found in rats, but their concentrations in human urine were much lower [8]. This and a similar finding with respect to MDA excretion are

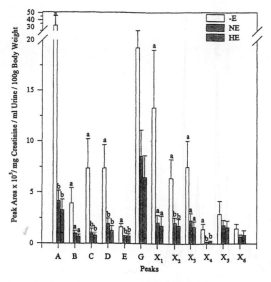

Fig. 7. Comparison of the 2,4-dinitrophenylhydrazine derivatives of urinary polar aldehydes and carbonyl compounds excreted by rats fed diets containing no vitamin E ($-$E), 30 mg/kg vitamin E (NE), or 300 mg/kg vitamin E (HE). E = 4-hydroxyhex-2-enal; X_4 = 4-hydroxyoct-2-enal; A, B, C, D, G, X_1, X_2, X_3, X_5, and X_6 unidentified. Footnotes as for Fig. 6.

lower, are both dominated by adducts with lysine, indicating a primary association of MDA with proteins in both the diet and the tissues. Fasting humans and rats excrete analogous adducts of MDA. The effect of fasting and a peroxide-free diet on excretion of lipophilic products of lipid peroxidation by humans has not been determined, nor has the identity of the conjugates.

Rats fed so-called "purified diets" excrete markedly higher amounts of MDA and lipophilic metabolites than fasting animals [4,10]. Vegetable oils used as a source of essential polyunsaturated fatty acids in these diets and inorganic salts used as a source of essential trace minerals provide a fertile milieu for lipid peroxidation. Short-term replacement of corn oil with hydrogenated coconut oil results in a marked decrease in MDA excretion, whereas prolonged fasting results in an increase [10]. The increase in the level of free fatty acids in blood serum induced by administration of the lipolytic hormones epinephrine and adrenocorticotropic hormone [10] and the lipolysis induced by strenuous exercise [11] are accompanied by an increase in MDA excretion. It is possible that the increase in MDA excretion caused by exercise is due to enhanced initiation of lipid peroxidation by hydroxyl radicals associated with a higher rate of oxygen inhalation, but this does not explain the MDA response to the administration of lipolytic hormones. Adipose tissue is low in PUFA, but these acids may be

attributable to a species difference in basal metabolic rate [9]. The profile of MDA adducts in the urine of rats fed a stock diet of natural ingredients and that of rats fed a purified diet free of peroxidizable fat, though much

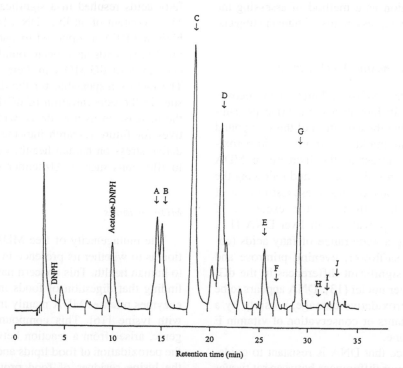

Fig. 8. Representative HPLC profile of 2,4-dinitrophenylhydrazine derivatives of urinary nonpolar aldehydes and carbonyl compounds excreted by a nonfasted human subject. A = butanal; B = butan-2-one; D = pentan-2-one; F = hex-2-enal; G = hexanal; H = hepta-2,4-dienal; I = hepta-2-enal; C, E, and J unidentified. Footnotes as for Fig. 6.

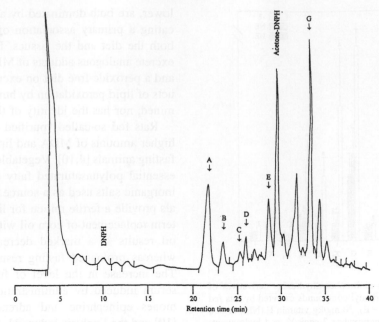

Fig. 9. HPLC profile of DNPH derivatives of urinary polar aldehydes and carbonyl compounds excreted by fasted male rats. E = 4-hydroxyhex-2-enal; A, B, C, D, and G unidentified.

more susceptible to peroxidation when mobilized in the free state. It is also possible that the increase in MDA excretion is simply due to mobilization of preformed fatty acid peroxides. In any event, normal fluctuations in lipolysis undermine the usefulness of urinalysis for products of lipid peroxidation as a method of assessing the steady state peroxidative stress status of human subjects.

Excretion of a deoxyguanosine MDA adduct

Identification of an MDA adduct with deoxyguanosine (dG-MDA) in human and rat urine [9] suggested the possibility that the excretion of this compound might serve as a specific marker for in vivo lipid peroxidation. However, in contrast to the increase in MDA excretion seen in vitamin E deficiency and following the administration of chemical oxidants, these instruments of oxidative stress had no effect on the excretion of dG-MDA by rats or its prevalence in liver DNA [12]. Feeding oils containing a wide range of fatty acids (coconut, olive, linseed, safflower, evening primrose and salmon) produced no significant differences in the dG-MDA content of rat liver nuclei [13]. DNA appears to be shielded from lipid peroxidative damage, possibly by a nuclear barrier to oxidants or conservation of vitamin E in the nuclear membrane.

Despite this evidence that DNA is resistant to oxidative stress, there are major differences between rat tissues in the concentration of dG-MDA in DNA (62 pmol/100 μg in brain vs. 10 pmol in liver and 2 pmol in kidney)

[12]. Furthermore, its frequency in DNA increases with age. The high concentration of dG-MDA in brain DNA suggests a relationship with the prevalence of highly unsaturated n-3 fatty acids in this organ. Adding 2% cod liver oil to a 10% distilled corn oil diet as a source of n-3 fatty acids resulted in a significant increase in the dG-MDA content of rat liver DNA [12]. Humans fed a diet high in PUFA as opposed to one high in monounsaturated fatty acids have been found to have a higher concentration of dG-MDA in their white blood cells [14]. The factors responsible for the differences between tissues in the concentration of dG-MDA in DNA and for the increase in its prevalence with age are prime objectives for future research into the effect of lipid peroxidative stress on human health, with particular reference to afflictions such as Alzheimer's disease.

MDA in foods

The mutagenicity of free MDA [15] raised the question as to whether its presence in the diet presents a risk to human health. This concern has been mitigated by the finding that digestion of foods in vitro with proteolytic enzymes releases MDA mainly in the form of an adduct with lysine [16]. This compound, which is nonmutagenic, arises from a reaction between MDA released in the peroxidation of food lipids and the ϵ-amino groups of the lysine residues of food proteins, from which it is released in the course of protein digestion. Under the acidic conditions of the thiobarbituric acid method for the determination of MDA, its bound forms are hydrolyzed to give a spuriously high value for the concentration of free MDA [16].

DISCUSSION

Halliwell [18] among others has discussed the difficulties involved in distinguishing between lipid peroxidation as a cause and as a result of disease. The list of clinical ailments reportedly associated with an increased concentration of MDA in the blood, urine, or tissues covers most of the major metabolic diseases (atherosclerosis, diabetes, congestive heart failure, cancer, arthritis, cold injury, alcoholic liver disease). This list could be extended to include perinatal hypoxia; infertility in men; tissue trauma resulting from physical injury, surgery, or burns; and the cachexia of cancer and other debilitative diseases. Subclinical disturbances and normal metabolic responses to stress that involve mobilization of adipose tissue fatty acids, including exposure to the cold and activities regarded as beneficial to human health, such as walking, jogging, and participation in sports, give rise to increases in MDA excretion in ostensibly healthy subjects. Added to these variables are wide fluctuations in the intake of peroxidized lipids in the diet. Consequently, short-term measurements of the urinary excretion of MDA and longer-chain aldehydes can not provide a reliable indication of the general state of oxidative stress.

The early perception of lipid peroxidation as a pathological condition that should be suppressed to the extent possible has given way to the realization that it occurs in the absence of any gross, histological, or biochemical evidence of disease. Experimental animals fed chemically defined diets adequate in antioxidant nutrients and devoid of lipid peroxides excrete significant, though reduced, quantities of lipid peroxidation products. Except for abnormalities of vitamin E metabolism such as abetalipoproteinemia and malabsorption, the intimate association between vitamin E and PUFA in higher plants and animals protects humans against dietary vitamin E deficiency.

Notwithstanding, there is strong evidence that raising the tissue level of vitamin E above the norm can have a protective effect against some diseases. This effect is attributable to the unique metabolic role of this vitamin. Most vitamins function as cofactors for enzymes involved in the metabolism of the major nutrients and in the synthesis and degradation of other compounds. Such reactions are seldom responsive to an excess of cofactors. In contrast, vitamin E functions as a reactant in a chemical process. Stripped of the enzymes that scavenge the oxygen radicals responsible for initiation of lipid peroxidation and those that degrade its products, the chemistry of lipid peroxidation in vivo is much the same as that which takes place in foods and plant oils. Indeed, certain synthetic antioxidants structurally unrelated to vitamin E are capable of replacing it in the tissues of animals. Rats have been raised through two generations on a vitamin E–free diet containing the synthetic rubber antioxidant N^1, N^1-diphenyl-p-phenylene diamine [19]. The same products of α-tocopherol oxidation are produced by its antioxidant action in natural products and in the body, as well as by mild oxidizing agents such as potassium ferricyanide and ferric chloride. This evidence that the efficacy of vitamin E in the prevention of lipid peroxidation in vivo follows chemical rather than enzyme kinetics suggests the possibility that tissue concentrations above the range sustainable by diet may be efficacious in the prevention of some human pathologies.

Persuasive evidence that this may be the case has been provided by studies on two large groups of human adults, one male and one female, who exhibited a substantially lower incidence of heart attack than a control group after consuming a vitamin E supplement equivalent to over ten times the median intake from a typical diet for two or more years [20,21]. There was no significant relationship between the incidence of heart attack and the intake of vitamin E in the diet. The efficacy of the supplement was attributed to decreased peroxidation of abetalipoprotein-B.

It is noteworthy in this connection that in the only experimental study of vitamin E deficiency carried out in humans, two years also were required to induce the first symptom of deficiency (a mild creatinuria), and then only after a high level of PUFA was incorporated into the depletion diet [22]. A low efficiency of vitamin E absorption at high intakes and a low binding capacity of transport lipoproteins preclude raising the serum level of vitamin E much above twice the normal concentration. A small differential between the concentration of vitamin E in the serum and the solid tissues further limits the impact of dietary vitamin E on its concentration in the tissues. Consequently, both vitamin E accumulation and depletion in human adults occur very slowly. This has long been known in the case of adult animals. Feeding a vitamin E–free diet to produce the classical sterility syndrome in female rats must begin with weanlings, as must the production of muscular dystrophy in rabbits, the deficiency disease in most animals.

The apparent capacity of pharmacological intakes of vitamin E to lower the risk of heart attack by inhibiting the peroxidation of apolipoprotein-B suggests that high concentrations of this vitamin may be efficacious in the prevention of oxidative pathology at other sites. The development of noninvasive methods for the identification and investigation of such sites is a worthy priority for future research.

Acknowledgements — This research was supported by the Natural Sciences and Engineering Research Council of Canada, the Minnesota Agricultural Experiment Station, the U.S Agency for International Development Contract No. 608-0160, and the North Dakota Agricultural Experiment Station.

REFERENCES

[1] Pryor, W. A.; Godber, S. S. Noninvasive measures of oxidative stress status in humans. *Free Radic. Biol. Med.* **10**:177–184; 1991.

[2] Lemoyne, M.; Van Gossum, A.; Kurian, R.; Ostro, M.; Axler, J.; Jeejeebhoy, K. Breath pentane analysis as an index of lipid peroxidation: a functional test of vitamin E status. *Am. J. Clin. Nutr.* **46**:267–270; 1987.

[3] Draper, H. H.; Polensek, L.; Hadley, M.; McGirr, L. G. Urinary malondialdehyde as an indicator of lipid peroxidation in the diet and in the tissues. *Lipids* **19**:836–843; 1984.

[4] Csallany, A. S.; Kim, S.-S.; Gallaher, D. D. Response of urinary aldehydes and related compounds to factors that stimulate lipid peroxidation in vivo. *Lipids* **35**:855–862; 2000.

[5] Mahmoodi, H.; Hadley, M.; Chang, Y.-X.; Draper, H. H. Increased formation and degradation of malondialdehyde-modified proteins under conditions of peroxidative stress. *Lipids* **30**:963–966; 1995.

[6] Davies, K. J. A.; Goldberg, A. L. Oxygen radicals stimulate intracellular proteolysis and lipid peroxidation by independent mechanisms in erythrocytes. *J. Biol. Chem.* **262**:8220–8226; 1987.

[7] Marcillat, O.; Zhang, Y.; Lin, S. W.; Davies, K. J. A. Mitochondria contain a proteolytic system which can recognize and degrade oxidatively-denatured proteins. *Biochem. J.* **254**:677–683; 1988.

[8] Kim, S.-S.; Gallaher, D. D.; Csallany, A. S. Lipophilic aldehydes and related compounds in rat and human urine. *Lipids* **34**:489–496; 1999.

[9] Agarwal, S.; Wee, J. J.; Hadley, M.; Draper, H. H. Identification of a deoxyguanosine-malondialdehyde adduct in rat and human urine. *Lipids* **29**:429–432; 1994.

[10] Dhanakoti, S. N.; Draper, H. H. Response of urinary malondialdehyde to factors that stimulate lipid peroxidation in vitro. *Lipids* **22**:643–646; 1987.

[11] Quintanilha, A. T. Oxidative effects of physical exercise. In: Quintanilha, A. T., ed. *Reactive oxygen species in chemistry, biology and medicine.* New York: Plenum Press; 1988:187–195.

[12] Draper, H. H.; Agarwal, S.; Voparil Nelson, D. E.; Wee, J. J.; Ghoshal, A. K.; Farber, E. Effects of peroxidative stress and age on the concentration of a deoxyguanosine-malondialdehyde adduct in rat DNA. *Lipids* **30**:959–961; 1995.

[13] Hadley, M.; Stangl, G. L.; Philbrick, D. J.; Kirchgessner, M.; Draper, H. H. Influence of dietary oils on the fatty acid composition of the nuclear lipids of rat brain and the oxidative stability of DNA. Unpublished results.

[14] Fang, J.-L.; Vaca, C. E.; Valsta, L. M.; Mutanen, M. Determination of DNA adducts of malondialdehyde in humans: effects of dietary fatty acid composition. *Carcinogenesis* **17**:1035–1040; 1996.

[15] Basu, A. K.; Marnett, L. J. Unequivocal demonstration that malondialdehyde is a mutagen. *Carcinogenesis* **4**:331–333; 1983.

[16] Piche, L. A.; Cole, P. D.; Hadley, M.; van den Bergh, R.; Draper, H. H. Identification of N-ε-(2-propenal)lysine as the main form of malondialdehyde in food digesta. *Carcinogenesis* **9**:473–477; 1988.

[17] Draper, H. H.; Squires, E. J.; Mahmoodi, H.; Wu, J.; Agarwal, S.; Hadley, M. A comparative evaluation of thiobarbituric acid methods for the determination of malondialdehyde in biological materials. *Free Radic. Biol. Med.* **15**:353–363; 1993.

[18] Halliwell, B.; Gutteridge, J. M. C. *Free radicals in biology and medicine.* Oxford: Clarendon Press; 1985.

[19] Draper, H. H.; Goodyear, S.; Barbee, K. D.; Johnson, B. C. A study of the role of antioxidants in the diet of the rat. *Br. J. Nutr.* **12**:89–97; 1952.

[20] Rimm, E. B.; Stampfer, M. J.; Ascherio, A.; Giovannucci, E.; Colditz, G. A.; Willett, W. C. Vitamin E consumption and the risk of coronary heart disease in men. *New Engl. J. Med.* **328**:1450–1456; 1993.

[21] Stampfer, M. J.; Hennekens, C. H.; Manson, J. E.; Colditz, G. A.; Rosner, B.; Willett, W. C. Vitamin E consumption and the risk of coronary heart disease in women. *New Engl. J. Med.* **328**:1444–1449; 1993.

[22] Horwitt, M. K. Interrelationships between vitamin E and polyunsaturated fatty acids in adult men. *Vita. Hormon.* **20**:541–558; 1962.

ABBREVIATIONS

dG—deoxyguanosine

MDA—malondialdehyde

PUFA—polyunsaturated fatty acids

Bioassays for Oxidative Stress Status (BOSS). Edited by W.A. Pryor

ACCURATE AND SENSITIVE MEASUREMENTS OF pO$_2$ IN VIVO USING LOW FREQUENCY EPR SPECTROSCOPY: HOW TO CONFER BIOCOMPATIBILITY TO THE OXYGEN SENSORS

BERNARD GALLEZ* and KARSTEN MÄDER†

*Laboratory of Medicinal Chemistry and Radiopharmacy and Laboratory of Biomedical Magnetic Resonance, Université Catholique de Louvain, Brussels, Belgium; and †Department of Pharmaceutical Sciences, Freie Universität, Berlin, Germany

(Received 16 December 1999; Revised 10 May 2000; Accepted 1 June 2000)

Abstract—Within the last few years, there has been a significant amount of progress using EPR oximetry, which has resulted in the availability of instrumentation and paramagnetic materials capable of measuring pO$_2$ in tissues with an accuracy and sensitivity comparable to or greater than that available by any other method. While the results obtained with EPR so far indicate that criteria for the measurements of pO$_2$—such as accuracy, sensitivity, repeatability, and noninvasiveness—can be met, some of the paramagnetic materials with optimum spectroscopic properties (i.e., strong simple signals which are appropriately responsive to changes in pO$_2$) may have some undesirable interactions with tissues, causing reactions with and/or losing responsiveness to oxygen. In this paper, several approaches are discussed, such as encapsulation procedures, which can result in the availability of oxygen-sensitive materials in a suitable configuration for long-term studies (absence of toxicity and preservation of the responsiveness to oxygen). © 2000 Elsevier Science Inc.

Keywords—EPR, ESR, In vivo, Oxygen, Encapsulation, Free radicals

INTRODUCTION

Oxidative stress primarily occurs in tissues as the result of an inappropriate oxygen environment for the cells. Small differences (decreases or increases) in the physiological partial pressure of oxygen (pO$_2$) can induce stress and/or the formation of reactive species with deleterious effects. In studies on oxidative stress, there is more and more recognition that there is a need for systems that can detect very subtle variations of pO$_2$ in tissues, especially in living systems, in order to predict the relevance of pathophysiological situations. Although the measurement of oxygen is a key factor for understanding oxidative stress, there is a lack of methodologies that are able to report continuously, noninvasively, and accurately the pO$_2$ in tissues. In this paper, we describe the principles of EPR oximetry, the types of oxygen sensors, and the recent advances to enhance the biocompatibility of these sensors. It is essential to demonstrate that the variations of pO$_2$ measured are related to physiological or pathophysiological processes, and are not related to disturbances coming from the presence of the oxygen sensor inside the tissue.

IN VIVO EPR OXIMETRY AND OXYGEN SENSORS

The term *EPR oximetry* encompasses a number of distinct techniques [1]. One class of techniques relies on the oxygen dependence of the rates of metabolism of EPR-sensitive probes (e.g., nitroxides). Other methods rely on the paramagnetic properties of molecular oxygen since oxygen has two unpaired electrons and is therefore effective in relaxing other paramagnetic species. Measurements that depend on T$_1$ and T$_2$ provide direct evidence of the local

Dr. Bernard Gallez is director of the Laboratory of Biomedical Magnetic Resonance at Catholic University of Louvain, Brussels, Belgium. He received his Ph.D. in 1993 in Pharmaceutical Sciences from the same university. His main research interests focus on developments of methodologies for measuring oxygenation and perfusion in vivo, as well as their applications in tumors and cerebral ischemia.

Dr. Karsten Mäder is a senior lecturer of pharmaceutics at Free University in Berlin. He obtained his Ph.D. in 1993 at Humboldt-University Berlin, Germany. His research is focused on the development and characterization of nano- and microparticulate biodegradable drug carriers.

Address correpondence to: Bernard Gallez, Ph.D., Laboratory of Medicinal Chemistry and Radiopharmacy, Avenue Mounier 73.40, B-1200 Brussels, Belgium; Tel: +32 (2) 7647348; Fax: +32 (2) 7647363; E-Mail: gallez@cmfa.ucl.ac.be.

Table 1. Stability of Several Paramagnetic Compounds When Implanted in Gastrocnemius
Muscle of Mice

Paramagnetic material	Stability in vivo	Reference
Lithium Phthalocyanine	< 3 d	[17]
India ink	> 4 months	[18]
Fusinite	> 4 months	[4]
Typical heated charcoal	< 1 d	[12]
Nitroxides	Few min	[11]
Aldrich activated carbon decolorizing 16155-1	> 3 months	[6]
Aldrich activated carbon darco 27809-2	< 7 d	[6]
Merck charcoal activated GR 99-002186/0250	< 7 d	[6]
Merck charcoal activated extra pure 99-021841/1000	< 3 d	[6]
Fluka activated charcoal 05105	< 1 d	[6]
Strem chemicals carbon powder 93-0601	< 2 d	[6]
EM Science charcoal wood CX0670-1	> 4 months	[6]
Carlo Erba charcoal powder 332658	> 3 months	[6]

concentration in oxygen. By far, the most common method is the measurement of the broadening of the principal hyperfine line (related to T_2). The line broadening is due to Heisenberg spin exchange between unpaired electrons of the probe and paramagnetic oxygen. The physical description is well established for soluble paramagnetic compounds (such as nitroxides). However, the characterization of the physical processes leading to the broadening of particulate materials is still being investigated [2]. Particulate materials offer a more sensitive response than do the soluble materials. Because of their intrinsic high sensitivity, variations of less than 1 mm Hg can be measured directly in vivo after implantation of such materials acting as oxygen reporters. Several paramagnetic probes have been described and used in vivo as oxygen sensors: lithium phthalocyanine [3], charcoals such as fusinite [4] or gloxy [5], activated charcoals [6], and synthetic carbohydrate chars [7]. The measurement in vivo can be performed using low-frequency EPR spectrometers. At low magnetic field, the wavelength to be used is compatible with deep penetration into the tissues, and EPR spectra can be recorded in living systems. A principal goal to further develop EPR oximetry is to achieve the following characteristics: increased sensitivity (ability to detect very subtle changes of pO₂), localization (ability to make the measurement from a defined volume), repeatability, rapidity, noninvasiveness, and biocompatibility [8].

SPECIFIC PROBLEMS OF THE METHODOLOGY

While the results obtained so far with EPR oximetry indicate that we can meet many of the criteria noted above, there are also a number of potential limitations of the current capabilities. We have found that some of the paramagnetic materials with optimum spectroscopic properties (i.e., strong simple signals which are appropriately responsive to changes in the pO₂) may have some undesirable interactions with tissues, causing reac-

tions with and/or losing responsiveness to oxygen. These problems are due to a loss of paramagnetism and/or loss of responsiveness to oxygen.

Loss of paramagnetism

Nitroxides are quickly converted into diamagnetic hydroxylamines, and the metabolism of the nitroxides is oxygen-dependent [9]. Therefore, it was previously suggested that the monitoring of the nitroxide decrease was a possible tool for measuring the pO₂ both in vitro and in vivo. Changes in nitroxide kinetics were also used in specific circumstances to detect the formation of oxidative radicals, thus reflecting the oxidative stress status of the tissue [10]. In fact, the decrease in the EPR signal coming from a specific tissue in vivo can be due to many other factors (the strongest dependence is from the washout of the compound and is related to the perfusion of the tissue; other mechanisms such as diffusion inside the tissue or necrosis of a tissue could also affect the decay of the EPR signal) [11]. Another way to use nitroxides as pO₂ reporters is to use their line width or the super hyperfine splitting (observed with some nitroxides such as carbamoyl-proxyl) as indicators of their environment. For this approach it is, of course, more convenient to get a stable signal from the sensor, and therefore to protect the nitroxide from this metabolism or from rapid washout. The inclusion of nitroxides inside microcapsules or microdevices that are still permeable to oxygen could in this way enhance their usefulness as oxygen sensors in tissues.

Loss of responsiveness to oxygen

There are several experiments that show the relative instability of the oxygen sensitivity under specific conditions. As illustrative examples, we present in Table 1 data on the stability of the oxygen response of several paramagnetic materials in the gastrocnemius muscles of mice. When a paramagnetic material has an EPR line width that is dependent on the pO_2, one of the simplest ways to evaluate its usefulness as an in vivo marker of pO_2 consists of injecting a small amount of this material into a leg muscle of mice or rats; after equilibrium is reached in the tissues, hypoxic conditions can be easily obtained by reversible restriction of the blood supply. To demonstrate the stability in the responsiveness, one can repeat these experiments over days, weeks, or months. In this context, instability in vivo means that EPR line widths in vivo tend to have the same values in both normal and hypoxic tissues. At the end of the experiment the paramagnetic material should be removed from its site after the sacrifice of the animal and the calibration curve (line width as a function of the pO_2) should be repeated to determine whether the calibration curve was modified. Both types of experiments (in vivo modification of the line width following a clear change in pO_2 and possible change of the calibration curve) are needed to determine the stability of the sensor to its environment. As shown in Table 1, the loss of oxygen sensitivity of the paramagnetic material can occur within a few days for several materials, potentially limiting their usefulness in this tissue.

Concerning the loss of oxygen-sensitivity with time, several considerations should be emphasized. (i) The loss of responsiveness to oxygen is material-dependent and tissue-dependent. For example, lithium phthalocyanine loses the responsiveness to oxygen within a few days in gastrocnemius muscle of mice, but has a stable response for months in the brain or in the spinal cord of rats. Similar findings were made using different types of charcoal. (ii) The time needed to reach the equilibrium for a sensor in its environment is variable from one material to another and from one tissue to another (this phenomenon should not be confused with the loss of responsiveness to oxygen). (iii) There is a need to demonstrate the stability for a specific batch of material; caution should be used when extrapolating results that were established using another batch of oxygen sensors. (iv) The loss of responsiveness to oxygen seems to be the consequence of a modification in the environment of the particle. Although these phenomena are not yet fully understood, it is likely that the instability of the response is related to "chemical" changes at the surface of the particles and/or physical limitation in the accessibility of oxygen. The first event generally occurs rapidly, and the change in the calibration also can be demonstrated in vitro (e.g., in tissue homogenates). In this case, the response will be quickly lost in vivo. The second type of event could be related to the deposit of proteins or other factors after a reaction inside the tissue: such biological components can form a small capsule. As the sensitivity of the response is strongly dependent on the collisions with oxygen, any barrier can lead to a change in the responsiveness to oxygen. In

this case, the loss of responsiveness will occur progressively in vivo. Such reactions were also described for some biosensors implanted in tissues. Some biosensors rely on a catalytic reaction for which oxygen is important; some biosensors progressively lose their activity because a tissue reaction occurs, leading to the formation of a capsule surrounding the sensor, which makes a barrier to oxygen [13].

CONCEPT OF BIOCOMPATIBILITY

The biocompatibility of an implanted material can be defined as follows. The performance of the material should not be affected by the host, and the host should not be affected by the implanted material. The potential modifications decreasing the performance of the oxygen sensors include both loss of paramagnetism and/or loss of responsiveness to oxygen. The reactions affecting the host include inflammatory and/or toxic reactions in tissues. It appears quite possible that the embedding of oxygen-sensitive paramagnetic materials in biopolymers that already are acceptable for use in humans would facilitate their approval for use in human subjects. The development of appropriate coating/embedding capabilities appears to be an effective means to reduce some of these potentially important limitations.

Some of the potential beneficial effects of the coatings include: (i) decreasing the probability of the paramagnetic materials having deleterious biological reactions either directly or by stimulating processes such as cytokine release or specific immune response; (ii) decreasing the probability of the paramagnetic materials being altered by the biological system such that they have a decreased signal intensity (due to a decrease in the number of spins and/or an increase in the intrinsic line width) and/or a change in their responsiveness to oxygen (either a loss of responsiveness or a change in the calibration curve); (iii) stabilizing small particles in tissues, preventing their unwanted aggregation or movement; (iv) enhancing their spectroscopic characteristics.

The coating material to be used in order to stabilize the responsiveness of the oxygen sensor as well as to avoid a deleterious tissue reaction should possess the following characteristics. (i) The material should have a good permeability to oxygen. (ii) The coating material should be stable in tissues and should isolate completely the sensor in order to preserve its configuration as long as needed for the studies. The polymers generally used for designing a drug delivery system are biodegradable (e.g., co-polymer of polylactide and polyglycolide). For the present purpose, there is a need to maintain the system in its initial configuration, and consequently to use a polymer without degradation within this time period. (iii) The material should be known to have good tolerance in tissues, and preferably already be accepted for human use. (iv) The material should be sufficiently versatile to lead to a reproducible and uniform coating.

APPLICATIONS: HOW TO CONFER THE BICOMPATIBILITY TO EPR OXYGEN SENSORS?

Soluble materials (e.g., nitroxides)

In order to avoid a direct contact of the nitroxide with the surrounding tissue, and consequently lose the paramagnetism, several approaches such as the inclusion of nitroxides in implants or the microencapsulation of the nitroxides in liposomes or microspheres have been described [14,16,17]. Interestingly, one of the first studies of in vivo EPR oximetry described the use of an oxygen-permeable capsule containing a nitroxide in solution implanted in the peritoneal cavity of a mouse [14]. Using such a capsule has several advantages: this physical barrier avoids the interaction between the nitroxide solution and the tissue components responsible for the bioreduction of the nitroxide into hydroxylamines; this plastic container also permits the exact localization of the region where oxygen tension is to be determined; and this system allows the use of a nonaqueous solvent with higher oxygen solubility, consequently increasing the sensitivity of the measurement. Oximetry using nitroxides is based on EPR line broadening caused by Heisenberg exchange between molecular oxygen dissolved in solution and the nitroxide. There is a direct relationship between the EPR line exchange broadening and the radical-radical collision rate [15]. The exchange rate, and consequently the EPR line width, is proportional to the solubility of oxygen in the solvent. In this pioneering experiment, the authors used an oxygen-permeable capsule in a plastic TPX, a methylpentene polymer. The nitroxide used was 15N-Tempone dissolved in light paraffin oil, and wax was used to seal the capillaries of PTX. This method had high sensitivity and the sensor remained intact during the period necessary for the measurement.

Another approach based on the same principle was carried out at the microscopic scale. In this approach, the nitroxide was encapsulated in liposomes [16] or proteinaceous microspheres [17]. Glockner et al. described the preparation of large unilamellar liposomes containing positively charged deuterated nitroxides [16]. Compared to unencapsulated nitroxides, these encapsulated nitroxides had several advantages. (i) The stability of the nitroxide encapsulated in liposomes was higher than free nitroxide after administration by intramuscular injection. (ii) This system provided stable levels of free nitroxides in vivo, which is needed for accurate oxygen measurements based on line broadening. However, as the nitroxides are dissolved in an aqueous phase inside the lipo-somes, the sensitivity of the nitroxide to changes of pO$_2$ remains low for in vivo applications. The approach of Liu et al. [17] overcame the problem of poor nitroxide sensitivity. In this study, proteinaceous microspheres filled with nitroxides dissolved in an organic liquid were synthesized with high intensity ultrasound. Due to the increase of sensitivity of the line width to oxygen because of the higher solubility of oxygen in organic solvents, reliable measurements of pO$_2$ were obtained in vivo. A potential concern is the release of the organic solvent inside the tissues as a potential source of toxicity. These authors suggested that the same approach using inert solvents such as perfluorocarbons might solve this problem. This system also protected the nitroxide from bioreduction. However, these microspheres are metabolized over extended periods, and the measurements of pO$_2$ using this system will be valid only for short-term studies. The particulate probes might be better for extended periods such as weeks or months.

Encapsulation of particulate oxygen sensors

All types of oxygen-sensitive particles (e.g., lithium phthalocyanine and charcoals) could potentially benefit from a convenient coating. Up to now, because of the low availability of lithium phthalocyanine, only studies describing the coating of carbonaceous particles have been reported.

Inks and synthetic inks

Since india ink is very well tolerated by living tissues [18,19], our group tried to prepare a special ink for in vivo EPR applications. The ink consisted of a carbon-based material (fusinite or a carbohydrate char) with optimal EPR properties, dispersed in a liquid solution containing gums or suspending agents [12]. Such a coating could have several advantages: preventing aggregation of small particles (<1 μm) due to electrostatic attraction; improving the tolerance by the tissues and decreasing immunological responses; facilitating systemic injection instead of a local injection with appropriate sized particles, and allowing liver oximetry (accumulation occurs in Kupffer cells). In the case of oxygen-sensitive materials that lose their oxygen sensitivity with time, the coating may prevent this loss of sensitivity to the pO$_2$. Suspensions of the paramagnetic material were made in water solutions containing 3% arabic gum. Very small particles were obtained using different homogenization systems. To select particles of different sizes, centrifugation and filtration on calibrated filters were used. No significant differences were observed among the calibration curves for the different water suspensions of carbon materials, which had either been coated, contained small particles, or contained large particles (within the range of size used in this study). Although no significant difference was observed in vitro in the pO_2 sensitivity for coated or uncoated materials, we observed that several coatings had a dramatic effect on the pO_2 sensitivity in vivo [12]. The coating with arabic gum preserved the pO_2 sensitivity of the fusinite particles. The pO_2 sensitivity (measured by observing the line width change in unrestricted or blood-restricted muscles) remained unchanged during the 6 weeks of the experiments. The chars used in that study lost their oxygen sensitivity in vivo and the coatings using these hydrosoluble polymers were unable to prevent this loss of oxygen sensitivity. The use of small particles of fusinite (300 nm diameter) coated with arabic gum made intravenous administration possible with eventual accumulation in the liver. The uncoated fusinite was toxic when administered by iv, due to the large size of the particles (apparently the particles formed aggregates that were trapped in the capillaries of the lungs). Although we demonstrated at that time the feasibility of synthesizing inks with optimal EPR properties, the approach of using hydosoluble polymers did not preserve the responsiveness to oxygen of unstable chars. Therefore, we have assumed that the preparation of sensors with a deposit of an insoluble thin film of oxygen-permeable biopolymer on its surface can preserve the responsiveness to pO_2 inside the tissue.

Film coatings of sensors

Many techniques of microencapsulation exist and are used in pharmaceutical technology. As a starting point, we used well-established methods that were already described in the literature for coating charcoals. Charcoal, due to its large adsorption capacity, can be used in the treatment of drug intoxication. The technique, called "charcoal hemoperfusion," relies on the passage of the blood through a column containing charcoal particles in order to remove the drugs or xenobiotics causing the intoxication. In this system, charcoal particles are directly in contact with blood elements. In order to overcome problems associated with platelet removal, excessive blood damage, and/or the release of fine carbon particles into the blood stream, activated charcoal has been successfully coated with many biocompatible polymers. These include albumin-collodion, nylon and collodion, cellulose derivatives, polyhydroxyethyl methacrylate, silicone, polyethyleneglycol, and polyacrylate-polymethacrylate. For our initial studies, we selected two charcoals which have the oxygen-sensing properties required for EPR oximetry combined with a tendency to lose responsiveness to oxygen when placed in tissues [6].

We prepared different batches of coated materials, varying the amount of pyroxylin (cellulose nitrate) used as the coating material [20]. Particles of defined sizes were selected using mechanical sieves to produce materials in increments of 25 μm, ranging from smaller than 25 μm to larger than 175 μm; the smallest and the largest particles were tested both as uncoated and coated materials. There were no significant differences in the calibration curve (line width as a function of pO_2) between the uncoated and coated materials. The thin layer film made by this cellulose derivative is sufficiently permeable to oxygen, which allows equilibrium with the external medium. Finally, we evaluated the performance of the coated particles as oxygen sensors by inducing hypoxia in the muscle of mice injected with charcoals and we repeated the experiments for 2 months to determine the reproducibility and the stability of the EPR line width of the paramagnetic material. As shown in Fig. 1, the uncoated charcoal used in this study lost its responsiveness to oxygen within 1 week; the responsiveness to changes of pO_2 decreased 1 week after the injection of the charcoal, and there was a complete loss of responsiveness by 9 d after the injection. Using the same animal model as for uncoated materials, we found no loss in the responsiveness to oxygen over 2 months (time of observation) when a sufficient amount of coating material was used (20 to 40% w/w). Using a lower content in pyroxylin, the responsiveness to oxygen was not preserved. Using a higher amount of pyroxylin (more than 40% w/w), we observed the formation of films that were difficult to handle and inject in animals. Coated particles can thus be used in long-term studies where an accurate measurement of the pO_2 in tissues is necessary. We concluded that these results demonstrate that an appropriate film coating on the surface of charcoals (i.e., using pyroxylin) is able to preserve the responsiveness to oxygen of paramagnetic materials used for in vivo EPR oximetry [20].

Biocompatible implants as oxygen reporters

Another approach of EPR oximetry is based on devices directly implanted into living tissue. This includes the preparation of very small implants and the development of coating/membranes for the attachment to insertable probe resonators. We are following strategies already used in the design of biosensors, especially devices requiring the use of biocompatible membranes permeable to oxygen. Most of them use thin silicone tubing. Other biocompatible oxygen-permeable materials could also be used for that purpose. One application we developed relied on the preparation of biocompatible silicon implants containing fusinite [21]. Finely ground fusinite was mixed with polydimethysiloxane oil and silicon paste, placed in a syringe, and extruded

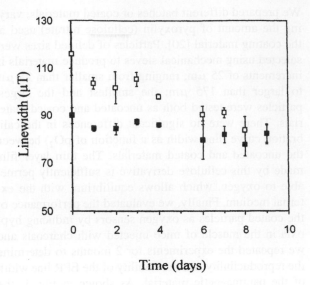

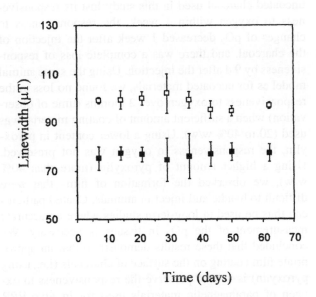

Fig. 1. Effect of the coating using pyroxylin on the responsiveness of a charcoal to changes of pO₂ in vivo. *Top*: Change of the responsiveness to oxygen in vivo observed on a typical unstable charcoal. The EPR line width was recorded using an L-Band spectrometer on anesthetized mice before (□) and after (■) the restriction of the blood flow in the muscle. Note the decrease in the responsiveness to changes of pO₂ one week after the injection, and the complete loss of responsiveness 9 d after the injection. *Bottom*: The same charcoal was coated using 30% (w/w) pyroxylin. Note that the coating preserved the responsiveness to oxygen for more than 2 months.

in a Silastic (Dow Corning, Midland, MI, USA) medical grade tubing. Small pieces of ± 2 mm were cut and sealed. The calibration curves of the fusinite in a slurry compared to those of the silicon implant indicated a larger slope value, indicating a higher sensitivity to pO₂ for the silicon implants than for the fusinite slurry. The reasons for this increase in sensitivity are not yet clear. In contrast to the soluble paramagnetic compounds (i.e., nitroxides) where

the EPR line width directly depends on the oxygen concentration, the EPR line width of particles is more independent of oxygen concentration, but more dependent on the partial pressure of oxygen. Therefore, it is likely that this particular observed behavior is not linked to the higher oxygen solubility in silicone than in aqueous phase, but rather is due to changes at the surface of the charcoal particles, leading to a change in sensitivity.

CONCLUSIONS AND PERSPECTIVES

The need for a special coating and/or embedding of paramagnetic materials is increasing. The interest for such studies started when it appeared that some paramagnetic materials were unstable in biological media and that a special coating can enhance their usefulness by stabilizing their responsiveness. Moreover, with the success of the in vivo EPR applications in living animals, specifically oximetry, it was apparent that it would be very useful to apply this technology in humans. Before this technique can be applied to humans, extensive research must be done to determine which paramagnetic compounds are biocompatible. The alternative is to incorporate these probes into biocompatible implants, a strategy that would greatly reduce the time required for approval for its use in human subjects. In this manuscript, we have presented several possible strategies. We should emphasize that this area of research is very new, and that the recent progress made in pharmaceutical technology should greatly help in finding suitable coatings of paramagnetic materials [22,23].

Acknowledgements — The authors wish to thank H.M. Swartz (Dartmouth Medical School) for continuous support and suggestions, and Ann Smith for revising the English. The research was supported by the Belgian National Fund for Scientific Research and the NIH grant PO1 GM 51630.

REFERENCES

[1] Swartz, H. M.; Glockner, J. F. Measurement of oxygen by EPRI and EPRS. In: Eaton, G. R.; Eaton, S. S.; Ohno, K., eds. *EPR imaging and in vivo EPR*. Boca Raton, FL: CRC Press; 1991: 261–290.

[2] Clarkson, R. B.; Ceroke, P. J.; Norby, S. W.; Odintsov, B. Water interactions in porous carbohydrate chars: multi-frequency DNP, EPR, and NMR studies. *International workshop on techniques and biomedical applications of in vivo EPR and PEDRI* (abstr.) 1999:29.

[3] Liu, K. J.; Gast, P.; Moussavi, M.; Norby, S. W.; Vahidi, N.; Walczak, T.; Wu, M.; Swartz, H. M. Lithium phthalocyanine: a probe for electron paramagnetic resonance oximetry in viable biological systems. *Proc. Natl. Acad. Sci. USA* **90**:5438–5442; 1993.

[4] Vahidi, N.; Clarkson, R. B.; Liu, K. J.; Norby, S. W.; Wu, M.; Swartz, H. M. In vivo and in vitro EPR oximetry with fusinite: a new coal-derived, particulate EPR probe. *Magn. Reson. Med.* **31**:139–146; 1994.

[5] James, P. E.; Grinberg, O. Y.; Goda, F.; Panz, T.; O'Hara, J. A.; Swartz, H. M. Gloxy: an oxygen-sensitive coal for accurate mea-

surement of low oxygen tensions in biological systems. *Magn. Reson. Med.* **38**:48–58; 1997.

[6] Jordan, B.; Baudelet, C.; Gallez, B. Carbon-centered radicals as oxygen sensors for in vivo electron paramagnetic resonance: screening for an optimal probe among commercially available charcoals. *MAGMA* **7**:121–129; 1998.

[7] Clarkson, R. B.; Odintsov, B.; Ceroke, P.; Ardenkjaer-Larsen, J. H.; Fruianu, M.; Belford, R. L. EPR and DNP of char suspensions: surface science and oximetry. *Phys. Med. Biol.* **43**:1907–1920; 1998.

[8] Swartz, H. M.; Clarkson, R. B. The measurement of oxygen in vivo using EPR techniques. *Phys. Med. Biol.* **43**:1957–1975; 1998.

[9] Chen, K.; Glockner, J. F.; Morse, P. D.; Swartz, H. M. Effects of oxygen on the metabolism of nitroxide spin labels in cells. *Biochemistry* **28**:2496–2501; 1989.

[10] Utsumi, H.; Takeshita, K.; Miura, Y.; Masuda, S.; Hamada, A. In vivo EPR measurement of radical reaction in whole mice. Influence of inspired oxygen and ischemia-reperfusion injury on nitroxide reduction. *Free Radic. Res.* **19**:S219–S225; 1993.

[11] Gallez, B.; Bacic, G.; Goda, F.; Jiang, J. J.; O'Hara, J. A.; Dunn, J. F.; Swartz, H. M. Use of nitroxides for assessing perfusion, oxygenation, and viability of tissues: in-vivo EPR and MRI studies. *Magn. Reson. Med.* **35**:97–106; 1996.

[12] Gallez, B.; Debuyst, R.; Liu, K. J.; Goda, F.; Walczak, T.; Demeure, R.; Taper, H.; Swartz, H. M. Small particles of fusinite and carbohydrate chars coated with aqueous soluble polymers: preparation and applications for in vivo EPR oximetry. *Magn. Reson. Med.* **40**:152–159; 1998.

[13] Park, H.; Park, K. Biocompatibility issues of implantable drug delivery systems. *Pharm. Res.* **13**:1770–1776; 1996.

[14] Subczinski, W. K.; Lukiewicz, S.; Hyde, J. S. Murine in vivo L-Band ESR spin-label oximetry with a loop-gap resonator. *Magn. Reson. Med.* **3**:747–754; 1986.

[15] Hyde, J. S.; Subczynski, W. K. Spin-label oximetry. In: Berliner, L. J.; Reuben, J., eds. *Biological magnetic resonance. Spin labeling: theory and applications.* New York: Plenum Press; 1989: 399–425.

[16] Glockner, J. F.; Chan, H. C.; Swartz, H. M. In vivo oximetry using a nitroxide-liposome system. *Magn. Reson. Med.* **20**:123–133; 1991.

[17] Liu, K. J.; Grinstaff, M. W.; Jiang, J.; Suslick, K. S.; Swartz, H. M.; Wang, W. In vivo measurement of oxygen concentration using sonochemically synthesized microspheres. *Biophys. J.* **67**:896–901; 1994.

[18] Goda, F.; Liu, K. J.; Walczak, T.; O'Hara, J. A.; Jiang, J.; Swartz, H. M. In vivo oximetry using EPR and india ink. *Magn. Reson. Med.* **33**:237–245; 1995.

[19] Goda, F.; Bacic, G.; O'Hara, J. A.; Gallez, B.; Swartz, H. M.; Dunn, J. F. The relationship between partial pressure of oxygen and perfusion in two murine tumors after X-ray irradiation: a combined gadopentetate dimeglumine dynamic magnetic resonance imaging and in vivo electron paramagnetic resonance oximetry study. *Cancer Res.* **56**:3344–3349; 1996.

[20] Gallez, B.; Jordan, B.; Baudelet, C. Microencapsulation of paramagnetic particles by pyrroxylin in order to preserve their responsiveness to oxygen when used as sensors for in vivo EPR oximetry. *Magn. Reson. Med.* **42**:193–196; 1999.

[21] Gallez, B.; Debuyst, R.; Liu, K. J.; Demeure, R.; Swartz., H. M. Development of biocompatible implants of fusinite for in vivo EPR oximetry. *MAGMA* **4**:71–75; 1996.

[22] Mäder, K. EPR and controlled drug delivery. Something to offer in both directions. *International workshop on techniques and biomedical applications of in vivo EPR and PEDRI* (abstr). 1999: 21.

[23] Gallez, B. Packaging of paramagnetic materials in oximetry and other applications. *International workshop on techniques and biomedical applications of in vivo EPR and PEDRI* (abstr). 1999: 31.

ABBREVIATIONS

EPR—Electron Paramagnetic Resonance
pO_2—partial pressure of oxygen
w/w—weight/weight

Bioassays for Oxidative Stress Status (BOSS). Edited by W.A. Pryor

LC-MS/MS DETECTION OF PEROXYNITRITE-DERIVED 3-NITROTYROSINE IN RAT MICROVESSELS

JOHN S. ALTHAUS,* KARI R. SCHMIDT,* SCOTT T. FOUNTAIN,† MICHAEL T. TSENG,‡ RICHARD T. CARROLL,*
PAUL GALATSIS,§ and EDWARD D. HALL *

*Neuroscience Therapeutics, Pfizer Global Research and Development, Ann Arbor, MI, USA; †Pharmacokinetics, Dynamics, and
Metabolism, Pfizer Global Research and Development, Ann Arbor, MI, USA; ‡Department of Anatomical Sciences and
Neurobiology, School of Medicine, University of Louisville, Louisville, KY, USA; and §Neuroscience Chemistry, Pfizer Global
Research and Development, Ann Arbor, MI, USA

(Received 28 March 2000; Revised 15 May 2000; Accepted 8 June 2000)

Abstract—3-Nitrotyrosine (3NT) is used as a biomarker of nitrative pathology caused by peroxynitrite (PN), myeloperoxidase (MPO)–, and/or eosinophil peroxidase (EPO)–dependent nitrite oxidation. 3NT measurements in biological materials are usually based on either antibody staining, HPLC detection, or GC detection methodologies. In this report, a procedure is described for the measurement of 3NT and tyrosine (TYR) by LC-MS/MS that is simple, direct, and sensitive. Though highly specialized in its use as an assay, LC-MS/MS technology is available in many research centers in academia and industry. The critical assay for 3NT was linear below 100 ng/ml and the limit of detection was below 100 pg/ml. Regarding protein digested samples, we found that MRM was most selective with 133.1 m/z as the daughter ion. In comparison, LC-ECD was 100 times less sensitive. Basal levels of 3NT in extracted digests of rat brain homogenate were easily detected by LC-MS/MS, but were below detection by LC-ECD. The LC-MS/MS assay was used to detect 3NT in rat brain homogenate that was filtered through a 180 micron nylon mesh. Three fractions were collected and examined by phase contrast microscopy. The mass ratio (3NT/TYR) of 3NT in fractions of large vessel enrichment, microvessel enrichment, and vessel depletion was 0.6 ng/mg, 1.2 ng/mg, and 0.2 ng/mg, respectively. Ultimately, we found that the basal 3NT/TYR mass ratio as determined by LC-MS/MS was six times greater in microvessel-enriched brain tissue vs. tissue devoid of microvessels. © 2000 Elsevier Science Inc.

Keywords—Mass spectrometry, LC-MS/MS, 3-nitrotyrosine, Electrochemical detection, LC-ECD, Peroxynitrite, Rat forebrain, Microvessels, Free radicals

John S. Althaus received a master's degree in biochemistry from the University of Maryland in 1980 and in business administration from Western Michigan University in 1989. His research interests include free radical mechanisms in disease pathology.

Kari R. Schmidt received a bachelor's degree in biology from Siena Heights University in 1999. She synthesized oligonucleotides at the University of Michigan and her research interests include mass spectrometry in biomarker analysis.

Scott T. Fountain received a bachelor's degree in chemistry from Central Michigan University in 1989 and a doctorate degree in analytical chemistry from The University of Michigan in 1994. His research interests include bioanalytical mass spectrometry in pharmaceutical research.

Michael T. Tseng received a doctorate degree in experimental pathology from the State University of New York in 1973, with postdoctoral training in reproductive endocrinology at the Oregon Regional Primate Research Center in 1974. His research interests include neuronal resuscitation in injured CNS.

Richard T. Carroll received a doctorate degree in chemistry from the University of Toledo in 1993 and a master's degree in chemistry from the University of Toledo in 1991. His research interests include chemical mechanisms involved in neuronal insult.

Paul Galatsis received a doctorate degree in organic chemistry from Indiana University in 1987 and completed postdoctoral studies at The Ohio State University in 1989. His research interests include the biological implications of free radical chemistry.

Edward D. Hall received a doctorate degree in pharmacology from the Cornell University in 1976. Currently, he is Director of Neurology Section, Neuroscience Therapeutics at Pfizer Global Research and Development. His research interests include free radical mechanisms of acute and chronic neurodegenerative disease.

Address correspondence to: Dr. John S. Althaus, Pfizer Global Research and Development, Neuroscience Therapeutics, 20N/111, Ann Arbor, MI 48105, USA; Tel: (734) 622-1119; Fax: (734) 622-7178; E-Mail: john.althaus@wl.com.

Reprinted from: *Free Radical Biology & Medicine, Vol. 29, No. 11, pp. 1085-1095, 2000*

INTRODUCTION

Peroxynitrite (PN) is a rich source of free radical production in vivo [1]. The theory of its formation, pathological activity, and nitration of tyrosine (TYR) to 3-nitrotyrosine (3NT) has been extensively reviewed [2–6]. The nitration of TYR to 3NT, by PN as well as by other nitrative pathways involving myeloperoxidase (MPO) and/or eosinophil peroxidase (EPO), may be an important mechanism regarding disease pathology. Increased 3NT formation is associated with Alzheimer's disease [7], Parkinson's disease [8], Huntington's disease [9], amyotrophic lateral sclerosis [10], as well as a number of pathologies associated with ischemia [11,12]. For this reason, the accurate and sensitive measurement of 3NT has important implications regarding mechanism, diagnosis, and drug development for diseases with significant oxidative components.

Detection of 3NT in biological samples has been reported extensively in the literature. The methods used fall into two basic categories: molecular analysis using 3NT antibody-staining techniques [13] and chemical analysis using HPLC and GC [14] with a wide range of detection techniques. In one case however, a caution was given by Kaur and colleagues [15] regarding the use of liquid chromatography with electrochemical detection (LC-EDC) or liquid chromatography with ultraviolet detection (LC-UV) for the detection of 3NT in diseased human brain. In this report they announced the discovery of an unknown artifact with chromatographic and detection properties similar to authentic 3NT.

We present a method for 3NT detection in biological samples using liquid chromatography with multiple-reaction monitoring mass spectrometry (LC-MS/MS), which is simple, direct, and sensitive. Sample processing involves enzymatic protein digestion and a single extraction step. Chromatographic results of biological material gave a single, clean, baseline-to-baseline resolved peak for 3NT at concentrations below 100 pg/ml within 12 min. We applied this method to the detection of 3NT in rat (naïve) forebrain homogenate. The 3NT to TYR ratio (3NT/TYR) was measured in homogenate that was filtered through a 180 micron nylon mesh and gave fractions containing material with varying degrees of microvessel enrichment.

METHODS

Tissue isolation

The isolation of tissue with microvessel enrichment was modified and based on a procedure by DeBault et al. [16]. The forebrain was removed from naïve adult Wistar rats. The forebrain was placed onto a petri dish and 10 ml of ice-cold basal eagle's medium (containing 0.1% BSA)

was added. Using two scalpels and a scissor-like cutting action, the brain was minced into about 1 mm pieces.

This material was transferred to a graduated 50 ml plastic tube with two additional 5 ml medium washes of the petri dish. Material in the tube was allowed to stand on ice for 5 min and fluid was removed from the top until a level of 5 ml was reached. The material remaining in the tube was transferred with a 2 ml wash to a 7 ml dounce homogenizer. Using a tightly fitting pestle, the material was homogenized with 10 stokes of the pestle. Swinnex holders (25 mm) (VWR S/P, Chicago, IL, USA) were fitted with 180 micron nylon mesh filters. An empty 60 ml plastic syringe was attached to the filter holder and the homogenate was poured into the empty syringe and allowed to drain by gravity. A 200 μl sample of homogenate that passed through the filter (the "through" fraction), was collected in a 1.5 ml Eppendorf centrifuge tube (Brinkmann Instruments, Inc., Westbury, NY, USA) for later analysis.

A 35 ml volume of medium was then poured into the empty syringe and allowed to drain by gravity. The syringe plunger was reattached and a volume of air was slowly pushed through the filter to remove moisture. The filter was removed from the holder and the outer edge of the filter was held in place between the lid and tube opening of a 1.5 ml Eppendorf tube. The Eppendorf tube was centrifuged at 5000 RPMs for 10 s in a tabletop Eppendorf centrifuge (model 5415C). With centrifugation, loosely held material on top of the filter (the "above" fraction) was "spun" to the bottom of the Eppendorf tube and formed a pellet. The filter was removed along with material that remained firmly imbedded in the filter and was placed in a separate tube (the "in" fraction). The three fractions ("through," "above," and "in") of the original homogenate were stored at −70°C.

Digestion and solid-phase extraction (SPE)

Digestion of tissue was performed according to the method of Hensley et al. [17]. A digestion buffer was prepared by dissolving 8 mg of pronase (Sigma P-5147; Sigma Chemical Co.; St. Louis, MO, USA) per ml of a stock buffer that was 10 mM sodium acetate, pH 6.5 (final pH of the digestion buffer was about 7.2). Samples were retrieved and 1 ml of the digestion buffer was added to each 1.5 ml Eppendorf tube. Samples were mixed using a 5 s burst from a Virsonic 100 probe sonicator (VWP S/P) at setting 10. Samples were then incubated at 55°C. After 1 h, all samples received a second 5 s burst of sonication, and then the samples were incubated at 55°C for about 18 more h. Under these conditions, the pronase enzyme appeared to be saturated because significant undigested material still remained after 18 h of incubation. Based on measures of free tyrosine, about

85% of the protein in a typical 40 mg sample of tissue was digested.

Following incubation, samples were frozen at −70°C and then centrifuged for 10 min using an Eppendorf centrifuge (model 5415C) set at 14,000 RPMs. Solid-phase extraction (SPE) columns (Waters Sep-Pak C18 cartridges WAT020805; Waters Corp., Milford, MA, USA) were activated by passing 5 ml of methanol followed by 5 ml of distilled water through each column. Following centrifugation, the supernatant from each sample was applied to an activated SPE column. Each column was washed with 1 ml of distilled water followed by 1 ml of 10% acetonitrile in water. A second ml of 10% acetonitrile was applied to the column and 1 ml of effluent was collected that contained TYR and 3NT. This 1 ml volume gave optimal recovery (> 90%) based on testing of representative samples that were spiked with standard 3NT and TYR (data not shown). We recommend that anyone wishing to repeat this procedure should perform his or her own optimization tests. The 1 ml of effluent was then dried to 250 μl using a Savant SpeedVac system (model AES2010; Savant, Holbrook, NY, USA). In some cases, SPE columns were preloaded with 100 μg of polyclonal antibody to 3NT (Upstate Biotechnology, Inc., Lake Placid, NY, USA).

Multiple-reaction monitoring by LC-MS/MS

Nitrotyrosine assay. LC-MS/MS was carried out using an Alliance HT model 2790 liquid chromatographic system (Waters Corp.), a Micromass Quattro Ultima mass spectrometry system (Micromass, Beverly, MA, USA), and MassLynx NT version 3.3 software (Micromass). Following a 20 μl injection, the sample was eluted at 200 μl per minute. Mobile phase A was 99.9% water and 0.1% acetic acid and mobile phase B was 99.9% acetonitrile and 0.1% acetic acid. For each 12 min run, the sample was eluted between 0 and 8 min with 93% A, between 8 and 9 min with 80% A, and between 9 and 12 min with 93% A. Separation was achieved with the tandem coupling of a Waters J'Sphere ODS-H80 (JH08S040502WTA) column (50 × 2 mm; 80 A°) (Waters Corp.) and a Phenomenex Luna C18 (00B-4041-BO) column (50 × 2 mm; 5μ) (Phenomenex, Torrence, CA, USA). The retention time for 3NT under these condition was about 8 min. Sample concentrations were calculated using a five-point standard curve and MassLynx software.

Eluent from the LC system was analyzed using an electrospray ionization (ESI) technique. The mode examined was positive ionization (ESP+) with multiple-reaction monitoring. The parent ion (M+1) had a mass of 227.1 m/z. One daughter ion selected had a mass of 181.1 m/z and the other daughter ion selected has a mass of 133.1 m/z. Detection at 181.1 m/z was more sensitive

but less selective than the other daughter ion. The cone voltage used was 25 V and the collision energy was 20 eV. The source block temperature was 150°C and the desolvation temperature was 270°C. The cone gas was set at 200 l/h while the desolvation gas was set at 590 l/h.

Tyrosine assay. LC-MS/MS was carried out using an Alliance HT model 2790 liquid chromatographic system, a Micromass Quattro Ultima mass spectrometry system, and MassLynx NT version 3.3 software (Micromass). Following a 5 μl injection, sample was eluted at 270 μl/min. Mobile phase A was 99.9% water and 0.1% acetic acid and mobile phase B was 99.9% Acetonitrile and 0.1% acetic acid. For each 5 min run the sample was eluted with 95% A. Separation was achieved with the tandem coupling of 2 TosoHaas TSK-GEL ODS-80TS (B0122-07AM) columns (150 × 2 mm; 5 μ) (TOSO-HAAS, Montgomeryville, PA, USA). The retention time for TYR under these conditions was about 4 min.

Because the TYR concentration was more than one million times greater than the 3NT concentration, all samples were diluted 1 to 1000 before assaying. The mode examined was positive ionization (ESP+) with MRM. The parent ion (M+1) had a mass of 182.1 m/z. The daughter ion selected had a mass of 136.1 m/z. The cone voltage used was 22 V and the collision energy was 13 eV. The source block temperature was 150°C and the desolvation temperature was 270°C. The cone gas was set at 200 l/h while the desolvation gas was set at 590 l/h. Sample concentrations were calculated using a five-point standard curve and MassLynx software.

3NT assay by electrochemical detection (LC-ECD)

The assay of 3NT by LC-ECD was slightly modified from a procedure reported by Hensley et al. [17]. A 60 μl aliquot of sample was injected onto a CoulArray system HPLC equipped with an eight-electrode detection platform (ESA, Chelmsfold, MA, USA). The column (250 × 4.6; 5 μ) used was a TosoHaas ODS-80TM (08149). Mobile phase A was 40 mM sodium phosphate at pH 3.4 and 215 μM sodium octyl sulfate in 100% distilled water. Mobile phase B was 20 mM sodium phosphate at pH 3.4 and 107.5 μM sodium octyl sulfate in 50% acetonitrile. For each 75 min run, the sample was eluted between 0 and 20 min with 100% A, between 20 and 50 min with 93%A, between 50 and 55 min with 0% A, and between 55 and 75 min with 100% A. Peak area was quantified using Coularray software (ESA), and these values were converted to masses per ml based on extrapolation from standard values best fit to a straight line.

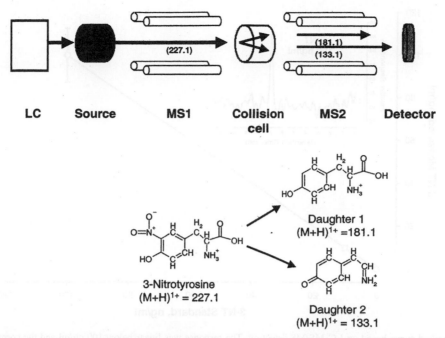

Fig. 1. Multiple-reaction monitoring (MRM) of 3NT. A sample containing the analyte 3NT is injected into an LC, which is coupled to a triple quadrupole mass spectrometer. Positive electrospray ionization is used to generate the $(M+H)^{1+}$ species of 3NT at the ion source. This ionized parent is filtered at the MS1 quadrupole, which is tuned to pass ions with a mass-to-charge (m/z) ratio equal to 227.1 m/z. All ions with this m/z pass into a collision cell where energy is applied, which causes fragmentation. These fragments are then filtered at the MS2 quadrupole that is programmed to pass ions with a m/z ratio equal to 181.1 m/z or 133.1 m/z. Finally, ions that reach the detector are counted and processed by the MassLynx software. Structures are shown that represent the $(M+H)^{1+}$ ion of the parent 3NT and proposed structures for the daughter ions with m/z ratios of 181.1 m/z and 133.1m/z.

Calculation of 3NT/TYR mass ratio

For each sample that was analyzed, the concentration of 3NT and TYR was determined in terms of mass per ml of sample assayed. These values were determined based on software employed as described above. Mass ratios were then calculated by dividing the concentration of 3NT in ng/ml by the concentration of TYR in mg/ml. The result gave a unit that was ng/mg. In other words, a 3NT/TYR mass ratio of 1 ng/mg means that 1 ng of 3NT was found for every 1 mg of TYR that was found in the same tissue being analyzed.

Micrographs of filtered tissue homogenate

Phase contrast micrographs were taken using an Olympus microscope system (model BX60; Olympus, Melville, NY, USA) equipped with a Diagnostics Instruments digital camera (model Spot using 40× magnification; Diagnostic Instruments, Sterling Heights, MI, USA).

RESULTS

In Fig. 1, a schematic of the LC-MS/MS instrumentation employed is shown. Proposed structures of daugh-

ter fragment ions with mass-to-charge (m/z) ratios of 181.1 m/z and 133.1 m/z are shown.

LC-MS/MS detection of 3NT

In Fig. 2, a standard curve of 3NT based on detection by LC-MS/MS is shown. Eluent from the LC system was analyzed using an electrospray ionization (ESI) technique. The mode examined was positive ionization (ESP+) ion selection and MRM. The parent ion (M+1) had a mass of 227.1 m/z and the daughter ion selected had a mass of 181.1 m/z. The correlation coefficient was 0.999 and the limit of detection was less than 2 pg on column. As a direct measure of 3NT, this is one of the most sensitive assays reported.

Representative chromatograms of extracted protein digests are shown in Fig. 3. A 20 μl injection of sample gave chromatograms that showed the selectivity and sensitivity of the assay. The concentration of 3NT in the samples was about 1 ng/ml and the on column amount detected was therefore 20 pg. In comparison, the same sample was injected onto an LC-ECD system and a representative chromatogram of 3NT is shown (see Fig. 7).

As a further test of the selectivity of 3NT detection by LC-MS/MS, samples were passed over C18 solid-phase extraction columns that were either loaded with poly-

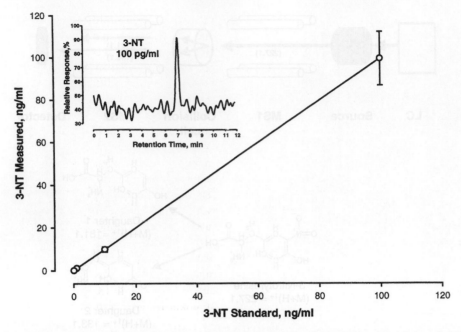

Fig. 2. 3NT standard curve based on LC-MS/MS detection. The response was linear below 100 ng/ml and the correlation coefficient was 0.999. The 3NT measured values were derived from the best-fitted line as determined by the MassLynx software. Data are means ± SD for three determinations per standard concentration. Inset: Representative chromatogram of 3NT at a standard concentration of 100 pg/ml. Eluent from the LC system was analyzed using an electrospray ionization (ESI) technique. The mode examined was positive ionization (ESP+) ion selection with MRM. The parent ion (M+1) had a mass of 227.1 m/z. The daughter ion selected had a mass of 181.1 m/z. The cone voltage used was 25 V and the collision energy was 22 eV. The source block temperature was 150°C and the desolvation temperature was 270°C. The cone gas was set at 200 l/h while the desolvation gas was set at 590 l/h.

clonal antibody to 3NT or buffer. Figure 4 shows that the peak identified as 3NT was removed by the polyclonal antibody treatment.

LC-MS/MS detection of TYR

Figure 5 shows a standard curve of TYR based on detection by LC-MS/MS. The parent ion (M+1) had a mass of 182.1 m/z and the daughter ion selected had a mass of 136.1 m/z. The correlation coefficient was 0.999 and the limit of detection was well below 50 pg on column. Because the concentration of TYR in the sample relative to 3NT was in the order of one million times greater, sensitivity was not an issue. In fact, samples were diluted 1000 times in order to prevent saturation of the detector.

LC-ECD detection of 3NT

A standard curve for 3NT by LC-ECD is shown in Fig. 6. The correlation coefficient for the straight line was 0.987. Detection was linear below 1 μg/ml and the limit of detection was 10 ng/ml. In Fig. 7, a representative chromatogram by LC-ECD of an extracted protein digest is shown. Detection of 3NT in the sample was based on peak detection at electrode 8 with a retention time of 48 min. The trace corresponding to the chromato-

gram at electrode 8 in Fig. 7 is viewed in isolation and appears in Fig. 8. In Fig. 8, the trace shows chromatographic activity between 44 and 52 min. In addition, Fig. 8 contains an overlay of a chromatographic trace of an identical sample that was spiked with 3NT at 10 ng/ml.

3NT/TYR mass ratio with cerebral microvessel enrichment

In a previous report, our data suggested that the microvasculature of the CNS may be a target tissue of peroxynitrite actions [18]. Here we show the fractionation of rat brain homogenate using a 180 micron nylon mesh filter. Tissues were analyzed for 3NT in homogenate that passed "through" the filter, that remained "above" the filter, and that remained firmly imbedded "in" the filter. In Fig. 9, micrographs show that in homogenate that passed "through" the filter, few microvessels are found. However, microvessel enrichment was seen in tissue that remained imbedded "in" the filter and larger vessels were seen in tissue "above" the filter that could be easily removed.

In Fig. 10, the 3NT/TYR mass ratio of the three filtered fractions is shown. The results show that the highest mass ratio of 3NT was associated with the enriched microvessel fraction found in tissue firmly imbed-

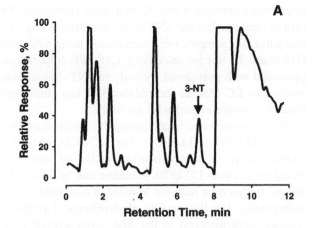

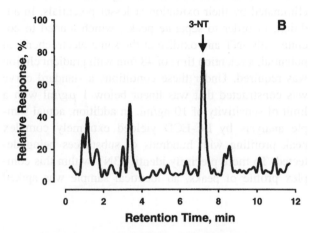

Fig. 3. Representative chromatograms of microvessel digests by LC-MS/MS. Microvessel digests were prepared as described in Methods. A 20 μl sample was injected onto the column. The concentration of 3NT measured in this particular digest sample was 0.84 ng/ml. As a fraction of the TYR concentration in the same sample (data not shown), the value for 3NT was 1.1 ng/mg. Chromatograms by LC-MS/MS detection were determined according to Methods. (A) The mode examined was positive ionization (ESP+) ion selection with MRM. The parent ion (M+1) had a mass of 227.2 m/z. The daughter ion selected had a mass of 181.1 m/z. (B) The mode examined was positive ionization (ESP+) ion selection with MRM. The parent ion (M+1) had a mass of 227.2 m/z. The daughter ion selected had a mass of 133.1 m/z.

ded "in" the filter. The mass ratio was moderate in loosely held tissue "above" the filter and was smallest in tissue that passed "through" the filter.

DISCUSSION

3NT measurements in biological materials are usually based on either antibody-staining methodology [13,19–21] or on HPLC detection analysis [18,22,23]. Results between these reports are varied perhaps stemming in part from differences in methodologies. For example, Coeroli et al. [19] reported 3NT immunoreactivity in the vasculature of neonatal rats following focal ischemia. In

this work, the left middle cerebral artery was coagulated and the left common carotid artery was occluded for 1 h. Brains were examined from rats that were allowed to recover from between 24 h and 7 d. Mesenge et al. [20], however, reported little 3NT staining in vessels of control as well as injured animals. In this study, adult mice were subjected to a concussive injury force of 50 g dropped from 22 cm. Recovery periods between 4 and 24 h were examined. Fukuyama et al. [22] used an HPLC method with UV detection and reported 3NT/TYR percentage ratios in a rat model of focal cerebral ischemia/reperfusion. In this experiment, two microaneurysm clips were applied to the left middle cerebral artery of adult rats. The left common carotid artery was then permanently ligated. After 2 h, the clips were removed and rats were subjected to at most 3 h of reperfusion. Ratios as high as 1% were observed. Kaur et al. [15], using an almost identical HPLC method, reported the detection of an unknown artifact that had chromatographic, UV, and electrochemical properties similar to 3NT. This artifact was observed in postmortem brain tissue from patents afflicted with Alzheimer's, Huntington's, or Parkinson's diseases but not in control tissue. The authors made an appeal for better methods of quantitative analysis of 3NT. Interestingly, Hensley et al. [17] reported that the artifact phenomenon was somehow observed with the acid hydrolysis of protein but not observed with enzymatic hydrolysis of protein.

In a recent review [14] of analytical methods, the authors endorsed LC-ECD as a versatile method for measuring 3NT in biological mixtures. They cautioned, however, that quantitation by LC-ECD might be limited to sample preparations required to achieve acceptable chromatography. In this report, we offer an LC-MS/MS analytical method for the detection of 3NT that is simple, direct, and extremely sensitive. Sample preparation involved enzymatic digestion of protein and was followed by a one-step extraction using SPE with C18 resin. The sample was then concentrated by 75% and 20 μl was injected onto an LC-MS/MS system. Detection below 100 ng/ml was linear with a limit of detection of less than 100 pg/ml.

When multiple-reaction monitoring was performed on a protein-digested sample, a parent/daughter ion combination of 227.1 m/z and 181.1 m/z gave the best-detectable response for 3NT. However, with this ion combination, other peaks of unknown identity were much more predominant. In fact, initially during assay development, the 3NT peak could not be separated from the very large peak appearing at a retention time of about 9 min (see Fig. 3A). A tandem combination of two different C18 columns was required for peak separation. Peak misidentification was avoided based on the discovery that multiple-reaction monitoring using a parent/daughter combi-

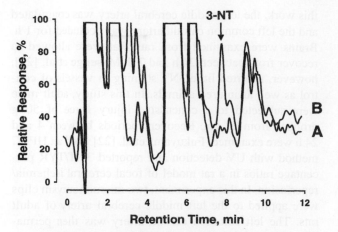

Fig. 4. Comparison of LC-MS/MS chromatograms of microvessel digest reisolated using SPE columns with and without 3NT polyclonal antibody preloading. The LC-MS/MS and the SPE procedures are described in greater detail in Methods. For trace A, a C18 SPE column was loaded with 3NT polyclonal antibody (100 μg) and the column was washed with 1 ml of distilled water. This SPE column containing 3NT polyclonal antibody was used to reisolate 3NT from a microvessel digest. For trace B, a C18 SPE column was loaded only with buffer and the column was washed with 1 ml of distilled water. This SPE column NOT containing 3NT polyclonal antibody was then used to reisolate 3NT from a microvessel digest.

nation of 227.1 m/z and 133.1 m/z though less sensitive (by ~ 70%) was much more selective. In Fig. 3B, it is clear that the 3NT peak is predominant and that the very

large peaks between 8 and 12 min were eliminated. To further confirm that the 3NT peak identified in Fig. 3A was authentic, samples were reprocessed using SPE with C18 resin. To test for selectivity, C18 SPE column was preloaded with polyclonal antibody to 3NT. Subsequent analysis by LC-MS/MS showed that 3NT was eliminated from the sample treated with antibody.

We were interested in comparing the detection of 3NT by LC-MS/MS vs. LC-ECD. For LC-ECD, an eight-channel array of electrochemical cells was used. As stated above, with LC-ECD, detection can be limited by sample preparation required to achieve acceptable chromatography. For this reason, electrochemical potential settings were assigned to the first seven-arrayed electrodes so that substances co-eluting with 3NT could be eliminated by their oxidation at lesser potentials. In addition, in order to separate peaks, which tended to co-elute with 3NT and oxidize at the same electrochemical potential, a retention time of 48 min with gradient elution was required. Under these conditions a standard curve was constructed that was linear below 1 μg/ml with a limit of sensitivity of 10 ng/ml. In addition, actual sample analysis by LC-ECD yielded extremely complex peak profiling with hundreds of substances being detected. To more precisely identify 3NT within this complex profile of peaks, an identical sample was spiked

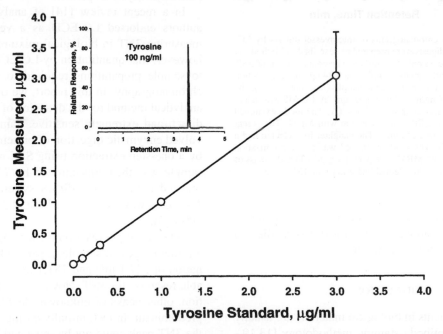

Fig. 5. TYR standard curve based on LC-MS/MS detection. The response was linear below 3 μg/ml and the correlation coefficient was 0.999. The TYR measured values were derived from the best-fitted line as determined by the MassLynx software. Data are means ± SD for three determinations per standard concentration. Inset: Representative chromatogram of TYR at a standard concentration of 100 ng/ml. Eluent from the LC system was analyzed using an electrospray ionization (ESI) technique. The mode examined was positive ionization (ESP+) ion selection with MRM. The parent ion (M+1) had a mass of 182.1 m/z. The daughter ion selected had a mass of 136.1 m/z. The cone voltage used was 22 V and the collision energy was 13 eV. The source block temperature was 150°C and the desolvation temperature was 270°C. The cone gas was set at 200 l/h while the desolvation gas was set at 590 l/h.

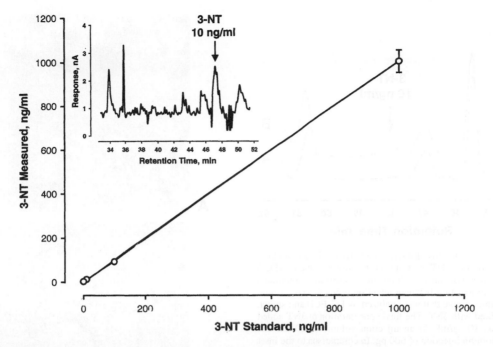

Fig. 6. 3NT standard curve based on LC-ECD detection. The response was linear below 1 μg/ml and the correlation coefficient was 0.987. The 3NT measured values were derived from the best-fitted line as determined by the linear regression analysis. Data are means ± SD for three determinations per standard concentration. Inset: Representative chromatogram of 3NT at a standard concentration at 10 ng/ml. Eluent from the LC system was analyzed using an eight-channel electrochemical detection system. The potential settings for channels 1 to 8 corresponded to 450, 450, 450, 450, 450, 600, 600, and 650 mV respectively. The representative chromatogram shown was determined at electrode 8 set at 650 mV. The amount of 3NT measured represents approximately 60% of the total amount injected based on losses occurring at upsteam electrodes.

with 3NT at 10 ng/ml. The result clearly showed that measurement of 3NT in the sample was below detection by LC-ECD. This was confirmed by LC-MS/MS, which indicated that the concentration of 3NT in the unspiked sample was less than 1 ng/ml.

In 1991, Beckman et al. [24] hypothesized that the

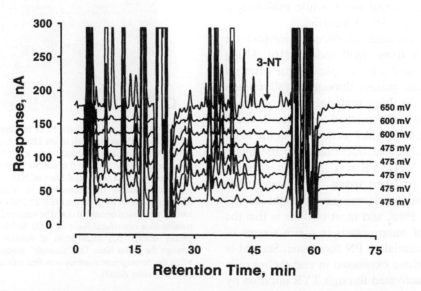

Fig. 7. An LC-ECD chromatogram of a microvessel digest sample. A 60 μl sample of the digest was injected onto the column. The sample passed through an eight-channel electrochemical detection system. Electrochemical potentials ranged from 450 mV at channel one to 650 mV at channel eight. Under these conditions, approximately 60% of 3NT injected was oxidized at channel 8 (data not shown). The retention time for 3NT was about 48 min. The assignment of electrochemical potentials to the eight electrodes was based on achieving the most effective elimination (by electrochemical oxidation) of extraneous peaks which co-eluted with 3NT.

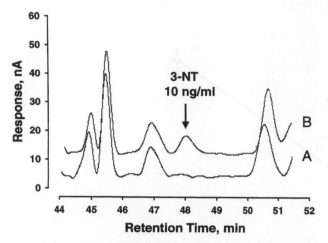

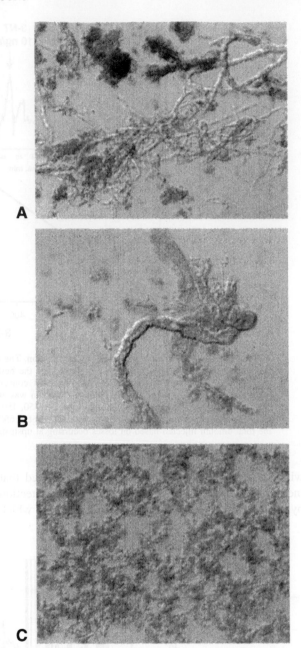

Fig. 8. LC-ECD chromatograms of microvessel digest with and without the exogenous addition of 3NT at 10 ng/ml. For trace A, the LC-ECD chromatographic trace associated with channel 8 (650 mV) illustrated in Fig. 7 is shown in isolation between retention times 44 and 52 min. For trace B, the same sample that was analyzed by LC-ECD and shown as trace A was spiked with 3NT. The final concentration of 3NT added to this sample was 10 ng/ml. At an injection volume of 60 μl, this represents an on column injection of 600 pg. In comparison to the inset in Fig. 6, at an injection volume of 20 μl, the peak area represents an on column injection of 200 pg.

apparent toxicity observed with NO was actually being mediated by its reaction with superoxide to form PN. In the first figure of that report, a schematic illustrated a potential pathological relationship between PN and the microvasculature. Consistent with a microvascular localization of PN-dependent pathology, we found it difficult to measure appreciable amounts of 3NT in brain parenchymal tissue of model injury [18]. We believed that the vasculature might be a target tissue while exhibiting a greater mass ratio toward 3NT formation.

A search of the literature produced a method for isolating microvessels from small rodents [16]. Using this method we confirmed microscopically that when rat brain homogenate was passed through a 180 micron nylon mess that the resulting material that was firmly attached to the filter was enriched with small blood vessels. We applied LC-MS/MS technology to the analysis of rat brain homogenate fractionated by the nylon filter and found that 3NT/TYR was indeed most abundant in microvessel enriched tissue. The greater 3NT mass ratio with microvessel enrichment might be explained in two ways. First, and most obvious is that the endothelium lining of microvessels is a rich source of NO, a component essential to PN formation. Second is that prostacyclin synthase expressed in endothelial cells [25] is selectively inactivated through TYR nitration by peroxynitrite [26]. A comparison of IC50 values indicated that prostacyclin synthase is one of the most sensitive macromolecules toward PN reactivity yet described [27].

Fig. 9. Phase contrast micrographs of rat forebrain homogenate fractionated with a 180 micron mesh nylon filter. (A) Micrograph of homogenate material that was firmly trapped by the nylon filter. This "in" material was considered trapped because it could not be removed from the nylon filter placed parallel to a centrifugal force of 5000 RPMs for 10 s (see Methods for greater detail). (B) Micrograph of homogenate material that was loosely associated by the nylon filter. The "above" material was considered loosely associated because it could be removed from the nylon filter placed parallel to a centrifugal force of 5000 RPMs for 10 s (see Methods for greater detail). (C) Micrograph of homogenate material that passed through the nylon filter. The "through" material was collected by gravity when the homogenate material was first subjected to filtration (see Methods for greater detail).

The relationship between peroxynitrite and the cerebral microvasculature may be important in acute traumatic brain injury (TBI) as well. In previous work [28,

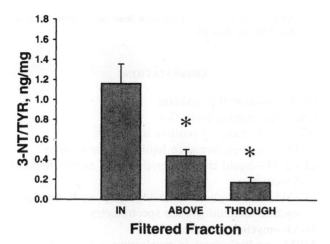

Fig. 10. Mass ratio of 3NT in rat forebrain homogenate fractionated by a 180 micron mesh nylon filter. Data are means ± SD for 6 animals. The "in" fraction refers to homogenate material that was trapped by the filter. The "above" fraction refers to homogenate material that could be removed from the filter by centifugation of the filter parallel to the centrifugal force. The "through" fraction refers to material that was collected by gravity when the homogenate was first subjected to filtration (see Methods for greater detail). The filtered fractions were subjected to pronase digestion, SPE purification and LC-MS/MS analysis (see Methods for greater detail). Statistically, *$p < .05$ vs. "in" by ANOVA.

29], we studied scavengers of peroxynitrite in a mouse model of TBI. In these experiments, mice were treated with doses of either penicillamine or the more brain-penetrating derivative, penicillamine methyl ester. Neurological recovery following injury was observed with both drugs, but there was no increased potency or efficacy with the more brain-penetrating derivative. These results are consistent with a role of PN in acute head injury, and suggest that microvascular scavenging of PN may be of primary therapeutic importance during the early posttraumatic period. In laboratories that have measured 3NT in rat focal cerebral ischemia/reperfusion, 3NT/TYR ratios as high as 1% with injury were found [22]. Regarding human disease, Hensley et al. [7] found 3NT/TYR in excess of 0.1% in the brains of Alzheimer's patents.

In conclusion, an LC-MS/MS assay was developed for 3NT detection in protein digests that was simple, direct, and sensitive. The detection limit for the assay was less than 100 pg/ml. In comparison, LC-ECD was 100 times less sensitive. Basal levels of 3NT in brain homogenate were easily detected by LC-MS/MS but were below detection by LC-ECD. The LC-MS/MS assay was used to analyze rat brain homogenate fractionated by passage through a 180 micron nylon mesh. The results showed that the mass ratio of 3NT was about six times greater in tissue enriched vs. tissues devoid of microvessels.

REFERENCES

[1] Squadrito, G. L.; Pryor, W. A. Oxidative chemistry of nitric oxide: the role of superoxide, peroxynitrite and carbon dioxide. *Free Radic. Biol. Med.* **25**:392–403; 1998.

[2] Richeson, C. E.; Muller, P.; Bowry, V. W.; Ingold, K. U. The complex chemistry of peroxynitrite decomposition: new insights. *J. Am. Chem. Soc.* **120**:7211–7219; 1998.

[3] Bonini, M. G.; Radi, R.; Ferrer-Sueta, G.; Ferreira, A. M. D.; Augusto, O. Direct EPR detection of the carbonate radical anion produced from peroxynitrite and carbon dioxide. *J. Biol. Chem.* **274**:10802–10806; 1999.

[4] Carroll, R. T.; Galatsis, P.; Borosky, S.; Kopec, K. K.; Kumar, V.; Althaus, J. S.; Hall, E. D. 4-Hydroxy-2,2,6,6,-tetramethyl-piper-idine-1-oxyl (tempol) inhibits peroxynitrite-mediated phenol nitration. *Chem. Res. Toxicol.* **13**:294–300; 2000.

[5] van der Vliett, A.; Eiserich, J. P.; Halliwell, B.; Cross, C. E. Formation of reactive nitrogen species during peroxidase-catalyzed oxidation of nitrite. *J. Biol. Chem.* **272**:7617–7625; 1997.

[6] Souza, J. M.; Giasson, B. I.; Chen, Q.; Lee, V. M.-Y.; Ischiropoulos, H. Dityrosine cross-linking promotes formation of stable α-synuclein polymers: implication of nitrative and oxidative stress in the pathogenesis of neurodegenerative synucleinopathies. *J. Biol. Chem.* **275**:18344–18349; 2000.

[7] Hensley, K.; Maidt, M. L.; Yu, Z.; Sang, H.; Markesbery, W. R.; Floyd, R. A. Electrochemical analysis of protein nitrotyrosine and dityrosine in the Alzheimer brain indicates region-specific accumulation. *J. Neurosci.* **18**:8126–8132; 1998.

[8] Good, P. F.; Hsu, A.; Werner, P.; Olanow, C. W. Protein nitration in Parkinson's disease. *J. Neuropathol. Exp. Neurol.* **57**:338–342; 1998.

[9] Browne, S. E.; Ferrante, R. J.; Beal, M. F. Oxidative stress in Huntington's disease. *Brain Pathol.* **9**:147–163; 1999.

[10] Wong, N. K.; Strong, M. J. Nitric oxide synthase expression in cervical spinal cord in sporadic amyotrophic lateral sclerosis. *Eur. J. Cell Biol.* **77**:338–343; 1998.

[11] Forster, C.; Clark, H. B.; Ross, M. E.; Iadecola, C. Inducible nitric oxide synthase expression in human cerebral infarcts. *Acta Neuropathol. Berl.* **97**:215–220; 1999.

[12] Skinner, K. A.; Crow, J. P.; Skinner, H. B.; Chandler, R. T.; Thomson, J. A.; Parks, D. A. Free and protein-associated nitrotyrosine formation following rat liver preservation and transplantation. *Arch. Biochem. Biophys.* **342**:282–288; 1997.

[13] Ischiropoulos, H. Biological tyrosine nitration: a pathophysiological function of nitric oxide and reactive oxygen species. *Arch. Biochem. Biophys.* **356**:1–11; 1998.

[14] Herce-Pagliai, C.; Kotecha, S.; Shuker, D. E. G. Analytical methods for 3-nitrotyrosine as a marker of exposure to reactive nitrogen species: a review. *Nitric Oxide: Biol. Chem.* **2**:324–336; 1998.

[15] Kaur, H.; Lyras, L.; Jenner, P.; Halliwell, B. Artifacts in HPLC detection of 3-nitrotyrosine in human brain tissue. *J. Neurochem.* **70**:2220–2223; 1998.

[16] DeBalut, L. E.; Kahn, L. E.; Frommes, S. P.; Cancilla, P. A. Cerebral microvessels and derived cells in tissue culture: isolation and preliminary characterization. *In Vitro* **15**:473–487; 1979.

[17] Hensley, K.; Maidt, M. L.; Pye, Q. N.; Stewart, C. A.; Mack, M.; Tabatabaie, T.; Floyd, R. A. Quantitation of protein-bound 3-nitrotyrosine and 3, 4-dihydroxyphenylalanine by high-performance liquid chromatography with electrochemical array detection. *Anal. Biochem.* **251**:187–195; 1997.

[18] Althaus, J. S.; Carroll, R. T.; Roof, R. L.; Tseng, M. T.; Hall, E. D. Eight-channel electrochemical detection of nitrotyrosine by HPLC: a survey of animal models of brain injury. *Nitric Oxide: Biol. Chem.* **3**:22; 1999.

[19] Coeroli, L.; Renolleau, S.; Arnaud, S.; Plotkine, D.; Cachin, N.; Plotkine, M.; Ben-Ari, Y.; Charriaut-Marlangue, C. Nitric oxide production and perivascular tyrosine nitration following focal ischemia in neonatal rat. *J. Neurochem.* **70**:2516–2525; 1998.

[20] Mesenge, C.; Charriaut-Marlangue, C.; Verrecchia, C.; Allix, M.; Boulu, R. R.; Plotkine, M. Reduction of tyrosine nitration after

N^w-nitro-L-arginine-methylester treatment of mice with traumatic brain injury. *Euro. J. Pharm.* **353**:53–57; 1998.

[21] Kamisaki, Y.; Wada, K.; Bian, K.; Balabanli, B.; Davis, K.; Martin, E.; Behbod, F.; Lee, Y. C.; Murad, F. An activity in rat tissues that modifies nitrotyrosine-containing proteins. *Proc. Natl. Acad. Sci. USA* **95**:11584–11589; 1998.

[22] Fukuyama, N.; Takizawa, S.; Ishida, H.; Hoshiai, K.; Shinoharra, Y.; Nakazawa, H. Peroxynitrite formation in focal cerebral ischemia/reperfusion in rats occurs predominately in the peri-infarct region. *J. Cereb. Blood Flow Metab.* **18**:123–129; 1998.

[23] Takizawa, S.; Fukuyama, N.; Hirabayashi, H.; Nakazawa, H.; Shinohara, Y. Dynamics of nitrotyrosine formation and decay in rat brain during focal ischemia/reperfusion. *J. Cereb. Blood Flow Metab.* **19**:667–672; 1999.

[24] Beckman, J. S. The double-edged role of nitric oxide in brain function and superoxide-mediated injury. *J. Dev. Physiol.* **15**:53–59; 1991.

[25] Mehl, M.; Bidmon, H.-J.; Hilbig, H.; Zilles, K.; Dringen, R.; Ullrich, V. Prostacyclin synthase is localized in rat, bovine, and human neuronal brain cells. *Neurosci. Lett.* **271**:143–146; 1999.

[26] Zou, M.-H.; Ullrich, V. Peroxynitrite formed by simultaneous generation of nitric oxide and superoxide selectively inhibits bovine aortic prostacyclin synthase. *FEBS Lett.* **382**:101–104; 1996.

[27] Zou, M.; Martin, C.; Ullrich, V. Tyrosine nitration as a mechanism of selective inactivation of prostacyclin synthase by peroxynitrite. *Biol. Chem.* **378**:707–713; 1997.

[28] Althaus, J. S.; Fici, G. J.; VonVoigtlander, P. F. The pharmacology of peroxynitrite-dependent neurotoxicity blockade. In: Rubanyi, G. M., ed. *Pathophysiology and clinical applications of nitric oxide.* Amsterdam: Harwood Academic Publishers; 1999:523–538.

[29] Hall, E. D.; Kupina, N. C.; Althaus, J. S. Peroxynitrite scavengers

for the acute treatment of traumatic brain injury. *Ann. NY Acad. Sci.* **890**:462–468; 1999.

ABBREVIATIONS

EPO—eosinophil peroxidase

ESI—electrospray ionization

ESP+—electrospray positive ionization

HPLC—high-performance liquid chromatography

LC-ECD—liquid chromatography with electrochemical detection

LC-MS/MS—liquid chromatography with multiple-reaction monitoring mass spectrometry

MPO—myeloperoxidase

MRM—multiple-reaction monitoring

m/z—mass-to-charge

PN—peroxynitrite

SPE—solid-phase extraction

TBI—traumatic brain injury

3NT—3-nitrotyrosine

3NT/TYR—tissue mass ratio of 3-nitrotyrosine to tyrosine

TYR—tyrosine

UV—ultraviolet

HISTOCHEMICAL VISUALIZATION OF OXIDANT STRESS

J. Frank,* A. Pompella,[†] and H. K. Biesalski*

*Institute of Biological Chemistry and Nutrition, University of Hohenheim, Stuttgart, Germany; and [†]Dept. of Pathophysiology and
Experimental Medicine, University of Siena, Siena, Italy

(Received 7 April 2000; Revised 7 July 2000; Accepted 12 July 2000)

Abstract—Free radicals induce oxidative modification in distinct components of the living matter (lipid, proteins, and DNA). For qualitative and quantitative determination of free radical–induced modifications, different, more or less sensitive biochemical methods are available. Because of the high reactivity and short life of free radicals, ongoing oxidative damage is generally analyzed by measurement of secondary products—such as H_2O_2, oxidized proteins, peroxidized lipids, and their breakdown products, oxidized DNA—or by fluorographic analysis in combination with fluorescent dyes such as dichlorofluorescin (DCFH). In addition, the determination of free radical–related oxidation products is usually carried out in plasma, urine, or, less frequently, in bioptic material. Consequently, biochemical data seldom reflect the effects of free radical insults in situ. The histochemical visualization of selected molecular markers of oxidative damage can often provide more valuable information concerning the in vivo distribution of oxidative processes. This review summarizes the methods currently available for histochemical detection and indirect visualization of free radical–induced alterations in tissues and isolated cells.   © 2000 Elsevier Science Inc.

Keywords—Free radicals, Histochemistry, Lipid peroxidation, Protein oxidation, Nitrosylated proteins

INTRODUCTION

Several sensitive biochemical procedures are available for the determination of even minimal levels of oxidant stress in vivo [1]. However, such approaches do not generally allow collection of information concerning the distribution of such phenomena in situ, even though this aspect is of great potential importance for the understanding of oxidative processes, especially in the case of tissues with a heterogeneous cell composition, such as brain, lung, or kidney. As a result, some laboratories—including our own—have approached the issue of determination of oxidant stress in vivo from a histochemical point of view [2–5]. The possibility of developing specific and sensitive procedures able to directly reveal some of the biochemical changes induced by oxidant stress has thus made feasible the discrimination of areas, cellular types, and sometimes the subcellular sites that are involved in the process. This review will present an updated survey of the available histochemical approaches, the problems encountered with some of them, and some of the most relevant applications.

Sites of production of reactive oxygen species

Increased production of reactive oxygen species (ROS) has been recognized to occur in a number of pathophysiologic conditions. The specific detection of superoxide anions in biological samples is made difficult by several methodological problems [6] and therefore requires the most accurate application of specificity con-

Jürgen Frank was born in 1960 and studied biology at the University of Hohenheim. In 1992, he finished his doctoral thesis at the University of Tübingen, Germany, where he received a scholarship from the state Baden-Württemberg from 1990 until 1992. In 1994, he joined the team of H.K. Biesalski in the position of an Assistant Lecturer, and he is working on oxidant stress and antioxidant defense systems of cells and tissues.

Alfonso Pompella, born in 1956, studied medicine at the Universities of Naples and Siena, Italy, from which he graduated with his M.D. in 1980. He subsequently worked on the role of lipid peroxidation in cell injury with the research group of Mario Comporti, and since 1992, he has held a position as Associate Professor of General Pathology at the University of Siena Medical School.

Hans Konrad Biesalski obtained his M.D. in 1979 at the University of Mainz, Germany. His research was focused on retinoids and signal transmission in the inner ear. He is head of the Department of Biological Chemistry and Nutrition, and he currently studies effects of oxidative stress, antioxidants, and retinoids on signal transduction.

Address correspondence to: Professor H.K. Biesalski, Department of Biological Chemistry and Nutrition, University of Hohenheim, Fruwirthstrassen 12, D-70593, Stuttgart, Germany; Tel: +49 711-459-4112; Fax: +49 711-459-3822; E-Mail: biesal@uni-hohenheim.de.

Reprinted from: *Free Radical Biology & Medicine, Vol. 29, No. 11, pp. 1096–1105, 2000*

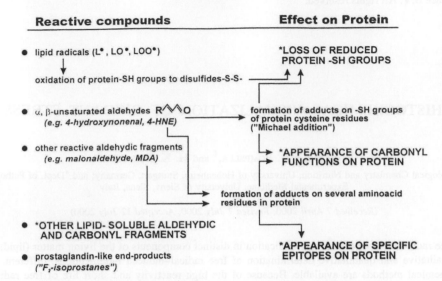

Reactive compounds

- lipid radicals (L•, LO•, LOO•)

  oxidation of protein-SH groups to disulfides-S-S-

- α, β-unsaturated aldehydes R⌇⌇O
  (e.g. 4-hydroxynonenal, 4-HNE)

- other reactive aldehydic fragments
  (e.g. malonaldehyde, MDA)

- *OTHER LIPID- SOLUBLE ALDEHYDIC
  AND CARBONYL FRAGMENTS

- prostaglandin-like end-products
  ("F₂-isoprostanes")

Effect on Protein

*LOSS OF REDUCED
PROTEIN -SH GROUPS

formation of adducts on -SH groups
of protein cysteine residues
("Michael addition")

*APPEARANCE OF CARBONYL
FUNCTIONS ON PROTEIN

formation of adducts on several aminoacid
residues in protein

*APPEARANCE OF SPECIFIC
EPITOPES ON PROTEIN

Fig. 1. Main products of lipid peroxidation and their modifying effects on proteins. Several classes of compounds are originated during the process of lipid peroxidation, some of which are directly detectable by histochemical procedures and/or lead to histochemically-detectable alterations in proteins (indicated with an asterisk).

trols. For many years, especially in studies using activated phagocytes, the reduction of nitroblue tetrazolium (NBT) to insoluble blue formazan has been widely used to evaluate superoxide production at the level of the light microscope, as well as with electron microscopy [7]. However, the cytochemical visualization of subcellular sites of ROS production has predominantly been made possible by the outstanding contributions of M.J. Karnowsky and his coworkers, who defined and optimized the procedures for detection of hydrogen peroxide (by the cerium chloride method) and superoxide (by the Mn^{2+}-diaminobenzidine method), additionally providing insights into the cellular production of singlet oxygen [8]. Using activated neutrophils, the cerium chloride method has also recently been used for detection of cellular sites of H_2O_2 production with the laser-scanning confocal reflectance microscope [9]. The same methods have successfully been employed for the demonstration of the production of ROS at the endothelial surface of cardiac vessels during the first moments of reperfusion following a period of anoxia [10,11].

Current studies aimed at revealing the production of ROS often employ a method involving the preloading of living cells with 2′,7′-dichlorofluorescin diacetate (DCF-DA), a compound whose fluorescence sharply increases in the presence of oxidizing agents. DCF-DA has thus been successfully employed in the investigation of whole organs and isolated cells, as well as in flow cytometry applications [12,13]. Nevertheless, the ability of DCF-DA to detect superoxide anion has been repeatedly questioned, and it has been shown that DCF-DA can serve as a substrate for xanthine oxidase and other cellular peroxidases [14,15]. Indeed, the latter process ap-

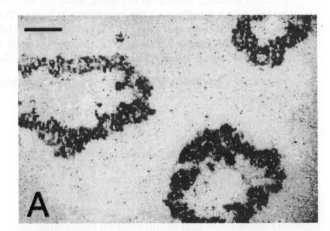

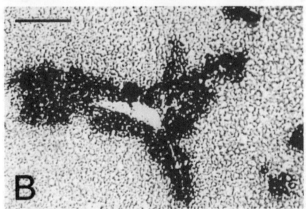

Fig. 2. Visualization by direct Schiff's reaction of tissue areas becoming involved by lipid peroxidation in vivo in mouse liver after the administration of glutathione-depleting toxins. (A) Mediolobular distribution of LPO (purple stain) caused by bromobenzene intoxication; (B) strict periportal localization of LPO induced by intoxication with allyl alcohol. See Pompella et al. [18] for details of the procedures employed. Bars correspond to 50 μm.

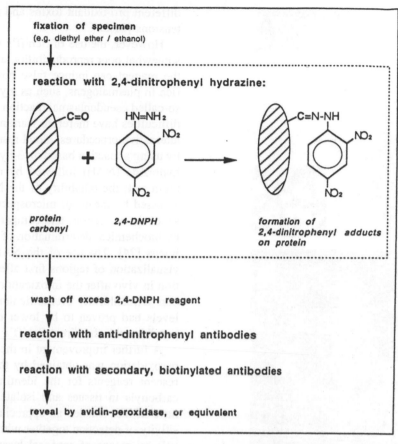

Fig. 3. The two-step procedure for the histochemical determination of protein carbonyl groups. See Pompella et al. [30] for details.

pears itself to be a source of superoxide, leading to artifactual amplification of DCF fluorescence and making the use of DCF-DA problematic [15].

Aldehydes and carbonyls derived from the peroxidation of unsaturated lipids

From a histochemical point of view, the most prominent aspect of LPO is the formation of a wide range of aldehyde and carbonyl compounds after the breakdown of polyunsaturated fatty acids. Of these LPO-derived products, the lipophilic ones will remain associated with the lipid phase, whereas others (e.g., malondialdehyde) will promptly diffuse into aqueous media. An important class of LPO products is the α,β-unsaturated aldehydes, which show variable degrees of reactivity with amino acid residues in protein. Altogether, the process of LPO will result in a marked increase in the amount of carbonyl and aldehyde groups in cellular lipids and proteins, along with a decrease in protein-reduced thiol groups (Fig. 1). Thus, the occurrence of LPO in a given specimen can be inferred by the histochemical identification of these biochemical alterations.

The direct Schiff's reaction has long been employed

for the identification of aldehydes in tissues, even though the exact mechanism for the generation of chromogen has never been convincingly elucidated, despite a long-lasting debate [16]. With respect to LPO-derived aldehydes, Schiff's reaction was first used for the visualization of areas with decreased sensitivity to induction of lipid peroxidation in vitro in cryostat sections obtained from the liver of rats after administration of a carcinogen [17]. Subsequently, the same procedure was applied to the detection of lipid peroxidation in vivo in the whole animal [18], using a model involving intoxication with bromobenzene, a glutathione-depleting agent with a strong pro-oxidant action on several rat organs. Figure 2 shows the distribution of LPO in mouse liver, as assessed by the direct Schiff's reaction, after treatment with bromobenzene and with allyl alcohol, another distinct glutathione-depleting pro-oxidant toxin. The direct Schiff's reaction was subsequently employed with success under other experimental conditions, allowing the demonstration of the selective involvement of rat substantia nigra during in vitro iron-induced lipid peroxidation [19] and of rat tubular proximal epithelium during in vivo lipid peroxidation induced by the nephrocarcinogen iron nitrilotriacetate [20]. Very interesting results were obtained

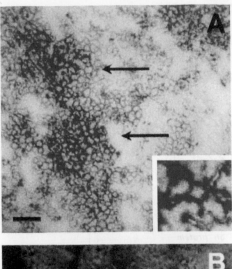

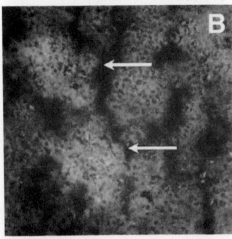

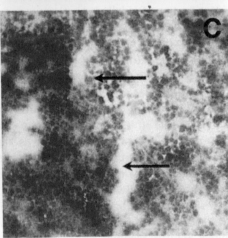

Fig. 4. Colocalization of protein oxidation (panel A, DNPH/anti-DNP reaction) and lipid peroxidation (panel B, immunofluorescence by polyclonal antiserum against protein-4-HNE adducts) at sites of cellular injury (panel C, hematoxylin/eosin). Rat DS-sarcoma cells treated in vivo with hyperthermia plus respiratory hyperoxia and xanthine oxidase. Bar corresponds to 50 μm. Reproduced from Frank et al. [34].

by Masuda and Yamamori, who studied the distribution of LPO in relation to that of cell injury in rat livers subjected to anterograde vs. retrograde perfusion with

different pro-oxidant toxins and under different oxygen tensions [21,22].

However, the use of Schiff's reaction is limited by a somewhat poor reproducibility, and the strong acidity of the reagent can induce false positive results in tissues rich in plasmalogens, such as myocardium, in which the so-called pseudoplasmal reaction can be seen [23]. These difficulties have therefore warranted the development of alternative procedures. Good results have been obtained by using a reaction based on 3-hydroxy-2-naphthoic acid hydrazide (NAH) followed by coupling with a diazonium salt; the reliability of the NAH reaction has been assessed by means of microspectrophotometrical analysis of tissue sections and comparison with data obtained by biochemical determination of LPO in the same specimens [24]. The use of the NAH reaction allowed the visualization of regions first affected by lipid peroxidation in vivo after the intoxication with haloalkanes (carbon tetrachloride, bromotrichloromethane); such LPO levels had proven to be lower than the detection limit possible with the direct Schiff's reaction.

A further improvement in the histochemistry of lipid peroxidation was obtained by the employment of fluorescent reagents for the identification of LPO-derived carbonyls in tissues and isolated cells. Fluorochromes have in fact enabled an appreciable increase in the sensitivity of detection together with the possibility of analysis by means of confocal laser scanning fluorescence microscopy with image videoanalysis. Interesting results with this procedure were obtained by exploiting the fluorescence of the NAH reagent itself [25]. An alternative approach was followed by others, using a biotin-labeled hydrazide coupled with fluorescent-conjugated streptavidin [26].

An additional tool for the cytochemical detection of LPO is the naturally fluorescent fatty acid, *cis*-parinaric acid. Once preloaded in living cells, *cis*-parinaric acid is readily consumed during lipid peroxidation, thus allowing the monitoring of the lipid peroxidation process in the form of a fluorescence decrease. To date, the procedure has found successful application in flow cytometry [27].

OXIDATIVE CHANGES IN CELLULAR PROTEIN (PROTEIN OXIDATION)

The oxidative modification of proteins by reactive oxygen species and other reactive compounds has been recognized as playing a role in the progression of several pathophysiologic processes, including a range of notable diseases and aging [28]. For the detection of protein-associated carbonyl functions, the method originally developed by Levine et al. [29] was adapted in our laboratories for histochemical application [30]. As outlined in

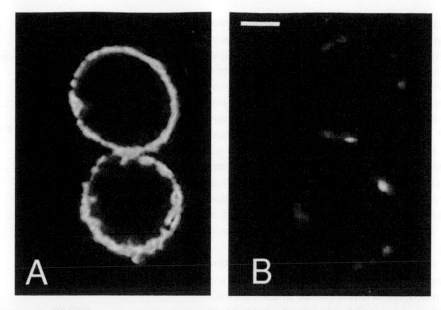

Fig. 5. Selective visualization of cell surface–reduced protein sulfhydryl groups by MPB-based immunofluorescence. Confocal laser scanning fluorescence imaging; see Pompella et al. [30] for details of the procedures employed. (A) Control HPBALL leukemia cells, showing a ring of reduced protein sulfhydryls at their surface. (B) Loss of surface protein sulfhydryls after the exposure of cells to the lipid peroxidation product, 4-hydroxynonenal (4-HNE). Bar corresponds to 10 μm.

Fig. 3, the procedure consists of a first step, in which protein carbonyls are derivativized by 2,4-dinitrophenyl hydrazine (2,4-DNPH) to yield the corresponding 2,4-dinitrophenyl hydrazones. In a second step, the dinitrophenyl (DNP) becoming associated with proteins is detected immunochemically by using a commercial anti-DNP antiserum. Finally, antibodies bound to specimens are identified with a conventional biotin-avidin system or equivalent [30].

In principle, the 2,4-DNPH/anti-DNP procedure should reveal all kinds of protein-associated carbonyls, irrespective of their origin. With this method, oxidized proteins have been visualized in several interesting studies, such as in activated neutrophil phagocytes [30,31], in brain tissue from Alzheimer patients [32,33], or in sarcoma cells exposed to pro-oxidant/hyperthermic (HT) treatments [34]. The latter study showed, under in vivo conditions, the effect of a pro-oxidant stimulus (HT), enhanced ATP hydrolysis in tissue (similar to the events during ischemia), and respiratory hyperoxia (RH: administration of 100% oxygen via a face mask). Such experimental conditions were meant to mimic the oxidative stress processes underlying the ischemia/reperfusion syndrome. HT is known to induce xanthine oxidase (XO) activation, and it is known that the interaction of XO with its substrate (hypoxanthine) is associated with the generation of superoxide anion. XO has been described in tumor tissues, in which hypoxanthine concentrations are also elevated because of impaired energy metabolism with accumulation of ADP, as well as to degradation of nucleic acids following cell death, thus prompting the

conditions for the production of significant amounts of superoxide during the XO reaction. In our study [34], significant increases in the amounts of malondialdehyde (MDA) equivalents (TBARS) were found in tumor tissue after the various treatments, compared with the case of control tumors. The highest levels were found in tumors treated with HT + RH + XO, these values being significantly higher than those found with HT alone. The occurrence of oxidative damage was confirmed by the detection of increasing levels of protein carbonyls in tumor tissue after the various treatments (Fig. 4). Carbonyl functions can originate on protein as the result of a direct attack by reactive oxygen species on amino acid side chains and/or from the covalent binding of reactive aldehyde and carbonyl products of lipid peroxidation, such as MDA and 4-hydroxynonenal. Increased levels of lipid peroxidation were indeed detected in situ in treated tumors, for which the focal accumulation on tumor cell protein of 4-hydroxynonenal was visualized (Fig. 4). As expected, areas presenting protein-bound 4-hydroxynonenal were found to correspond strictly to areas with increased protein carbonyls (Fig. 4). Finally, increased protein oxidation and lipid peroxidation were associated with the occurrence of cellular damage and death; as shown in Fig. 4, acidophilia and nuclear condensation were detectable in cells in strict correspondence with areas with higher levels of oxidative alterations. The immunohistochemical detection of oxidized proteins matched well with the extent of the lipoperoxidative process measured as MDA equivalents.

Another important parameter that can be used as a

marker of oxidant insult to protein is the loss of protein-SH groups, a phenomenon that has mainly been
investigated with reference to toxic cell injury and lipid
peroxidation [35,36]. Modern histochemistry of protein
thiols was developed and optimized by G. Nöhammer,
who laid down the foundations for semiquantitative mi-
crospectrophotometry of protein-reduced thiols, protein
disulfides, and protein-mixed disulfides [37,38]. These
procedures found applications in the histochemical evaluation of protein thiol redox status in neoplastic and
preneoplastic cells [39,40]. Fluorescent-labeling procedures have recently been developed for the visualization
of both total and cell surface protein–reduced thiols at
the single-cell level by laser scanning confocal microscopy [30] (Fig. 5).

A further pathway leading to oxidative modification
of protein is given by glycation reactions. Glycation and
glycoxidation reactions, first recognized in diabetes mellitus, can produce irreversible structural alteration in
protein, which eventually result in the accumulation of
the so-called advanced glycation end products (AGEs) in
tissue. Immunohistochemical procedures are currently in
use for detection of these compounds, as detailed below.

Immunodetection of oxidant stress-induced epitopes in proteins and nucleic acids

The immunohistochemical approach to oxidant stress
has expanded rapidly over the past few years and today
represents the tool of choice for the specific detection of
oxidative changes in tissue and cells. After the structural
alterations introduced by an oxidant insult, proteins in
fact acquire new antigenic properties because of the
appearance of new specific epitopes on the polypeptide
chain. This is primarily the case with reactive aldehydes
derived from lipid peroxidation, which are able to bind to
several amino acid residues, as outlined in Fig. 1. By
means of specific polyclonal or monoclonal antibodies,
the occurrence of MDA and 4-hydroxynonenal (4-HNE)
bound to cellular protein have thus been documented
under a number of experimental and clinical conditions.
Lipid peroxidation has been demonstrated in this way in
collagen-producing fibroblasts [41,42], in the liver of
human alcoholics [43], hepatitis C patients and other
chronic liver diseases [44,45], in the arterial wall during
experimentally induced atherosclerosis [46], in activated
neutrophils [30,31,47], in nigral neurons of Parkinson
patients [48], in ferric nitrilotriacetate–induced renal carcinogenesis [49], as well as in human renal [50, 51] and
colon [52] carcinoma. Most of these studies were largely
made possible by the availability of thoroughly characterized monoclonal antibodies [53,54].

Specific epitopes are also present in oxidized low-
density lipoproteins, a distinctive class of oxidized pro-

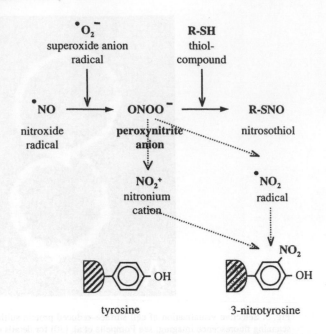

protein nitration

Fig. 6. Reactions of the reactive nitrogen species, nitroxide radical and
peroxynitrite anion, leading to the formation of adducts on protein
tyrosine residues (protein nitration). Reaction of reactive nitrogen species formed from NO and superoxide anion radicals. The formation of
peroxynitrite anion leads to the formation of adducts on protein tyrosine residues (protein nitration), leaving a marker for nitrosative
stress detectable in vivo.

teins probably involved in the pathogenesis of atherosclerosis. The exact nature of such epitopes is a matter of
debate, although it seems certain that the antigenicity of
oxidized LDL can be at least partially accounted for by
the binding of LPO-derived aldehydes, such as MDA and
4-HNE, to the LDL apoprotein moiety [55–57]. By
means of polyclonal and monoclonal antibodies raised
against in vitro oxidized LDL, the immunohistochemical
visualization of oxidized LDL has been repeatedly reported in atherosclerotic lesions [58,59]; in a recent work
from our laboratories, oxidized LDL were found to colocalize with active gamma-glutamyl transpeptidase in
the subendothelial space, thus providing the basis a possible role of this enzyme activity in atherosclerosis [60].

With respect to AGE-modified proteins, immunohistochemical studies aimed at determining the sites of
accumulation of these products have generally employed
antibodies specific for N-ε-(carboxymethyl) lysine, the
main antigenic structure produced during protein glycoxidation [61]. In this way, AGEs have recently been
detected in several disease conditions [62–65].

Besides protein oxidation, resulting in increased levels of protein carbonyls, a different type of oxidative
attack on protein must also be considered; in other
words, that caused by reactive nitrogen species (Fig. 6).

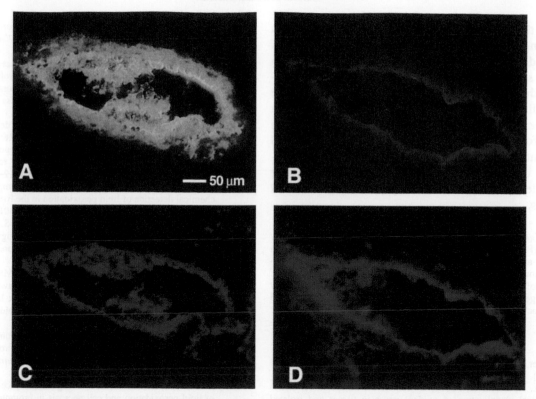

Fig. 7. Colocalization of nitrosylated proteins (A) green fluorescence and von Willebrand factor (C,D) red fluorescence in vessels of rat DS-sarcoma tissues treated in vivo with hyperthermia and photodynamic therapy. The treatment resulted in an induction of nitrosylated proteins within the vessel wall (A). To demonstrate the specificity of binding, the primary antibody (antinitro-tyrosin) was blocked for 1 h by mixing with 10 mM nitrotyrosine at room temperature. The consecutive section showed no staining for nitrotyrosine (B).

The chemistry of NO in biological systems is extensive and complex, and there has been confusion as to what role(s) NO has in different physiological and pathophysiological circumstances. The toxicity of NO is likely to result from the diffusion-limited reaction of nitric oxide with superoxide anion to produce the toxic oxidant peroxynitrite anion ($ONOO^-$), which in turn is capable of oxidizing a wide variety of biomolecules, including proteins, lipids, carbohydrates, and nucleic acids. Peroxynitrite has been implicated in the pathophysiology of a variety of diseases, including inflammation, atherosclerosis, arthritis, endotoxemia, ischaemia-reperfusion injury, or acute respiratory distress syndrome [66–69]. Peroxynitrite can react with aromatic amino acids, probably through the formation of the nitronium cation NO_2^+ and of the $^•NO_2$ radical (Fig. 6). This will primarily result in the addition of nitrate groups to the ortho position of tyrosines, a process referred to as *protein nitration,* leaving a marker detectable in vivo. Histochemically, protein nitration can be documented by means of immunomethods employing anti-nitrotyrosine antibodies [70–73]. The availability of antibodies that recognize nitrosative stress in vivo provides, in fact, a simple method to localize the sites of reactive nitrogen-related damage in diseased tissue of different origin (Fig. 7).

Finally, recent studies have explored the possibility of extending histochemical investigations to the detection of oxidized DNA. DNA damage by both reactive oxygen and reactive nitrogen species is in fact a prominent aspect of a number of oxidant stress conditions. Quantitative immunohistochemical procedures have been developed for determination of 8-hydroxy-2′-deoxyguanosine—one of the main new epitopes on oxidized DNA—by means of specific monoclonal antibodies [74, 75]. Oxidized DNA was thus detected in human mucosal cells of smokers [74] and—in association with protein-bound 4-hydroxynonenal and protein nitrosylation—in human colon carcinoma tissue [52].

CONCLUDING REMARKS

Through the development of numerous methods and specific agents as outlined in this review, histochemistry has now become a useful approach for researchers investigating oxidant stress. Although oxidative alterations have long been known to be associated with several disease conditions, the actual involvement of oxidant stress in the pathogenesis of a given condition has been convincingly documented only in a small number of cases to date. In this respect, when compared with even the most sophisticated and specific biochemical determinations, the histochemical approach offers the advantage of providing a certification that this process is occurring at a given time and at a given site, either in vitro or in vivo. Information obtained in this way can thus add significant details to the investigation of oxidative alterations in biological samples and help the understanding of the role of oxidant stress in experimental and clinical processes.

Acknowledgements — The present study was supported by C.R.U.I./ D.A.A.D. (Programma Vigoni) and the Science Center of Nutrition in Prevention and Therapy (FEP-Life Science Center, Esslingen, Germany). Additional funds were obtained from A.I.R.C. (Italy) and A.I.C.R. (U.K.). Prof. M. Comporti (University of Siena, Italy) is warmly acknowledged for continued scientific support to researchers involved in the histochemistry of oxidant stress.

REFERENCES

[1] Packer, L., ed. Oxygen radicals in biological systems [special issue]. *Methods Enzymol.* **233**:1994.

[2] Frank, J.; Biesalski, H. K.; Dominici, S.; Pompella, A. Histochemical visualization of oxidant stress in tissues and isolated cells. *Histol. Histopathol.* **15**:173–184; 2000.

[3] Frank, J.; Pompella, A.; Biesalski, H. K. Immunohistochemical detection of protein oxidation. In: Armstrong, D., ed. *Methods in ultrastructural and molecular biology protocols.* Totowa, NJ: Humana Press; 2000. In press.

[4] Pompella, A.; Dominici, S.; Frank, J.; Biesalski, H. K. Indirect immunofluorescence detection of protein-bound 4-hydroxynonenal in tissue sections and isolated cells. In: Armstrong, D., ed. *Methods in ultrastructural and molecular biology protocols.* Totowa, NJ: Humana Press; 2000. In press.

[5] Pompella, A.; Dominici, S.; Cambiaggi, C.; Frank, J.; Biesalski, H. K. Cytofluorescence techniques for the visualization of distinct pools of protein thiols at the single cell level. In: Armstrong, D., ed. *Methods in ultrastructural and molecular biology protocols.* Totowa, NJ: Humana Press; 2000. In press.

[6] Fridovich, I. Superoxide anion radical, superoxide dismutases, and related matters. *J. Biol. Chem.* **272**:18515–18517; 1997.

[7] Hirai, K.-I.; Moriguchi, K.; Wang, G.-Y. Real time histochemistry of oxygen free radical production in human leukocytes. *Acta Histochem. Cytochem.* **25**:319–323; 1992.

[8] Karnowsky, M. J. Cytochemistry and reactive oxygen species: a retrospective. *Histochemistry* **102**:15–27; 1994.

[9] Robinson, J. M.; Batten, B. E. Localization of cerium-based reaction products by scanning laser reflectance confocal microscopy. *J. Histochem. Cytochem.* **38**:315–318; 1990.

[10] Shlafer, M.; Brosamer, K.; Froder, J. R.; Simon, R. H.; Ward, P. A.; Grum, C. M. Cerium chloride as a histochemical marker of hydrogen peroxide in reperfused hearts. *J. Mol. Cell Cardiol.* **22**:83–97; 1990.

[11] Babbs, C. F.; Cregor, M. D.; Turek, J. J.; Badylak, S. F. Endothelial superoxide production in the isolated rat heart during early reperfusion after ischemia—a histochemical study. *Am. J. Pathol.* **139**:1069–1080; 1991.

[12] Royall, J. A.; Ischiropoulos, H. Evaluation of 2′,7′-dichlorofluorescin and dihydrorhodamine 123 as fluorescent probes for intracellular H_2O_2 in cultured endothelial cells. *Arch. Biochem. Biophys.* **302**:348–355; 1993.

[13] Tsuchiya, M.; Suematsu, M.; Suzuki, H. *In vivo* visualization of oxygen radical-dependent photoemission. *Methods Enzymol.* **233**:128–140; 1994.

[14] Zhu, H.; Bannenberg, G. L.; Moldéus, P.; Schertzer, H. G. Oxidation pathways for the intracellular probe 2′,7′-dichlorofluorescin. *Arch. Toxicol.* **68**:582–587; 1994.

[15] Rota, C.; Chignell, C. F.; Mason, R. P. Evidence for free radical formation during the oxidation of 2′-7′-dichlorofluorescin to the fluorescent dye 2′-7′-dichlorofluorescein by horseradish peroxidase: possible implications for oxidative stress measurements. *Free Radic. Biol. Med.* **27**:873–881; 1999.

[16] Hörmann, H.; Grassmann, W.; Fries, G. Über den mechanismus der Schiffschen reaktion. *Liebigs Ann. Chem.* **616**:125–147; 1958.

[17] Benedetti, A.; Malvaldi, G.; Fulceri, R.; Comporti, M. Loss of lipid peroxidation as a histochemical marker for preneoplastic hepatocellular foci of rats. *Cancer Res.* **44**:5712–5717; 1984.

[18] Pompella, A.; Maellaro, E.; Casini, A. F.; Comporti, M. Histochemical detection of lipid peroxidation in the liver of bromobenzene-poisoned mice. *Am. J. Pathol.* **129**:295–301; 1987.

[19] Tanaka, M.; Sotomatsu, A.; Kanai, H.; Hirai, S. Combined histochemical and biochemical demonstration of nigral vulnerability to lipid peroxidation induced by DOPA and iron. *Neurosci. Lett.* **140**:42–46; 1992.

[20] Toyokuni, S.; Okada, S.; Hamazaki, S.; Minamiyama, Y.; Yamada, Y.; Liang, P.; Fukunaga, Y.; Midorikawa, O. Combined histochemical and biochemical analysis of sex hormone dependence of ferric nitrilotriacetate-induced renal lipid peroxidation in ddY mice. *Cancer Res.* **50**:5574–80; 1990.

[21] Masuda, Y.; Yamamori, Y. Histological detection of lipid peroxidation following infusion of tert–butyl hydroperoxide and ADP–iron complex in perfused rat livers. *Jpn. J. Pharmacol.* **56**:133–142; 1991.

[22] Masuda, Y.; Yamamori, Y. Histological evidence for dissociation of lipid peroxidation and cell necrosis in bromotrichloromethane hepatotoxicity in the perfused rat liver. *Jpn. J. Pharmacol.* **56**:143–150; 1991.

[23] Pearse, A. G. E. *Histochemistry—theoretical and applied.* London: Churchill Livingstone; 1985.

[24] Pompella, A.; Comporti, M. The use of 3-hydroxy-2-naphthoic acid hydrazide and Fast Blue B for the histochemical detection of lipid peroxidation in animal tissues—a microphotometric study. *Histochemistry* **95**:255–262; 1991.

[25] Pompella, A.; Comporti, M. Imaging of oxidative stress at subcellular level by confocal laser scanning microscopy after fluorescent derivativization of cellular carbonyls. *Am. J. Pathol.* **142**:1353–1357; 1993.

[26] Harris, M. E.; Carney, J. M.; Hua, D. H.; Leedle, R. A. Detection of oxidation products in individual neurons by fluorescence microscopy. *Exp. Neurol.* **129**:95–102; 1994.

[27] Hedley, D.; Chow, S. Flow cytometric measurement of lipid peroxidation in vital cells using parinaric acid. *Cytometry* **13**:686–692; 1992.

[28] Davies, M. J.; Dean, R. T., eds. *Radical-mediated protein oxidation—from chemistry to medicine.* New York: Oxford University Press; 1997.

[29] Levine, R. L.; Williams, J. A.; Stadtman, E. R.; Schachter, E. Carbonyl assays for determination of oxidatively modified proteins. *Methods Enzymol.* **233**:346–357; 1994.

[30] Pompella, A.; Cambiaggi, C.; Dominici, S.; Paolicchi, A.; Tongiani, R.; Comporti, M. Single-cell investigation by laser scanning confocal microscopy of cytochemical alterations resulting from extracellular oxidant challenge. *Histochem. Cell Biol.* **105**:173–178; 1996.

[31] Cambiaggi, C.; Dominici, S.; Comporti, M.; Pompella, A. Modulation of human T lymphocyte proliferation by 4-hydroxynonenal, the bioactive product of neutrophil-dependent lipid peroxidation. *Life Sci.* **61**:777–785; 1997.

[32] Smith, M. A.; Perry, G.; Richey, P. L.; Sayre, L. M.; Anderson, V. E.; Beal, M. F.; Kowall, N. Oxidative damage in Alzheimer's. *Nature* **382**:120–121; 1996.

[33] Smith, M. A.; Sayre, L. M.; Anderson, V. E.; Harris, P. L.; Beal, M. F.; Kowall, N.; Perry, G. Cytochemical demonstration of oxidative damage in Alzheimer disease by immunochemical en-

hancement of the carbonyl reaction with 2,4-dinitrophenylhydrazine. *J. Histochem. Cytochem.* **46:**731–735; 1998.

[34] Frank, J.; Kelleher, D. K.; Pompella, A.; Thews, O.; Biesalski, H. K.; Vaupel, P. Enhancement of the antitumor effect of localized 44°C-hyperthermia by combination with xanthine oxidase and respiratory hyperoxia. *Cancer Res.* **58:**2693–2698; 1998.

[35] Bellomo, G.; Orrenius, S. Altered thiol and calcium homeostasis in oxidative hepatocellular injury. *Hepatology* **5:**876–882; 1985.

[36] Pompella, A.; Romani, A.; Benedetti, A.; Comporti, M. Loss of membrane protein thiols and lipid peroxidation in allyl alcohol hepatotoxicity. *Biochem. Pharmacol.* **41:**1255–1259; 1991.

[37] Nöhammer, G.; Desoye, G.; Khoschsorur, G. Quantitative cytospectrophotometrical determination of the total protein thiols with "Mercurochrom". *Histochemistry* **71:**291–300; 1981.

[38] Nöhammer, G. Quantitative microspectrophotometrical determination of protein thiols and disulfides with 2,2′-dihydroxy-6,6′-dinaphthyldisulfide (DDD). *Histochemistry* **75:**219–250; 1982.

[39] Nöhammer, G.; Bajardi, F.; Benedetto, C.; Schauenstein, E.; Slater, T. F. Quantitative cytospectrophotometric studies on protein thiols and reactive protein disulfides in samples of normal human uterine cervix and on samples obtained from patients with dysplasia or carcinoma *in situ. Br. J. Cancer* **53:**217–222; 1986.

[40] Pompella, A.; Paolicchi, A.; Dominici, S.; Comporti, M.; Tongiani, R. Selective colocalization of lipid peroxidation and protein thiol loss in chemically induced hepatic preneoplastic lesions: the role of γ-glutamyl transpeptidase activity. *Histochem. Cell Biol.* **106:**275–282; 1996.

[41] Chojkier, M.; Houglum, K.; Solis Herruzo, J.; Brenner, D. A. Stimulation of collagen gene expression by ascorbic acid in cultured human fibroblasts—a role for lipid peroxidation? *J. Biol. Chem.* **264:**16957–16962; 1989.

[42] Bedossa, P.; Houglum, K.; Trautwein, C.; Holstege, A.; Chojkier, M. Stimulation of collagen β_1(I) gene expression is associated with lipid peroxidation in hepatocellular injury: a link to tissue fibrosis? *Hepatology* **19:**1262–1271; 1994.

[43] Niemelä, O.; Parkkila, S.; Ylä-Herttuala, S.; Halsted, C.; Witztum, J. L.; Lanca, A.; Israel, Y. Covalent protein adducts in the liver as a result of ethanol metabolism and lipid peroxidation. *Lab. Invest.* **70:**537–546; 1994.

[44] Paradis, V.; Mathurin, P.; Kollinger, M.; Imbert-Bismut, F.; Charlotte, F.; Piton, A.; Opolon, P.; Holstege, A.; Poynard, T.; Bedossa, P. *In situ* detection of lipid peroxidation in chronic hepatitis C: correlation with pathological features. *J. Clin. Pathol.* **50:**401–406; 1997.

[45] Paradis, V.; Kollinger, M.; Fabre, M.; Holstege, A.; Poynard, T.; Bedossa, P. In situ detection of lipid peroxidation by-products in chronic liver disease. *Hepatology* **26:**135–142; 1997.

[46] Palinski, W.; Rosenfeld, M. E.; Ylä-Herttuala, S.; Gurtner, G. C.; Socher, S. S.; Butler, S. W. Parthasarathy, S.; Carew, T. E.; Steinberg, D; Witztum, J. L. Low density lipoprotein undergoes oxidative modification in vivo. *Proc. Natl. Acad. Sci. USA* **86:**1372–1376; 1989.

[47] Quinn, M. T.; Linner, J. G.; Siemsen, D.; Dratz, E. A.; Buescher, E. S.; Jesaitis, A. J. Immunocytochemical detection of lipid peroxidation in phagosomes of human neutrophils: correlation with expression of flavocytochrome b. *J. Leukocyte Biol.* **57:**415–421; 1995.

[48] Yoritaka, A.; Hattori, N.; Uchida, K.; Tanaka, M.; Stadtman, E. R.; Mizuno, Y. Immunohistochemical detection of 4-hydroxynonenal protein adducts in Parkinson disease. *Proc. Natl. Acad. Sci. USA* **93:**2696–2701; 1996.

[49] Uchida, K.; Fukuda, A.; Kawakishi, S.; Hiai, H.; Toyokuni, S. A renal carcinogen ferric nitrilotriacetate mediates a temporary accumulation of aldehyde-modified proteins within cytosolic compartment of rat kidney. *Arch. Biochem. Biophys.* **317:**405–411; 1995.

[50] Okamoto, K.; Toyokuni, S.; Uchida, K.; Ogawa, O.; Takenewa, J.; Kakehi, Y.; Kinoshita, H.; Hattori-Nakakuki, Y.; Hiai, H.; Yoshida, O. Formation of 8-hydroxy-2′-deoxyguanosine and 4-hydroxy-2-nonenal-modified proteins in human renal cell carcinoma. *Int. J. Cancer* **58:**825–829; 1994.

[51] Oberley, T. D.; Toyokuni, S.; Szweda, L. J. Localization of hydroxynonenal protein adducts in normal human kidney and selected human kidney cancers. *Free Radic. Biol. Med.* **27:**695–703; 1999.

[52] Kondo, S.; Toyokuni, S.; Iwasa, Y.; Tanaka, T.; Onodera, H.; Hiai, H.; Imamura, M. Persistent oxidative stress in human colorectal carcinoma, but not in adenoma. *Free Radic. Biol. Med.* **27:**401–410; 1999.

[53] Toyokuni, S.; Miyake, N.; Hiai, H.; Hagiwara, M.; Kawakishi, S.; Osawa, T.; Uchida, K. The monoclonal antibody specific for the 4-hydroxy-2-nonenal histidine adduct. *FEBS Lett.* **359:**189–191; 1995.

[54] Waeg, G.; Dimsity, G.; Esterbauer, H. Monoclonal antibodies for detection of 4-hydroxynonenal-modified proteins. *Free Radic. Res.* **25:**149–159; 1996.

[55] Chen, Q.; Esterbauer, H.; Jürgens, G. Studies on epitopes on low-density lipoprotein modified by 4-hydroxynonenal. Biochemical characterization and determination. *Biochem. J.* **288:**249–254; 1992.

[56] O'Brien, K. D.; Alpers, C. E.; Hokanson, J. E.; Wang, S.; Chait, A. Oxidation-specific epitopes in human coronary atherosclerosis are not limited to oxidized low-density lipoprotein. *Circulation* **94:**1216–1225; 1996.

[57] Requena, J. R.; Fu, M. X.; Ahmed, M. U.; Jenkins, A. J.; Lyons, T. J.; Baynes, J. W.; Thorpe, S. R. Quantification of malondialdehyde and 4-hydroxynonenal adducts to lysine residues in native and oxidized human low-density lipoprotein. *Biochem. J.* **322:**317–325; 1997.

[58] Palinski, W.; Rosenfeld, M. E.; Yla-Herttuala, S.; Gurtner, G. C.; Socher, S. S.; Butler, S. W.; Parthasarathy, S.; Carew, T. E.; Steinberg, D.; Witztum, J. L. Low density lipoprotein undergoes oxidative modification in vivo. *Proc. Natl. Acad. Sci. USA* **86:**1372–1376; 1989.

[59] Ylä-Herttuala, S.; Rosenfeld, M. E.; Parthasarathy, S.; Glass, C. K.; Sigal, E.; Witztum, J. L.; Steinberg, D. Colocalization of 15-lipoxygenase mRNA and protein with epitopes of oxidized low density lipoprotein in macrophage-rich areas of atherosclerotic lesions. *Proc. Natl. Acad. Sci. USA* **87:**6959–6963; 1990.

[60] Paolicchi, A.; Minotti, G.; Tonarelli, P.; Tongiani, R.; De Cesare, D.; Mezzetti, A.; Dominici, S.; Comporti, M.; Pompella, A. Gamma-glutamyl transpeptidase-dependent iron reduction and LDL oxidation—a potential mechanism in atherosclerosis. *J. Invest. Med.* **47:**151–160; 1999.

[61] Ikeda, K.; Nagai, R.; Sakamoto, T.; Sano, H.; Araki, T.; Sakata, N.; Nakayama, H.; Yoshida, M.; Ueda, S.; Horiuchi S. Immunochemical approaches to AGE-structures: characterization of anti-AGE antibodies. *J. Immunol. Methods* **215:**95–104; 1998.

[62] Takayama, F.; Aoyama, I.; Tsukushi, S.; Miyazaki, T.; Miyazaki, S.; Morita, T.; Hirasawa, Y.; Shimokata, K.; Niwa, T. Immunohistochemical detection of imidazolone and N-ε-(carboxymethyl)lysine in aortas of hemodialysis patients. *Cell Mol. Biol.* **44:**1101–1109; 1998.

[63] Matsuse, T.; Ohga, E.; Teramoto, S.; Fukayama, M.; Nagai, R.; Horiuchi, S.; Ouchi, Y. Immunohistochemical localisation of advanced glycation end products in pulmonary fibrosis. *J. Clin. Pathol.* **51:**515–519; 1998.

[64] Sasaki, N.; Fukatsu, R.; Tsuzuki, K.; Hayashi, Y.; Yoshida, T.; Fuji, N.; Koike, T.; Wakayama, I.; Yanagihara, R.; Garruto, R.; Amano, N.; Makita, Z. Advanced glycation end products in Alzheimer's disease and other neurodegenerative diseases. *Am. J. Pathol.* **153:**1149–1155; 1998.

[65] Sun, M.; Yokoyama, M.; Ishiwata, T.; Asano, G. Deposition of advanced glycation end products (AGE) and expression of the receptor for AGE in cardiovascular tissue of the diabetic rat. *Int. J. Exp. Pathol.* **79:**207–222; 1998.

[66] Wink, D. A.; Mitchell, J. B. Chemical biology of nitric oxide: insights into regulatory, cytotoxic, and cytoprotective mechanisms of nitric oxide. *Free Radic. Biol. Med.* **25:**434–456; 1998.

[67] Beckman, J. S.; Koppenol, W. H. Nitric oxide, superoxide, and peroxynitrite: the good, the bad, and the ugly. *Am. J. Physiol.* **271:**C1424–C1437; 1996.

[68] Demiryurek, A. T.; Cakici, I.; Kanzik, I. Peroxynitrite: a putative cytotoxin. *Pharmacol. Toxicol.* **82**:113–117; 1998.

[69] Darley-Usmar, V.; Wiseman, H.; Halliwell, B. Nitric oxide and oxygen radicals: a question of balance. *FEBS Lett.* **369**:131–135; 1995.

[70] Viera, L.; Ye, Y. Z.; Estévez, A. G.; Beckman, J. S. Immunohistochemical methods to detect nitrotyrosine. *Methods Enzymol.* **301**:373–381; 1999.

[71] Beckman, J. S.; Ye, Y. Z.; Anderson, P. G.; Chen, J.; Accavitti, M. A.; Tarpey, M. M.; White, C. R. Excessive nitration of protein tyrosines in human atherosclerosis detected by immunohistochemistry. *Biol. Chem. Hoppe-Seyler* **375**:81–88; 1994.

[72] Good, P. F.; Werner, P.; Hsu, A.; Warren Olanow, C.; Perl, D. P. Evidence for neuronal oxidative damage in Alzheimer's disease. *Am. J. Pathol.* **149**:21–28; 1996.

[73] Virág, L.; Scott, G. S.; Cuzzocrea, S.; Marmer, D.; Salzman, A. L.; Szabó, C. Peroxynitrite-induced thymocyte apoptosis: the role of caspases and poly (ADP-ribose) synthetase (PARS) activation. *Immunology* **94**:345–355; 1998.

[74] Yarborough, A.; Zhang, Y.-J.; Hsu, T.-M.; Santella, R. M. Immunoperoxidase detection of 8-hydroxydeoxyguanosine in aflatoxin B1-treated rat liver and human oral mucosal cells. *Cancer Res.* **56**:683–688; 1996.

[75] Toyokuni, S.; Tanaka, T.; Hattori, Y.; Nishiyama, Y.; Yoshida, A.; Uchida, K.; Hiai, H.; Ochi, H.; Osawa, T. Quantitative immunohistochemical determination of 8-hydroxy-2'-deoxyguanosine by a monoclonal antibody N45.1: its application to ferric nitrilitriacetate-induced renal carcinogenesis model. *Lab. Invest.* **76**:365–374; 1997.

ABBREVIATIONS

DCF-DA—2′,7′-dichlororfluorescin-diacetate
DCFH—dichlorofluorescin
DNP—dinitrophenyl
2,4-DNPH—2,4-dinitrophenyl hydrazine
4-HNE—4-hydroxynonenal
HT—hyperthermia
LPO—lipidperoxidation
MDA—malondialdehyde
NAH—3-hydroxy-2-naphtoic acid hydrazide
NBT—nitroblue tetrazolium
NO—nitric oxide
RH—respiratory hyperoxia
ROS—reactive oxygen species
TBARS—thiobarbituric acid reactive substances
XO—xanthine oxidase

Bioassays for Oxidative Stress Status (BOSS). Edited by W.A. Pryor

TOTAL ANTIOXIDANT CAPACITY AS A TOOL TO ASSESS REDOX STATUS: CRITICAL VIEW AND EXPERIMENTAL DATA

ANDREA GHISELLI, MAURO SERAFINI, FAUSTA NATELLA, and CRISTINA SCACCINI

National Institute for Food and Nutrition Research (Istituto Nazionale per la Ricerca su Alimenti e Nutrizione), 546 Via Ardeatina, 00178 Rome, Italy

(Received 15 November 1999; Revised 27 June 2000; Accepted 12 July 2000)

Abstract—The measure of antioxidant capacity (AC) considers the cumulative action of all the antioxidants present in plasma and body fluids, thus providing an integrated parameter rather than the simple sum of measurable antioxidants. The capacity of known and unknown antioxidants and their synergistic interaction is therefore assessed, thus giving an insight into the delicate balance in vivo between oxidants and antioxidants. Measuring plasma AC may help in the evaluation of physiological, environmental, and nutritional factors of the redox status in humans. Determining plasma AC may help to identify conditions affecting oxidative status in vivo (e.g., exposure to reactive oxygen species and antioxidant supplementation). Moreover, changes in the plasma AC after supplementation with galenic antioxidants or with antioxidant-rich foods may provide information on the absorption and bioavailability of nutritional compounds. Consequently, this review discusses the rationale, interpretation, confounding factors, measurement limits, and human applications of the measure of plasma AC. © 2000 Elsevier Science Inc.

Keywords—Free radical, Antioxidant capacity, Oxidative stress, Plasma, Uric acid, Plant phenols

INTRODUCTION

The increasing evidence that reactive oxygen species (ROS) and oxidative damage is involved in several inflammatory and degenerative diseases [1] has recently stimulated much interest and concern. Oxidative damage can originate from an increase in free radical production either by exogenous radicals such as radiation, pollution, and cigarette smoking, or by endogenous sources, such as inflammation, the respiratory burst, and xenobiotic killing [1].

Mammals have evolved complex antioxidant strategies to utilize oxygen and to minimize the noxious effects of its partially reduced species [2]. Antioxidants within cells, cell membranes, and extracellular fluids can be upregulated and mobilized to neutralize excessive and inappropriate ROS formation. Within the strategy to maintain redox balance against oxidant conditions (e.g., chronic inflammation, cigarette smoking, and diets poor in antioxidants and/or rich in pro-oxidants) [3,4], blood has a central role because it transports and redistributes antioxidants to every part of the body. For example, plasma can scavenge long-lived ROS, such as the super-

Andrea Ghiselli, M.D., specialized in Internal Medicine, has been a researcher at the Istituto Nazionale di Ricerca su Alimenti e Nutrizione (INRAN) since 1989. He works in the field of antioxidant modulation of platelet function and cardiovascular diseases and currently conducts studies in humans on oxidative stress and dietary antioxidants. In 1996, he actively contributed to the establishment of the Free Radical Research Group (http://inn.ingrm.it/frrg.htm).

Mauro Serafini, Ph.D. in Experimental Physiopathology, is a researcher at the INRAN. Previously, he was a post-doctoral fellow at the faculty of Pharmacy of the University of Coimbra, and a visiting scientist at the Nutritional Immunology Lab of the Human Nutrition Research Center on Aging (HNRC) at Tufts University. His interests are in the dietary modulation of oxidative stress in humans and on the redox modulation of cell-mediated immune function.

Fausta Natella is a post-doctoral fellow in the Free Radical Research Group at INRAN. She received her Degree in Biological Sciences at the University of Rome (1996). She is currently involved in research on the oxidative modification of LDL and the antioxidant activity of phenolic compounds in different model systems.

Cristina Scaccini has been a senior scientist at the INRAN since 1993. Prior to this she held a research scientist position at the Center for Human Nutrition—Southwestern Medical School, University of Texas at Dallas (USA). Her scientific interests in the Free Radical Research Group currently range from the in vitro antioxidant activity of single molecules to the modulation by natural antioxidants of oxidative damage, and apoptotic and proliferative response in cellular systems.

Address correspondence to: Andrea Ghiselli, Free Radical Research Group, Istituto Nazionale di Ricerca su Alimenti e Nutrizione, 546 Via Ardeatina, 00178 Rome, Italy; Tel: +(39)065032412; Fax: +(39)065031592; E-Mail: ghiselli@inn.ingrm.it.

Reprinted from: *Free Radical Biology & Medicine*, Vol. 29, No. 11, pp. 1106-1114, 2000

oxide anion or hydrogen peroxide, thus preventing reactions with catalytic metal ions to produce more harmful species [1,5,6]. It can also reduce oxidized ascorbic acid back to ascorbate [7–10]. Hence, plasma antioxidant status is the result of the interaction of many different compounds and systemic metabolic interactions.

The cooperation among the different antioxidants provides greater protection against attack by ROS, than any compound alone. A typical example of synergism between antioxidants is glutathione regenerating ascorbate [11], and ascorbate regenerating α-tocopherol [12]. Thus, the overall antioxidant capacity (AC) may give more biologically relevant information than that obtained from measuring concentrations of individual antioxidants. In addition, the AC of the cell is mainly attributable to the enzyme system, whereas that of plasma is mostly accounted for by low molecular weight antioxidants of dietary origin. These "sacrificial" compounds, rapidly consumed during the scavenging of ROS, need to be regenerated or replaced by new dietary-derived compounds. Thus, plasma AC is modulated either by radical overload or by the intake of dietary antioxidants and can therefore be regarded as more representative of the in vivo balance between oxidizing species and antioxidant compounds (known and unknown, measurable and not measurable) than the concentration of single, selected antioxidants.

Appropriate application of AC measurement requires a clear understanding and description of what is really being measured as follows:

1) Although the terms *antioxidant capacity* and *antioxidant activity* are often used interchangeably [13–16], their real meanings are quite distinct. The *antioxidant activity* corresponds to the rate constant of a single antioxidant against a given free radical. The *antioxidant capacity* is the measure of the moles of a given free radical scavenged by a test solution, independently from the antioxidant activity of any one antioxidant present in the mixture. In biological samples such as plasma, a number of heterogeneous compounds displaying diverse antioxidant activity are present, and therefore it can be argued that the antioxidant status is better represented by its total antioxidant capacity or activity alone.

2) A particular antioxidant under testing would present different activities, depending on the species used to initiate the oxidation reaction. Superoxide anion, hydroxyl radical, and hydrogen peroxide are the most frequently used compounds, because they are known to occur in vivo. However, water-soluble peroxyl radicals generated through the thermal decomposition of azoinitiators (mostly 2,2'-diazobis(2-amidinopropane) hydrochloride (ABAP) and 2,2'-diazobis(2-

amidinopropane) dihydrochloride (AAPH) have also been widely used [13,17–29]. Catalytic transition metal ions are also powerful initiators of oxidative reactions, but they produce a wide range of oxidant species [30], and their presence in vivo in the catalytic form is questionable. Oxidative reactions initiated by transition metals are prevented by chelating agents, but not by chain-breaking antioxidants. As single antioxidants generally have different activities against different ROS, AC of a complex mixture depends on the radical to be scavenged. For this reason, the radical to be used to initiate the oxidation reaction must be always specified and carefully selected.

3) The overall AC of plasma is a combination of the effect of all of the chain-breaking antioxidants, including the thiol groups of proteins and uric acid. "Paradoxical" results can, therefore, occur. For example, because the concentration of uric acid in plasma is very high, it contributes substantially to plasma AC. As plasma uric acid increases in clinical conditions where oxidative stress is also implicated (e.g., kidney failure, metabolic disorders), and after strenuous physical exercise, a significant increase in AC may occur, although the opposite effect might have been anticipated [31,32].

Consequently, it is perhaps best to define the overall plasma AC as a "concept" rather than a simple analytical determination, because it results from the simultaneous presence of many poorly defined pro-oxidant and antioxidant compounds. Although a large number of methodologies have been developed in the last few years [22,29,33–39], the quantification of AC remains troublesome. Thus, the aim of this paper is to critically analyze the method used in our laboratory and to arrive at a general methodological approach to the measurement of plasma AC. Some observation from in vitro and in vivo applications of the method will be also reported.[1]

THE EVOLUTION OF THE METHOD: THE TRAP HISTORY

The original Total Radical-trapping Antioxidant Parameter (TRAP) assay, proposed by Wayner et al. in 1985 [28], has been the most widely used method for evaluating plasma AC. The method is based on the

---

[1]In the present paper we describe and discuss only the method used in our laboratory. In the last decade several methods have been proposed. These methods are mostly based on the measure of inhibition of artificially generated oxidative reactions [19,22,29,36,37,39–43]. Although differences exist in the choice of free radical generator, target molecule, endpoint, and biological matrix, we hope that our work will provide some general indications, independent from the methodological approach adopted.

property of "azo-initiators," such as ABAP, to decompose, producing a peroxyl radical flow at a constant temperature-dependent rate. These peroxyl radicals have enough energy to abstract hydrogen from a (lipid) substrate, thus initiating a (lipid) peroxidation chain. In the Wayner assay, the consumption of dissolved oxygen is the marker of the rate of lipid peroxidation and, therefore, an indirect measure of plasma's ability to inhibit the reaction. The lag phase induced by plasma on the rate of oxygen consumption is compared with the lag phase induced by a known amount of trolox, the water-soluble analog of α-tocopherol. In this assay, TRAP is expressed as micromoles of peroxyl radicals scavenged by a liter of plasma. This method is time-consuming (2 h per sample), and therefore only a limited number of samples can be handled daily.

Another, more relevant, problem in the TRAP assay originates from the high dilution of plasma required to produce a suitable lag phase. This dilution makes the propagation chain reaction between fatty acids "physically" difficult [44]. In fact, the TRAP value proportionally increases with dilution. The authors overcame this problem by adding a small amount of linoleic acid to the reaction mixture [44], potentially introducing an additional source of error.

A few years later, DeLange and colleagues [23] proposed the utilization of an external probe (*R*-phycoerythrin (R-PE), a protein extracted from *Corallina officinalis*) to measure AC. ROS produce a decrease in R-PE fluorescence, which is slowed down by antioxidants, allowing the monitoring of oxidative reactions. This assay measures directly the attack of peroxyl radical upon the target molecule, rather than oxygen consumed during chain reactions of lipid peroxidation. Although this innovative aspect rendered the methodology more rapid, reliable, and sensitive, the authors could only measure the effect of single antioxidants on R-PE oxidation (reported as percentage of inhibition) rather than the overall AC of solutions or biological fluids.

We merged the approaches of Wayner and De-Lange, utilizing R-PE as an external probe and the comparison between the lag-phase produced by plasma to that produced by trolox, to calculate the concentration of peroxyl radicals quenched. Details of the analytical conditions are reported in the footnote.[2]

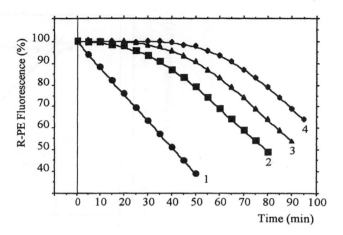

Fig. 1. Kinetics of oxidation reaction initiated by AAPH (5 mM). In the absence of antioxidants (sample 1) a linear decrease of R-PE fluorescence is observed. The addition of antioxidants (4, 6, and 8 mM trolox, samples 2, 3, and 4, respectively) induces a lag phase proportional to the amount of the antioxidant added.

The method

Based on the original TRAP assay [28], we proposed an assay [24] based on the ability of plasma to trap a flow of water-soluble peroxyl radicals produced at constant rate, through thermal decomposition of AAPH.

The addition of AAPH produces a linear decrease of R-PE fluorescence (Sample 1 of Fig. 1), and addition of plasma or single antioxidants induces a lag proportional to the amount of the antioxidant added (Samples 2–4 of Fig. 1). In a recent review [45], Prior and Cao raised concerns on the linearity of fluorescence decrease on the basis of their personal results. However, R-PE has been utilised successfully by other researchers before we started working in this topic and a linear decrease of R-PE with time was clearly reported by Glazer [23,46, 47], Packer [48], Stocker [49], and others [50,51].

Chain-breaking antioxidants react with peroxyl radicals 100-fold faster than does R-PE, thus representing the fast-reacting substances in the reaction [23]. At the end of the lag phase, when all these fast-reacting antioxidants are completely used up, R-PE begins to be oxidized and to lose its fluorescence properties. The results are quantified by standardizing the lag phase of plasma by the lag phase of a known amount of trolox 30–40 min after the initiation of the reaction. When R-PE fluorescence is about 50% of the initial value, trolox is added and the reaction is followed until the loss of fluorescence is linear again (Fig. 2). The lag phase is then calculated by extrapolating the slope of maximal R-PE oxidation before and after trolox addition, up to the intersection of the slopes of induction phases of plasma and trolox. Both

[2]Human plasma is collected after an overnight fast in a prechilled EDTA Vacutainer. The blood is immediately centrifuged at 1200 × *g* for 5 min and plasma is separated and kept at 4°C. Samples are assayed for AC within 5–10 min. The reaction mixture consists in 15 nM R-PE in 75 mM phosphate buffer, pH 7.0. After addition of plasma, trolox, or other compounds, the oxidation reaction is started by adding AAPH to a final concentration of 5.0 mM. The loss of R-PE fluorescence is monitored every 5 min on a luminescence spectrometer equipped with a thermostated cell-holder; monochromators are operating at an exci-

tation wavelength of 495 nm/10 nm slit width and an emission wavelength of 575 nm/5 nm slit width.

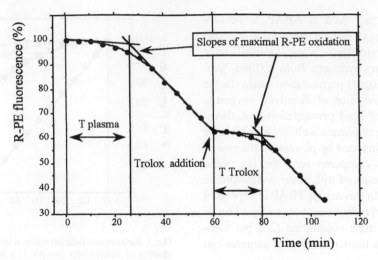

Fig. 2. The kinetics of R-PE oxidation initiated by 5 mM AAPH in the presence of plasma (8 μl) before and after trolox addition (1.8 μM final solution). The antioxidant capacity of each plasma sample is calculated by comparing the two lag phases obtained in the presence and in the absence of trolox. Redrawn from Ghiselli et al. [24].

slopes of maximal R-PE oxidation in any assay are the same. Plasma AC value is obtained by comparing the lag phase of plasma to that of trolox:

$$C_{\text{trolox}}/T_{\text{trolox}} = X/T_{\text{plasma}} \qquad (1)$$

where C_{trolox} is the concentration of the trolox added, T_{trolox} the lag phase induced by trolox, X the AC of plasma (or other matrix), and T_{plasma} the lag phase induced by plasma. The resulting value of X is then multiplied by 2.0 (the stoichiometric factor of trolox) and by plasma dilution factor. Values are expressed as μM (micromoles of peroxyl radicals trapped by 1 liter of plasma).

Methodological problems: tips & tricks, unanswered questions

Plasma or serum? The instability of antioxidants in biological samples, especially when they are exposed to air and light at room temperature, suggests that plasma and serum cannot be interchangeably used for total AC measurement. Some authors reported to have used plasma [19,23,24,28,37,52–60], and some others serum [40,61–70]. No differences were seen between plasma and serum TRAP by some authors [28], whereas others found higher values in serum than in plasma [71]. This last observation is quite surprising, because serum is obtained after clotting blood at room temperature. Time and temperature are indeed crucial when dealing with antioxidants such as ascorbic acid, glutathione, and ubiquinol, which are relatively unstable at neutral pH [72,73]. Moreover, during aggregation, platelets release ROS [74,

75], also suggesting that plasma and not serum is to be preferred. To attempt to clarify the disparity in the literature, we collected two different blood samples after an overnight fast from 10 healthy adults (5 smokers, 5 non-smokers) in prechilled Vacutainers with or without EDTA as anticoagulant. Serum was obtained by clotting blood samples at 22°C for 30 min, while EDTA-treated blood samples were immediately centrifuged and the harvested plasma maintained in an ice-bath in the dark. Plasma and serum AC values were significantly different (1500 ± 455 μM and 1061 ± 514 μM, respectively, $p <$ .05). However, within groups analysis revealed that this difference was due to the presence of smokers in the group. Smoker's AC was significantly higher in plasma than in serum (1480 ± 183 μM and 700 ± 391 μM, $p <$ .01, respectively). Plasma and serum AC values in non-smokers were slightly, though not significantly different (6%).

This observation indicates that serum and plasma AC can be significantly different, according to the specific conditions of the subjects. Preparation of serum may expose the blood of smokers to higher pro-oxidant conditions than non-smokers, causing a more marked loss of antioxidants. The reason for this smoking-related effect is unclear, although the hemorheological consequences of chronic cigarette smoking (higher hematocrit, platelet hyperfunction, higher white cell count) [76] may give rise to "oxidative conditions" during blood clotting. Therefore, plasma should be used to assess AC rather than serum, and the assay should be performed immediately after blood collection. The use of a refrigerated microcentrifuge is also needed to rapidly prepare plasma to avoid any thermal stress.

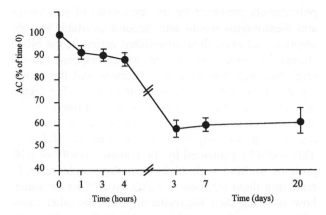

Fig. 3. Stability of plasma samples ($n = 6$) stored at $-80°C$ for different times. Values are expressed as percentage from T_0 (100%). Each analysis requires about 1 h to reach its completion, thus the storage time "1 hour" is referred to as the beginning of the assay (1 h from blood withdrawal).

Storage condition. As a consequence of the short lifetime of some antioxidant compounds, the effect of storage on plasma AC must also be considered. In fact, the measure of AC at different timepoints (0–4 h and 3–20 d) of the same sample stored at $-80°C$ indicates a modest drop in the first 4 h, followed by a significant and rapid loss within the first 3 d of storage (Fig. 3). On this basis, AC assay must be run immediately after blood collection and plasma separation. When laboratory conditions do not allow this procedure (such as during a field study), quick plasma separation should be followed by storage at $-80°C$ and samples should be analyzed for AC within 3 d.

Protein. Protein interference is another important confounder in the measurement of total plasma AC. Protein sulphydryl groups participate physiologically to the overall redox balance and modulate oxidative stress generating reversible semi-oxidized species (mixed disulfides with non-protein low-molecular-weight thiols [77, 78]. Protein sulfydryl groups are responsible for part of plasma AC, because they react with peroxyl radicals during the very first steps of the oxidation reaction and contribute to the duration of the lag phase. The protein skeleton only reacts slowly with peroxyl radicals in the R-PE oxidation phase (60-fold less than R-PE [23]). However, plasma protein concentration (more than 50 g/liter) is high enough to be competitive with R-PE oxidation, thus producing interference. When all fast-reacting compounds are completely exhausted, a competition for peroxyl radicals between proteins and R-PE begins (Fig. 2). Therefore, different plasma samples produce different slopes of maximal R-PE oxidation. These differences depend on their protein content (lower protein concentration corresponding to steeper slopes). The slope obtained from deproteinized plasma is comparable

to a slope determined in the absence of plasma [24]. The protein interference represents a major drawback in methods, such as ORAC [22], in which protein contribution to total plasma AC is as high as 86%. It would be very difficult (if not impossible) to discriminate between different plasma samples with similar levels of fast reactive antioxidants (14% of total AC) and different levels of protein (86%). To avoid this kind of interference, our method uses trolox as an internal standard, so that the slopes of both plasma and plasma + trolox share the same "noise" from the proteins (1).

Uric acid. Uric acid has powerful antioxidant activity, and its concentration in plasma is almost 10-fold higher than other antioxidants, such as vitamin C or vitamin E. Plasma levels of uric acid can range from 200 to 450 μM, which corresponds to about 400 μM AC.

Gender and metabolic differences, as well as some pathological conditions (kidney diseases, metabolic disorders) [79,80], diet [80,81], and strenuous exercise [59], may be associated with an increase of plasma uric acid, thus introducing another possible confounding factor in the measure of plasma AC. For example, a significant AC increase has been observed in rats fed an ethanol-supplemented diet (1.1 g/100 g) for 6 weeks [82]. This AC increase was explained by the simultaneous increase of plasma uric acid concentration due to ethanol-induced purine degradation. Thus, the toxic effect of chronic ethanol consumption [83] resulted in a paradoxical increase on plasma AC. Similar results have been reported by MacKinnon and colleagues [32] in patients with a renal dysfunction, in which the increase in uric acid levels gave a parallel increase in total antioxidant capacity.

In conclusion, the potential confounding effects of uric acid indicates that AC assays should be accompanied by the measure of plasma uric acid concentration.

APPLICATIONS: WHEN AND WHERE TO UTILIZE AC

An assay should be ideally adaptable to every substrate and condition. However, AC assays are subject to artefactual confounding (see above) and results have to be interpreted with caution. Below we report and discuss some application using data coming exclusively from our laboratory.

In vitro and in vivo antioxidant capacity of complex mixtures: the case of a single dose of phenolic-rich beverages

Some dietary plant constituents, such as flavonoids and related phenolics, are powerful antioxidants in vitro, and may also have a protective role in several human

Table 1. Differences Between the in Vitro Antioxidant Capacity (in the Presence and in the Absence of Milk) and in the in Vivo Antioxidant Capacity Before and After Supplementation with Black and Green Tea

	In vitro AC (mM)	In vivo AC (μM) Before	After
Black tea	3.54 ± 0.15	1300 ± 190	1680 ± 130
Black tea and milk	3.42 ± 0.16	1270 ± 170	1170 ± 160
Green tea	17.85 ± 0.13	1330 ± 170	1790 ± 210
Green tea and milk	17.20 ± 0.17	1240 ± 190	1360 ± 90

From Serafini et al. [90].

pathologies [84,85]. However, their absorption, metabolic fate, and availability for antioxidant protection in humans are not fully understood [86–88]. Moreover, the definition "plant phenolics" includes thousand of compounds with different activities and different chemical structure. The chemical structure (number of phenolic rings, aromatic substitution, glycosylation, conjugation with other phenolics or organic acids) can be an important determinant in their bioavailability. However, the profile of phenolic compounds in plasma can potentially be quite different from that of the original dietary source due to metabolization and biotransformation.

In this contest, the measure of plasma AC can represent a tool for indirectly investigating bioavailability of compounds present in plant foods. Monitoring changes in plasma AC after administration of polyphenol-rich foods may represent a kinetic marker of phenolic bioavailability. In the studies reported below, beverages rich in phenolic compounds but almost free from other antioxidants (green and black tea, white and red wine) were used.

Tea represents an excellent example of the lack of correspondence between the in vitro AC of a phenolic-rich beverage and its capacity to induce an increase in plasma AC after supplementation (Table 1). Phenolic patterns of black and green teas are very different, and black tea may be considered the oxidized form of green tea. During the processes to obtain black tea, oxidation of polyphenols occurs, thus decreasing the levels of simple compounds and increasing condensed compounds. In agreement with other authors [89], we showed that tea processing profoundly affects the in vitro AC of the infusion. In fact, green tea displayed an AC 5 times higher than black tea [90] (Table 1). In vivo green and black tea beverages (300 ml) induced a rapid increase in plasma AC; however, the relative increase was the same for both teas (40% at 30 min and 48% at 50 min, respectively, for green and black tea).

This discrepancy between in vitro and in vivo experiments suggests that acid gastric juice might rapidly break down the condensed phenols of black tea. Simple polyphenols produced by the hydrolysis of theaflavins and thearubigins would thus become available for absorption and exert their antioxidant action in the blood stream. The time course of antioxidant response (30–50 min) indicates that these modifications and the subsequent polyphenol absorption occur in the higher region of the gastrointestinal system, probably starting from the stomach. Interestingly, the addition of milk to black tea decreased its capacity to increase plasma AC (Table 1). This could be explained by the known capacity of milk proteins to bind with high-molecular-weight phenols, rendering them resistant to gastric hydrolysis, or somehow reducing their bioavailability. On the other hand, milk proteins did not affect tea AC in the in vitro tests [90].

Based on these experiments, we proceeded to the evaluation of the in vitro and in vivo AC of red and white wine [91]. To avoid any interference from acute ethanol ingestion, dealcoholized wine was used both for the in vitro and for the in vivo experiment. Also in this case, there was a significant difference between the in vitro and in vivo AC of the two wines. The in vitro AC of the wines was of 40.0 and 1.9 mM for red and white wine, respectively, reflecting their total phenol contents (3630 and 31 mg/l, respectively). Only the ingestion of red wine induced a significant increase in plasma AC in humans, whereas white wine did not produce any appreciable change.

The results described here indicate that polyphenol-rich beverages are able to transfer their antioxidant capacity to body fluids. However, the phenolic content itself is not a comprehensive index of antioxidant capacity; the chemical nature of phenols must be taken into account when evaluating the AC of polyphenol-rich beverages.

These data corroborate the concept that simple in vitro AC analysis of food can be misleading, as different factors, such as metabolic transformation and dietary interactions, might affect plant phenols availability and activity in vivo.

Long-term counteraction of oxidative stress induced by cigarette smoking

In vivo oxidative stress can be counteracted by therapeutic interventions aimed at either decreasing the risk of the patient of being exposed to ROS or at supplementing the patient with antioxidants. The measurement of changes in AC may represent a suitable indicator of the success of either intervention.

To assess this possibility, we measured plasma AC of smokers before and after 4 weeks of abstention from smoking (A. Ghiselli and M. Serafini, unpublished observations). Smokers were used as a model of chronic exposition to high ROS flux, mostly peroxyl radicals, and to a lesser extent, carbon-centered radicals [1,4]. Sixteen subjects (smoking >250 mg of tar/day and not taking any drug or vitamin supplementation) were recruited in a group attending a "stop smoking program" organized by the Italian League Against Cancer. A group of 10 non-smokers was used as control. The abstinence from smoking corresponded to a 42% increase in plasma AC (963 ± 369 μM before stopping smoking vs. 1363 ± 473 μM after stopping smoking, $p < .05$); these values were close to those found in non-smokers (1288 ± 350 μM). The rise in plasma AC was partially due to an increase in thiol groups (34%), whereas no changes in α-tocopherol, ascorbic acid, and uric acid were detected.

Another group of smokers (16 subjects) was divided in two subgroups, which were given either a placebo (sucrose tablets) or an antioxidant supplement (1 g of vitamin C, 600 mg of vitamin E, and 25 mg of β-carotene) daily for 4 weeks. Supplementation significantly increased plasma AC (1349 ± 249 vs. 1721 ± 345 μM; $p < .001$). This increase was mainly, but not totally, ascribed to the rise of plasma antioxidant levels. In fact, vitamin C, α-tocopherol, and β-carotene would have had to increase by 10- to 20-fold to achieve the observed 400 μM increase of AC. These results further demonstrate that the measure of AC represents the body redox status better than does the measure of the single circulating antioxidants.

CONCLUSIONS

AC is a sensitive and reliable marker to detect changes of in vivo oxidative stress, which may not be detectable through the measure of single "specific" antioxidants. The method can be used for evaluating the effect of different treatments on plasma redox status in healthy subjects when the results are expressed as change with respect to the basal value. However, several clinical, metabolic, or physiological conditions, not necessarily related to oxidative stress, may result in different plasma values of AC. For this reason the comparison of plasma AC of different population groups should be performed following strictly controlled protocols and accompanied by analysis of single antioxidant compounds.

REFERENCES

[1] Halliwell, B.; Gutteridge, J. M. C., eds. *Free radicals in biology and medicine.* Oxford: Clarendon Press; 1989.
[2] Halliwell, B.; Gutteridge, J. M. The antioxidants of human extracellular fluids. *Arch. Biochem. Biophys.* 280:1–8; 1990.
[3] Halliwell, B.; Cross, C. E. Oxygen-derived species: their relation to human disease and environmental stress. *Environ. Health Perspect.* 102(Suppl. 10):5–12; 1994.
[4] Davies, K. J. Oxidative stress: the paradox of aerobic life. *Biochem. Soc. Symp.* 61:1–31; 1995.

[5] Lynch, R. E.; Fridovich, I. Permeation of the erythrocyte stroma by superoxide radical. *J. Biol. Chem.* 253:4697–4699; 1978.
[6] Mao, G. D.; Poznansky, M. J. Electron spin resonance study on the permeability of superoxide radicals in lipid bilayers and biological membranes. *FEBS Lett.* 305:233–236; 1992.
[7] Wagner, E. S.; White, W.; Jennings, M.; Bennett, K. The entrapment of [^{14}C]ascorbic acid in human erythrocytes. *Biochim. Biophys. Acta* 902:133–136; 1987.
[8] Okamura, M. Uptake of L-ascorbic acid and L-dehydroascorbic acid by human erythrocytes and HeLa cells. *J. Nutr. Sci. Vitaminol. (Tokyo)* 25:269–279; 1979.
[9] Iheanacho, E. N.; Stocker, R.; Hunt, N. H. Redox metabolism of vitamin C in blood of normal and malaria-infected mice. *Biochim. Biophys. Acta* 1182:15–21; 1993.
[10] May, J. M.; Qu, Z. C.; Whitesell, R. R. Ascorbic acid recycling enhances the antioxidant reserve of human erythrocytes. *Biochemistry* 34:12721–12728; 1995.
[11] Packer, J. E.; Slater, T. F.; Willson, R. L. Direct observation of a free radical interaction between vitamin E and vitamin C. *Nature* 278:737–738; 1979.
[12] Stocker, R.; Weidemann, M. J.; Hunt, N. H. Possible mechanisms responsible for the increased ascorbic acid content of *Plasmodium vinckei*-infected mouse erythrocytes. *Biochim. Biophys. Acta* 881:391–397; 1986.
[13] Stratford, N.; Murphy, P. Antioxidant activity of propofol in blood from anaesthetized patients. *Eur. J. Anaesthesiol.* 15:158–160; 1998.
[14] Tissié, G.; Flangakis, S.; Missotten, L.; D'Hermies, F.; de Laey, J. J.; Bourgeois, H.; Zenatti, C.; Hermet, J. R.; Rigeade, M. C.; Bonne, C. Antioxidant activity of plasma from subjects with and without senile cataract. *Doc. Ophthalmol.* 83:357–361; 1993.
[15] Toivonen, H. J.; Ahotupa, M. Free radical reaction products and antioxidant capacity in arterial plasma during coronary artery bypass grafting. *J. Thorac. Cardiovasc. Surg.* 108:140–147; 1994.
[16] van Zoeren-Grobben, D.; Lindeman, J. H.; Houdkamp, E.; Moison, R. M.; Wijnen, J. T.; Berger, H. M. Markers of oxidative stress and antioxidant activity in plasma and erythrocytes in neonatal respiratory distress syndrome. *Acta Paediatr.* 86:1356–1362; 1997.
[17] Kahl, R.; Hildebrandt, A. G. Methodology for studying antioxidant activity and mechanisms of action of antioxidants. *Food Chem. Toxicol.* 24:1007–1014; 1986.
[18] Aruoma, O. I.; Halliwell, B.; Hoey, B. M.; Butler, J. The antioxidant action of taurine, hypotaurine and their metabolic precursors. *Biochem. J.* 256:251–255; 1988.
[19] Lissi, E.; Salim-Hanna, M.; Pascual, C.; del Castillo, M. D. Evaluation of total antioxidant potential (TRAP) and total antioxidant reactivity from luminol-enhanced chemiluminescence measurements. *Free Radic. Biol. Med.* 18:153–158; 1995.
[20] Fauré, M.; Lissi, E. A.; Videla, L. A. Antioxidant capacity of allopurinol in biological systems. *Biochem. Int.* 21:357–366; 1990.
[21] Abella, A.; Messaoudi, C.; Laurent, D.; Marot, D.; Chalas, J.; Breux, J.; Claise, C.; Lindenbaum, A. A method for simultaneous determination of plasma and erythrocyte antioxidant status. Evaluation of the antioxidant activity of vitamin E in healthy volunteers. *Br. J. Clin. Pharmacol.* 42:737–741; 1996.
[22] Cao, G.; Alessio, H. M.; Cutler, R. G. Oxygen-radical absorbance capacity assay for antioxidants. *Free Radic. Biol. Med.* 14:303–311; 1993.
[23] DeLange, R. J.; Glazer, A. N. Phycoerythrin fluorescence-based assay for peroxy radicals: a screen for biologically relevant protective agents. *Anal. Biochem.* 177:300–306; 1989.
[24] Ghiselli, A.; Serafini, M.; Maiani, G.; Azzini, E.; Ferro-Luzzi, A. A fluorescence-based method for measuring total plasma antioxidant capability. *Free Radic. Biol. Med.* 18:29–36; 1995.
[25] Lotito, S. B.; Fraga, C. G. (+)-Catechin prevents human plasma oxidation. *Free Radic. Biol. Med.* 24:435–441; 1998.
[26] Tamura, A.; Sato, T.; Fujii, T. Antioxidant activity of indapamide

and its metabolite. *Chem. Pharm. Bull. (Tokyo)* **38**:255–257; 1990.

[27] Vaya, J.; Belinky, P. A.; Aviram, M. Antioxidant constituents from licorice roots: isolation, structure elucidation and antioxidative capacity toward LDL oxidation. *Free Radic. Biol. Med.* **23**:302–313; 1997.

[28] Wayner, D. D.; Burton, G. W.; Ingold, K. U.; Locke, S. Quantitative measurement of the total, peroxyl-radical trapping antioxidant capability of human blood plasma by controlled peroxidation. *FEBS Lett.* **187**:33–37; 1985.

[29] Winston, G. W.; Regoli, F.; Dugas, A. J. J.; Fong, J. H.; Blanchard, K. A. A rapid gas chromatographic assay for determining oxyradical scavenging capacity of antioxidants and biological fluids. *Free Radic. Biol. Med.* **24**:480–493; 1998.

[30] Patel, R. P.; Svistunenko, D.; Wilson, M. T.; Darley-Usmar, V. M. Reduction of Cu(II) by lipid hydroperoxides: implications for the copper-dependent oxidation of low-density lipoprotein. *Biochem. J.* **322**:425–433; 1997.

[31] Sahlin, K.; Ekberg, K.; Cizinsky, S. Changes in plasma hypoxanthine and free radical markers during exercise in man. *Acta Physiol. Scand.* **142**:275–281; 1991.

[32] MacKinnon, K. L.; Molnar, Z.; Lowe, D.; Watson, I. D.; Shearer, E. Measures of total free radical activity in critically ill patients. *Clin. Biochem.* **32**:263–268; 1999.

[33] Wang, H.; Joseph, J. A. Quantifying cellular oxidative stress by dichlorofluorescein assay using microplate reader. *Free Radic. Biol. Med.* **27**:612–616; 1999.

[34] Re, R.; Pellegrini, N.; Proteggente, A.; Pannala, A.; Yang, M.; Rice-Evans, C. Antioxidant activity applying an improved ABTS radical cation decolorization assay. *Free Radic. Biol. Med.* **26**:1231–1237; 1999.

[35] Lussignoli, S.; Fraccaroli, M.; Andrioli, G.; Brocco, G.; Bellavite, P. A microplate-based colorimetric assay of the total peroxyl radical trapping capability of human plasma. *Anal. Biochem.* **269**:38–44; 1999.

[36] Chevion, S.; Berry, E. M.; Kitrossky, N.; Kohen, R. Evaluation of plasma low molecular weight antioxidant capacity by cyclic voltammetry. *Free Radic. Biol. Med.* **22**:411–421; 1997.

[37] Tubaro, F.; Ghiselli, A.; Rapuzzi, P.; Maiorino, M.; Ursini, F. Analysis of plasma antioxidant capacity by competition kinetics. *Free Radic. Biol. Med.* **24**:1228–1234; 1998.

[38] van den Berg, R.; Haenen, G. R. M. M.; van den Berg, H.; Bast, A. Applicability of an improved Trolox equivalent antioxidant capacity (TEAC) assay for evaluation of antioxidant capacity measurements of mixtures. *Food Chem.* **66**:511–517; 1999.

[39] Benzie, I. F. F.; Strain, J. J. The ferric reducing ability of plasma (FRAP) as a measure of "antioxidant power": the FRAP assay. *Anal. Biochem.* **239**:70–76; 1996.

[40] Chapple, I. L.; Mason, G. I.; Garner, I.; Matthews, J. B.; Thorpe, G. H.; Maxwell, S. R.; Whitehead, T. P. Enhanced chemiluminescent assay for measuring the total antioxidant capacity of serum, saliva and crevicular fluid. *Ann. Clin. Biochem.* **34**(Pt 4):412–421; 1997.

[41] Valkonen, M.; Kuusi, T. Spectrophotometric assay for total peroxyl radical-trapping antioxidant potential in human serum. *J. Lipid Res.* **38**:823–833; 1997.

[42] Miller, N. J.; Rice-Evans, C.; Davies, M. J.; Gopinathan, V.; Milner, A. A novel method for measuring antioxidant capacity and its application to monitoring the antioxidant status in premature neonates. *Clin. Sci.* **84**:407–412; 1993.

[43] Whitehead, T. P.; Thorpe, G. H. C.; Maxwell, S. R. J. Enhanced chemiluminescence assay for antioxidant capacity in biological fluids. *Anal. Chim. Acta* **266**:265–277; 1992.

[44] Wayner, D. D.; Burton, G. W.; Ingold, K. U.; Barclay, L. R. C.; Locke, S. The relative contribution of vitamin E, urate, ascorbate and proteins to the total peroxyl radical-trapping antioxidant activity of human blood plasma. *Biochim. Biophys. Acta* **924**:408–419; 1987.

[45] Prior, R. L.; Cao, G. In vivo total antioxidant capacity: comparison of different analytical methods. *Free Radic. Biol. Med.* **27**:1173–1181; 1999.

[46] Glazer, A. N. Fluorescence-based assay for reactive oxygen species: a protective role for creatinine. *FASEB. J.* **2**:2487–2491; 1988.

[47] Glazer, A. N. Phycoerythrin fluorescence-based assay for reactive oxygen species. *Methods Enzymol.* **186**:161–168; 1990.

[48] Kagan, V. E.; Shvedova, A.; Serbinova, E.; Khan, S.; Swanson, C.; Powell, R.; Packer, L. Dihydrolipoic acid—a universal antioxidant both in the membrane and in the aqueous phase. Reduction of peroxyl, ascorbyl and chromanoxyl radicals. *Biochem. Pharmacol.* **44**:1637–1649; 1992.

[49] Christen, S.; Peterhans, E.; Stocker, R. Antioxidant activities of some tryptophan metabolites: possible implication for inflammatory diseases. *Proc. Natl. Acad. Sci. USA* **87**:2506–2510; 1990.

[50] Sugihara, T.; Rao, G.; Hebbel, R. P. Diphenylamine: an unusual antioxidant. *Free Radic. Biol. Med.* **14**:381–387; 1993.

[51] Wehmeier, K. R.; Mooradian, A. D. Autoxidative and antioxidative potential of simple carbohydrates. *Free Radic. Biol. Med.* **17**:83–86; 1994.

[52] Ceriello, A.; Bortolotti, N.; Falleti, E.; Taboga, C.; Tonutti, L.; Crescentini, A.; Motz, E.; Lizzio, S.; Russo, A.; Bartoli, E. Total radical-trapping antioxidant parameter in NIDDM patients. *Diabetes Care* **20**:194–197; 1997.

[53] Cowley, H. C.; Bacon, P. J.; Goode, H. F.; Webster, N. R.; Jones, J. G.; Menon, D. K. Plasma antioxidant potential in severe sepsis: a comparison of survivors and nonsurvivors. *Crit. Care Med.* **24**:1179–1183; 1996.

[54] Cao, G.; Booth, S. L.; Sadowski, J. A.; Prior, R. L. Increases in human plasma antioxidant capacity after consumption of controlled diets high in fruit and vegetables. *Am. J. Clin. Nutr.* **68**:1081–1087; 1998.

[55] Aguirre, F.; Martin, I.; Grinspon, D.; Ruiz, M.; Hager, A.; De Paoli, T.; Ihlo, J.; Farach, H. A.; Poole, C. P. Jr. Oxidative damage, plasma antioxidant capacity, and glucemic control in elderly NIDDM patients. *Free Radic. Biol. Med.* **24**:580–585; 1998.

[56] Aejmelaeus, R. T.; Holm, P.; Kaukinen, U.; Metsä-Ketelä, T. J.; Laippala, P.; Hervonen, A. L.; Alho, H. E. Age-related changes in the peroxyl radical scavenging capacity of human plasma. *Free Radic. Biol. Med.* **23**:69–75; 1997.

[57] Erhola, M.; Nieminen, M. M.; Kellokumpu-Lehtinen, P.; Metsä-Ketelä, T.; Poussa, T.; Alho, H. Plasma peroxyl radical trapping capacity in lung cancer patients: a case-control study. *Free Radic. Res.* **26**:439–447; 1997.

[58] Jendryczko, A.; Tomala, J. The total free radical trapping ability of blood plasma in eclampsia. *Zentrbl. Gynaekol.* **117**:126–129; 1995.

[59] Maxwell, S. R.; Jakeman, P.; Thomason, H.; Leguen, C.; Thorpe, G. H. Changes in plasma antioxidant status during eccentric exercise and the effect of vitamin supplementation. *Free Radic. Res. Commun.* **19**:191–202; 1993.

[60] Mulholland, C. W.; Strain, J. J. Total radical-trapping antioxidant potential (TRAP) of plasma: effects of supplementation of young healthy volunteers with large doses of alpha-tocopherol and ascorbic acid. *Int. J. Vitam. Nutr. Res.* **63**:27–30; 1993.

[61] Ahmad, S.; Singh, V.; Rao, G. S. Antioxidant potential in serum and liver of albino rats exposed to benzene. *Indian J. Exp. Biol.* **32**:203–206; 1994.

[62] Asayama, K.; Nakane, T.; Uchida, N.; Hayashibe, H.; Dobashi, K.; Nakazawa, S. Serum antioxidant status in streptozotocin-induced diabetic rat. *Horm. Metab. Res.* **26**:313–315; 1994.

[63] Cao, G.; Prior, R. L. Comparison of different analytical methods for assessing total antioxidant capacity of human serum. *Clin. Chem.* **44**:1309–1315; 1998.

[64] Child, R. B.; Wilkinson, D. M.; Fallowfield, J. L.; Donnelly, A. E. Elevated serum antioxidant capacity and plasma malondialdehyde concentration in response to a simulated half-marathon run. *Med. Sci. Sports Exerc.* **30**:1603–1607; 1998.

[65] Jones, A. F.; Winkles, J. W.; Jennings, P. E.; Florkowski, C. M.; Lunec, J.; Barnett, A. H. Serum antioxidant activity in diabetes mellitus. *Diabetes Res.* **7**:89–92; 1988.

[66] Karmazsin, L.; Oláh, V. A.; Balla, G.; Makay, A. Serum antiox-

idant activity in premature babies. *Acta Paediatr. Hung.* **30**:217–224; 1990.

[67] Mulholland, C. W.; Strain, J. J. Serum total free radical trapping ability in acute myocardial infarction. *Clin. Biochem.* **24**:437–441; 1991.

[68] Pinzani, P.; Petruzzi, E.; Orlando, C.; Stefanescu, A.; Antonini, M. F.; Serio, M.; Pazzagli, M. Reduced serum antioxidant capacity in healthy centenarians [letter]. *Clin. Chem.* **43**:855–856; 1997.

[69] Whitehead, T. P.; Robinson, D.; Allaway, S.; Syms, J.; Hale, A. Effect of red wine ingestion on the antioxidant capacity of serum [see comments]. *Clin. Chem.* **41**:32–35; 1995.

[70] Woodford, F. P.; Whitehead, T. P. Is measuring serum antioxidant capacity clinically useful? *Ann. Clin. Biochem.* **35**(Pt 1):48–56; 1998.

[71] Rea, C. A.; Maxwell, S. R.; Maslin, D. J.; Thomason, H. L.; Thorpe, G. H. Anticoagulant effects of antioxidant capacity [letter; comment]. *Ann. Clin. Biochem.* **33**(Pt 2):174; 1996.

[72] Lee, W.; Davis, K. A.; Rettmer, R. L.; Labbe, R. F. Ascorbic acid status: biochemical and clinical considerations. *Am. J. Clin. Nutr.* **48**:286–290; 1988.

[73] Beutler, E.; Gelbart, T. Plasma glutathione in health and in patients with malignant disease. *J. Lab. Clin. Med.* **105**:581–584; 1985.

[74] Ghiselli, A. Aromatic hydroxylation. In: Armstrong, D., ed. *Methods in molecular biology: free radical and antioxidant protocols.* Totowa, NJ: Humana Press Inc.; 1998:89–99.

[75] Leo, R.; Iuliano, L.; Ghiselli, A.; Colavita, A. R.; Pulcinelli, F.; Praticò, D.; Fitzgerald, G. A.; Violi, F. Platelet activation by superoxide anion and hydroxyl radical generated by platelet exposed to anoxia and reoxygenation. *Circulation* **95**:885–891; 1997.

[76] Ernst, E. Haemorheological consequences of chronic cigarette smoking. *J. Cardiovasc. Risk* **2**:435–439; 1995.

[77] Grant, C. M.; Quinn, K. A.; Dawes, I. W. Differential protein S-thiolation of glyceraldehyde-3-phosphate dehydrogenase isoenzymes influences sensitivity to oxidative stress. *Mol. Cell Biol.* **19**:2650–2656; 1999.

[78] Di Simplicio, P.; Cacace, M. G.; Lusini, L.; Giannerini, F.; Giustarini, D.; Rossi, R. Role of protein-SH groups in redox homeostasis—the erythrocyte as a model system. *Arch. Biochem. Biophys.* **355**:145–152; 1998.

[79] Puig, J. G.; Mateos, F. A.; Ramos, T. H.; Capitán, C. F.; Michán, A. A.; Mantilla, J. M. Sex differences in uric acid metabolism in adults: evidence for a lack of influence of estradiol-17 beta (E2). *Adv. Exp. Med. Biol.* **195**(Pt A):317–323; 1986.

[80] Maesaka, J. K.; Fishbane, S. Regulation of renal urate excretion: a critical review. *Am. J. Kidney Dis.* **32**:917–933; 1998.

[81] Yamashita, S.; Matsuzawa, Y.; Tokunaga, K.; Fujioka, S.; Tarui, S. Studies on the impaired metabolism of uric acid in obese subjects: marked reduction of renal urate excretion and its improvement by a low-calorie diet. *Int. J. Obes.* **10**:255–264; 1986.

[82] Gasbarrini, A.; Addolorato, G.; Simoncini, M.; Gasbarrini, G.; Fantozzi, P.; Mancini, F.; Montanari, L.; Nardini, M.; Ghiselli, A.; Scaccini, C. Beer affects oxidative stress due to ethanol in rats. *Dig. Dis. Sci.* **43**:1332–1338; 1998.

[83] Faller, J.; Fox, I. H. Ethanol induced alterations of uric acid metabolism. *Adv. Exp. Med. Biol.* **165**(Pt A):457–462; 1984.

[84] Rice-Evans, C. Plant polyphenols: free radical scavengers or chain-breaking antioxidants? *Biochem. Soc. Symp.* **61**:103–116; 1995.

[85] Croft, K. D. The chemistry and biological effects of flavonoids and phenolic acids. *Ann. N. Y. Acad. Sci.* **854**:435–442; 1998.

[86] Hollman, P. C.; Katan, M. B. Bioavailability and health effects of dietary flavonols in man. *Arch. Toxicol. Suppl.* **20**:237–248; 1998.

[87] Hollman, P. C. Bioavailability of flavonoids. *Eur. J. Clin. Nutr.* **51**(Suppl. 1):S66–S69; 1997.

[88] Hollman, P. C.; Katan, M. B. Absorption, metabolism and health effects of dietary flavonoids in man. *Biomed. Pharmacother.* **51**:305–310; 1997.

[89] Robinson, E. E.; Maxwell, S. R.; Thorpe, G. H. An investigation of the antioxidant activity of black tea using enhanced chemiluminescence. *Free Radic. Res.* **26**:291–302; 1997.

[90] Serafini, M.; Ghiselli, A.; Ferro-Luzzi, A. In vivo antioxidant effect of green and black tea in man. *Eur. J. Clin. Nutr.* **50**:28–32; 1996.

[91] Serafini, M.; Maiani, G.; Ferro-Luzzi, A. Alcohol-free red wine enhances plasma antioxidant capacity in humans. *J. Nutr.* **128**:1003–1007; 1998.

ABBREVIATIONS

AC—antioxidant capacity

AAPH—2,2'-diazobis(2-amidinopropane) dihydrochloride

ABAP—2,2'-diazobis(2-amidinopropane) hydrochloride

EDTA—ethylenediaminetetraacetic acid

TRAP—Total Radical-trapping Antioxidant Parameter

R-PE—*R*-phycoerythrin

ROS—reactive oxygen species

FORUM ON OXIDATIVE STRESS STATUS (OSS)—THE SIXTH SET

WILLIAM A. PRYOR

Biodynamics Institute, Louisiana State University, Baton Rouge, LA, USA

(*Received* 20 *November* 2000; *Accepted* 21 *November* 2000)

This is the sixth installment in our ongoing Forum on Oxidative Stress Status (OSS). The first five groups of papers (27 articles in all) on oxidative stress and its measurement can be found in *FRBM* as follows:

- The first set of seven articles in *FRBM* 27:11/12 (December, 1999), pp. 1135–1196.
- The second set of four articles in *FRBM* 28:4 (February, 2000), pp. 503–536.
- The third set of six articles in *FRBM* 28:6 (March, 2000), pp. 837–886.

Address correspondence to: Dr. William A. Pryor, Biodynamics Institute, 711 Choppin Hall, Louisiana State University, Baton Rouge, LA 70803, USA.

- The fourth set of four articles in *FRBM* 29:5 (September, 2000), pp. 387–415.
- The fifth set of six articles in *FRBM* 29:11 (December, 2000), pp. 1063–1114.

The first of the three articles in this sixth set of OSS Forum articles is "Profiles of antioxidants in human plasma" by M. C. Polidori, W. Stahl, O. Eichler, I. Niestroj, and H. Sies. This is followed by "Unraveling peroxynitrite formation in biological systems" by R. Radi, G. Peluffo, M. N. Alvarez, M. Naviliat, and A. Cayota. The final article in this segment, by L. J. Berliner, V. Khramisov, H. Fujii, and T. L. Clanton, is entitled, "Unique *in vivo* applications of spin traps".

PROFILES OF ANTIOXIDANTS IN HUMAN PLASMA

Maria Cristina Polidori,* Wilhelm Stahl,* Olaf Eichler,* Irmgard Niestroj,[†] and Helmut Sies*

*Institut für Physiologische Chemie I, Heinrich-Heine-Universität, Düsseldorf, Germany; and [†]Schwarzwald Privatklinik Obertal, Baiersbronn, Germany

(Received 1 March 2000; Revised 6 June 2000; Accepted 8 June 2000)

Abstract—The profile of antioxidants in biological fluids and tissues may be helpful in assessing oxidative stress in humans. Plasma antioxidants can be decreased as compared to established normal values, in abnormal or subnormal conditions, for instance as a consequence of disease-related free radical production. Alternatively, plasma antioxidants may be below the normal range due to insufficient dietary supply. Therefore, the profile of antioxidants can be of use only in conjunction with other parameters of the oxidative stress status. This article examines the profiles of plasma antioxidants in oxidative stress–related conditions, e.g., diabetes and some other diseases, as well as smoking and smoking cessation. © 2001 Elsevier Science Inc.

Keywords—Free radicals, Antioxidants, Humans, Vitamins, Aging, Smoking, Health, Assay

INTRODUCTION

Oxidative stress, defined as "the imbalance between oxidants and antioxidants in favor of the oxidants, potentially leading to damage" [1,2], has been associated with a number of disease states in humans [3]. As part of physiological defense mechanisms, small molecular weight antioxidant compounds such as ascorbate (vitamin C), α-tocopherol (vitamin E), and carotenoids are consumed and may fall below normal ranges. The assessment of circulating antioxidants can be used, together with other information, to evaluate conditions of oxidative stress in humans and to disclose nutritional needs of the subjects studied. Plasma antioxidant measurement in humans may be influenced by a variety of parameters related to disease, nutrition, and lifestyle.

The aim of this article is to discuss the potential use of plasma antioxidant profiles as markers of oxidative stress in health and disease. The article is concerned with antioxidant patterns rather than absolute concentrations of antioxidants in plasma. This is because sampling, sample preparation, storage conditions, and operator competence can influence the accuracy of measurement

Maria Cristina Polidori received her M.D. degree at the University of Perugia and did a postdoctoral research visit with Dr. Balz Frei at Boston University in 1995 and 1997, followed by a second postdoctoral position in 1999 at the Institute of Physiological Chemistry of the Heinrich-Heine-University in Düsseldorf, Germany. Currently, she is a last-year resident in Geriatrics at the University Hospital of Perugia, Italy.

Wilhelm Stahl received his Ph.D. degree in chemistry at the University of Kaiserslautern and thereafter worked on drug pharmacokinetics in the pharmaceutical industry. At present, he is a Dozent at the Heinrich-Heine-University Düsseldorf and leads a research group working on antioxidant micronutrients at the Institute of Physiological Chemistry. His research interests are carotenoids, retinoids, and vitamin E, their biological properties and role as antioxidants, as well as their uptake and metabolism in the human.

Olaf Eichler received his diploma in chemistry from the University of Cologne and is presently a doctoral student at the Institute of Physiological Chemistry at the University of Düsseldorf. He is a Stipendiat of the "Graduiertenkolleg Toxikologie und Umwelthygiene" of the Deutsche Forschungsgemeinschaft, Bonn. His research interests are biokinetics of carotenoids and protective effects of these compounds towards photooxidative stress.

Irmgard Niestroj received her M.D. degree from the University of Heidelberg. Thereafter she worked at the Center for Rheumatology at Bad Wildbad (FRG). In 1989, she joined the Schwarzwald Privatklinik Obertal in Baiersbronn where she developed new therapy concepts. Her research interests are nutritional and environmental medicine.

Helmut Sies received a M.D. degree from the University of Munich and an honorary Ph.D. from the University of Buenos Aires. He is professor and chairman of the Institute of Physiological Chemistry I at the Heinrich-Heine-University Düsseldorf. Currently, he is president of the Society for Free Radical Research (SFRR) International. He is Fellow of the National Foundation for Cancer Research, Bethesda, MD, USA. His research interests include oxidative stress, oxidants, and antioxidants.

Address correspondence to: Dr. Helmut Sies, Heinrich-Heine-Universität, Institut für Physiologische Chemie I, Postfach 101007, D-40001 Düsseldorf, Germany; Tel: +49 211-811-2707; Fax: +49 211-811-3029; E-Mail: helmut.sies@uni-duesseldorf.de.

Reprinted from: *Free Radical Biology & Medicine*, Vol. 30, No. 5, pp. 456-462, 2001

Table 1. Antioxidant Profiles in Human Plasma

	A-Healthy women,[a] $n = 64$, this paper (Europe)	B-Healthy women,[b] $n = 408$, ref. [5] (USA)	C-Healthy controls,[a] $n = 75$, ref. [6]	D-Diabetes,[a] $n = 72$, ref. [6]	E-Percent decrease[c]
α-tocopherol (μM)	44.7 ± 12.3	33.2 ± 24.2	26.8 ± 11.1	18.6 ± 10.2*	30.6
γ-tocopherol (μM)	2.15 ± 0.85	6.4 ± 3.2	...	...	...
β-carotene (μM)	0.41 ± 0.31	0.51 ± 0.33	0.58 ± 0.3	0.18 ± 0.1*	69.0
α-carotene (μM)	0.10 ± 0.13	0.12 ± 0.08	0.06 ± 0.05	0.026 ± 0.015*	56.7
Lycopene (μM)	0.19 ± 0.15	0.62 ± 0.34	0.75 ± 0.25	0.11 ± 0.045*	85.4
β-cryptoxanthin (μM)	0.25 ± 0.17	0.27 ± 0.17	0.32 ± 0.18	0.04 ± 0.02*	87.5
Lutein (μM)	0.28 ± 0.12	...	0.32 ± 0.11	0.11 ± 0.5*	65.6
Zeaxanthin (μM)	0.09 ± 0.04	...	0.12 ± 0.05	0.028 ± 0.02*	76.7
Retinol (μM)	1.83 ± 0.58	1.7 ± 0.5	2.56 ± 1.1	1.75 ± 0.5*	31.7
Ascorbic acid (μM)	67.0 ± 19.2	...	...	...	...
Total cholesterol (mg/dl)	214 ± 44	218.9 ± 37.9	199 ± 35	223 ± 41	...
HDL cholesterol (mg/dl)	65.7 ± 17.4	54.5 ± 15.9	49.6 ± 15.6	50.1 ± 14.9	...
LDL cholesterol (mg/dl)	134 ± 43	165 ± 39	149 ± 38	161 ± 38	...
Triglycerides (mg/dl)	99.4 ± 45.6	138 ± 79	128 ± 51	133 ± 52	...

[a] Plasma concentrations of antioxidants and of other constituents in healthy German women (this paper).

[b] Plasma concentrations of lipophilic antioxidants and plasma lipid profile in healthy American women (ref. [5]).

[c] Plasma antioxidant profile in Italian patients with type 2 diabetes as compared to 75 Italian healthy subjects (ref. [6]) and percentages of antioxidant decrease. Data are shown as means ± SD.

* Significantly different from controls (for p values, see text).

and lead to substantial interlaboratory variation. To compare concentrations, participation in a quality assurance scheme is worthwhile, particularly for nutritional and epidemiological studies involving analytical laboratories in different locations.

PROFILES OF PLASMA ANTIOXIDANTS

The measurement of ascorbic acid, α-tocopherol, β-carotene, and other carotenoids usually referred to as "antioxidant vitamins" provides means of evaluating the antioxidant status of a population. In addition to these nonenzymatic antioxidants, there is a variety of other compounds in plasma with free radical–scavenging capability, including glutathione, ubiquinol-10, flavonoids, as well as micronutrient elements such as zinc and selenium, which are incorporated into antioxidant enzymes. Furthermore, it is important to emphasize that plasma is endowed with other substances exhibiting various grades

of antioxidant activity, such as protein sulfhydryl groups, glucose, albumin, and albumin-bound bilirubin.

The antioxidant defense system represents a complex network with interactions, synergisms, and specific tasks for a given antioxidant. It is for this reason that the measurement of the total antioxidant potential of plasma may not be considered an alternative to the assay of individual plasma antioxidants [4].

We measured α- and γ-tocopherol, β-carotene, α-carotene, lycopene, β-cryptoxanthin, lutein, zeaxanthin, retinol, and ascorbate in 64 healthy German women with an average age of 66 years (Table 1). Part of the plasma from these subjects was used to evaluate the lipid profile (total cholesterol, LDL- and HDL-bound cholesterol, and triglycerides). The results of these measurements are shown in Table 1 (column A) in comparison with those from the Framingham Heart Study [5], in which 408 women with average age 76 years were included (Table 1, column B). Plasma concentrations of lipid-soluble

Table 2. Literature Citing Diseases in Which Specific Plasma Antioxidants, Among Antioxidants Measured, Have Been Shown Significantly Lower in Patients as Compared to Controls

Disease	Significantly depleted antioxidants	Reference
Cystic fibrosis	β-carotene	Benabdeslam et al. [8]
Wilson's disease	Ascorbic acid and uric acid; α-tocopherol	Ogihara et al. [9]; von Herbay et al. [10]
Lung cancer	β-carotene, β-cryptoxanthin, lutein/zeaxanthin	Comstock et al. [11]
Prostate cancer	Lycopene	Rao et al. [13]
Cataract	Ascorbic acid	Simon et al. [14]
Age-related macular degeneration	α-tocopherol	Delcourt et al. [16]
Coronary heart disease	α-carotene, γ-tocopherol	Kontush et al. [17]

Table 3. Antioxidant Profiles in Human Plasma

	A-Smokers, $n = 50$, ref. [19]	B-Nonsmokers, $n = 50$, ref. [19]	C-Smokers $n = 31$, ref. [20]	D-Nonsmokers $n = 38$, ref. [20]	E-Smokers $n = 7$, this paper	F-Quitters $n = 7$, this paper
α-tocopherol (μM)	26.4 ± 0.9	26.3 ± 0.8	17.2 ± 1.1	18.9 ± 1.3	19.8 ± 1.4*	26.2 ± 2.9
γ-tocopherol (μM)	2.11 ± 0.09	2.04 ± 0.17	1.0 ± 0.1	1.2 ± 0.1	1.6 ± 0.08	2.3 ± 0.1
β-carotene (μM)	0.48 ± 0.03*	0.60 ± 0.04	0.12 ± 0.03*	0.23 ± 0.04	0.19 ± 0.04	0.26 ± 0.05
α-carotene (μM)	0.08 ± 0.01*	0.11 ± 0.01	...	...	0.03 ± 0.0005	0.036 ± 0.005
Lycopene (μM)	0.50 ± 0.04	0.52 ± 0.04	...	...	0.76 ± 0.05*	0.99 ± 0.07
β-cryptoxanthin	0.08 ± 0.01*	0.11 ± 0.01	...	...	0.11 ± 0.008	0.12 ± 0.007
Lutein (μM)	...	...	...	...	0.36 ± 0.07	0.51 ± 0.08
Lutein/zeaxanthin	0.42 ± 0.04	0.51 ± 0.04	...	...	...	...
Zeaxanthin (μM)	...	...	...	...	0.08 ± 0.008	0.1 ± 0.01
Retinol (μM)	1.92 ± 0.03	1.88 ± 0.07	1.2 ± 0.1	1.3 ± 0.1	2.4 ± 0.3	3.1 ± 0.5
Ascorbic acid (μM)	25.6 ± 3.3*	37.6 ± 2.4	...	...	44.2 ± 4.3*	59.3 ± 5.3
Total cholesterol (mg/dl)	220 ± 4	209 ± 8	247 ± 15	228 ± 11	191 ± 6	192 ± 4
HDL cholesterol (mg/dl)	41.7 ± 1.5	45.2 ± 1.5	...	...	51.2 ± 2.8	51.8 ± 2.7
LDL cholesterol (mg/dl)	150 ± 4	139 ± 4	...	...	128 ± 6	126 ± 5
Triglycerides (mg/dl)	118 ± 6	115 ± 7	202 ± 24	186 ± 32	93.2 ± 5.3	91.6 ± 5.9

[a] Plasma antioxidant profile (μM) and lipid profile (mg/dl) in smokers as compared to nonsmokers (ref. [19]).

[b] Plasma lipophilic antioxidant profile (μM) in smokers as compared to nonsmokers (ref. [20]).

[c] Plasma concentrations of lipophilic and water-soluble antioxidants (μM) and plasma lipid profile (mg/dl) in 7 subjects before and after smoking cessation.

Data are shown as means ± SEM.

* Significantly different values as compared to nonsmokers or to smoking quitters (for *p* values, see text).

antioxidants as well as the lipid profile of 75 Italian healthy subjects (48 female, 27 male, mean age 77 years) (Table 1, column C) included as a control group for 72 elderly patients with type 2 diabetes (40 female, 32 male, mean age 76 years) (Table 1, column D) are shown [6]. The profiles of antioxidants in normal subjects are rather consistent between studies, showing some variation likely due to geographical and dietary differences, especially γ-tocopherol and lycopene.

However, the concentrations of antioxidants were significantly lower in patients with type 2 diabetes as compared to controls ($p < .0001$) (Table 1) [6]. The depletion of antioxidants in the diabetic patients was independent of body mass index (BMI) and dietary intake as assessed by a food frequency questionnaire [6]. Thus, as far as this disease is concerned, all lipophilic antioxidants studied appear to be depleted in plasma more or less in parallel.

Other disease states, however, have been shown to be associated with specific patterns of antioxidants (Table 2). A more marked decrease of plasma β-carotene levels than retinol and α-tocopherol [7] has been reported in patients with cystic fibrosis [8] (Table 2). Another disease associated with free radical–induced damage to tissues and organs, Wilson's disease, has been correlated with a particular plasma antioxidant pattern, consisting in the preferential depletion of ascorbic acid and uric acid as compared to other antioxidants [9] (Table 2). Plasma vitamin E was found to be lowered in patients with Wilson's disease (showing high free serum copper) in a

study where similar findings were obtained also in patients with acute or chronic ethanol intoxication and in patients with hemochromatosis (and high serum iron/low iron-binding capacity) as compared to healthy controls [10].

In certain types of cancer, some plasma antioxidants have been shown to be lowered more than others. When plasma concentrations of ascorbic acid, α- and β-carotene, cryptoxanthin, lutein/zeaxanthin, lycopene, α-tocopherol, and selenium were measured in 258 patients with lung cancer as compared to 515 controls, significant differences between groups were found for cryptoxanthin, lutein/zeaxanthin, and β-carotene levels; small nonsignificant differences between patients and controls were found for α-carotene and ascorbic acid; whereas lycopene, α-tocopherol, and selenium were similar between lung cancer patients and controls [11] (Table 2). In a further study, plasma concentrations of α-tocopherol and β-carotene were evaluated in women with uterine cervix dysplasias and cancer, and were both found to be significantly lower in patients as compared to controls [12]. In a case-control study of serum and tissue carotenoids in prostate cancer patients, significantly lower serum and tissue lycopene levels (44% and 78% less, respectively) were found in patients as compared to controls, whereas the other major carotenoids were not different between groups both in serum and prostate tissue [13] (Table 2).

Data regarding cataract suggest that low vitamin C levels might be a better indicator of risk of disease than

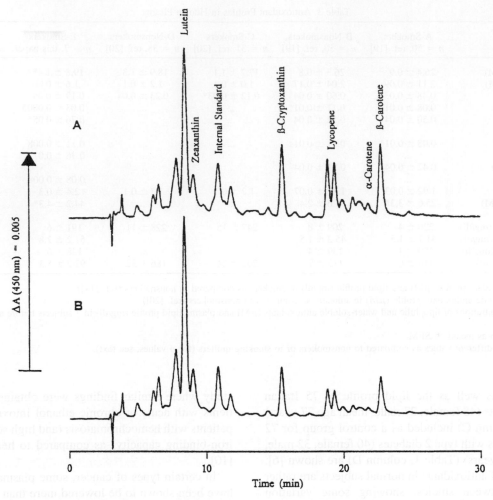

Fig. 1. Carotenoid profile in human plasma. HPLC chromatogram of a subject 4 weeks after (A) and before (B) smoking cessation.

vitamin E and carotenoids, but again few studies have explored a range of different antioxidants in patients vs. controls [14,15] (Table 2). A recent study has shown that lipid-standardized α-tocopherol levels, but not levels of other antioxidants measured, were significantly negatively associated with late age-related macular degeneration [16] (Table 2).

Oxidative stress and antioxidants have been repeatedly studied in atherosclerosis, whose pathogenesis is thought to involve oxidative modification of LDL, but there is a substantial lack of studies investigating whether plasma antioxidants may be used to assess oxidative stress in atherosclerosis or its risk. Kontush et al. [17], however, have shown that among a variety of lipophilic antioxidants—α-tocopherol, γ-tocopherol, α-carotene, β-carotene, and ubiquinol-10—only α-carotene and γ-tocopherol were significantly lower in plasma from patients with coronary heart disease as compared to controls both before and after normalization

to plasma lipids, whereas normalized β-carotene to lipids but not absolute levels were lower in patients as compared to controls [17] (Table 2).

Smoking is condition associated with free radical–induced damage to tissues and organs [18]. When plasma concentrations of carotenoids and antioxidant vitamins were assayed in 50 Scottish male smokers as compared to a group of 50 subjects who had never smoked (mean age 54 years), plasma levels of α-carotene, β-carotene, β-cryptoxanthin, and ascorbic acid were shown to be significantly lower in smokers as compared to nonsmokers ($p = .0006$, $p = .026$, $p = .041$, and $p = .03$, respectively); whereas the levels of lycopene, lutein, zeaxanthin, phytofluene, and vitamin E were similar between the two groups [19] (Table 3, columns A and B). Similarly, al Senaidy et al. [20] found significantly lower levels of β-carotene ($p < .05$) in 31 smokers as compared to 38 nonsmokers (mean age 21 years), but they did not find significant differences for retinol, α-tocoph-

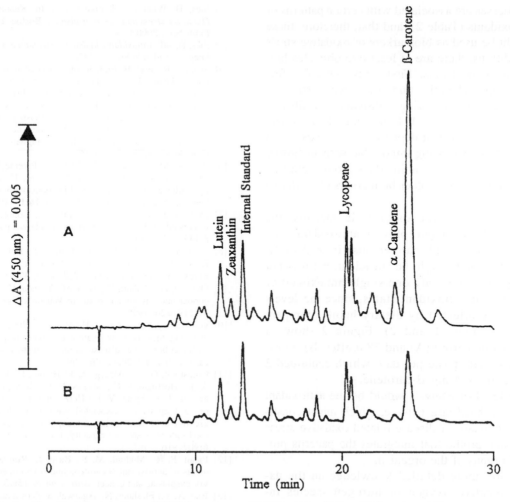

Fig. 2. Carotenoid profile in human plasma. HPLC chromatogram 28 d after (A) and before (B) ingestion of 20 g tomato paste per day. The carotenoid intake with the paste was 2 mg lycopene and 0.4 mg β-carotene per day.

erol, γ-tocopherol, and total vitamin E (Table 3, columns C and D).

We recruited by advertisement 7 young, normolipidemic volunteers (mean age 33 years) who were smoking more than 7 cigarettes per day during at least 6 years and who wanted to quit smoking. We collected blood from these subjects the day before smoking cessation and 4 weeks thereafter, for plasma measurements of α- and γ-tocopherol, β- and α-carotene, lycopene, β-cryptoxanthin, lutein, zeaxanthin, retinol, and ascorbate, as well as of total and bound cholesterol and triglycerides. All antioxidants were shown to be higher after smoking cessation than before, but differences reached statistical significance only for ascorbic acid ($p < .05$), α-tocopherol ($p < .04$), and lycopene ($p < .03$) (Table 3, columns E and F). These results, unrelated to the plasma lipid profile of the subjects nor to their diet, suggest that specific antioxidants might be lowered more than others by smoking, and that therefore some plasma antioxidant patterns might be typically different between smokers

and ex-smokers or nonsmokers. Figure 1 shows a carotenoid HPLC trace from a smoker 4 weeks after smoking cessation (A) and of the same subject at zero time (B); this carotenoid profile was selected because it also shows that the particularly high level of lutein appears to be unchanged.

PRACTICAL CONSIDERATIONS

The steady state balance between pro-oxidants and antioxidants in the human organism may be disturbed in case of depletion of the antioxidants, which may occur endogenously or as a consequence of a diminished dietary intake. The disequilibrium between free radicals and scavenging compounds, however, may be also due to an increase in the pro-oxidant load related to habits (such as smoking) or diseases, and the resulting condition of oxidative stress may be assessed by analysis of the profile of antioxidants. Although there is literature maintaining that particular so-called oxidative stress–related states and illnesses are associated with certain patterns of plasma antioxidants (Table 2), and that, therefore, these patterns might be used as biomarkers of oxidative stress in these conditions, there are at least two obstacles hindering tenable conclusions. First, only relatively few studies have approached the study of oxidative stress in distinct diseases by measuring antioxidant profiles in comparison to controls [7–17]. Second, disease- and nutrition-related antioxidant depletion may coexist, with a separation of the two being impossible without further information. Thus, plasma antioxidant profiles may be useful in conjunction with other biomarkers of oxidative stress.

When studying specific diseases in humans, the measurement of the major plasma antioxidants (α-tocopherol, ascorbic acid, carotenoids) instead of all scavenging molecules might be sufficient, but it is not recommended to study only one single antioxidant or the so-called total antioxidant status, since the levels may be strongly affected by dietary habits (compare lutein peaks in Figs. 1 and 2). Figure 2 shows a carotenoid profile before (A) and 28 d after (B) ingestion of 20 g tomato paste per day, which contained 2 mg lycopene and 0.4 mg β-carotene.

A further level of analysis would be the antioxidant profile in an afflicted target tissue, which might constitute a better indicator of disease-related oxidative stress than the plasma profile that integrates the patterns provided by all organs of the organism.

Furthermore, more detailed knowledge on the nature of the reactive oxygen or nitrogen species involved in a disease process will facilitate insight gained from changes in antioxidant patterns. For example, it has been shown that γ-tocopherol traps nitric oxide more efficiently than α-tocopherol [21]. However, in LDL, α- and β-carotene are the preferential targets for nitric oxide–induced oxidation [22] and are consumed more than α- and γ-tocopherol. Such preferences would be reflected in the antioxidant pattern, so that conversely the antioxidant pattern may be of use obtaining clues on the nature of reactive oxygen species involved in a disease state.

Finally, plasma levels of products generated from interaction of antioxidants with pro-oxidants will be a future area of interest, in analogy to the use of DNA damage products such as 8oxodG for assessing DNA damage. Among these, the tocopherol metabolites α- and γ-CEHC [23] or tocopherol quinones [24] or metabolites of lycopene or bioflavonoids could serve as examples.

REFERENCES

[1] Sies, H. Oxidative stress: introductory remarks. In: Sies, H., ed. *Oxidative stress*. London: Academic Press; 1985:1–8.

[2] Sies, H. What is oxidative stress? In: Keaney, J. F. Jr., ed. *Oxidative stress and vascular disease*. Boston: Kluwer Academic Publishers; 2000:1–8.

[3] Sies, H., ed. *Antioxidants in disease mechanisms and therapy*. San Diego: Academic Press; 1997.

[4] Stocker, R.; Frei, B. Endogenous antioxidant defenses in human blood plasma. In: Sies, H., ed. *Oxidative stress: oxidants and antioxidants*. London: Academic Press; 1991:213–243.

[5] Vogel, S.; Contois, J. H.; Tucker, K. L.; Wilson, P. W. F.; Schaefer, E. J.; Lammi-Keefe, C. J. Plasma retinol and plasma and lipoprotein tocopherol and carotenoid concentrations in healthy elderly participants of the Framingham heart study. *Am. J. Clin. Nutr.* **66**:950–958; 1997.

[6] Polidori, M. C.; Mecocci, P.; Stahl, W.; Parente, B.; Cecchetti, R.; Cherubini, A.; Cao, P.; Sies, H.; Senin, U. Plasma levels of lipophilic antioxidants in very old patients with noninsulin-dependent diabetes. *Diab. Metab. Res. Rev.* **16**:15–19; 2000.

[7] Dominguez, C.; Gartner, S.; Linan, S.; Cobos, N.; Moreno, A. Enhanced oxidative damage in cystic fibrosis patients. *Biofactors* **8**:149–153; 1998.

[8] Benabdeslam, H.; Abidi, H.; Garcia, I.; Bellon, G.; Gilly, R.; Revol, A. Lipid peroxidation and antioxidant defenses in cystic fibrosis patients. *Clin. Chem. Lab. Med.* **37**:511–516; 1999.

[9] Ogihara, H.; Ogihara, T.; Miki, M.; Yasuda, H.; Mino, M. Plasma copper and antioxidant status in Wilson's disease. *Pediatr. Res.* **37**:219–226; 1995.

[10] von Herbay, A.; de Groot, H.; Hegi, U.; Stremmel, W.; Strohmeyer, G.; Sies, H. Low vitamin E content in plasma of patients with alcoholic liver disease, hemochromatosis and Wilson's disease. *J. Hepatol.* **20**:41–46; 1994.

[11] Comstock, G. W.; Alberg, A. J.; Huang, H. Y.; Wu, K.; Burke, A. E.; Hoffman, S. C.; Norkus, E. P.; Gross, M.; Cutler, R. G.; Morris, J. S.; Spate, V. L.; Helzlsouer, K. J. The risk of developing lung cancer associated with antioxidants in the blood: ascorbic acid, carotenoids, α-tocopherol, selenium, and total peroxyl radical–absorbing capacity. *Cancer Epidemiol. Biom. Prev.* **6**:907–916; 1997.

[12] Palan, P. R.; Mikhail, M. S.; Basu, J.; Romney, S. L. Plasma levels of antioxidant β-carotene and α-tocopherol in uterine cervix dysplasias and cancer. *Nutr. Cancer* **15**:13–20; 1991.

[13] Rao, A. V.; Fleshner, N.; Agarwal, S. Serum and tissue lycopene and biomarkers of oxidation in prostate cancer patients: a case control study. *Nutr. Cancer* **33**:159–164; 1999.

[14] Simon, J. A.; Hudes, E. S. Serum ascorbic acid and other correlates of self-reported cataract among older Americans. *J. Clin. Epidemiol.* **52**:1207–1211; 1999.

[15] Taylor, A.; Nowell, T. Oxidative stress and antioxidant function in relation to risk for cataract. In: Sies, H., ed. *Antioxidants in disease mechanisms and therapy*. London: Academic Press; 1997: 515–536.

[16] Delcourt, C.; Cristol, J. P.; Tessier, F.; Leger, C. L.; Descomps, B.; Papoz, L. Age-related macular degeneration and antioxidant status in the POLA study. POLA study group. Pathologies Oculaires Liees a l'Age. *Arch. Ophthalmol.* **117**:1384–1390; 1999.

[17] Kontush, A.; Spranger, T.; Reich, A.; Baum, K.; Beisiegel, U. Lipophilic antioxidants in blood plasma as markers of atherosclerosis: the role of α-carotene and γ-tocopherol. *Atherosclerosis* **144**:117–122; 1999.

[18] Pryor, W. A.; Stone, K. Oxidants in cigarette smoke. Radicals, hydrogen peroxide, peroxynitrate, and peroxynitrite. *Ann. NY Acad. Sci.* **686**:12–27; 1993.

[19] Ross, M. A.; Crosley, L. K.; Brown, K. M.; Duthie, S. J.; Collins, A. C.; Arthur, J. R.; Duthie, G. G. Plasma concentrations of carotenoids and antioxidant vitamins in Scottish males: influences of smoking. *Eur. J. Clin. Nutr.* **49**:861–865; 1995.

[20] al Senaidy, A. M.; Zahrany, Y. A.; Faqeeh, M. B. Effects of smoking on serum levels of lipid peroxides and essential fat-soluble antioxidants. *Nutr. Health* **12**:55–65; 1997.

[21] Christen, S.; Woodall, A. A.; Shigenaga, M. K.; Southwell-Keely, P. T.; Duncan, M. W.; Ames, B. N. Gamma-tocopherol traps mutagenic electrophiles such as NO(X) and complements α-to-

copherol: physiological implications. *Proc. Natl. Acad. Sci. USA* **94:**3217–3222; 1997.

[22] Kontush, A.; Weber, W.; Beisiegel, U. Alpha- and β-carotenes in low-density lipoprotein are the preferred target for nitric oxide–induced oxidation. *Atherosclerosis* **148:**87–93; 2000.

[23] Stahl, W.; Graf, P.; Brigelius-Flohé, R.; Wechter, W.; Sies, H. Quantification of the α- and γ-tocopherol metabolites 2,5,7,8-tetramethyl-2-(2′-carboxyethyl)-6-hydroxychroman and 2,7,8-trimethyl-2-(2′-carboxyethyl)-6-hydroxychroman in human serum. *Anal. Biochem.* **275:**254–259; 2000.

[24] Murphy, M. E.; Kolvenbach, R.; Aleksis, M.; Hansen, R.; Sies, H. Antioxidant depletion in aortic crossclamping ischemia: increase of the plasma α-tocopheryl quinone/α-tocopherol ratio. *Free Radic. Biol. Med.* **13:**95–100; 1992.

ABBREVIATIONS

BMI—body-mass index

HDL—high-density lipoprotein

HPLC—high-performance liquid chromatography

LDL—low-density lipoprotein

α-CEHC—2,5,7,8-tetramethyl-2-(2′-carboxyethyl)-6-hydroxychroman

γ-CEHC—2,7,8-trimethyl-2-(2′-carboxyethyl)-6-hydroxychroman

Bioassays for Oxidative Stress Status (BOSS). Edited by W.A. Pryor

UNRAVELING PEROXYNITRITE FORMATION IN BIOLOGICAL SYSTEMS

RAFAEL RADI,* GONZALO PELUFFO,* MARÍA NOEL ALVAREZ,* MERCEDES NAVILIAT,[†] and ALFONSO CAYOTA[‡]

Departamentos de *Bioquímica, [†]Reumatología, y [‡]Medicina, Facultad de Medicina, Universidad de la República, Montevideo, Uruguay

(Received 2 May 2000; Revised 29 June 2000; Accepted 29 June 2000)

Abstract—Peroxynitrite promotes oxidative damage and is implicated in the pathophysiology of various diseases that involve accelerated rates of nitric oxide and superoxide formation. The unambiguous detection of peroxynitrite in biological systems is, however, difficult due to the combination of a short biological half-life, limited diffusion, multiple target molecule reactions, and participation of alternative oxidation/nitration pathways. In this review, we provide the conceptual framework and a comprehensive analysis of the current experimental strategies that can serve to unequivocally define the existence and quantitation of peroxynitrite in biological systems of different levels of organization and complexity. © 2001 Elsevier Science Inc.

Keywords—Nitric oxide, Superoxide, Peroxynitrite, Hydroxyl radical, Nitrogen dioxide, 3-nitrotyrosine, Free radicals, Detection, Methods

INTRODUCTION

Peroxynitrite,[1] the product of the combination reaction between nitric oxide ($^{\bullet}$NO) and superoxide ($O_2^{\bullet-}$), is a reactive and short-lived species that promotes oxidative molecular and tissue damage [1–7]. In addition to the generation of a pro-oxidant species, the formation of peroxynitrite results in decreased bioavailability of $^{\bullet}$NO, therefore diminishing both its salutary physiological functions [8–10] and its strong antioxidant actions over free radical and metal-mediated processes [11–13]. Peroxynitrite formation and reactions are proposed to contribute to the pathogenesis of a series of diseases including acute and chronic inflammatory processes, sepsis, ischemia-reperfusion, and neurodegenerative disorders, among others [14–27].

The detection of peroxynitrite in biological systems has been a challenge over the past decade because of the (i) elusive nature of peroxynitrite which precludes its direct isolation and detection, (ii) necessity to find detector molecules that can efficiently outcompete the multiple reactions that peroxynitrite can undergo, (iii) nonexistence of footprints totally specific of peroxynitrite reactions, and (iv) the difficulty to discriminate between

Rafael Radi, M.D., Ph.D., obtained his doctoral degree at the Universidad de la República, Montevideo, Uruguay in 1989 and performed posdoctoral studies at the University of Alabama at Birmingham (1989–1991). He returned to Uruguay in 1992 to a faculty position at the Departamento de Bioquímica, Facultad de Medicina, Universidad de la República, where he initiated a research group that investigates the biochemistry and cell biology of nitric oxide and peroxynitrite. He is at present a Professor of Biochemistry, an International Research Scholar of the Howard Hughes Medical Institute, and the current Secretary General of the Oxygen Society.

Gonzalo Peluffo, M.D., and María Noel Alvarez, M.S., obtained their degrees at the Universidad de la República in 1998 and are currently performing Ph.D. studies on aspects referred to the biological formation, detection, and diffusion of oxygen radicals, nitric oxide, and peroxynitrite in biochemical and cell systems (MNA) or humans (GP). They are both Assistant Professors of Biochemistry at the Facultad de Medicina, Universidad de la República.

Mercedes Naviliat, M.D., obtained her degree at the Universidad de la República in 1985 and has just completed a Ph.D. thesis on the role of nitric oxide and peroxynitrite in inflammatory disease. She is an Assistant Professor of Rheumatology at the Facultad de Medicina, Universidad de la República.

Alfonso Cayota, M.D., Ph.D., obtained his M.D. degree at Universidad de la República in 1986 and performed Ph.D. studies at the Université Paris VI-Pasteur Institute, Paris, France from 1990–1995. He is currently an Associate Professor of Medicine at the Facultad de Medicina, Universidad de la República, where he investigates the role of reactive oxygen and nitrogen species during normal and pathological immune responses.

Address correspondence to: Dr. Rafael Radi, Departamento de Bioquímica, Facultad de Medicina, Avda. General Flores 2125, 11800 Montevideo, Uruguay; Fax: +598 (2) 9249563; E-Mail: rradi@fmed.edu.uy.

[1]The term peroxynitrite refers to the sum of peroxynitrite anion (ONOO$^-$) and peroxynitrous acid (ONOOH). IUPAC recommended names for peroxynitrite anion, peroxynitrous acid, nitroso-peroxocarboxylate (ONOOCO$_2^-$), and nitric oxide are oxoperoxynitrate (1$-$), hydrogen oxoperoxynitrate, 1-carboxylato-2-nitrosodioxidane, and nitrogen monoxide, respectively.

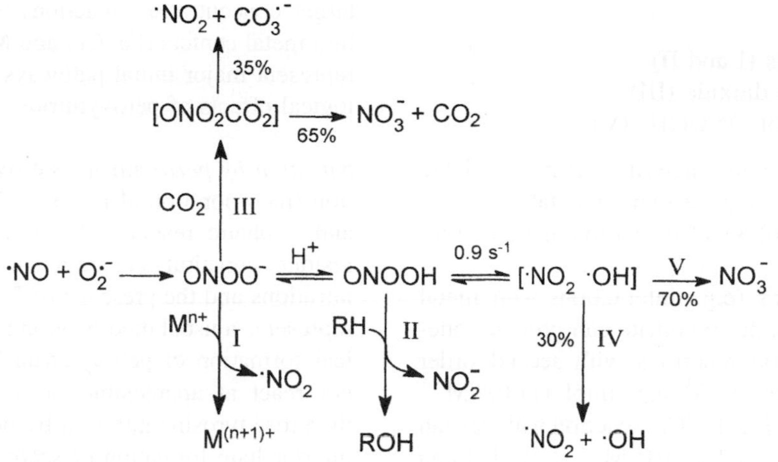

Fig. 1. Peroxynitrite reaction pathways. Numbers I to V indicate possible fates of peroxynitrite: direct reactions include the one-electron oxidation of transition metal centers (Fe, Mn, Cu) (I); the two-electron oxidation with a target substrate (RH) (II) and the formation of nitroso-peroxocarboxylate (III), that rapidly decomposes to secondary radicals in 35% yield. Peroxynitrous acid undergoes homolysis at $0.9 \ s^{-1}$ to yield free radicals in 30% yields (IV) or rearrange to nitrate (V).

the biological effects of peroxynitrite versus that of its precursors, $^{\bullet}NO$ and $O_2^{\bullet-}$, and other $^{\bullet}NO$-derived oxidants.

In spite of the fact that the biological formation and reactions of peroxynitrite are kinetically and thermodynamically favored, the importance of the peroxynitrite pathway in biology has been occasionally questioned [28–30], in part due to the difficulties related to its detection. Even though there is solid evidence supporting the formation of peroxynitrite in vivo and its contribution to biomolecular damage and cell and tissue pathology, unambiguous detection and quantitation is not trivial and requires a subtle knowledge of its biological chemistry and a multifaceted approach. In this work, we will (i) provide biochemical and physico-chemical foundation needed to search for peroxynitrite, (ii) analyze current methodologies used in the detection of peroxynitrite and (iii) establish criteria that must be fulfilled to unravel the formation of peroxynitrite in biological systems of different levels of organization and complexity.

PEROXYNITRITE BIOCHEMISTRY

Formation reactions

The biological formation of peroxynitrite anion (ONOO$^-$) is mainly due to the fast reaction between $^{\bullet}NO$ and $O_2^{\bullet-}$ (Eqn. 1). This radical-radical combination reaction undergoes with a second order rate constant that has been independently determined as 4.3, 6.7, and 19×10^9 $M^{-1} \ s^{-1}$ [31–33], and therefore one can safely assume a value of $\sim 10^{10} \ M^{-1}s^{-1}$, which indicates a diffusion-controlled reaction.

$$^{\bullet}NO + O_2^{\bullet-} \rightarrow ONOO^- \qquad v = k[^{\bullet}NO][O_2^{\bullet-}] \tag{1}$$

Since both precursor radical species, $^{\bullet}NO$ and $O_2^{\bullet-}$, are transient in nature, the biological formation of peroxynitrite requires the simultaneous generation of both radicals which, in addition, must approach and react within the same compartment. However, while $^{\bullet}NO$ has a biological half-life in the range of seconds and readily diffuses across membranes [34,35], $O_2^{\bullet-}$ lasts less than milliseconds and permeates membranes only via anion channels [36]. Thus, due to both the greater half-life and facile diffusion of $^{\bullet}NO$ compared to $O_2^{\bullet-}$, peroxynitrite formation will predominantly occur nearer to the $O_2^{\bullet-}$ formation sites.

Peroxynitrite anion exists in protonation equilibrium with peroxynitrous acid (ONOOH, pKa = 6.8) [2]. Thus, under biological conditions both ONOO$^-$ and ONOOH will be present, the ratio depending on local pH (i.e., at pH 7.4, 80% of peroxynitrite will be in the anionic form). This is relevant because both species have different reactivities and diffusional properties [7].

In addition to the $^{\bullet}NO$ plus $O_2^{\bullet-}$ reactions, other processes such as the reaction of nitroxyl anion with molecular oxygen [37,38], $^{\bullet}NO$ oxidation of oxy-, hemo-, and myoglobin [39], and turnover of L-arginine-depleted nitric oxide synthase [40] may also contribute to the biological formation of peroxynitrite.

Reactivity

Peroxynitrite promotes biological effects via different types of reactions (Fig. 1), which could be classified in three main groups:

1) direct redox reactions (I and II)
2) reaction with carbon dioxide (III)
3) homolytic cleavage of ONOOH (IV)

Recognizing the different reactivities of peroxynitrite is critical for planning adequate experimental strategies directed to assess its biological formation and quantitation.

In the direct reactions (e.g., interactions with metal centers, thiol oxidation), peroxynitrite can promote one- or two-electron oxidation reactions with second order rate constants in the order of 10^3 (e.g., thiols) to 10^6 M^{-1} s^{-1} (e.g., metal centers) [7,41]. The reaction with carbon dioxide [42] is fast ($k = 5.7 \times 10^4$ M^{-1} s^{-1}) [43] and yields a short-lived intermediate (estimated half-life <1 μs), nitroso-peroxocarboxylate ($ONOOCO_2^-$), which homolyses to carbonate radical ($CO_3^{\bullet -}$) and nitrogen dioxide ($^{\bullet}NO_2$) in ~35% yields [44–46]. Carbonate radical is a relatively strong one-electron oxidant and $^{\bullet}NO_2$ is a more moderate oxidant and also a nitrating agent; therefore the radical products arising from $ONOOCO_2^-$ decomposition promote secondary oxidation events. Finally, ONOOH can undergo homolysis to $^{\bullet}OH$ and $^{\bullet}NO_2$, with a first order rate constant of 0.9 s^{-1} at pH 7.4 and 37°C in ~30% yields, while the rest of ONOOH isomerizes directly to nitrate (NO_3^-) [1,47,48] (Fig. 1, route V). Hydroxyl radical is a more powerful oxidant than $CO_3^{\bullet -}$ and $^{\bullet}NO_2$, but it is significantly less selective in target molecule reactions and addition reactions predominate over one-electron abstractions. In the absence of targets, the proton- or carbon dioxide-catalyzed decomposition of peroxynitrite mainly yields nitrate, due to recombination of the radical intermediates arising from homolysis. However, in the presence of targets, most peroxynitrite yields nitrite (NO_2^-).

A wide variety of biomolecules can be oxidized by peroxynitrite in vitro either by direct reactions or by the secondary radicals ($CO_3^{\bullet -}$, $^{\bullet}NO_2$, $^{\bullet}OH$). However, in biological systems where multiple direct targets of peroxynitrite are present, the $^{\bullet}OH$-pathway is slow in comparison with the direct bimolecular reactions of peroxynitrite and a relatively small number of reactions predominate [41]. In fact, in vivo, more than 95% of all peroxynitrite formed will be consumed by direct reactions, with less than 5% evolving to $^{\bullet}OH$ and $^{\bullet}NO_2$. Unlike $^{\bullet}OH$ that typically reacts with most biomolecules with rate constants in the order of 10^9 $M^{-1}s^{-1}$, peroxynitrite is a much more selective oxidant and reacts at significantly slower rates. The rate constants and preferential pathways of peroxynitrite decomposition in biology have been reviewed elsewhere [7,41], as the contribution of a biomolecule to the overall fate of peroxynitrite will be a function of both rate constant and target concentration, reactions with sulfhydryls, transition metal centers (Fe, Cu, and Mn), and carbon dioxide represent major initial pathways accounting for the biological effects of peroxynitrite.

Nitration by peroxynitrite. Peroxynitrite promotes nitration (incorporation of a nitro $-NO_2$ group) of aromatic and aliphatic residues. Most notably, protein tyrosine residues constitute key targets for peroxynitrite-mediated nitrations and the presence of 3-nitrotyrosine in proteins represents a usual modification introduced by the biological formation of peroxynitrite [49]. Peroxynitrite does not react at appreciable rates with tyrosine [50] and therefore tyrosine nitration by peroxynitrite requires the intermediate formation of secondary species. The nitration process involves free radical mechanisms in which one-electron oxidants derived from peroxynitrite attack the aromatic ring leading to the formation of tyrosyl radical, which then rapidly combines with $^{\bullet}NO_2$ to yield 3-nitrotyrosine [51,52] (Fig. 2).

In addition to 3-nitrotyrosine, the reactions of peroxynitrite-derived oxidants with tyrosine, also yield 3,3'-dityrosine [53] (Fig. 2). Small quantities of 3-hydroxytyrosine may be formed, following the addition reaction of $^{\bullet}OH$ with tyrosine [54] (Fig. 2).

Peroxynitrite diffusion

Due to target molecule reactions, the biological half-life of peroxynitrite is estimated to be less than 100 ms [41,55]. This half-life is long enough for peroxynitrite to potentially travel some distances (e.g., 5–20 μm) across extra- and/or intracellular compartments. However, in addition to the estimated diffusion in aqueous environments, the biological effects and detection of peroxynitrite will be influenced by its ability to permeate cell membranes. In this regard, both $ONOO^-$ and ONOOH can cross biological membranes, via anion channels and passive diffusion, respectively [56,57].

The diffusion distances of peroxynitrite will critically influence the distribution of oxidative modifications within a tissue and the reaction yields with reporter molecules.

AFFIRMING THE BIOLOGICAL FORMATION OF PEROXYNITRITE

General considerations

The detection of peroxynitrite relies on either (i) modification of exogenously added probes, or (ii) footprinting reactions on endogenous molecules. However, these are not straightforward procedures; at present there are no totally specific modifications of either probe or bi-

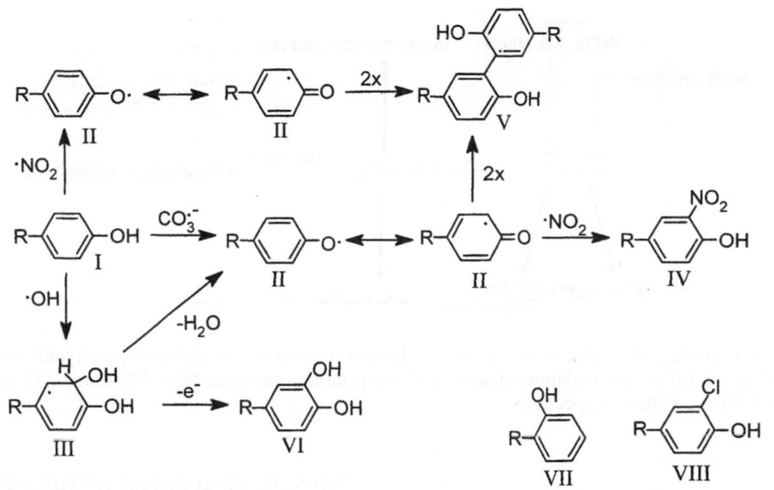

Fig. 2. Tyrosine oxidation pathways by peroxynitrite. Represented compounds are tyrosine (I), tyrosyl radical (II), tyrosine-hydroxyl radical adduct (III), 3-nitrotyrosine (IV), 3-3'-dityrosine (V), 3-hydroxytyrosine (VI), all of which can be yielded during reactions of peroxynitrite-derived oxidants with tyrosine. Both $CO_3^{\cdot-}$ and $^{\cdot}NO_2$ can perform a one-electron abstraction of tyrosine to yield tyrosyl radical, while $^{\cdot}OH$ predominantly (>95%) leads to the formation of a radical adduct that can either decay by water elimination to tyrosyl radical, or be oxidized to 3-hydroxy-tyrosine, among other reactions. o-Tyrosine (VII) and 3-chlorotyrosine (VIII), products of the reaction of $^{\cdot}OH$ or HOCl with tyrosine, respectively, are also shown. Final product distribution will largely depend on type and flux of the different radical/oxidant species present as well as the amount of available free and protein-bound tyrosine.

omolecules that can directly and unambiguously assure the formation of peroxynitrite. Probe modification and/or footprinting reactions require additional criteria to constitute sufficient evidence for affirming peroxynitrite formation. Some of these additional criteria involve pharmacological modulation, while others rely on the appreciation of the unique biochemistry of peroxynitrite versus that of $^{\cdot}NO$ and other $^{\cdot}NO$-derived species such as nitrogen dioxide ($^{\cdot}NO_2$), S-nitrosothiols (RSNO), and the species arising from catalytic action of hemeperoxidases in the presence of hydrogen peroxide (H_2O_2) and nitrite (NO_2^-).

Differential reactivities of $^{\cdot}NO$ and $ONOO^-$

An important aspect to unravel the formation of peroxynitrite in biology is to recognize the differential reactivities (and effects) of $^{\cdot}NO$ and peroxynitrite over cell and tissues. In particular, $^{\cdot}NO$ is neither a strong oxidant nor a nitrating agent; it mostly participates in reversible interactions with iron centers, radical-radical combination reactions (e.g., with lipid radicals to terminate lipid oxidation chain reactions, with $O_2^{\cdot-}$ to form peroxynitrite) and nitrosylation reactions via intermediate formation of dinitrogen trioxide (N_2O_3). On the other hand, peroxynitrite is a strong oxidant and nitrating agent and a poor nitrosylating agent. Thus, several oxidation and nitration reactions measured in probes or biomolecules secondary to $^{\cdot}NO$ formation reflect the presence of peroxynitrite.

It has become apparent [6,58–61] that a series of

"$^{\cdot}NO$-dependent" biological processes, are not *directly* mediated by $^{\cdot}NO$, but they rather depend on the formation of secondary species. For instance, the $^{\cdot}NO$-dependent inactivation of mitochondrial electron transport complexes I and II cannot be readily accomplished by $^{\cdot}NO$ itself, but is efficiently mediated by peroxynitrite. Thus, in this case protection from inactivation by blocking $^{\cdot}NO$ production infers that secondary species derived from $^{\cdot}NO$ are participating in the process, and then orient for the search of peroxynitrite as the proximal oxidant.

Probe oxidation

Oxidizable probes can be conveniently used to monitor peroxynitrite formation in biochemical and cell systems. For quantitation purposes, probes would be ideally located close to the peroxynitrite forming sites and outcompete peroxynitrite reactions with various other biological targets. Also, the selected probe should be readily oxidized by peroxynitrite, but not by $^{\cdot}NO$ and $O_2^{\cdot-}$.

There are caveats in the use of probes which are to be defined in each case: (i) intra- versus extracellular distribution of the probe and detection of reactive intermediates, (ii) contribution of competing reactions to overall oxidation yields, (iii) potential redox-cycling and/or secondary reactions of probe radical intermediates, which may artifactually modify oxidation yields, (iv) limited knowledge of the mechanisms of probe oxidation by peroxynitrite, which restrains quantitative interpretation of the data.

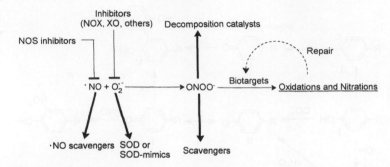

Fig. 3. Peroxynitrite pharmacology. Potential sites of pharmacological intervention to diminish peroxynitrite formation and reactions are indicated and aimed to inhibit probe oxidation/nitration and footprinting reactions. NOS, NOX and XO are nitric oxide synthase, NADPH oxidase and xanthine oxidase, respectively.

Footprinting

Peroxynitrite formation and reactions can be evidenced through the detection of oxidative modifications that peroxynitrite promotes in target biomolecules. Oxidative modifications can be performed by different oxidants and/or may be readily reversed by appropriate repair systems, therefore ideal modifications to measure would be those that are (i) more specific for peroxynitrite versus other oxidant systems, and (ii) relatively permanent and stable. These modifications involve oxidation reactions in proteins, DNA or lipids, and most notably, protein tyrosine nitration.

Protein tyrosine nitration deserves special discussion. As ${}^{\bullet}NO$ is incapable of directly promoting nitration reactions, nitration was initially thought to constitute a specific footprint of peroxynitrite reactions in biology [17]. However, it has become apparent that other mechanisms of biological nitration may exist in addition to peroxynitrite, including (i) the H_2O_2-NO_2^--hemeperoxidase [62] (ii) the reactions of ${}^{\bullet}NO_2$ (formed from the aerobic oxidation of ${}^{\bullet}NO$) with tyrosine residues [63], and (iii) the oxidation of unstable nitrosotyrosine [64]. Importantly, however, these alternative mechanisms of nitration appear to be more restricted than the nitration mediated by peroxynitrite. Indeed, the peroxidase mechanism requires the presence of a specific enzyme (e.g., myeloperoxidase or eosinophil peroxidase) in the site of formation of oxygen radicals and ${}^{\bullet}NO$. This may be limited to those tissue regions or compartments and processes involving an important participation of activated inflammatory cells. The ${}^{\bullet}NO_2$ mechanism appears to be of minor relevance because under normal or low tissue oxygen tensions, the oxidation of ${}^{\bullet}NO$ to ${}^{\bullet}NO_2$ is a rather slow reaction, and also, because the initial reaction of ${}^{\bullet}NO_2$ with tyrosine is not too fast and other reactions such as ${}^{\bullet}NO_2$-mediated thiol oxidation will be kinetically favored. Finally, the rapid reaction of tyrosyl radical with ${}^{\bullet}NO$ transiently yields nitrosotyrosine, in a reaction that appears to be reversible. In the case of free tyrosine, most nitrosotyrosine dissociates back to tyrosyl radical and ${}^{\bullet}NO$, to ultimately predominantly yield 3,3'-dityrosine [52]. However, in some proteins containing adjacent redox centers (e.g., prostaglandin synthase) [64], protein-bound nitrosotyrosine may be further oxidized to 3-nitrotyrosine by a "site-specific" one-electron oxidation to iminoxyl radical, followed by a second one-electron oxidation to nitro-tyrosine. The mechanism of 3-nitrotyrosine formation involving the transient formation of nitrosotyrosine would be limited to a small number of proteins.

Thus, nitration of biomolecules reveals the participation of ${}^{\bullet}NO$-derived oxidants during oxidative damage to tissues. In this context, peroxynitrite is a central contributor to protein tyrosine nitration.

Peroxynitrite pharmacology

Pharmacological strategies are available to unravel peroxynitrite formation. Indeed, pharmacological modulation by NOS inhibitors and ${}^{\bullet}NO$ scavengers, inhibitors of $O_2^{\bullet-}$ formation and SOD/SOD-mimics and peroxynitrite decomposition catalysts and scavengers is directed to attenuate peroxynitrite-mediated oxidative modifications in biomolecules and probes as well as biological effects such as cell death and tissue injury (Fig. 3).

Compounds used to interfere on the peroxynitrite pathway and their mechanism of action will be discussed in more detail at the end of the chapter.

METHODOLOGIES FOR PEROXYNITRITE DETECTION

In this section we will describe current methodologies for peroxynitrite detection. The analysis will concentrate on those techniques that have been more widely used and validated. We will provide the biochemical background for each methodology, its potency and limitations. The field is in progress and awaits further application and development.

Probe oxidation/nitration

Probes must ideally combine simplicity, sensitivity, and specificity for their application. In this regard, a series of compounds have been successfully used to detect peroxynitrite. None of them being totally specific, knowledge of their reactivity with different oxidants and proper use of scavengers and inhibitors is required.

Oxidation of fluorescent probes

Two fluorescent probes, dichlorofluorescin (DCFH) and dihydrorhodamine (DHR), have been frequently used for assaying the formation of cell and tissue-derived oxidants. Indeed, the two-electron oxidation of dichlorofluorescin to dichlorofluorescein (DCF; λ_{ex} = 502 nm, λ_{em} = 523 nm) and dihydrorhodamine to rhodamine (RH; λ_{ex} = 500 nm, λ_{em} = 536 nm), results in the formation of highly fluorescent products and is promoted by strong oxidants such as $^\bullet OH$, oxo-iron complexes, and peroxynitrite [65–69].

This method allows the detection of submicromolar levels (e.g., 50 nM) of peroxynitrite. Additionally, since DCF and RH are strong chromophores at 500 nm (ε_{500} = 59,500 $M^{-1} cm^{-1}$ and 78,800 $M^{-1} cm^{-1}$, respectively), larger amounts of peroxynitrite can be also detected spectrophotometrically. Importantly, neither $^\bullet NO$ nor $O_2^{\bullet-}$ and only to a marginal extent $^\bullet NO_2$, are able to oxidize either probe at significant yields.

The oxidation of DCFH or DHR by peroxynitrite results in ~35 and 42% oxidation yields (molecules of oxidized probe per molecules of peroxynitrite) at pH 7.4, 37°C, in potassium phosphate buffer, respectively [69]. Oxidation yields are highly influenced by environmental conditions including pH and buffer components as well as by the presence of molecules that compete for peroxynitrite including bicarbonate, thiols, and urate that may inhibit probe oxidation (Table 1). Changes in pH may not only affect oxidation efficiency by peroxynitrite (which is the highest in the pH 7.4 region for both probes) [66,67], but it may also cause subtle effects on fluorescent intensity yields. Thus, it is critical to measure probe oxidation yields with authentic peroxynitrite under the conditions of the assay, as this will permit a more accurate determination of the levels produced.

The mechanisms of DCFH and DHR oxidation by peroxynitrite are largely unknown; this remains as a limitation for interpreting the probe oxidation data under conditions in which other biological targets for peroxynitrite exist, since the probes used in the biological assays are never present in large excess (typical concentration ~10 μM), due to their high cost and relatively low solubility. For both DCFH and DHR, hydroxyl radical scavengers are unable to inhibit oxidation, suggesting

Table 1. Probe Oxidation Yields and Influence of Media Components

| Probe[a] | Yield | Influence of components | | | |
		CO_2	Thiols	Uric acid	Desferrioxamine
DCFH [67,69]	35	↓	↓ ↓	↓	↓
DHR [66,69]	42	↓	↓ ↓	↓ ↓	↓
Tyrosine [4,43,200]	6	↑	↓ ↓	↓ ↓	↓
Cyt c²⁺ [43,119, 120]	>50	—	↓	—	—
Luminol[b] [78,83, 133]	(~20%)[c]	↑ ↑	↓ ↓	↓ ↓	↓ ↓

[a] Data provided were obtained at pH 7.4, 37°C in phosphate buffer and from bolus addition of authentic peroxynitrite. Symbols indicate: ↓, partial inhibition; ↓ ↓, total inhibition; —, without influence; ↑, moderate enhancement; ↑ ↑, large enhancement.
[b] Referred as light emission.
[c] Non applicable, as luminol oxidation yield is not utilized for detection/quantitation purposes.
DCFH = dichlorofluorescin; DHR = dihydrorhodamine.

that peroxynitrite itself may be mediating the oxidation process (Fig. 4). However, evidence for direct reactions between peroxynitrite and the probes is at the moment lacking.

A separate note deserves the influence of excess $^\bullet NO$ or $O_2^{\bullet-}$ on probe oxidation. Nitric oxide/superoxide formation ratios different to one will result in peroxynitrite formation but also in remaining quantities of one of the precursor radical species. Both excess $^\bullet NO$ and $O_2^{\bullet-}$ may decrease probe oxidation yields [68,70] due to reactions with radical intermediates arising from one-electron oxidation of the probes (Fig. 4). Thus, excess radical formation may result in underestimation of peroxynitrite

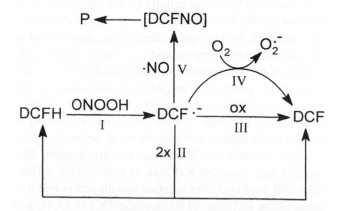

Fig. 4. Dichlorofluorescin oxidation by peroxynitrite. Dichlorofluorescin (DCFH) is postulated to be oxidized by one-electron by peroxynitrite (I) to yield the corresponding radical anion (DCF$^{\bullet-}$), which is either dismutated (II) or further oxidized (III) to yield dichlorofluorescein (DCF). Secondary relevant reactions include the reaction of DCF$^{\bullet-}$ with molecular oxygen to yield $O_2^{\bullet-}$ (IV), or with $^\bullet NO$ to yield nonfluorescent products (P) (V). Analogous reactions are expected to occur with dihydrorhodamine.

formation rates due to secondary, $^{\bullet}NO/O_2^{\bullet-}$-mediated termination reactions.

It has been recently pointed out that the one-electron oxidation product of DCFH, the DCF semiquinone radical ($DCF^{\bullet-}$) may subsequently react with molecular oxygen to yield $O_2^{\bullet-}$ and DCF [71,72] (Fig. 4) and then artifactually enhance probe oxidation yields. This mechanism of probe-dependent $O_2^{\bullet-}$ formation may also be operative for other fluorescent compounds of similar structure to that of DCFH, such as DHR. Indeed, $DCF^{\bullet-}$-dependent $O_2^{\bullet-}$ formation could, in principle, contribute to generating peroxynitrite in $^{\bullet}NO$-generating systems. This phenomenon requires study but may not be critical during peroxynitrite-mediated DCFH oxidation, since excess $^{\bullet}NO$ over $O_2^{\bullet-}$ leads to *inhibition* instead of augmentation of probe oxidation [68], indicating that, if any, $O_2^{\bullet-}$ formation by $DCF^{\bullet-}$ does not significantly contribute to further formation/detection of peroxynitrite. In any event, experiments designed to assess (i) cell $O_2^{\bullet-}$ formation in the presence of DCFH, and (ii) DCFH oxidation with and without NOS activity, will provide valuable information to account for or rule out the detection of potentially high values of peroxynitrite.

DCFH and DHR oxidation may serve to indicate either intra- or extracellular formation of peroxynitrite. Both probes are able to enter cells and participate in intracellular oxidation reactions [65]. Thus, extracellular formation of peroxynitrite may be preferentially studied by adding the probes right before the beginning of the experiment. On the other hand, preincubation of cells with DCFH or DHR for ~1 h leads to significant probe accumulation intracellularly, and both intra- and extracellular-associated fluorescence can be studied. However, there are potential pitfalls in the interpretation of the data for the following reasons: probes oxidized in a compartment, for example, extracellular, may diffuse to intracellular compartments and vice versa, with the exception of the extracellular diffusion of RH; indeed, RH is lipophilic, positively charged, and concentrates in mitochondria and other negatively charged cell compartments, and will mostly remain inside the cell as long as mitochondria maintain electrochemical potential. However, peroxynitrite itself may lead to mitochondrial dysfunction and therefore RH leak to extracellular milieu. Also, cell washings after incubation with probes result in extracellular diffusion of intracellular DCFH, DCH, and DHR [65,69]. Thus, the dynamics of parent and oxidized probe permeation through cell membranes must be specifically studied. In any event, fluorescence of cell supernatants, cell homogenates, and even fluorescence microscopy has been applied successfully for peroxynitrite detection with DCFH and DHR [73–77].

Chemiluminescence probes

Luminol (5-amino-2,3-dihydro-1,4-phthalazinedione) is a chemiluminescent probe widely use for detection of oxygen radical intermediates produced in biological systems. The general mechanism of chemiexcitation includes a two-electron oxidation process and the production of an unstable endoperoxide intermediate decomposes yielding excited 3-aminophtalate, which decays to ground state aminophtalate with the emission of light with a maxima at 425 nm [78].

Peroxynitrite is capable of inducing luminol chemiluminescence (LCL), especially in the presence of bicarbonate-carbon dioxide, which results in high quantum yields [78]. Indeed, this technique has been successfully used in a number of studies for peroxynitrite detection at the cellular and organ level [79–82]. The mechanism of chemiexcitation of luminol by peroxynitrite in the presence of bicarbonate-carbon dioxide, involves the formation of $CO_3^{\bullet-}$ from the homolytic decomposition of $ONOOCO_2^-$, which readily performs a one-electron oxidation of luminol, initiating the pathway for light emission (Fig. 5). Peroxynitrite-mediated light emission is inhibited by thiols and urate (Table 1), and quantum yields increase with pH, especially because only decomposition of luminol monoanion (pKa = 8.2) leads to the light-emitting route. Chemiluminescence makes it possible to follow the time course of peroxynitrite formation, and with highly sensitive photon counting techniques [83], peroxynitrite fluxes down to ~1 nM/min can be accurately detected.

LCL can be induced by other biologically relevant oxidants such as $^{\bullet}OH$ and $O_2^{\bullet-}$, which may operate sequentially during luminol oxidation; however, $^{\bullet}NO$ is not capable of inducing LCL, and the potential contribution of $^{\bullet}NO_2$ is, at most, marginal [83]. Thus, inhibition of LCL by NOS inhibitors is compatible with the formation of peroxynitrite. Nevertheless, it is important to note that excess $^{\bullet}NO$ may lead to inhibition of luminol chemiluminescence, most likely via termination reactions with luminol radical intermediates [83] (Fig. 5). In this regard, quantum yields from authentic peroxynitrite are somewhat higher than those obtained with equimolar fluxes of $^{\bullet}NO$ and $O_2^{\bullet-}$.

In addition to luminol, recently another highly sensitive probe, coelenterazine (2-(4-hydroxybenzyl-)-6-(4-hydroxyphenyl)-8-benzyl-3,7-dihydroimidazo[1,2-α]pyrazin-3-one), has been shown to efficiently produce light emission during its reaction with reactive oxygen species and peroxynitrite [84]. Further studies characterizing peroxynitrite-induced coelenterazine chemiluminescence in chemical and biological systems are being carried out in our laboratories.

While $O_2^{\bullet-}$-producing systems also lead to the che-

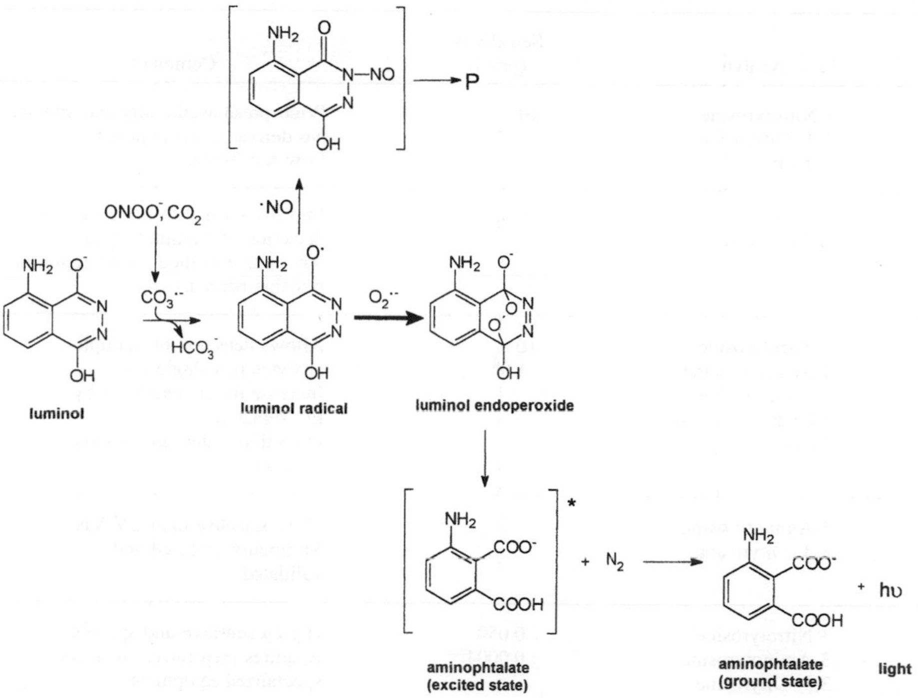

Fig. 5. Luminol chemiluminescence by peroxynitrite. Carbonate radicals arising from the decomposition of $ONOOCO_2^-$ promote a one-electron oxidation to luminol radical, which then reacts with $O_2^{\cdot-}$ to follow the light emitting pathway. Under excess $^\cdot$NO, luminol radicals yield intermediates and products (P) that led to a dark route.

miexcitation of another widely used probe, lucigenin (10,10-dimethyl-bis-9,9-bisacridinium nitrate) via a reductive dioxygenation process [85], lucigenin is unable to yield light by reactions with oxidants. Thus, peroxynitrite is *unable* to promote lucigenin chemiluminescence as described in an early report [78]. Then, while $^\cdot$NO may enhance luminol and coelenterazine-dependent chemiluminescence in $O_2^{\cdot-}$-producing systems through the formation of peroxynitrite, it will inhibit lucigenin chemiluminescence. Combined utilization of luminol and lucigenin may provide further insights with regard to the relative contributions of oxygen radicals and peroxynitrite to observed chemiluminescence.

Nitration of phenolic compounds

Peroxynitrite promotes nitration and oxidation of phenolic compounds such as tyrosine or p-hydroxyphenylacetic acid (p-HPA). Nitration leads to the formation of 3-nitrotyrosine or 3-nitro-pHPA, while oxidation results in dimerization and hydroxylation reactions, which can also be accomplished by other oxidant systems such as $^\cdot$OH and oxo-iron complexes. Thus, nitration of phenolic probes has been utilized as a more specific modification induced by peroxynitrite and potentially other $^\cdot$NO-derived oxidants.

Nitration of the aromatic compound p-hydroxyphenyl-

acetic acid (p-HPA) in the extracellular milieu was first used to detect peroxynitrite release from activated macrophages [86] and more recently for the detection of nitrating species generated by activated neutrophils [62]. In addition, free tyrosine has been used to trap nitrating species arising from peroxynitrite formed from chemical/biochemical fluxes of $^\cdot$NO and $O_2^{\cdot-}$ [52,87].

3-Nitro-pHPA and 3-nitrotyrosine are stable products and can be directly assessed spectrophotometrically, since they have a characteristic absorbance in the 350–450 nm region that is strongly pH-dependent [88]. While at low pH (<5.5) 3-nitrotyrosine has a peak absorbance at 360 nm ($\varepsilon = 2,790$ M^{-1} cm^{-1}), at alkaline pH maximum absorption is at 430 nm ($\varepsilon = 4,400$ M^{-1} cm^{-1}). Since dissociation of the aromatic hydroxyl group in 3-nitrotyrosine is responsible for the shift in absorbance to 430 nm (pKa = 7.5), samples can be quantitated at 430 nm by the difference between absorbance values at pH <5.5 and pH ~10. The limit of sensitivity is ~1 μM and potentially subject to interference from other compounds that may absorb in the same region, therefore it is only useful for relatively concentrated and pure samples.

More sensitive and specific detection of 3-nitro-PHPA or 3-nitrotyrosine typically requires high performance liquid chromatography (HPLC) or gas chromatography (GC) separation and different alternative methods for

Table 2. Selected Methods for Detection of 3-Nitrotyrosine

Method	Analyte	Sensitivity (pmol)	Comments	Reference
HPLC/UV-VIS	3-Nitrotyrosine 3-3′-Dityrosine Tyrosine	10	Widespread availability and simple. No derivatization required. Low sensitivity.	[88]
HPLC/EC	N-acetyl-3-aminotyrosine	0.02	Improved signal to noise ratio by detection of 3-aminotyrosine. Alternative method for detection of tyrosine needed.	[89]
HPLC/EC array system	3-Nitrotyrosine 3-Aminotyrosine 3,3′-Dityrosine 3-Hydroxytyrosine Tyrosine	10 1.0[a]	Allows detection of multiple analytes in a single run. Increase direct specificity by EC signature. Detection system not readily available.	[91]
HPLC/Fluo	3-Aminotyrosine 3-3′-Dityrosine	–	More sensitive than UV-Vis. Minimally explored and validated.	[95]
GC/MS[b] (NICI)	3-Nitrotyrosine 3-Aminotyrosine 3,3′-Dityrosine 3-Hydroxytyrosine Tyrosine	0.050 0.0004[a,c]	Highly sensitive and specific. Requires preparative steps and specialized equipment.	[97]

[a] Sensitivity values for 3-nitrotyrosine and 3-aminotyrosine are independently indicated.

[b] Chromatographed and measured as n-propyl heptafluorobutyryl-derivatives.

[c] The limit of detection of the n-propyl heptafluorobutyryl-derivative of 3-aminotyrosine is 400 amol.

HPLC = high performance liquid chromatography; EC = electrochemical; GC = gas chromatography; MS = mass spectrometry; NICI = negative ion chemical ionization.

detection, including UV/VIS, electrochemical (EC), fluorescent, and mass spectrometry (MS), depending on required detection limits and equipment availability and expertise (Table 2). This methodology also allows the concomitant detection of the parent aromatic compound (i.e., p-HPA and tyrosine) and other oxidized forms including dimeric and hydroxylated products. Several HPLC methods have been developed to determine these analytes, the vast majority relying in reverse phase HPLC with different mobile phase setups [21,86–93] (Table 2). UV-VIS detection can be accomplished by the use of acidified samples and simultaneous detection at 280 and 360 nm. The A_{280}/A_{360} ratio for 3-nitrotyrosine is 2.3, while tyrosine and other oxidation products such as 3-hydroxytyrosine and 3,3′-dityrosine do not absorb at 360 nm. Standards of 3-nitrotyrosine are commercially available. Electrochemical detection can achieve over 100-fold higher sensitivity than UV-VIS (see Table 2) and various EC methods have been developed. An intrinsic problem of EC-detection of 3-nitrotyrosine relies on the high voltage required for a response ($\sim$800 mV), which results in instability of baseline values and decreased specificity, since other compounds that may coelute with 3-nitrotyrosine and even components of the

mobile phase may also provide an electrochemical signal. Various approaches have been used to overcome this problem, including the use of array detector systems and/or reduction of 3-nitrotyrosine to 3-aminotyrosine, a treatment that substantially reduces the potential for oxidation ($\sim$70–100 mV). 3-Nitrotyrosine is resistant to reduction by common reductants such as ascorbate and dithiothreitol, but can be readily reduced by addition of sodium dithionite under alkaline conditions. 3-Aminotyrosine is colorless and fluoresces at pH = 3.0–3.5 with $\lambda_{exc} = 277$ nm and λ_{em} of 308 and 350 nm, which can serve for fluorescence detection purposes. Since tyrosine also fluoresces at 308 nm, the λ_{em} at 350 nm is the one of choice, when assessing 3-aminotyrosine by fluorescent detection methods. 3-Aminotyrosine typically has a poor retention in reverse-phase HPLC columns, thus either derivatization (e.g., acetylation) [89,94] or use of ion pairing mobile phases [90] have been designed to shift 3-aminotyrosine retention to a convenient elution time. Commercial standards of 3-aminotyrosine are available; however, the reduction of 3-nitrotyrosine to 3-aminotyrosine is in many cases not complete and 3-aminotyrosine can progressively be air-oxidized back to 3-nitrotyrosine.

Recent studies have also used gas chromatography separation methods followed by mass spectrometry [95–100] (Table 2). Of these, isotope dilution GC-MS method in the negative ion chemical ionization (NICI) mode for the n-propyl, heptaflurobutyryl-derivative of 3-nitrotyrosine [97] results in a limit of detection (signal to noise ratio >10) of ~50 fmol and reduction to 3-aminotyrosine of the derivatized amino acid allows detection down to ~400 amol (attomol) (Table 2). This method provides specific structural information on the analyte, minimizing potential confounding effects of compounds that coelute with the target analyte. It represents a highly sensitive technique, which has been successfully applied for the detection of 3-nitrotyrosine from protein hydrolyzates of tissue samples [96] (Table 2). Recently, another highly sensitive method for detection of plasma *free* 3-nitrotyrosine was developed [99]. In this, free 3-nitrotyrosine was separated from protein-bound 3-nitrotyrosine by ultrafiltration, then excluded from nitrate and nitrite by HPLC and recovered by solid phase extraction, derivatized, and analyzed by GC-tandem MS in NICI mode. By this several steps methodology, the authors obtain an average *basal* concentration of plasma *free* 3-nitrotyrosine of 2.8 nM and indicate an overall recovery of 3-nitrotyrosine of 50% and a limit of detection as low as 4 amol.

As seen in the previous paragraphs, several potentially useful methods for nitro-PHPA and 3-nitrotyrosine detection are available. Each laboratory must adapt or develop the appropriate methodology depending on the experimental models under study. Given that the utilized method will be specific for 3-nitrotyrosine, the decision should be made based on expected detection levels, the intrinsic complexity of the method, and the availability of special equipment such as multielectrode array detector systems or GS-MS technology, not readily accessible in all laboratories.

Nitration yields by peroxynitrite in biological systems are typically low due to the intrinsic reaction chemistry and competing reactions (Table 1). Some compounds can enhance and catalyze nitration, including bicarbonate-carbon dioxide [43], Fe-edta, hemeperoxidases, or inactivated SOD [4,49]. However, the modulation of biological nitration by these compounds has been marginally explored and requires further specific investigation.

Specificity of tyrosine nitration as a footprint of peroxynitrite. Nitration is not totally specific of peroxynitrite, since other nitrating species may be released from cells and lead phenolic nitration as well, as recently indicated for the case of activated human neutrophils [62] and monocytes [101], which can promote extracellular nitration via a myeloperoxidase mechanism. Thus, specific pharmacological experiments to define whether

the peroxynitrite or the nitrite-hemeperoxidase pathway lead to phenolic nitration may be necessary to perform in systems in which both mechanisms may coexist, such as in the case of neutrophils, monocytes, or eosinophils (in this latter case due to eosinophil peroxidase, (EPO) [102]. In addition, myeloperoxidase- and eosinophil peroxidase-dependent mechanisms of oxidation lead to the formation of halogenated derivatives of tyrosine [102–104] (e.g., 3-chlorotyrosine, 3-bromotyrosine), which can also be detected by some of the techniques described above.

Nitration yields by $^\bullet NO$ *plus* $O_2^{\bullet-}$ *fluxes.* An important issue is whether fluxes of $^\bullet NO$ and $O_2^{\bullet-}$ would result in comparable nitration yields to that of authentic peroxynitrite [4,52,87,105]. It is becoming apparent that nitration yields by fluxes of $^\bullet NO$ and $O_2^{\bullet-}$ are highly dependent on free radical formation rates and ratios, with lower nitration yields obtained at low $^\bullet NO$ plus $O_2^{\bullet-}$ fluxes [87]. Dimerization of tyrosyl radicals to 3,3'-dityrosine predominate at low fluxes, while the tyrosyl radical combination reaction with $^\bullet NO_2$ to yield 3-nitrotyrosine becomes more relevant at higher fluxes. Thus, the ratio of oxidized over nitrated products (Fig. 2) may vary depending on the rate of formation of peroxynitrite and even the concentration of reporter molecules. The relevance of these findings to nitration (and oxidation) yields in reporter molecules by cell or tissue-derived $^\bullet NO$ and $O_2^{\bullet-}$ remains to be established.

Other methods

Conceptually, any molecule that can be oxidized or nitrated by peroxynitrite leading to a stable and measurable change could potentially be used for peroxynitrite detection. In this regard, other methods relying on oxidation reactions have been used, including aromatic hydroxylation, EPR-spin trapping detection, and oxidation or formation of chromophores.

Aromatic hydroxylation. Aromatic compounds such as benzoate, phenol, phenyalanine, and salicylate become hydroxylated and nitrated by peroxynitrite via $^\bullet OH$ and $^\bullet NO_2$ reactions formed from ONOOH homolysis, and can be used to follow the formation of peroxynitrite [43,54,106,107]. Benzoate hydroxylation has been conveniently used in simple biochemical systems by direct fluorescence spectroscopy ($\lambda_{exc} = 300$ nm, $\lambda_{em} = 410$ nm) for competition assays [43]. Phenylalanine reactions with peroxynitrite leads to the formation of p-, m- and o-tyrosine, specific products of $^\bullet OH$ radical attack, as well as nitrated products, including possibly 4-nitrophenylalanine among other products [106]. Salicylate reacts with peroxynitrite-derived species leading to the formation of various products including 2,3-dihydroxybenzoate, 2,5-dihydroxybenzoate, and 2-hydroxy-5-benzoate [54]. The two hydroxylated derivatives are typical of ˙OH reaction with salicylate and the nitro-derivative is specific for ˙NO-derived oxidants. The combined detection of hydroxylated and nitrated products arising from peroxynitrite reactions with phenylalanine and/or salicylate, could be applied for the in vitro and even in vivo detection of peroxynitrite. In vivo, however, free 3-nitrotyrosine formation may be a more sensitive marker than salicylate because the yield of peroxynitrite reaction with salicylate is much smaller than with tyrosine [54]. This is supported by the fact that increased levels of free 3-nitrotyrosine were not accompanied with elevated ˙OH trapping by exogenous addition of salicylate, in spinal cord and brain stem of a mice model of amyotrophic lateral sclerosis, a neurodegenerative disorder in which peroxynitrite is proposed to play a pathogenic role [108].

EPR-spin trapping of peroxynitrite-derived oxidants. The EPR-spin trapping technique has been proved to be useful to study peroxynitrite decomposition pathways and target molecule reactions and mechanisms [46,47,109–112]. Spin traps also have some potential for detection of peroxynitrite derived-oxidants or secondary radicals formed in biochemical/cellular systems. The formation of ˙OH from peroxynitrite decomposition can be detected using different spin traps such as the cell permeable hydrophilic DMPO (5,5-dimethylpyroline N-oxide), that yield stable spin trap-OH adducts [47,109]. The reaction mechanism of peroxynitrite with DMPO involves the formation of ˙OH after homolysis of peroxynitrous acid, and therefore it is unable to compete efficiently under conditions in which other targets that react with second order kinetics with peroxynitrite are present. Other frequently used spin traps such as the cell permeable (lipophilic) PBN (C-phenyl N-*tert*-butylnitrone) and (hydrophilic) POBN ($\propto$-4-pyridiyl-1-oxide N-*tert*-butylnitrone) provide more stable signals with carbon-centered radicals and can be used as long as a ˙OH scavenger such as ethanol or DMSO is added in excess to the system, in which case the radical character is transferred to yield α-hydroxy-ethyl or methyl radicals that then form spin adducts with the traps [47].

Other compounds have been proposed for peroxynitrite detection by EPR, such as the cell-permeable TEMPONE (1-hydroxy-2,2,6,6-tetramethyl-4-oxo-piperidine) and CP-H (1-hydroxy-3-carboxy-2,2,5,5-tetramethylpirrolidine) or non-cell-permeable PP-H (1-hydroxy-4-phosphono-oxy-2,2,6,6-tetramethyl pyridine) [113–115]. These compounds are not spin traps, because they do not participate in addition (trapping) reactions with radical intermediates, and therefore do not provide structural information on the attacking reactive species; they are, rather, prone to one-electron oxidation reactions by several oxidants including $O_2^{˙-}$ or ˙OH derived from homolysis of ONOOH, to yield stable nitroxides.

These methods could potentially be used for biochemical as well as more complex biological systems but have been only minimally explored. In particular, their use in cell (and even animal) systems remains to be explored and validated. A major problem with several spin traps is that spin adducts may decay in the presence of excess peroxynitrite or ˙NO$_2$ (e.g., DMPO-OH), and then the signal may be formed and "silenced" due to the same oxidants. Low concentrations of thiols such as GSH (e.g., 1 mM) may help to protect the signal from excess oxidant, but small amounts of thiyl radical may be formed. Another problem can arise from the reduction of the spin adducts by cellular reductants such as glutathione and ascorbate [109]. Thus, all these variables must be considered when using and validating spin traps for peroxynitrite studies.

Current EPR-spin trapping methodology does not provide precise quantitation of peroxynitrite formed since the radicals trapped are always a small fraction of those formed and spin adducts can decay, be reduced, or further oxidized. However, this method may serve to establish qualitative differences among different experimental conditions and the use of cell-permeable or nonpermeable probes may provide data in regards to peroxynitrite formation sites, all of which needs to be established.

Oxidation or formation of chromophores. Use of compounds that undergo significant absorbance changes during oxidation reactions could be potentially applied to follow peroxynitrite formation in biochemical systems. Among these, compounds such as carminic acid, gallocyanine, pyrogallol red [116], 2,2′-azinobis(3-ethylbenzothiazoline-6-sulfonate) (ABTS) [117] and ferrocyanide [118], react with the ˙OH radical arising from decomposition of ONOOH. Others such as iodide, Ni^{2+} cyclam [118], and reduced cytochrome c (cytochrome c^{2+}) [119] react directly at variable rates with peroxynitrite. Many of these compounds have been applied in competition kinetic studies and their use as peroxynitrite detectors require further characterization, as it may be hampered by the fact that some will be toxic to cells and others can react and/or undergo redox cycles with intermediates other than peroxynitrite.

To illustrate some of the complexities and potential pitfalls behind the use of oxidizable chromophores for peroxynitrite detection, we will briefly describe the use of cytochrome c^{2+}, which we have successfully applied to follow peroxynitrite formation by biochemical systems [119,120], and also to determine second order rate constants of targets via competition kinetics [121,122].

Cytochrome c^{2+} oxidation. While the reduction of oxidized cytochrome c (cytochrome c^{3+}) is widely employed to follow O$_2^{•-}$ formation by biochemical and cellular systems [123,124], the cogeneration of •NO and O$_2^{•-}$ usually limits O$_2^{•-}$ detection by this technique since the formation of peroxynitrite outcompetes the O$_2^{•-}$-dependent cytochrome c^{3+} reduction. Moreover, cytochrome c^{2+} is readily oxidized by peroxynitrite at pH 7.4 and 37°C (k = 2.5 × 10^4 M^{-1}s^{-1}) and at ~50 μM results in >50% oxidation yields [119]. The method has the following advantages: (i) fast-second order-kinetics of cytochrome c^{2+} with peroxynitrite, (ii) cytochrome c^{2+} can be conveniently prepared in concentrated (mM) stock solutions and added in excess (e.g., 50–100 μM) to the reaction to outcompete most other reactions of peroxynitrite, and (iii) oxidation to cytochrome c^{3+} can be readily followed spectrophotometrically (ε_{550} = 21,000 M^{-1} cm^{-1}), either as a continuous or as end-point assay. The main limitations with the method are: (i) air oxidation to yield O$_2^{•-}$, (ii) cytochrome c^{2+} oxidation by excess hydrogen peroxide (k = ~2 M^{-1} s^{-1}) [125], (iii) reduction of cytochrome c^{3+} by O$_2^{•-}$ (k = 3 × 10^6 M^{-1} s^{-1}) [126], (iv) reactions of excess •NO with either reduced (k = ~10 M^{-1} s^{-1}) or oxidized (k ~10^3 M^{-1} s^{-1}) cytochrome c [12]. Air oxidation is slow and can be discounted by using appropriate controls. The addition of catalase minimizes the potential contribution of H$_2$O$_2$-dependent oxidation; the reduction reaction of cytochrome c^{3+} back to cytochrome c^{2+} becomes relevant only if the oxidized form is accumulated significantly and under conditions of excess O$_2^{•-}$ over •NO. The reactions of •NO with cytochrome c, especially cytochrome c^{2+}, are slow, and its potential participation can be diagnosed by the formation of cytochrome c-nitrosyl complexes, of characteristic absorption spectra. Cautious use of this method has been successfully applied in vitro both for measuring peroxynitrite fluxes formed from chemical/biochemical fluxes of •NO plus O$_2^{•-}$, and for defining rate constants and reaction mechanisms of competing targets. Importantly, due to its fast reaction rate, peroxynitrite-mediated cytochrome c^{2+} oxidation is not easily affected by low molecular weight scavengers such as uric acid that affect many peroxynitrite-mediated oxidative processes [119,120] (Table 1).

Indirect methods

These methods rely on the independent detection of •NO (directly, end products, or by bioassays) and O$_2^{•-}$. Direct detection of •NO (either by electrochemical detection or by EPR techniques) or its bioactivity (i.e., vasodilation, inhibition of platelet aggregation, cGMP accumulation) can be performed and evaluate whether significant interactions with O$_2^{•-}$ exist. Indeed, if this were the case SOD or SOD mimics should enhance the concentration of •NO and its bioactivity [127,128]. A commonly used method for the detection of •NO in biology is the quantitation of NO$_2^-$ plus NO$_3^-$ (NO$_x^-$). In biochemical and cellular systems, most •NO preferentially decays to NO$_2^-$, with NO$_3^-$ being typically less abundant. The interactions of •NO with O$_2^{•-}$ increase the formation of NO$_3^-$, due to the proton-catalyzed peroxynitrous acid (Fig. 1). Thus, at the same output of •NO and then total amount of NO$_x^-$, the formation of O$_2^{•-}$ leading to peroxynitrite increases the nitrate/nitrite ratio [86]. On the other hand, O$_2^{•-}$ detection by methods such as the SOD-inhibitable reduction of cytochrome c^{2+} under conditions in which peroxynitrite is being formed may be marginal, but will be greatly enhanced by the addition of NOS inhibitors [86].

Reaction yields and quantitation

In order to quantitate peroxynitrite, it is critical to know the reaction yield of the probe used (i.e., DHR, DCF, cytochrome c^{2+}, p-HPA) with authentic peroxynitrite (Table 1). However, it is important to recognize that bolus addition of peroxynitrite to probes does not mimic the low flux formation of peroxynitrite that may occur under biologically relevant conditions. In addition, since most reactions involving probe oxidation by peroxynitrite occur through free radical intermediates, and the fate of free radicals (including probe radicals) is typically dependent on steady-state concentrations, oxidation yields obtained with acute addition of peroxynitrite can be, at most, an approximation to what is expected to occur in biology. In particular, phenolic nitration by peroxynitrite infusion over an extended period of time or independent fluxes of •NO and O$_2^{•-}$ results in lower nitration yields than those observed during bolus addition of peroxynitrite; however, peroxynitrite fluxes may be slightly more efficient in other assays such as DHR (unpublished observations).

Peroxynitrite quantitation is affected by various other factors including (i) competing reactions that peroxynitrite may have with cell-tissue components, (ii) •NO and O$_2^{•-}$ reactions with probe radical intermediates, (iii) redox reactions of probe radical intermediates with molecular oxygen or reductants, among other reactions. Notwithstanding, the independent detection of •NO and O$_2^{•-}$ provides critical information regarding the existence of potential confounding factors during peroxynitrite detection and help to obtain accurate values in both biochemical and biological (cell-tissue-organ level) systems.

Potential pitfalls and artifacts

The detection of peroxynitrite by oxidizable probes is highly affected by conditions such as pH, temperature, metals, buffer composition, presence of carbon dioxide-bicarbonate, and biomolecules such as thiols.

The reaction with carbon dioxide. The reaction of peroxynitrite with carbon dioxide is critical to consider because carbon dioxide is ubiquitous in biological systems and also because sometimes carbon dioxide levels are poorly controlled in reaction systems. Indeed, variable levels of carbon dioxide are present on air-equilibrated solutions and buffer systems, depending on composition, ionic strength, and temperature. In air-equilibrated 50 mM phosphate buffer, pH 7.4 a typical value of contaminating carbon dioxide is in the order of 5–10 μM, unless special measures are taken for preparation of the solutions [129]. Moreover, stock solutions of alkaline peroxynitrite may provide significant amounts of carbon dioxide, arising from equilibration of trapped carbonate with neutral pH buffers. Experiments can be specifically designed in the presence of known amounts of carbon dioxide-bicarbonate. In addition, cells and tissues actively produce carbon dioxide and cell cultures are typically equilibrated with 5% carbon dioxide that translates to a dissolved concentration of ~1.0–1.5 mM carbon dioxide in equilibrium with bicarbonate at pH 7.4. Carbon dioxide has the following net consequences on peroxynitrite biochemistry: (i) decreases peroxynitrite half-life and therefore its distance of diffusion, (ii) decreases one-electron oxidations and hydroxylation of probes, (iii) enhances nitration of phenolic probes. Therefore, in biochemical, cell, tissue, and organ levels, the presence of carbon plays a critical role in modulating peroxynitrite biochemistry and affecting reaction yields of the different probes.

Buffer systems, media components, and metals. Phosphate buffer does not interfere in peroxynitrite reactions, but other commonly used buffer systems including Tris and Hepes react with peroxynitrite inhibiting oxidative processes [130] and may secondarily lead to the formation of radical intermediates and $^\bullet$NO-donors [131,132]. Similarly, the presence of thiols, glucose, proteins, and/or plasma will decrease probe oxidation yields.

Transition metals also react with peroxynitrite and may affect probe oxidation yields. In biochemical systems, the use of metal chelators such as dtpa (diethylenetriamine-N-N-N′-N″-pentaacetate) allows a more precise detection of peroxynitrite, minimizing poorly controlled metal-catalyzed oxidation reactions. However, dtpa reacts with $^\bullet$OH and $CO_3^{\bullet -}$, and therefore large concentrations (>100 μM) should not be used. Importantly, desferrioxamine is not recommended because it has been shown to inhibit peroxynitrite-mediated oxida-

tive processes, independently of its capacity to bind transition metals [3,133]. In a separate approach, addition of known amounts of transition metal-complexes may serve to significantly increase nitration yields.

Footprinting

The reactions of peroxynitrite with cell and tissue components leave oxidative modifications that may serve as "footprints" of peroxynitrite formations and reactions. Of these, the most frequently used is the nitration of protein tyrosine residues. Protein nitration is a rather stable chemical modification and metabolic systems able to repair or reduce nitration are either nonexistent or operate at low rates [134–136]. Indeed, at present there is no solid evidence for the presence of specific metabolic systems that remove or reduce 3-nitrotyrosine. However, proteolysis of oxidized proteins may contribute to 3-nitrotyrosine turnover [137], and can affect 3-nitrotyrosine stability in stored samples over extended periods of time or in samples that undergo repeated freeze-thawing cycles. Detection and quantitation of 3-nitrotyrosine in biological samples (cells and tissues) have been detected by one of two main strategies: immunochemical detection of nitrated proteins, and quantitation of 3-nitrotyrosine after protein hydrolysis. Although detection of protein 3-nitrotyrosine can increase several-fold in diverse pathophysiological situations, nitrated proteins exist under physiological conditions, supporting the concept of a low flux of oxidants being formed and causing molecular damage, even under basal/normal conditions.

Immunochemistry for 3-nitrotyrosine

Immunochemical-based methods rely on the use of anti-nitrotyrosine antibodies [138]. Polyclonal and monoclonal antibodies have been raised and purified by different laboratories and used as immunological probes in different experimental methods [138–140]. Immunochemical methods are useful tools for studying biological samples and the only to be applied to whole tissues. They have the advantage of being simple and they require only routine laboratory equipment, and therefore can be performed in most research and clinical laboratories.

The preparation and characterization of antinitrotyrosine antibodies have been described recently elsewhere [138,139]. As there are no controlled studies comparing both specificity and sensitivity of the available antibodies, precise characterization must be performed before utilization. The specificity of these antibodies for nitrated proteins in the samples under study have to be confirmed by two types of experiments. First, antibody binding must be displaced by either free 3-nitrotyrosine or, more efficiently, by smaller amounts of 3-nitrotyrosine-containing peptides such as Gly-NO_2Tyr-Ala or Gly-Gly-NO_2-Tyr-Ala. Secondly, samples must be reduced by sodium dithionite to 3-aminotyrosine, therefore preventing antibody binding. While monoclonal antibodies are more specific than the polyclonal, they are less sensitive for detection of protein-bound 3-nitrotyrosine.

Enzyme-linked immunosorbent assays (ELISA)

ELISA is a very versatile method to detect and measure nitrated proteins. In a simple ELISA, a standard nitrated protein (e.g., nitrated bovine serum albumin) is coated on the plate and the antinitrotyrosine antibody is used as the first antibody. When ELISA assays are performed by using secondary antibodies conjugated to horseradish peroxidase (HRP) and o-Phenylenediamine dihydrochloride as chromogen, levels of 3-nitrotyrosine down to 200 fmol can be detected using 0.5 μg/ml of antinitrotyrosine polyclonal antibody.

In order to detect 3-nitrotyrosine present in biological fluids, and as nitrated proteins can be poorly represented in these samples, two different approaches can be performed: (i) competitive ELISA, where the antibody binding to the standard nitrated protein coated on the ELISA plate is inhibited by nitrated proteins contained in the sample [141], or (ii) a capture ELISA, in which the nitrated protein is first captured by the anti-nitrotyrosine antibody coated on the plate and then alternatively detected by secondary antibodies directed against (i) the native protein, (ii) the nitrotyrosine epitope with either antinitrotyrosine antibodies raised in a different animal species than that used for coating or the same antibody used for coating conjugated to HRP on an alternative detection system [140]. In the competitive ELISA, detection and quantitation is determined as a function of the percentage of antibody binding. In the capture ELISA, measurement requires comparison with standard nitrated proteins.

However, useful for detection purposes, results obtained with ELISA assays are not fully extrapolable in quantitative terms, and different factors may be taken into account for their analysis. For example, the antibody avidity and specificity may change related to the place of the 3-nitrotyrosine epitope in different proteins, the antibody may not detect some internal 3-nitrotyrosine moieties, and peroxynitrite may not nitrate the same tyrosine residues in vitro and in vivo.

Western-blot and dot-blot assays

Western blot is a descriptive and semiquantitative method that can be used to detect immunoreactivity of proteins carrying 3-nitrotyrosine residues with antinitrotyrosine antibodies including isolated proteins [57], and protein extracts from cells or tissue samples [138,139,142]. Immunoreactive proteins can be further characterized by their molecular weight, specific recognition by a second western blot developed with antibodies directed against specific proteins or, preferably, after immunoprecipitation [143]. Additionally, western-blot assays can be used for competition and specificity analysis of antinitrotyrosine antibodies [138,139] and for semiquantitative purposes in conjunction with digital imaging analysis.

As a more simple and faster alternative to western blots to assess total nitrated proteins, irrespective of specific nitration profiles, and also for antibody characterization, dot-blot analysis can be performed. Detection limits must be defined in each particular system; for instance, concentrations as low as 10 ng/ml of nitrated keyhole limpet hemocyanin representing 430 pmol/ml 3-nitrotyrosine, can be detected by dot blot using 0.1 μg/ml of polyclonal antinitrotyrosine antibody associated to a chemiluminescent detection system.

Recently, it has been shown that during sample boiling under reducing conditions (β-mercaptoethanol) an important fraction of 3-nitrotyrosine can be reduced to 3-aminotyrosine, if hemeproteins are present [136]. This results in a significant decrease and even disappearance of the 3-nitrotyrosine signal and may explain why western-blot methodology has been less used and successful than immunohistochemistry for detection of nitrated proteins in complex biological samples. In future studies using western blot analysis of nitrated proteins it will be critical to assess the influence of sample reduction on 3-nitrotyrosine yields to avoid artifactually low signals.

Immunohistochemistry and immunocytochemistry

Peroxynitrite-mediated nitration has been shown to occur in different human and animal conditions by immunohistochemistry, including human atherosclerosis [17], acute lung injury [14,15], intestinal inflammation [144], and physiological immune responses [139,145], among several others (for a review see [49]). Indeed, immunohistochemistry has been widely used to detect 3-nitrotyrosine from tissues and the only one that permits localization of 3-nitrotyrosine residues in specific areas, with the maintenance of the histoarchitecture. Immunohistochemistry has the advantage of permitting the recognition of cells as well as subcellular structures or compartments implicated in peroxynitrite and ·NO-derived oxidants-induced nitration (Fig. 6). As peroxynitrite can diffuse in tissues up to ~20 μm from the formation site, a diffuse pattern of anti-nitrotyrosine

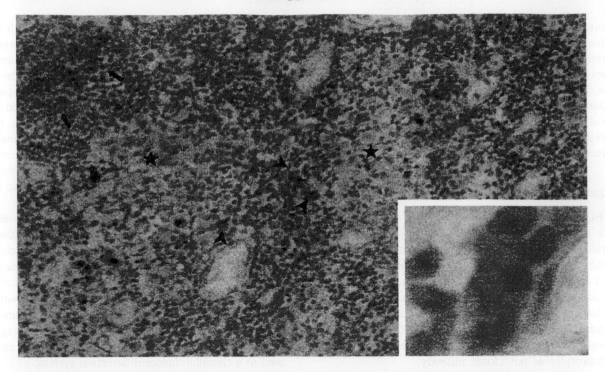

Fig. 6. Immunocytochemistry of protein-bound 3-nitrotyrosine. Nitrotyrosine staining of a human lymph node undergoing nonspecific immune activation is shown. Human lymph nodes were obtained from surgical resections for colonic cancer. Lymph nodes were fixed in 10% formalin, paraffin embedded and 5 μm sections were mounted in silanized microscope slides. Tissue sections were probed with antinitrotyrosine polyclonal antibodies (working dilution 50 μg/ml) and developed with a secondary antibody coupled to biotin using a streptavidin-peroxidase kit (Sigma) and diaminobenzidine as chromogen. Histological sections were counterstained with hemalum. Lymph nodes were free of metastasic invasion and showed mild to strong reactive follicular hyperplasia and sinusoidal histiocytosis. The sinuses of the medullar zone with histiocytosis show a diffuse staining (asterisk). A weak reactivity is seen at the interstitial level, between lymphocytes (arrow). The strongest 3-nitrotyrosine immunoreactivity is observed in macrophages (arrowhead), (100 ×). Insert: a macrophage with strong intracytoplasmic staining, (1000 ×). See also [139].

binding is frequently observed; that should not be considered as unspecific binding.

It is recommended to evaluate successive slices to perform the appropriate controls. Again, the technique is relatively simple to perform and can be carried out with paraffin-embedded or frozen tissues, as has been recently reviewed elsewhere [146].

3-Nitrotyrosine in cultured cells can be detected by immunofluorescence techniques [146]. This technique has been successfully applied, for instance, to the study of cell nitration during degeneration of motoneurons in culture by either growth-factor deprivation [26] or the action of intracellular Zn-deficient SODs [27].

Flow cytometry

Flow cytometry is an analytical method widely used in cell biology that has not been significantly applied for the detection of nitrated proteins. Initial work in our laboratory was performed in order to assess its validity to detect nitrated proteins in peripheral blood mononuclear cells (PBMC). We associated our polyclonal antinitrotyrosine antibodies [139] to

phycoerythrine (PE) or fluorescein isothyocyanate (FITC)-conjugated secondary antibody as fluorescent probes and evaluated the fluorescence intensity of intact (membrane-associated) versus permeabilized PBMC (membrane- and intracellular-associated) after peroxynitrite exposure (Fig. 7).

Flow Cytometry can be also performed by two- or three-color analysis using different fluorochromes. Figure 8 shows a two-color cytometry analysis that was performed with anexin V conjugated to FITC to identify apoptotic cells and with anti-nitrotyrosine antibodies associated to a PE-secondary antibody to detect cells bearing surface-associated nitrated proteins. These results indicate that levels of protein nitration can be successfully detected in cells by flow cytometry using PE- or FITC-conjugated secondary antibodies as fluorescent probes.

Protein hydrolysis and 3-nitrotyrosine quantitation

Tyrosine is a relatively abundant amino acid in proteins, with most mammalian proteins containing an average of ~3.5 mol% tyrosine residues, having a

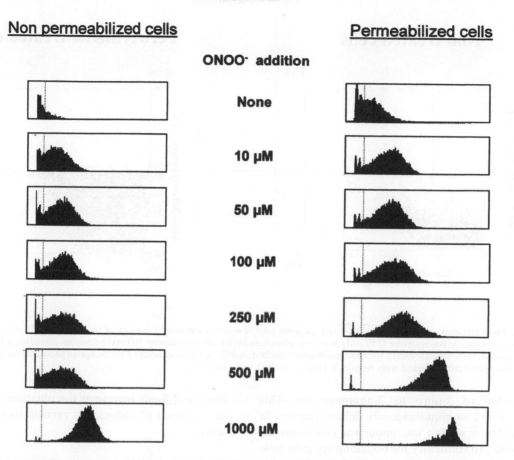

Fig. 7. Nitrotyrosine detection by flow cytometry. Nitrated proteins analyzed in normal peripheral blood mononuclear cells (PBMC) by monoparametric flow cytometry analysis. After peroxynitrite treatment to 5×10^6 cells/ml in isotonic phosphate buffer saline containing 100 mM potassium phosphate, pH 7.4, cells were either directly incubated with anti-nitrotyrosine antibodies followed by a secondary FITC-conjugated secondary antibody (nonpermeabilized cells) or fixed (permeabilized cells) prior to antibody incubations with 2% paraformaldehyde in PBS and permeabilized with Triton X-100 at 0.5% in PBS.

range of 1–8 mol%. Protein 3-nitrotyrosine can be quantitated after total protein hydrolysis of tissue or cell homogenate and biological fluids. Usual methods used for hydrolysis have been either vapor phase acid hydrolysis or protease treatment (e.g., pronase) [92, 94], with both methods resulting in high 3-nitrotyrosine recovery yields. During acid hydrolysis substantial care must be taken to minimize artifactual nitration (6 M HCl, 120°C) due to contaminating NO_2^- and NO_3^- present in the sample. Extensive wash and/or dialysis is critical to eliminate NO_x^- before protein. Also, addition of 1% phenol to the samples prior to hydrolysis helps to diagnose whether NO_2^- and NO_3^- contribute to artifactual nitration and also trap significantly (but not totally) the reactive nitrogen intermediates that may arise during sample workup and analysis. The hydrolyzed samples are dried under vacuum and neutralized before initiating the detection of 3-nitrotyrosine. Pronase treatment reduces the potential interference of undesirable secondary nitrations, but proteolysis of the sample may not be complete and

3-nitrotyrosine may not be recovered at pronase-resistant sites while, on the other hand, autoproteolysis may contribute to enhance tyrosine levels of the samples. Finally, a third alternative has been just published [100] involving alkaline hydrolysis, which may help to circumvent some of the problems observed with the other two methods.

Detection and quantitation of 3-nitrotyrosine. The resulting free 3-nitrotyrosine is separated from the rest of the amino acids using reverse phase-based HPLC methodology [92] and quantitated by one of the various methods shown in Table 2. Initial studies detected 3-nitrotyrosine by amino acid analysis after derivatization of all amino acids with phenyl isothiocyanate (PITC) [147], using appropriate mobile phase gradient and the PITC-derivatized amino acids in hydrolyzates detected by optical absorption at 254 nm. 3-Nitrotyrosine elutes between leucine and phenylalanine and can be detected at levels in the range of ~4 pmol. Controls can be performed by prereducing the sample with dithionite, which results in

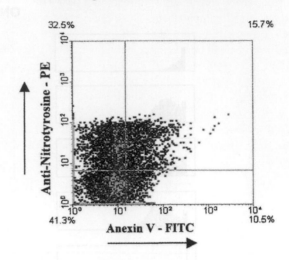

Non treated cells

Peroxynitrite treated cells

Fig. 8. Flow cytometry detection of nitrated and apoptotic cells. Two-color cytometric analysis of normal PBMC (5×10^6 cells/ml) exposed to authentic peroxynitrite (250 μM) in isotonic phosphate buffer saline containing 100 mM potassium phosphate, pH 7.4. After peroxynitrite, cells were incubated with anti-nitrotyrosine antibodies followed by a secondary PE-conjugated secondary antibody. After this, cells were further stained with Anexin V-FITC to identify apoptotic cells.

the conversion of 3-nitro to 3-aminotyrosine. This method is useful to simultaneously follow changes in different amino acids but has important limitations: (i) low sensitivity, (ii) difficulty for obtaining accurate 3-nitrotyrosine values, since 3-nitrotyrosine is usually represented as a small shoulder in the chromatogram between peaks of significantly more abundant vicinal eluting amino acids, and (iii) limited specificity in complex biological samples.

The current trend is to assess 3-nitrotyrosine in conjunction with tyrosine using more sensitive and specific methods, such as those described before in this chapter. Quantitation of 3-nitrotyrosine can be expressed on a molar basis respect to total tyrosine or per mg protein, thus normalizing for the variable protein content in different samples. However, it is recommended to express the 3-nitrotyrosine value as a function of total protein-bound tyrosine, as this ratio takes into account not only protein content, but also efficiency of protein hydrolysis [92]. For instance, 3-nitrotyrosine in protein hydrolyzates of RAW 264.7 macrophages under control (no stimulation) condition using EC-detection techniques [89] yielded values of ~one 3-nitrotyrosine (as N-acetyl-3-aminotyrosine) per 10^6 tyrosines (as N-acetyltyrosine) and increased ~8-fold after treatment with zymosan and IFN-γ.

Free 3-nitrotyrosine has been detected and used as a marker of $^\bullet$NO-derived oxidant formation in biological fluids (i.e., plasma, synovial fluid, cerebrospinal fluid, bronchoalveolar lavage fluid) under basal [99,100,148] and a variety of pathological conditions [18,19,149,150]. It is not defined at present whether free 3-nitrotyrosine in

biological fluids represents the nitration of free tyrosine or is a result of endogenous proteolysis of nitrated proteins.

Detection of nitrated peptides. Tyrosine nitration of peptides and proteins can be also studied by LC-MS techniques [151–153]. Nitration will increase the molecular weight of the control peptide by 45 Da; if more than one tyrosine residue is susceptible to nitration di-, tri-, or poly-nitrations can be observed. Importantly, when changes other than nitration occur, increases in molecular weights may not be a straightforward method for affirming nitration. As an additional or alternative approach, peptides can be followed by UV/VIS absorbance taking advantage of the characteristic absorption properties of 3-nitrotyrosine.

Partial proteolysis of nitrated proteins with trypsin or other proteases has been successfully used for peptide mapping of 3-nitrotyrosine residues in proteins [153].

Compartmentalization and determinants in protein nitration

Peroxynitrite formed extracellularly could react with either extra- or intracellular proteins, due to its capacity to diffuse and permeate membranes. Thus, potentially, a target cell protein can be nitrated from peroxynitrite arising from another source cell or compartment. For instance, inflammatory cells forming peroxynitrite may cause nitration in membrane or intracellular compartments of target cells. On the other hand, intracellularly formed peroxynitrite tends to react within the same or an adjacent cell compartment. Thus, the pattern of tissue or cell protein nitration observed in immunohistological and western blots studies can only suggest the site(s) where peroxynitrite had been formed. For instance, activated macrophages release most of peroxynitrite to the extracelullar milieu, but become nitrated themselves intracelullarly [86,89, see also Fig. 9 inset]. In a tissue, nitration is preferentially localized around the areas in which peroxynitrite-producing cells are more abundant, which is consistent with the concept that peroxynitrite can diffuse one to two cell diameters; indeed, strong 3-nitrotyrosine immunoreactivity in inflammatory processes is principally observed in the macrophage or neutrophil-rich areas [15,139,154].

Peroxynitrite reactions in a compartment results in nitration of a limited number of proteins. Selectivity of protein nitration depends on a series of factors [153], some of which are still unknown or partially defined, and include protein concentration, abundance and localization of tyrosine residues within a protein, presence of transition metal centers close to nitration sites, and influence of neighboring amino acids and secondary structure.

Other modifications

Various other oxidative modifications may be used for peroxynitrite footprinting in biological systems. In proteins, peroxynitrite may lead to the formation of low levels of o-tyrosine and 3,3'-dityrosine [96]. Formation of protein carbonyls in cells have been proposed to arise from peroxynitrite reactions with proteins [155,156], but peroxynitrite reactions with pure proteins do not significantly yield carbonyls [157], thus cell formation of protein carbonyls may depend on secondary oxidative processes. Other modifications that may contribute to unravel peroxynitrite which have been minimally explored are: (i) lipid oxidation and formation of nitrosylated and nitrated lipid derivatives [11,158]. These compounds are unstable, and methods for improvement of extraction techniques and stabilization are being developed; they are of particular interest in LDL and atherosclerotic plaques, where peroxynitrite may play a critical prooxidant role. (ii) DNA strand breaks and base modification. Peroxynitrite is able to cause DNA strand breaks and oxidation and nitration of DNA bases such as guanine to oxo- and 8-nitroguanine [159–161]. The formation of these modified bases by the actions of peroxynitrite in vivo awaits future research. (iii) modifications of ubiquitous low molecular weight compounds, may also serve as biomarkers of peroxynitrite formation and reactions. Some of these include recently characterized mod-

ifications, including the formation on nitrated-derivatives of glucose [162], the potential formation of nitrated thiols (such as nitroglutathione) [163] and decomposition products of uric acid reaction with peroxynitrite [164, 165]. It remains to be established whether these modifications are quantitatively sufficient and measurable to be used as potential footprints of peroxynitrite.

It is important to recognize that all of the modifications described in this section are not a consequence of direct bimolecular reactions of peroxynitrite with the target molecules, but rather of free radical intermediates arising from secondary reactions, including $CO_3^{\cdot-}$ radicals, and $^{\cdot}NO_2$. Moreover, in complex biological systems, as yet undescribed modifications may arise, products of free radical processes initiated by peroxynitrite.

Importantly, in spite of all the potential oxidative modifications indicated, there is no available single chemical modification 100% specific to peroxynitrite. Thus, unraveling peroxynitrite with the highest degree of certainty may require the detection of a panel of characteristic footprints, such as, for example, the simultaneous detection of 3-nitrotyrosine and 3,3'-dityrosine in biological samples. At the same time, the lack of detection of other biomarkers of oxidative damage such as o-tyrosine and chlorotyrosine, which most likely reflect reactions of $^{\cdot}OH$ and MPO-dependent reactions, respectively, will provide even stronger evidence of peroxynitrite-mediated oxidations.

PHARMACOLOGY TO UNRAVEL PEROXYNITRITE

Due to the lack of techniques and chemical modifications completely specific for peroxynitrite, additional experimental evidence is obtained by the use of: (i) drugs that decrease $^{\cdot}NO$ and $O_2^{\cdot-}$ levels, (ii) treatments that modify the concentration of biomolecules that critically influence peroxynitrite reactivity, (iii) peroxynitrite decomposition catalysts and scavengers, and (iv) compounds that interfere with alternative oxidation/nitration pathways. Finally, in some models it may be of interest to promote or increase peroxynitrite formation/detection by pharmacological modulation.

Drugs that decrease $^{\cdot}NO$ and $O_2^{\cdot-}$ levels—inhibition of $^{\cdot}NO$ and $O_2^{\cdot-}$ formation

Nitric oxide. The inhibition of $^{\cdot}NO$ formation is obtained with the use of a wide variety of NOS inhibitors of various degrees of selectivity and potency over the three different isoforms of NOS (nNOS, iNOS, and eNOS) [166,167]. Also, NOS-deficient cells and knockout models of one or more than one isoforms of NOS [21,168, 169] are potential tools to define the contribution of $^{\cdot}NO$ and $^{\cdot}NO$-derived oxidants.

Superoxide. Pharmacological inhibition of $O_2^{\cdot-}$ formation is more complex than that of $^{\cdot}NO$, because the sources of $O_2^{\cdot-}$ are variable depending on cell types and metabolic status, and also because typically more than one cellular source contributes to overall cell $O_2^{\cdot-}$ formation. Thus, inhibition of $O_2^{\cdot-}$ formation requires a previous definition of the main cell or tissue sources of $O_2^{\cdot-}$ under the specific conditions of the study. In the more straightforward case of $O_2^{\cdot-}$ formation from activated macrophages, neutrophils, and other cells of the immune system via NADPH oxidase activation, inhibition of this enzyme may prove effective to inhibit peroxynitrite formation. However, there is a relatively short group of NADPH oxidase inhibitors, most of which are relative unspecific or their mechanism of action largely unknown. For instance, the commonly used inhibitor of NADPH oxidase, diphenyliodonium [170], is a flavoprotein inhibitor that can also influence NOS activity, depending on the concentrations used [171]. In systems in which $O_2^{\cdot-}$ is principally formed due to the catalytic action of xanthine oxidase, enzyme inhibitors such as allopurinol or oxypurinol may be useful [172,173]. NADPH oxidase deficient cells and knockout models are also available, and may be of great help to define the contribution of $O_2^{\cdot-}$ (and therefore peroxynitrite) to $^{\cdot}NO$-mediated oxidations. Interestingly, double knockout models for NADPH oxidase and iNOS have been developed recently [168,169,174].

Enhanced consumption of $^{\cdot}NO$ and $O_2^{\cdot-}$

Nitric oxide. The consumption of $^{\cdot}NO$ may be promoted by the use of $^{\cdot}NO$ scavengers, such as derivatives of phenyl-4,4,5,5,-tetramethylimidazoline-1-oxyl 3-oxide (PTIO) [25]. PTIO reacts stoichometrically with $^{\cdot}NO$ to yield $^{\cdot}NO_2$ and 2-phenyl-4,4,5,5,-tetramethylimidazoline-1-oxyl (PTI), as seen in Eqn. 2:

$$PTIO + {^{\cdot}NO} \rightarrow {^{\cdot}NO_2} + PTI \qquad (2)$$

This reaction rapidly consumes $^{\cdot}NO$ and may partially compete with $O_2^{\cdot-}$.

Superoxide. Enhanced elimination of $O_2^{\cdot-}$ is obtained by overexpression of SOD (cytosolic or mitochondrial) [175–177] or supplementation with cell-permeable SOD mimics (e.g., manganese-complexes) [178,179], or with liposomal entrapped SOD [180]; these manipulations result in fast consumption of $O_2^{\cdot-}$ by pathways that do not yield peroxynitrite.

It is important to appreciate that since the combination reaction between $^{\cdot}NO$ and $O_2^{\cdot-}$ is diffusion-controlled, only those traps that react fast enough with either $^{\cdot}NO$ or $O_2^{\cdot-}$ and/or can approach relatively high concentrations in the sites where peroxynitrite is being produced, will be able to prevent peroxynitrite formation.

Endogenous components that modulate peroxynitrite reactivity

Modulation of cellular thiol levels. Glutathione and thiols are critical endogenous intracellular antioxidants against peroxynitrite and derived species, and therefore changes in thiol and glutathione levels will have a direct influence on peroxynitrite detection and biological outcome. Cellular glutathione depletion by treatment with buthionine sulfoximine has been shown to increase the detection (e.g., intracellular nitration) and biological effects of peroxynitrite [142]. Enhancement of intracellular thiol levels by supplementation with thiol or thiolesters may also decrease the detection and effects of peroxynitrite.

Modulation of carbon dioxide levels. The presence of carbon dioxide significantly affects peroxynitrite reactivity and diffusion. Thus, variation of carbon dioxide levels in biochemical systems, cells, and tissues could have a significant effect of detection reactions (e.g., carbon dioxide inhibits oxidations but promotes nitrations).

Use of peroxynitrite decomposition catalysts and scavengers

Peroxynitrite or its secondary products can be readily decomposed by the use of decomposition catalysts and scavengers, aimed to inhibit biological oxidations and/or nitrations and therefore protect from peroxynitrite-induced oxidative damage. However, to be effective in complex biological systems, both peroxynitrite decomposition catalysts and scavengers must be able to outcompete peroxynitrite target molecule reactions. Therefore, it is important to appreciate how much and how efficiently these molecules can be delivered to the sites where peroxynitrite is being formed and/or react when interpreting protection and detection data.

Among the first group of compounds are included those that catalyze peroxynitrite isomerization (e.g., iron porphyrins) [181] and those that catalytically reduce peroxynitrite by redox cycles that consume endogenous reductants such as thiols, ascorbate, or urate (e.g., manganese-porphyrins, selenocompounds) [182–187] (Fig. 9). Various organo-transition metal complexes have been lately used as peroxynitrite decomposition catalysts in biochemical, cell, and animal models and shown to protect against the effects of exogenous and endogenous peroxynitrite [188,189] and inhibit protein nitration [26,

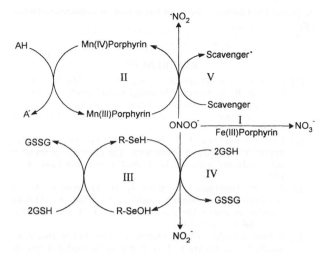

Fig. 9. Peroxynitrite decomposition catalysts and scavengers. Peroxynitrite can be catalytically isomerized to nitrate with iron-porphyrins (I), or be reduced by one or two electrons by the action of manganese-porphyrins (II), or seleno-compounds (III), respectively, at the expense of reducing equivalents of endogenous compounds. Scavengers such as glutathione react with peroxynitrite in a two-electron oxidation process to yield nitrite (IV); other scavengers can be oxidized by one-electron to yield a scavenger-derived free radical and nitrogen dioxide (V).

139]. Some compounds such as the manganese porphyrins (Mn-tmpyp) can potentially have a dual action by acting both as SOD mimics as well as peroxynitrite decomposition catalysts. Still, although promising, current compounds could have secondary effects [190] and short- and long-term toxicity in cell and animal models requires further investigation.

There are some caveats as regards the use of peroxynitrite decomposition catalysts as a way to confirm the participation of peroxynitrite by using probe oxidation or footprinting, because these compounds can undergo secondary nondesirable redox reactions under particular conditions. For instance, during the catalytic decomposition of peroxynitrite by manganese-porphyrins, cellular reductants such as glutathione are consumed. Therefore, the inhibitory effects on peroxynitrite detection will be highly dependent on the glutathione as well as other reductants concentrations in the cell. Moreover, the compounds may not be inhibitory of peroxynitrite-mediated processes or may even enhance oxidative modifications under conditions of cellular glutathione depletion, since the oxo-manganese complexes formed during the catalytic cycle may themselves promote oxidation and/or nitration of cellular targets [191].

The scavengers include a variety of endogenous or synthetic molecules that decompose peroxynitrite at the expense of their sacrificial oxidation. Endogenous scavengers that have been used (in addition to glutathione and cysteine) are methionine [101,192] and uric acid [25,193]. These compounds do not react fast with peroxynitrite but they can be added in relatively high con-

centrations and can react with secondary intermediates of peroxynitrite reactions. Synthetic scavengers include hydroxamates and substituted guanidines. Desferrioxamine and other hydroxamates may be able to inhibit peroxynitrite-mediated oxidations and nitrations in vitro [133] and in vivo [194], leading to the formation of the corresponding nitroxide. Guanidines are also NOS inhibitors and may also provide a dual mechanism of protection against peroxynitrite-mediated modifications. In particular, mercaptoethylguanidine has been shown to be particularly effective in the inhibition of peroxynitrite-mediated oxidative processes in vitro and in vivo [121,195].

Compounds that interfere with alternative oxidation/nitration pathways

Compounds can be used to define the contribution of H_2O_2, NO_2^-, and hemeperoxidases to probe oxidation and footprinting. Hydrogen peroxide rapidly equilibrates between extra- and intracellular compartments and can be removed by the use of exogenous catalase. Myeloperoxidase can be inhibited by azide and aminobenzoic acid hydrazide [62,101]. Importantly, while the hemeperoxidase-dependent mechanism of nitration will be inhibited by catalase, SOD may *promote* nitration due to a more efficient conversion of $O_2^{\bullet-}$ to H_2O_2. These pharmacological tests and others could be useful for discriminating between contributing pathways to $^{\bullet}NO$-mediated oxidative processes [101].

Facilitation of peroxynitrite formation

Peroxynitrite formation in most cases requires both $^{\bullet}NO$ and $O_2^{\bullet-}$ and therefore, maximal production is limited by the radical precursor produced at lower rates. Thus, peroxynitrite formation may be enhanced or promoted by conditions that enhance the generation of the limiting radical. In this regard, experimental approaches exist to enhance exogenous or endogenous production of either $^{\bullet}NO$ or $O_2^{\bullet-}$. Increased $^{\bullet}NO$ levels can be obtained by extracellular addition of $^{\bullet}NO$-donors such as NONOates, which can provide controlled and constant fluxes of $^{\bullet}NO$ over various periods of time [196], or by overexpression and/or activation of endogenous NOS. On the other hand, enhanced extracellular $O_2^{\bullet-}$ generation may be accomplished by the xanthine oxidase system, extra- and intracellular generation by quinone-redox cycling, and intramitochondrial generation by the use of respiratory inhibitors and uncouplers [142]. Enhanced steady-state concentrations of $O_2^{\bullet-}$ can also be accomplished by cellular Cu-Zn or Mn-SOD-depletion using oligonucleotide antisense technology [197–199].

These approaches have been minimally explored and

applied, but they can result in an increased formation/
detection of peroxynitrite, opening a new experimental
strategy to define biological processes mediated by peroxynitrite.

CONCLUSIONS

• Nitric oxide-superoxide interactions readily occur
in vivo leading to the formation of peroxynitrite.

• Peroxynitrite is short-lived, therefore its detection
relies on modification of exogenous detector (probe
oxidation) or endogenous target (footprinting) molecules.

• Peroxynitrite-induced modifications, most notably
oxidations and nitrations, can be followed in biological systems of different levels of complexity including biochemical, cellular, tissue, and organ levels.

• Various methods are available of different specific-
ities, sensitivities, and complexity, the choice of
which should be decided based on the experimental
design under study.

• At present no single chemical modification completely specific to peroxynitrite has been discovered,
and therefore its unambiguous detection requires a
multifaceted approach.

• Additional criteria include the combined detection
of a panel of oxidative modifications consistent with
peroxynitrite reactions and appropriate intervention
on the peroxynitrite pathway.

• Peroxynitrite pharmacology involves the modula-
tion of $^{\bullet}NO$ and $O_2^{\bullet-}$ levels, modification of tissue
thiol and carbon dioxide content, use of peroxynitrite
decomposition catalysts and scavengers. Experiments
using NADPH oxidase deficient cells or animals may
provide further evidence.

• Unsolved issues remain: (i) the relative contribution
of peroxynitrite versus other alternative mechanisms
to $^{\bullet}NO$-mediated oxidative modifications, (ii) precise
quantitation of peroxynitrite formed from cells and
tissues, and (iii) peroxynitrite formation in specific
intracellular compartments.

• Detecting and defining the participation of peroxynitrite during cell and tissue pathology is an active
and rapidly evolving area of research. More extensive
and judicious application of current methodologies
and development of more specific ones will further
contribute to unravel the presence of this elusive
intermediate in animal and human biology and disease.

Acknowledgements — This work was supported by grants from ICGEB
(Trieste), SAREC (Sweden), Fogarty-NIH (USA), CONICYT and
CSIC (Uruguay) to R.R. We thank Drs. Ohara Augusto and Jay W.
Heinecke for helpful comments during the preparation of the manu-
script and Dr. Gabriela Gualco for her advice on immunohistopathol-
ogy.

REFERENCES

[1] Beckman, J. S.; Beckman, T. W.; Chen, J.; Marshall, P. A.;
Freeman, B. A. Apparent hydroxyl radical production by peroxynitrite: implications for endothelial injury from nitric oxide
and superoxide. *Proc. Natl. Acad. Sci. USA* **87**:1620–1624;
1990.
[2] Radi, R.; Beckman, J. S.; Bush, K. M.; Freeman, B. A. Peroxynitrite oxidation of sulfhydryls. The cytotoxic potential of
superoxide and nitric oxide. *J. Biol. Chem.* **266**:4244–4250;
1991.
[3] Radi, R.; Beckman, J. S.; Bush, K. M.; Freeman, B. A. Peroxynitrite-induced membrane lipid peroxidation: the cytotoxic
potential of superoxide and nitric oxide. *Arch. Biochem. Biophys.* **288**:481–487; 1991.
[4] Ischiropoulos, H.; Zhu, L.; Chen, J.; Tsai, M.; Martin, J. C.;
Smith, C. D.; Beckman, J. S. Peroxynitrite-mediated tyrosine
nitration catalyzed by superoxide dismutase. *Arch. Biochem.
Biophys.* **298**:431–437; 1992.
[5] Pryor, W. A.; Squadrito, G. L. The chemistry of peroxynitrite: a
product from the reaction of nitric oxide with superoxide. *Am. J.
Physiol.* **268**:L699–L722; 1995.
[6] Beckman, J. S.; Koppenol, W. H. Nitric oxide, superoxide, and
peroxynitrite: the good, the bad, and ugly. *Am. J. Physiol.*
271:C1424–1437; 1996.
[7] Radi, R.; Denicola, A.; Alvarez, B.; Ferrer-Sueta, G.; Rubbo, H.
The biological chemistry of peroxynitrite. In: Ignarro, L., ed.
Nitric oxide. San Diego, CA: Academic Press; 2000:57–82.
[8] Ignarro, L. J. Biosynthesis and metabolism of endothelium-
derived nitric oxide. *Annu. Rev. Pharmacol. Toxicol.* **30**:535–
560; 1990.
[9] Moncada, S.; Higgs, A. The L-arginine-nitric oxide pathway.
N. Engl. J. Med. **329**:2002–2012; 1993.
[10] Beckman, J. S. The physiological and pathological chemistry of
nitric oxide. In: Lancaster, J., ed. Nitric oxide: principles and
actions. New York: Academic Press, 1996:1–82.
[11] Rubbo, H.; Radi, R.; Trujillo, M.; Telleri, R.; Kalyanaraman, B.;
Barnes, S.; Kirk, M.; Freeman, B. A. Nitric oxide regulation of
superoxide and peroxynitrite-dependent lipid peroxidation. Formation of novel nitrogen-containing oxidized lipid derivatives.
J. Biol. Chem. **269**:26066–26075; 1994.
[12] Radi, R. Reactions of nitric oxide with metalloproteins. *Chem.
Res. Toxicol.* **9**:828–835; 1996.
[13] Rubbo, H.; Radi, R. Antioxidant properties of nitric oxide. In:
Cadenas, E.; Packer, L., eds. *Handbook of antioxidants: Biochemical, nutritional and clinical aspects,* 2nd ed. New York:
Marcel Dekker Inc.; 2000:in press.
[14] Haddad, I. Y.; Pataki, G.; Hu, P.; Galliani, C.; Beckman, J. S.;
Matalon, S. Quantitation of nitrotyrosine levels in lung sections
of patients and animals with acute lung injury. *J. Clin. Invest.*
94:2407–2413; 1994.
[15] Kooy, N. W.; Royall, J. A.; Ye, Y. Z.; Kelly, D. R.; Beckman,
J. S. Evidence for in vivo peroxynitrite production in human
acute lung injury. *Am. J. Respir. Crit. Care Med.* **151**:1250–
1254; 1995.
[16] Akaike, T.; Noguchi, Y.; Ijiri, S.; Setoguchi, K.; Suga, M.;
Zheng, Y. M.; Dietzschold, B.; Maeda, H. Pathogenesis of
influenza virus-induced pneumonia: involvement of both nitric
oxide and oxygen radicals. *Proc. Natl. Acad. Sci. USA* **93**:2448–
2453; 1996.
[17] Beckmann, J. S.; Ye, Y. Z.; Anderson, P. G.; Chen, J.; Accaviti,
M. A.; Tarpey, M. M.; White, C. R. Extensive nitration of
protein tyrosines in human atherosclerosis detected by immunohistochemistry. *Biol. Chem. Hoppe Seyler* **375**:81–88; 1994.
[18] Kaur, H.; Halliwell, B. Evidence for nitric oxide-mediated oxidative damage in chronic inflammation. Nitrotyrosine in serum
and synovial fluid from rheumatoid patients. *FEBS Lett.* **350**:9–
12; 1994.

[19] Fukuyama, N.; Takebayashi, Y.; Hida, M.; Ishida, H.; Ichimori, K.; Nakazawa, H. Clinical evidence of peroxynitrite formation in chronic renal failure patients with septic shock. *Free Radic. Biol. Med.* **22:**771–774; 1997.

[20] Szabo, C. The pathophysiological role of peroxynitrite in shock, inflammation, and ischemia-reperfusion injury. *Shock* **6:**79–88; 1996.

[21] Eliasson, M. J.; Huang, Z.; Ferrante, R. J.; Sasamata, M.; Molliver, M. E.; Snyder, S. H.; Moskowitz, M. A. Neuronal nitric oxide synthase activation and peroxynitrite formation in ischemic stroke linked to neural damage. *J. Neurosci.* **19:**5910–5918; 1999.

[22] Beckman, J. S.; Carson, M.; Smith, C. D.; Koppenol, W. H. ALS, SOD and peroxynitrite. *Nature* **364:**584; 1993.

[23] Beal, M. F.; Ferrante, R. J.; Browne, S. E.; Matthews, R. T.; Kowall, N. W.; Brown, R. H. Jr. Increased 3-nitrotyrosine in both sporadic and familial amyotrophic lateral sclerosis. *Ann. Neurol.* **42:**644–654; 1997.

[24] Ara, J.; Przedborski, S.; Naini, A. B.; Jackson-Lewis, V.; Trifiletti, R. R.; Horwitz, J.; Ischiropoulos, H. Inactivation of tyrosine hydroxylase by nitration following exposure to peroxynitrite and 1-methyl-4-phenyl-1,2,3,6-tetrahydropyridine (MPTP). *Proc. Natl. Acad. Sci. USA* **95:**7659–7663; 1998.

[25] Hooper, D. C.; Bagasra, O.; Marini, J. C.; Zborek, A.; Ohnishi, S. T.; Kean, R.; Champion, J. M.; Sarker, A. B.; Bobroski, L.; Farber, J. L.; Akaike, T.; Maeda, H.; Koprowski, H. Prevention of experimental allergic encephalomyelitis by targeting nitric oxide and peroxynitrite: implications for the treatment of multiple sclerosis. *Proc. Natl. Acad. Sci. USA* **94:**2528–2533; 1997.

[26] Estevez, A. G.; Spear, N.; Manuel, S. M.; Radi, R.; Henderson, C. E.; Barbeito, L.; Beckman, J. S. Nitric oxide and superoxide contribute to motor neuron apoptosis induced by trophic factor deprivation. *J. Neurosci.* **18:**923–931; 1998.

[27] Estevez, A. G.; Crow, J. P.; Sampson, J. B.; Reiter, C.; Zhuang, Y.; Richardson, G. J.; Tarpey, M. M.; Barbeito, L.; Beckman, J. S. Induction of nitric oxide-dependent apoptosis in motor neurons by zinc-deficient superoxide dismutase. *Science* **286:**2498–2500; 1999.

[28] Wink, D. A.; Mitchell, J. B. Chemical biology of nitric oxide: insights into regulatory, cytotoxic and cytoprotective mechanisms of nitric oxide. *Free Radic. Biol. Med.* **25:**434–456; 1998.

[29] Fukuto, J.; Ignarro, L. In vivo aspects of nitric oxide (NO) chemistry: Does peroxinitrite (ONOO−) play a major role in cytotoxicity? *Acc. Chem. Res.* **30:**149–152; 1997.

[30] Brockhaus, F.; Brune, B. Overexpression of CuZn superoxide dismutase protects RAW 264.7 macrophages against nitric oxide cytotoxicity. *Biochem. J.* **338:**295–303; 1999.

[31] Goldstein, S.; Czapski, G. The reaction of NO• with O_2^{-} and HO_2•: a pulse radiolysis study. *Free Radic. Biol. Med.* **19:**505–510; 1995.

[32] Huie, R. E.; Padmaja, S. The reaction of NO with superoxide. *Free Radic. Res. Commun.* **18:**195–199; 1993.

[33] Kissner, R.; Nauser, T.; Bugnon, P.; Lye, P. G.; Koppenol, W. H. Formation and properties of peroxynitrite as studied by laser flash photolysis, high-pressure stopped-flow technique, and pulse radiolysis. *Chem. Res. Toxicol.* **10:**1285–1292; 1997.

[34] Lancaster, J. R. Jr. Simulation of the diffusion and reaction of endogenously produced nitric oxide. *Proc. Natl. Acad. Sci. USA* **91:**8137–8141; 1994.

[35] Denicola, A.; Souza, J. M.; Radi, R.; Lissi, E. Nitric oxide diffusion in membranes determined by fluorescence quenching. *Arch. Biochem. Biophys.* **328:**208–212; 1996.

[36] Fridovich, I. Superoxide radical and superoxide dismutases. *Annu. Rev. Biochem.* **64:**97–112; 1995.

[37] Hughes, M. N. Relationships between nitric oxide, nitroxyl ion, nitrosonium cation and peroxynitrite. *Biochim. Biophys. Acta* **1411:**263–272; 1999.

[38] Sharpe, M. A.; Cooper, C. E. Reactions of nitric oxide with mitochondrial cytochrome c: a novel mechanism for the formation of nitroxyl anion and peroxynitrite. *Biochem. J.* **332:**9–19; 1998.

[39] Wade, R. S.; Castro, C. E. Reactions of oxymyoglobin with NO, NO2, and NO2-under argon and in air. *Chem. Res. Toxicol.* **9:**1382–1390; 1996.

[40] Xia, Y.; Zweier, J. L. Superoxide and peroxynitrite generation from inducible nitric oxide synthase in macrophages. *Proc. Natl. Acad. Sci. USA* **94:**6954–6958; 1997.

[41] Radi, R. Peroxynitrite reactions and diffusion in biology. *Chem. Res. Toxicol.* **11:**720–721; 1998.

[42] Lymar, S. V.; Hurst, J. K. Rapid reaction between peroxynitrite ion and carbon dioxide: implications for biological activity. *J. Am. Chem. Soc.* **117:**8867–8868; 1995.

[43] Denicola, A.; Freeman, B. A.; Trujillo, M.; Radi, R. Peroxynitrite reaction with carbon dioxide/bicarbonate: kinetics and influence on peroxynitrite-mediated oxidations. *Arch. Biochem. Biophys.* **333:**49–58; 1996.

[44] Goldstein, S.; Czapski, G. The effect of bicarbonate on oxidation by peroxynitrite: implication for its biological activity. *Inorg. Chem.* **36:**5113–5117; 1997.

[45] Lymar, S. V.; Hurst, J. K. CO_2-Catalyzed one-electron oxidations by peroxynitrite: properties of the reactive intermediate. *Inorg. Chem.* **37:**294–301; 1998.

[46] Bonini, M. G.; Radi, R.; Ferrer-Sueta, G.; Ferreira, A. M. D. C.; Augusto, O. Direct epr detection of the carbonate radical anion produced from peroxynitrite and carbon dioxide. *J. Biol. Chem.* **274:**10802–10806; 1999.

[47] Augusto, O.; Gatti, R. M.; Radi, R. Spin-trapping studies of peroxynitrite decomposition and of 3-morpholinosydnonimine N-ethylcarbamide autooxidation: direct evidence for metal-independent formation of free radical intermediates. *Arch. Biochem. Biophys.* **310:**118–125; 1994.

[48] Merenyi, G.; Lind, J. Free radical formation in the peroxynitrous acid (ONOOH)/peroxynitrite (ONOO−) system. *Chem. Res. Toxicol.* **11:**243–246; 1998.

[49] Ischiropoulos, H. Biological tyrosine nitration: a pathophysiological function of nitric oxide and reactive oxygen species. *Arch. Biochem. Biophys.* **356:**1–11; 1998.

[50] Alvarez, B.; Ferrer-Sueta, G.; Freeman, B. A.; Radi, R. Kinetics of peroxynitrite reaction with amino acids and human serum albumin. *J. Biol. Chem.* **274:**842–848; 1999.

[51] Lymar, S. V.; Jiang, Q.; Hurst, J. K. Mechanism of carbon dioxide-catalyzed oxidation of tyrosine by peroxynitrite. *Biochemistry* **35:**7855–7861; 1996.

[52] Goldstein, S.; Czapski, G.; Lind, J.; Merenyi, G. Tyrosine nitration by simultaneous generation of (•)NO and O-(2) under physiological conditions. How the radicals do the job. *J. Biol. Chem.* **275:**3031–3036; 2000.

[53] van der Vliet, A.; Eiserich, J. P.; O'Neill, C. A.; Halliwell, B.; Cross, C. E. Tyrosine modification by reactive nitrogen species: a closer look. *Arch. Biochem. Biophys.* **319:**341–349; 1995.

[54] Ramezanian, M. S.; Padmaja, S.; Koppenol, W. H. Nitration and hydroxylation of phenolic compounds by peroxynitrite. *Chem. Res. Toxicol.* **9:**232–240; 1996.

[55] Romero, N.; Denicola, A.; Souza, J. M.; Radi, R. Diffusion of peroxynitrite in the presence of carbon dioxide. *Arch. Biochem. Biophys.* **368:**23–30; 1999.

[56] Marla, S. S.; Lee, J.; Groves, J. T. Peroxynitrite rapidly permeates phospholipid membranes. *Proc. Natl. Acad. Sci. USA* **94:**14243–14248; 1997.

[57] Denicola, A.; Souza, J. M.; Radi, R. Diffusion of peroxynitrite across erythrocyte membranes. *Proc. Natl. Acad. Sci. USA* **95:**3566–3571; 1998.

[58] Radi, R.; Rodriguez, M.; Castro, L.; Telleri, R. Inhibition of mitochondrial electron transport by peroxynitrite. *Arch. Biochem. Biophys.* **308:**89–95; 1994.

[59] Cassina, A.; Radi, R. Differential inhibitory action of nitric oxide and peroxynitrite on mitochondrial electron transport. *Arch. Biochem. Biophys.* **328:**309–316; 1996.

[60] Lizasoain, I.; Moro, M. A.; Knowles, R. G.; Darley-Usmar, V.; Moncada, S. Nitric oxide and peroxynitrite exert distinct effects on mitochondrial respiration which are differentially blocked by glutathione or glucose. *Biochem. J.* **314:**877–880; 1996.

[61] Clementi, E.; Brown, G. C.; Feelisch, M.; Moncada, S. Persistent inhibition of cell respiration by nitric oxide: crucial role of S-nitrosylation of mitochondrial complex I and protective action of glutathione. *Proc. Natl. Acad. Sci. USA* **95**:7631–7636; 1998.

[62] Eiserich, J. P.; Hristova, M.; Cross, C. E.; Jones, A. D.; Freeman, B. A.; Halliwell, B.; van der Vliet, A. Formation of nitric oxide-derived inflammatory oxidants by myeloperoxidase in neutrophils. *Nature* **391**:393–397; 1998.

[63] Squadrito, G. L.; Pryor, W. A. Oxidative chemistry of nitric oxide: the roles of superoxide, peroxynitrite and carbon dioxide. *Free Radic. Biol. Med.* **25**:392–403; 1998.

[64] Gunther, M. R.; Hsi, L. C.; Curtis, J. F.; Gierse, J. K.; Marnett, L. J.; Eling, T. E.; Mason, R. P. Nitric oxide trapping of the tyrosyl radical of prostaglandin H synthase-2 leads to tyrosine iminoxyl radical and nitrotyrosine formation. *J. Biol. Chem.* **272**:17086–17090; 1997.

[65] Royall, J. A.; Ischiropoulos, H. Evaluation of 2′,7′-dichlorofluorescin and dihydrorhodamine 123 as fluorescent probes for intracellular H2O2 in cultured endothelial cells. *Arch. Biochem. Biophys.* **302**:348–355; 1993.

[66] Kooy, N. W.; Royall, J. A.; Ischiropoulos, H.; Beckman, J. S. Peroxynitrite-mediated oxidation of dihydrorhodamine 123. *Free Radic. Biol. Med.* **16**:149–156; 1994.

[67] Kooy, N. W.; Royall, J. A.; Ischiropoulos, H. Oxidation of 2′,7′-dichlorofluorescin by peroxynitrite. *Free Radic. Res.* **27**:245–254; 1997.

[68] Crow, J. P. Dichlorodihydrofluorescein and dihydrorhodamine 123 are sensitive indicators of peroxynitrite in vitro: implications for intracellular measurement of reactive nitrogen and oxygen species. *Nitric Oxide* **1**:145–157; 1997.

[69] Ischiropoulos, H.; Gow, A.; Thom, S. R.; Kooy, N. W.; Royall, J. A.; Crow, J. P. Detection of reactive nitrogen species using 2,7-dichlorodihydrofluorescein and dihydrorhodamine 123. *Methods Enzymol.* **301**:367–373; 1999.

[70] Miles, A. M.; Bohle, D. S.; Glassbrenner, P. A.; Hansert, B.; Wink, D. A.; Grisham, M. B. Modulation of superoxide-dependent oxidation and hydroxylation reactions by nitric oxide. *J. Biol. Chem.* **271**:40–47; 1996.

[71] Rota, C.; Chignell, C. F.; Mason, R. P. Evidence for free radical formation during the oxidation of 2′-7′-dichlorofluorescin to the fluorescent dye 2′-7′-dichlorofluorescein by horseradish peroxidase: possible implications for oxidative stress measurements. *Free Radic. Biol. Med.* **27**:873–881; 1999.

[72] Rota, C.; Fann, Y. C.; Mason, R. P. Phenoxyl free radical formation during the oxidation of the fluorescent dye 2′,7′-dichlorofluorescein by horseradish peroxidase. Possible consequences for oxidative stress measurements. *J. Biol. Chem.* **274**:28161–28168; 1999.

[73] Possel, H.; Noack, H.; Augustin, W.; Keilhoff, G.; Wolf, G. 2,7-Dihydrodichlorofluorescein diacetate as a fluorescent marker for peroxynitrite formation. *FEBS Lett.* **416**:175–178; 1997.

[74] Tannous, M.; Rabini, R. A.; Vignini, A.; Moretti, N.; Fumelli, P.; Zielinski, B.; Mazzanti, L.; Mutus, B. Evidence for iNOS-dependent peroxynitrite production in diabetic platelets. *Diabetologia* **42**:539–544; 1999.

[75] Thom, S. R.; Xu, Y. A.; Ischiropoulos, H. Vascular endothelial cells generate peroxynitrite in response to carbon monoxide exposure. *Chem. Res. Toxicol.* **10**:1023–1031; 1997.

[76] Gagnon, C.; Leblond, F. A.; Filep, J. G. Peroxynitrite production by human neutrophils, monocytes and lymphocytes challenged with lipopolysaccharide. *FEBS Lett.* **431**:107–110; 1998.

[77] Lopez-Garcia, M. P.; Sanz-Gonzalez, S. M. Peroxynitrite generated from constitutive nitric oxide synthase mediates the early biochemical injury in short-term cultured hepatocytes. *FEBS Lett.* **466**:187–191; 2000.

[78] Radi, R.; Cosgrove, T. P.; Beckman, J. S.; Freeman, B. A. Peroxynitrite-induced luminol chemiluminescence. *Biochem. J.* **290**:51–57; 1993.

[79] Kooy, N. W.; Royall, J. A. Agonist-induced peroxynitrite production from endothelial cells. *Arch. Biochem. Biophys.* **310**:352–359; 1994.

[80] Catz, S. D.; Carreras, M. C.; Poderoso, J. J. Nitric oxide synthase inhibitors decrease human polymorphonuclear leukocyte luminol-dependent chemiluminescence. *Free Radic. Biol. Med.* **19**:741–748; 1995.

[81] Davidson, C. A.; Kaminski, P. M.; Wolin, M. S. NO elicits prolonged relaxation of bovine pulmonary arteries via endogenous peroxynitrite generation. *Am. J. Physiol.* **273**:L437–L444; 1997.

[82] Kudoh, S.; Suzuki, K.; Yamada, M.; Liu, Q.; Nakaji, S.; Sugawara, K. Contribution of nitric oxide synthase to human neutrophil chemiluminescence. *Luminescence* **14**:335–339; 1999.

[83] Castro, L.; Alvarez, M. N.; Radi, R. Modulatory role of nitric oxide on superoxide-dependent luminol chemiluminescence. *Arch. Biochem. Biophys.* **333**:179–188; 1996.

[84] Tarpey, M. M.; White, C. R.; Suarez, E.; Richardson, G.; Radi, R.; Freeman, B. A. Chemiluminescent detection of oxidants in vascular tissue. Lucigenin but not coelenterazine enhances superoxide formation. *Circ. Res.* **84**:1203–1211; 1999.

[85] Faulkner, K.; Fridovich, I. Luminol and lucigenin as detectors for O2$^{•-}$. *Free Radic. Biol. Med.* **15**:447–451; 1993.

[86] Ischiropoulos, H.; Zhu, L.; Beckman, J. S. Peroxynitrite formation from macrophage-derived nitric oxide. *Arch. Biochem. Biophys.* **298**:446–451; 1992.

[87] Pfeiffer, S.; Schmidt, K.; Mayer, B. Dityrosine formation outcompetes tyrosine nitration at low steady-state concentrations of peroxynitrite. Implications for tyrosine modification by nitric oxide/superoxide in vivo. *J. Biol. Chem.* **275**:6346–6352; 2000.

[88] Crow, J. P.; Ischiropoulos, H. Detection and quantitation of nitrotyrosine residues in proteins: in vivo marker of peroxynitrite. *Methods Enzymol.* **269**:185–194; 1996.

[89] Shigenaga, M. K.; Lee, H. H.; Blount, B. C.; Christen, S.; Shigeno, E. T.; Yip, H.; Ames, B. N. Inflammation and NO(X)-induced nitration: assay for 3-nitrotyrosine by HPLC with electrochemical detection. *Proc. Natl. Acad. Sci. USA* **94**:3211–3216; 1997.

[90] Hensley, K.; Maidt, M. L.; Yu, Z.; Sang, H.; Markesbery, W. R.; Floyd, R. A. Electrochemical analysis of protein nitrotyrosine and dityrosine in the Alzheimer brain indicates region-specific accumulation. *J. Neurosci.* **18**:8126–8132; 1998.

[91] Hensley, K.; Maidt, M. L.; Pye, Q. N.; Stewart, C. A.; Wack, M.; Tabatabaie, T.; Floyd, R. A. Quantitation of protein-bound 3-nitrotyrosine and 3,4-dihydroxyphenylalanine by high-performance liquid chromatography with electrochemical array detection [published erratum appears in *Anal. Biochem.* 1998 Feb 1;256:148], *Anal. Biochem.* **251**:187–195; 1997.

[92] Crow, J. P. Measurement and significance of free and protein-bound 3-nitrotyrosine, 3-chlorotyrosine, and free 3-nitro-4-hydroxyphenylacetic acid in biologic samples: a high-performance liquid chromatography method using electrochemical detection. *Methods Enzymol.* **301**:151–160; 1999.

[93] Herce-Pagliai, C.; Kotecha, S.; Shuker, D. E. Analytical methods for 3-nitrotyrosine as a marker of exposure to reactive nitrogen species: a review. *Nitric Oxide* **2**:324–336; 1998.

[94] Shigenaga, M. K. Quantitation of protein-bound 3-nitrotyrosine by high-performance liquid chromatography with electrochemical detection. *Methods Enzymol.* **301**:27–40; 1999.

[95] van der Vliet, A.; Eiserich, J. P.; Kaur, H.; Cross, C. E.; Halliwell, B. Nitrotyrosine as biomarker for reactive nitrogen species. *Methods Enzymol.* **269**:175–184; 1996.

[96] Pennathur, S.; Jackson-Lewis, V.; Przedborski, S.; Heinecke, J. W. Mass spectrometric quantification of 3-nitrotyrosine, ortho-tyrosine, and o,o′-dityrosine in brain tissue of 1-methyl-4-phenyl-1,2,3, 6-tetrahydropyridine-treated mice, a model of oxidative stress in Parkinson's disease. *J. Biol. Chem.* **274**:34621–34628; 1999.

[97] Crowley, J. R.; Yarasheski, K.; Leeuwenburgh, C.; Turk, J.; Heinecke, J. W. Isotope dilution mass spectrometric quantification of 3-nitrotyrosine in proteins and tissues is facilitated by reduction to 3-aminotyrosine. *Anal. Biochem.* **259**:127–135; 1998.

[98] Jiang, H.; Balazy, M. Detection of 3-nitrotyrosine in human

platelets exposed to peroxynitrite by a new gas chromatography/mass spectrometry assay. *Nitric Oxide* **2**:350–359; 1998.

[99] Schwedhelm, E.; Tsikas, D.; Gutzki, F. M.; Frolich, J. C. Gas chromatographic-tandem mass spectrometric quantification of free 3-nitrotyrosine in human plasma at the basal state. *Anal. Biochem.* **276**:195–203; 1999.

[100] Frost, M. T.; Halliwell, B.; Moore, K. P. Analysis of free and protein-bound nitrotyrosine in human plasma by a gas chromatography/mass spectrometry method that avoids nitration artifacts. *Biochem. J.* **345**(Pt. 3):453–458; 2000.

[101] Hazen, S. L.; Zhang, R.; Shen, Z.; Wu, W.; Podrez, E. A.; MacPherson, J. C.; Schmitt, D.; Mitra, S. N.; Mukhopadhyay, C.; Chen, Y.; Cohen, P. A.; Hoff, H. F.; Abu-Soud, H. M. Formation of nitric oxide-derived oxidants by myeloperoxidase in monocytes: pathways for monocyte-mediated protein nitration and lipid peroxidation in vivo. *Circ. Res.* **85**:950–958; 1999.

[102] Wu, W.; Chen, Y.; Hazen, S. L. Eosinophil peroxidase nitrates protein tyrosyl residues. Implications for oxidative damage by nitrating intermediates in eosinophilic inflammatory disorders. *J. Biol. Chem.* **274**:25933–25944; 1999.

[103] van der Vliet, A.; Eiserich, J. P.; Halliwell, B.; Cross, C. E. Formation of reactive nitrogen species during peroxidase-catalyzed oxidation of nitrite. A potential additional mechanism of nitric oxide-dependent toxicity. *J. Biol. Chem.* **272**:7617–7625; 1997.

[104] Wu, W.; Chen, Y.; d'Avignon, A.; Hazen, S. L. 3-Bromotyrosine and 3,5-dibromotyrosine are major products of protein oxidation by eosinophil peroxidase: potential markers for eosinophil-dependent tissue injury in vivo. *Biochemistry* **38**:3538–3548; 1999.

[105] Beckman, J. S.; Ischiropoulos, H.; Zhu, L.; van der Woerd, M.; Smith, C.; Chen, J.; Harrison, J.; Martin, J. C.; Tsai, M. Kinetics of superoxide dismutase- and iron-catalyzed nitration of phenolics by peroxynitrite. *Arch. Biochem. Biophys.* **298**:438–445; 1992.

[106] van der Vliet, A.; O'Neill, C. A.; Halliwell, B.; Cross, C. E.; Kaur, H. Aromatic hydroxylation and nitration of phenylalanine and tyrosine by peroxynitrite. Evidence for hydroxyl radical production from peroxynitrite. *FEBS Lett.* **339**:89–92; 1994.

[107] Daiber, A.; Mehl, M.; Ullrich, V. New aspects in the reaction mechanism of phenol with peroxynitrite: the role of phenoxy radicals [In Process Citation]. *Nitric Oxide* **2**:259–269; 1998.

[108] Bruijn, L. I.; Beal, M. F.; Becher, M. W.; Schulz, J. B.; Wong, P. C.; Price, D. L.; Cleveland, D. W. Elevated free nitrotyrosine levels, but not protein-bound nitrotyrosine or hydroxyl radicals, throughout amyotrophic lateral sclerosis (ALS)-like disease implicate tyrosine nitration as an aberrant in vivo property of one familial ALS-linked superoxide dismutase 1 mutant. *Proc. Natl. Acad. Sci. USA* **94**:7606–7611; 1997.

[109] Augusto, O.; Radi, R.; Gatti, R. M.; Vasquez-Vivar, J. Detection of secondary radicals from peroxynitrite-medicated oxidations by electron spin resonance. *Methods Enzymol.* **269**:346–354; 1996.

[110] Gatti, R. M.; Radi, R.; Augusto, O. Peroxynitrite-mediated oxidation of albumin to the protein-thiyl free radical. *FEBS Lett.* **348**:287–290; 1994.

[111] Gatti, R. M.; Alvarez, B.; Vasquez-Vivar, J.; Radi, R.; Augusto, O. Formation of spin trap adducts during the decomposition of peroxynitrite. *Arch. Biochem. Biophys.* **349**:36–46; 1998.

[112] Karoui, H.; Hogg, N.; Frejaville, C.; Tordo, P.; Kalyanaraman, B. Characterization of sulfur-centered radical intermediates formed during the oxidation of thiols and sulfite by peroxynitrite. ESR-spin trapping and oxygen uptake studies. *J. Biol. Chem.* **271**:6000–6009; 1996.

[113] Dikalov, S.; Grigor'ev, I. A.; Voinov, M.; Bassenge, E. Detection of superoxide radicals and peroxynitrite by 1-hydroxy-4-phosphonooxy-2,2,6,6-tetramethylpiperidine: quantification of extracellular superoxide radicals formation. *Biochem. Biophys. Res. Commun.* **248**:211–215; 1998.

[114] Dikalov, S.; Skatchkov, M.; Fink, B.; Bassenge, E. Quantification of superoxide radicals and peroxynitrite in vascular cells

using oxidation of sterically hindered hydroxylamines and electron spin resonance. *Nitric Oxide* **1**:423–431; 1997.

[115] Dikalov, S.; Skatchkov, M.; Bassenge, E. Spin trapping of superoxide radicals and peroxynitrite by 1-hydroxy-3-carboxypyrrolidine and 1-hydroxy-2,2,6,6-tetramethyl-4-oxo-piperidine and the stability of corresponding nitroxyl radicals towards biological reductants. *Biochem. Biophys. Res. Commun.* **231**:701–704; 1997.

[116] Balavoine, G. G.; Geletii, Y. V. Peroxynitrite scavenging by different antioxidants. Part I: convenient assay. *Nitric Oxide* **3**:40–54; 1999.

[117] Crow, J. P.; Spruell, C.; Chen, J.; Gunn, C.; Ischiropoulos, H.; Tsai, M.; Smith, C. D.; Radi, R.; Koppenol, W. H.; Beckman, J. S. On the pH-dependent yield of hydroxyl radical products from peroxynitrite. *Free Radic. Biol. Med.* **16**:331–338; 1994.

[118] Goldstein, S.; Czapski, G. Direct and indirect oxidations by peroxynitrite. *Inorg. Chem.* **34**:4041–4048; 1995.

[119] Thomson, L.; Trujillo, M.; Telleri, R.; Radi, R. Kinetics of cytochrome c2+ oxidation by peroxynitrite: implications for superoxide measurements in nitric oxide-producing biological systems. *Arch. Biochem. Biophys.* **319**:491–497; 1995.

[120] Trujillo, M.; Alvarez, M. N.; Peluffo, G.; Freeman, B. A.; Radi, R. Xanthine oxidase-mediated decomposition of S-nitrosothiols. *J. Biol. Chem.* **273**:7828–7834; 1998.

[121] Szabo, C.; Ferrer-Sueta, G.; Zingarelli, B.; Southan, G. J.; Salzman, A. L.; Radi, R. Mercaptoethylguanidine and guanidine inhibitors of nitric-oxide synthase react with peroxynitrite and protect against peroxynitrite-induced oxidative damage. *J. Biol. Chem.* **272**:9030–9036; 1997.

[122] Souza, J. M.; Radi, R. Glyceraldehyde-3-phosphate dehydrogenase inactivation by peroxynitrite. *Arch. Biochem. Biophys.* **360**:187–194; 1998.

[123] Lynch, R. E.; Fridovich, I. Permeation of the erythrocyte stroma by superoxide radical. *J. Biol. Chem.* **253**:4697–4699; 1978.

[124] Kelm, M.; Dahmann, R.; Wink, D.; Feelisch, M. The nitric oxide/superoxide assay. Insights into the biological chemistry of the NO/O-2. interaction. *J. Biol. Chem.* **272**:9922–9932; 1997.

[125] Radi, R.; Thomson, L.; Rubbo, H.; Prodanov, E. Cytochrome c-catalyzed oxidation of organic molecules by hydrogen peroxide. *Arch. Biochem. Biophys.* **288**:112–117; 1991.

[126] Butler, J.; Koppenol, W. H.; Margoliash, E. Kinetics and mechanism of the reduction of ferricytochrome c by the superoxide anion. *J. Biol. Chem.* **257**:10747–10750; 1982.

[127] Gryglewski, R. J.; Palmer, R. M.; Moncada, S. Superoxide anion is involved in the breakdown of endothelium-derived relaxing factor. *Nature* **320**:454–456; 1986.

[128] Ignarro, L. J.; Buga, G. M.; Wood, K. S.; Byrns, R. E.; Chaudhuri, G. Endothelium-derived relaxing factor produced and released from artery and vein is nitric oxide. *Proc. Natl. Acad. Sci. USA* **84**:9265–9269; 1987.

[129] Radi, R.; Denicola, A.; Freeman, B. A. Peroxynitrite reactions with carbon dioxide-bicarbonate. *Methods Enzymol.* **301**:353–367; 1999.

[130] Gadelha, F. R.; Thomson, L.; Fagian, M. M.; Costa, A. D.; Radi, R.; Vercesi, A. E. Ca2+-independent permeabilization of the inner mitochondrial membrane by peroxynitrite is mediated by membrane protein thiol cross-linking and lipid peroxidation. *Arch. Biochem. Biophys.* **345**:243–250; 1997.

[131] Kirsch, M.; Lomonosova, E. E.; Korth, H. G.; Sustmann, R.; de Groot, H. Hydrogen peroxide formation by reaction of peroxynitrite with HEPES and related tertiary amines. Implications for a general mechanism. *J. Biol. Chem.* **273**:12716–12724; 1998.

[132] Schmidt, K.; Pfeiffer, S.; Mayer, B. Reaction of peroxynitrite with HEPES or MOPS results in the formation of nitric oxide donors. *Free Radic. Biol. Med.* **24**:859–862; 1998.

[133] Denicola, A.; Souza, J. M.; Gatti, R. M.; Augusto, O.; Radi, R. Desferrioxamine inhibition of the hydroxyl radical-like reactivity of peroxynitrite: role of the hydroxamic groups. *Free Radic. Biol. Med.* **19**:11–19; 1995.

[134] Gow, A. J.; Duran, D.; Malcolm, S.; Ischiropoulos, H. Effects of

peroxynitrite-induced protein modifications on tyrosine phosphorylation and degradation. *FEBS Lett.* **385**:63–66; 1996.

[135] Kamisaki, Y.; Wada, K.; Bian, K.; Balabanli, B.; Davis, K.; Martin, E.; Behbod, F.; Lee, Y. C.; Murad, F. An activity in rat tissues that modifies nitrotyrosine-containing proteins. *Proc. Natl. Acad. Sci. USA* **95**:11584–11589; 1998.

[136] Balabanli, B.; Kamisaki, Y.; Martin, E.; Murad, F. Requirements for heme and thiols for the nonenzymatic modification of nitrotyrosine. *Proc. Natl. Acad. Sci. USA* **96**:13136–13141; 1999.

[137] Grune, T.; Blasig, I. E.; Sitte, N.; Roloff, B.; Haseloff, R.; Davies, K. J. Peroxynitrite increases the degradation of aconitase and other cellular proteins by proteasome. *J. Biol. Chem.* **273**: 10857–10862; 1998.

[138] Ye, Y. Z.; Strong, M.; Huang, Z. Q.; Beckman, J. S. Antibodies that recognize nitrotyrosine. *Methods Enzymol.* **269**:201–209; 1996.

[139] Brito, C.; Naviliat, M.; Tiscornia, A. C.; Vuillier, F.; Gualco, G.; Dighiero, G.; Radi, R.; Cayota, A. M. Peroxynitrite inhibits T lymphocyte activation and proliferation by promoting impairment of tyrosine phosphorylation and peroxynitrite-driven apoptotic death. *J. Immunol.* **162**:3356–3366; 1999.

[140] ter Steege, J. C.; Koster-Kamphuis, L.; van Straaten, E. A.; Forget, P. P.; Buurman, W. A. Nitrotyrosine in plasma of celiac disease patients as detected by a new sandwich ELISA. *Free Radic. Biol. Med.* **25**:953–963; 1998.

[141] Khan, J.; Brennand, D. M.; Bradley, N.; Gao, B.; Bruckdorfer, R.; Jacobs, M. 3-Nitrotyrosine in the proteins of human plasma determined by an ELISA method. *Biochem. J.* **332**:807–808; 1998.

[142] Castro, L. A.; Robalinho, R. L.; Cayota, A.; Meneghini, R.; Radi, R. Nitric oxide and peroxynitrite-dependent aconitase inactivation and iron-regulatory protein-1 activation in mammalian fibroblasts. *Arch. Biochem. Biophys.* **359**:215–224; 1998.

[143] MacMillan-Crow, L. A.; Crow, J. P.; Kerby, J. D.; Beckman, J. S.; Thompson, J. A. Nitration and inactivation of manganese superoxide dismutase in chronic rejection of human renal allografts. *Proc. Natl. Acad. Sci. USA* **93**:11853–11858; 1996.

[144] ter Steege, J.; Buurman, W.; Arends, J. W.; Forget, P. Presence of inducible nitric oxide synthase, nitrotyrosine, CD68, and CD14 in the small intestine in celiac disease. *Lab. Invest.* **77**: 29–36; 1997.

[145] Virag, L.; Scott, G. S.; Cuzzocrea, S.; Marmer, D.; Salzman, A. L.; Szabo, C. Peroxynitrite-induced thymocyte apoptosis: the role of caspases and poly (ADP-ribose) synthetase (PARS) activation. *Immunology* **94**:345–355; 1998.

[146] Viera, L.; Ye, Y. Z.; Estevez, A. G.; Beckman, J. S. Immunohistochemical methods to detect nitrotyrosine. *Methods Enzymol.* **301**:373–381; 1999.

[147] Haddad, I. Y.; Zhu, S.; Ischiropoulos, H.; Matalon, S. Nitration of surfactant protein A results in decreased ability to aggregate lipids. *Am. J. Physiol.* **270**:L281–L288; 1996.

[148] Kamisaki, Y.; Wada, K.; Nakamoto, K.; Kishimoto, Y.; Kitano, M.; Itoh, T. Sensitive determination of nitrotyrosine in human plasma by isocratic high-performance liquid chromatography. *J. Chromatogr. B. Biomed. Appl.* **685**:343–347; 1996.

[149] Lamb, N. J.; Quinlan, G. J.; Westerman, S. T.; Gutteridge, J. M.; Evans, T. W. Nitration of proteins in bronchoalveolar lavage fluid from patients with acute respiratory distress syndrome receiving inhaled nitric oxide. *Am. J. Respir. Crit. Care Med.* **160**:1031–1034; 1999.

[150] Tohgi, H.; Abe, T.; Yamazaki, K.; Murata, T.; Ishizaki, E.; Isobe, C. Remarkable increase in cerebrospinal fluid 3-nitrotyrosine in patients with sporadic amyotrophic lateral sclerosis. *Ann. Neurol.* **46**:129–131; 1999.

[151] Yi, D.; Smythe, G. A.; Blount, B. C.; Duncan, M. W. Peroxynitrite-mediated nitration of peptides: characterization of the products by electrospray and combined gas chromatography-mass spectrometry. *Arch. Biochem. Biophys.* **344**:253–259; 1997.

[152] Yamakura, F.; Taka, H.; Fujimura, T.; Murayama, K. Inactivation of human manganese-superoxide dismutase by peroxynitrite

is caused by exclusive nitration of tyrosine 34 to 3-nitrotyrosine. *J. Biol. Chem.* **273**:14085–14089; 1998.

[153] Souza, J. M.; Daikhin, E.; Yudkoff, M.; Raman, C. S.; Ischiropoulos, H. Factors determining the selectivity of protein tyrosine nitration. *Arch. Biochem. Biophys.* **371**:169–178; 1999.

[154] Evans, T. J.; Buttery, L. D.; Carpenter, A.; Springall, D. R.; Polak, J. M.; Cohen, J. Cytokine-treated human neutrophils contain inducible nitric oxide synthase that produces nitration of ingested bacteria. *Proc. Natl. Acad. Sci. USA* **93**:9553–9558; 1996.

[155] Ischiropoulos, H.; al-Mehdi, A. B. Peroxynitrite-mediated oxidative protein modifications. *FEBS Lett.* **364**:279–282; 1995.

[156] Szabo, C.; O'Connor, M.; Salzman, A. L. Endogenously produced peroxynitrite induces the oxidation of mitochondrial and nuclear proteins in immunostimulated macrophages. *FEBS Lett.* **409**:147–150; 1997.

[157] Tien, M.; Berlett, B. S.; Levine, R. L.; Chock, P. B.; Stadtman, E. R. Peroxynitrite-mediated modification of proteins at physiological carbon dioxide concentration: pH dependence of carbonyl formation, tyrosine nitration, and methionine oxidation. *Proc. Natl. Acad. Sci. USA* **96**:7809–7814; 1999.

[158] O'Donnell, V.; Eiserich, J. P.; Chumley, P. H.; Jablonsky, M. J.; Krishna, N. R.; Kirk, M.; Barnes, S.; Darley-Usmar, V. M.; Freeman, B. A. Nitration of unsaturated fatty acids by nitric oxide-derived reactive nitrogen species peroxynitrite, nitrous acid, nitrogen dioxide, and nitronium ion. *Chem. Res. Toxicol.* **12**:83–92; 1999.

[159] Yermilov, V.; Rubio, J.; Ohshima, H. Formation of 8-nitroguanine in DNA treated with peroxynitrite in vitro and its rapid removal from DNA by depurination. *FEBS Lett.* **376**:207–210; 1995.

[160] Yermilov, V.; Yoshie, Y.; Rubio, J.; Ohshima, H. Effects of carbon dioxide/bicarbonate on induction of DNA single-strand breaks and formation of 8-nitroguanine, 8-oxoguanine and basepropenal mediated by peroxynitrite. *FEBS Lett.* **399**:67–70; 1996.

[161] Spencer, J. P.; Wong, J.; Jenner, A.; Aruoma, O. I.; Cross, C. E.; Halliwell, B. Base modification and strand breakage in isolated calf thymus DNA and in DNA from human skin epidermal keratinocytes exposed to peroxynitrite or 3-morpholinosydnonimine. *Chem. Res. Toxicol.* **9**:1152–1158; 1996.

[162] Moro, M. A.; Darley-Usmar, V. M.; Lizasoain, I.; Su, Y.; Knowles, R. G.; Radomski, M. W.; Moncada, S. The formation of nitric oxide donors from peroxynitrite. *Br. J. Pharmacol.* **116**:1999–2004; 1995.

[163] Balazy, M.; Kaminski, P. M.; Mao, K.; Tan, J.; Wolin, M. S. S-Nitroglutathione, a product of the reaction between peroxynitrite and glutathione that generates nitric oxide. *J. Biol. Chem.* **273**:32009–32015; 1998.

[164] Skinner, K. A.; White, C. R.; Patel, R.; Tan, S.; Barnes, S.; Kirk, M.; Darley-Usmar, V.; Parks, D. A. Nitrosation of uric acid by peroxynitrite. Formation of a vasoactive nitric oxide donor. *J. Biol. Chem.* **273**:24491–24497; 1998.

[165] Santos, C. X.; Anjos, E. I.; Augusto, O. Uric acid oxidation by peroxynitrite: multiple reactions, free radical formation, and amplification of lipid oxidation. *Arch. Biochem. Biophys.* **372**: 285–294; 1999.

[166] Babu, B. R.; Griffith, O. W. Design of isoform-selective inhibitors of nitric oxide synthase. *Curr. Opin. Chem. Biol.* **2**:491–500; 1998.

[167] Griffith, O. W.; Kilbourn, R. G. Nitric oxide synthase inhibitors: amino acids. *Methods Enzymol.* **268**:375–392; 1996.

[168] Shiloh, M. U.; MacMicking, J. D.; Nicholson, S.; Brause, J. E.; Potter, S.; Marino, M.; Fang, F.; Dinauer, M.; Nathan, C. Phenotype of mice and macrophages deficient in both phagocyte oxidase and inducible nitric oxide synthase. *Immunity* **10**:29–38; 1999.

[169] Murray, H. W.; Nathan, C. F. Macrophage microbicidal mechanisms in vivo: reactive nitrogen versus oxygen intermediates in the killing of intracellular visceral Leishmania donovani. *J. Exp. Med.* **189**:741–746; 1999.

[170] Jones, S. A.; O'Donnell, V. B.; Wood, J. D.; Broughton, J. P.; Hughes, E. J.; Jones, O. T. Expression of phagocyte NADPH oxidase components in human endothelial cells. *Am. J. Physiol.* **271**:H1626–H1634; 1996.

[171] Stuehr, D. J.; Fasehun, O. A.; Kwon, N. S.; Gross, S. S.; Gonzalez, J. A.; Levi, R.; Nathan, C. F. Inhibition of macrophage and endothelial cell nitric oxide synthase by diphenyleneiodonium and its analogs. *FASEB J.* **5**:98–103; 1991.

[172] Ellis, A.; Li, C. G.; Rand, M. J. Effect of xanthine oxidase inhibition on endothelium-dependent and nitrergic relaxations. *Eur. J. Pharmacol.* **356**:41–47; 1998.

[173] White, C. R.; Darley-Usmar, V.; Berrington, W. R.; McAdams, M.; Gore, J. Z.; Thompson, J. A.; Parks, D. A.; Tarpey, M. M.; Freeman, B. A. Circulating plasma xanthine oxidase contributes to vascular dysfunction in hypercholesterolemic rabbits. *Proc. Natl. Acad. Sci. USA* **93**:8745–8749; 1996.

[174] Nicholson, S. C.; Grobmyer, S. R.; Shiloh, M. U.; Brause, J. E.; Potter, S.; MacMicking, J. D.; Dinauer, M. C.; Nathan, C. F. Lethality of endotoxin in mice genetically deficient in the respiratory burst oxidase, inducible nitric oxide synthase, or both. *Shock* **11**:253–258; 1999.

[175] Fujimura, M.; Morita-Fujimura, Y.; Noshita, N.; Sugawara, T.; Kawase, M.; Chan, P. H. The cytosolic antioxidant Copper/Zinc-superoxide dismutase prevents the early release of mitochondrial cytochrome c in ischemic brain after transient focal cerebral ischemia in mice [In Process Citation]. *J. Neurosci.* **20**:2817–2824; 2000.

[176] Ying, W.; Anderson, C. M.; Chen, Y.; Stein, B. A.; Fahlman, C. S.; Copin, J. C.; Chan, P. H.; Swanson, R. A. Differing effects of copper, zinc superoxide dismutase overexpression on neurotoxicity elicited by nitric oxide, reactive oxygen species, and excitotoxins. *J. Cereb. Blood Flow Metab.* **20**:359–368; 2000.

[177] Gonzalez-Zulueta, M.; Ensz, L. M.; Mukhina, G.; Lebovitz, R. M.; Zwacka, R. M.; Engelhardt, J. F.; Oberley, L. W.; Dawson, V. L.; Dawson, T. M. Manganese superoxide dismutase protects nNOS neurons from NMDA and nitric oxide-mediated neurotoxicity. *J. Neurosci.* **18**:2040–2055; 1998.

[178] Salvemini, D.; Wang, Z. Q.; Zweier, J. L.; Samouilov, A.; Macarthur, H.; Misko, T. P.; Currie, M. G.; Cuzzocrea, S.; Sikorski, J. A.; Riley, D. P. A nonpeptidyl mimic of superoxide dismutase with therapeutic activity in rats. *Science* **286**:304–306; 1999.

[179] Batinic-Haberle, I.; Benov, L.; Spasojevic, I.; Fridovich, I. The ortho effect makes manganese(III) meso-tetrakis(N-methylpyridinium-2-yl)porphyrin a powerful and potentially useful superoxide dismutase mimic. *J. Biol. Chem.* **273**:24521–24528; 1998.

[180] Estevez, A. G.; Sampson, J. B.; Zhuang, Y.; Spear, N.; Richardson, G. J.; Crow, J. P.; Tarpey, M. M.; Barbeito, L.; Beckman, J. S. Liposome-delivered superoxide dismutase prevents nitric oxide-dependent motor neuron death induced by trophic factor withdrawal. *Free Radic. Biol. Med.* **28**:437–446; 2000.

[181] Lee, J.; Hunt, J. A.; Groves, J. T. Mechanism of iron porphyrin reactions with peroxynitrite. *J. Am. Chem. Soc.* **120**:7493–7501; 1998.

[182] Lee, J.; Hunt, J. A.; Groves, J. T. Manganese porphyrins as redox-coupled peroxynitrite reductases. *J. Am. Chem. Soc.* **120**:6053–6061; 1998.

[183] Lee, J.; Hunt, J.; Groves, J. T. Rapid decomposition of peroxynitrite by manganese porphyrin-antioxidant redox couples. *Bioorg. Med. Chem. Lett.* **7**:2913–2918; 1997.

[184] Ferrer-Sueta, G.; Batinic-Haberle, I.; Spasojevic, I.; Fridovich, I.; Radi, R. Catalytic scavenging of peroxynitrite by Isomeric Mn(III) N-methylpyridylporphyrins in the presence of reductants. *Chem. Res. Toxicol.* **12**:442–449; 1999.

[185] Sies, H.; Masumoto, H. Ebselen as a glutathione peroxidase mimic and as a scavenger of peroxynitrite. *Adv. Pharmacol.* **38**:229–246; 1997.

[186] Masumoto, H.; Kissner, R.; Koppenol, W. H.; Sies, H. Kinetic study of the reaction of ebselen with peroxynitrite. *FEBS Lett.* **398**:179–182; 1996.

[187] Masumoto, H.; Sies, H. The reaction of 2-(methylseleno)benzanilide with peroxynitrite. *Chem. Res. Toxicol.* **9**:1057–1062; 1996.

[188] Misko, T. P.; Highkin, M. K.; Veenhuizen, A. W.; Manning, P. T.; Stern, M. K.; Currie, M. G.; Salvemini, D. Characterization of the cytoprotective action of peroxynitrite decomposition catalysts. *J. Biol. Chem.* **273**:15646–15653; 1998.

[189] Salvemini, D.; Wang, Z. Q.; Stern, M. K.; Currie, M. G.; Misko, T. P. Peroxynitrite decomposition catalysts: therapeutics for peroxynitrite-mediated pathology. *Proc. Natl. Acad. Sci. USA* **95**:2659–2663; 1998.

[190] Pfeiffer, S.; Schrammel, A.; Koesling, D.; Schmidt, K.; Mayer, B. Molecular actions of a Mn(III)Porphyrin superoxide dismutase mimetic and peroxynitrite scavenger: reaction with nitric oxide and direct inhibition of NO synthase and soluble guanylyl cyclase. *Mol. Pharmacol.* **53**:795–800; 1998.

[191] Ferrer-Sueta, G.; Ruiz-Ramirez, L.; Radi, R. Ternary copper complexes and manganese (III) tetrakis(4-benzoic acid) porphyrin catalyze peroxynitrite-dependent nitration of aromatics. *Chem. Res. Toxicol.* **10**:1338–1344; 1997.

[192] Alvarez, B.; Rubbo, H.; Kirk, M.; Barnes, S.; Freeman, B. A.; Radi, R. Peroxynitrite-dependent tryptophan nitration. *Chem. Res. Toxicol.* **9**:390–396; 1996.

[193] Hooper, D. C.; Spitsin, S.; Kean, R. B.; Champion, J. M.; Dickson, G. M.; Chaudhry, I.; Koprowski, H. Uric acid, a natural scavenger of peroxynitrite, in experimental allergic encephalomyelitis and multiple sclerosis. *Proc. Natl. Acad. Sci. USA* **95**:675–680; 1998.

[194] Oury, T. D.; Piantadosi, C. A.; Crapo, J. D. Cold-induced brain edema in mice. Involvement of extracellular superoxide dismutase and nitric oxide. *J. Biol. Chem.* **268**:15394–15398; 1993.

[195] Zingarelli, B.; Ischiropoulos, H.; Salzman, A. L.; Szabo, C. Amelioration by mercaptoethylguanidine of the vascular and energetic failure in haemorrhagic shock in the anesthetised rat. *Eur. J. Pharmacol.* **338**:55–65; 1997.

[196] Keefer, L. K.; Nims, R. W.; Davies, K. M.; Wink, D. A. "NONOates" (1-substituted diazen-1-ium-1,2-diolates) as nitric oxide donors: convenient nitric oxide dosage forms. *Methods Enzymol.* **268**:281–293; 1996.

[197] Sugino, N.; Takiguchi, S.; Kashida, S.; Takayama, H.; Yamagata, Y.; Nakamura, Y.; Kato, H. Suppression of intracellular superoxide dismutase activity by antisense oligonucleotides causes inhibition of progesterone production by rat luteal cells. *Biol. Reprod.* **61**:1133–1138; 1999.

[198] Raineri, I.; Huang, T. T.; Epstein, C. J.; Epstein, L. B. Antisense manganese superoxide dismutase mRNA inhibits the antiviral action of interferon-gamma and interferon-alpha. *J. Interferon Cytokine Res.* **16**:61–68; 1996.

[199] Troy, C. M.; Derossi, D.; Prochiantz, A.; Greene, L. A.; Shelanski, M. L. Downregulation of Cu/Zn superoxide dismutase leads to cell death via the nitric oxide-peroxynitrite pathway. *J. Neurosci.* **16**:253–261; 1996.

[200] Gow, A.; Duran, D.; Thom, S. R.; Ischiropoulos, H. Carbon dioxide enhancement of peroxynitrite-mediated protein tyrosine nitration. *Arch. Biochem. Biophys.* **333**:42–48; 1996.

Bioassays for Oxidative Stress Status (BOSS). Edited by W.A. Pryor

UNIQUE IN VIVO APPLICATIONS OF SPIN TRAPS

Lawrence J. Berliner,* Valery Khramtsov,†‡ Hirotada Fujii,§ and Thomas L. Clanton†

*Department of Chemistry, The Ohio State University, Columbus, OH, USA; †Department of Internal Medicine, Davis Heart & Lung Institute/Pulmonary and Critical Care, Columbus, OH, USA; ‡Institute of Chemical Kinetics and Combustion, Novosibirsk, Russia; and §Sapporo Medical University, School of Health Science, Sapporo, Hokkaido, Japan

(Received 2 August 2000; Revised 25 October 2000; Accepted 21 November 2000)

Abstract—The ultimate goal of in vivo electron spin resonance (ESR) spin trapping is to provide a window to the characterization and quantification of free radicals with time within living organisms. However, the practical application of in vivo ESR to systems involving reactive oxygen radicals has proven challenging. Some of these limitations relate to instrument sensitivity and particularly to the relative stability of these radicals and their nitrone adducts, as well as toxicity limitations with dosing. Our aim here is to review the strengths and weaknesses of both traditional and in vivo ESR spin trapping and to describe new approaches that couple the strengths of spin trapping with methodologies that promise to overcome some of the problems, in particular that of radical adduct decomposition. The new, complementary techniques include: (i) NMR spin trapping, which monitors new NMR lines resulting from diamagnetic products of radical spin adduct degradation and reduction, (ii) detection of •NO by ESR with dithiocarbamate: Fe(II) "spin trap-like" complexes, (iii) MRI spin trapping, which images the dithiocarbamate: Fe(II)-NO complexes by proton relaxation contrast enhancement, and (iv) the use of ESR to follow the reactions of sulfhydryl groups with dithiol biradical spin labels to form "thiol spin label adducts," for monitoring intracellular redox states of glutathione and other thiols. Although some of these approaches are in their infancy, they show promise of adding to the arsenal of techniques to measure and possibly "image" oxidative stress in living organisms in real time. © 2001 Elsevier Science Inc.

Keywords—EPR, NMR, MRI, DEPMPO, Nitric oxide, Disulfide biradical, Free radical

INTRODUCTION

In the first section of this article, we critically evaluate the utilization of spin traps with electron spin resonance

Address correspondence to: Lawrence J. Berliner, Ph.D., Department of Chemistry and Biochemistry, Olin 202, 2190 E. Iliff Avenue, University of Denver, Denver, CO 80208, USA; Tel: (303) 871-2436; Fax: (303) 871-2254; E-Mail: berliner@du.edu.

Professor Lawrence J. Berliner, Ph.D., was a graduate student at Stanford during the genesis of EPR spin labeling. He was one of the first scientists involved in the development of in vivo EPR. His current interests are free radical intermediates of drugs.

Valery Khramtsov received the Ph.D. from the Institute Chemical Kinetics and Combustion, Russian Academy of Sciences, Novosibirsk, where he is currently a senior researcher. He has worked on novel spin probes, labels and traps.

Dr. Hirotada Fujii received his Ph.D. in polymer science from the Tokyo Institute of Technology. He developed in vivo ESR with Prof. Berliner, continuing at the Tokyo Metropolitan Medical Research Institute before becoming Professor at Sapporo Medical University.

Professor Thomas Clanton, Ph.D., is also with Physiology and Cell Biology at Ohio State. His Ph.D. was in Physiology from the University of Nebraska. His current interests are acute redox-mediated responses to stress stimuli, skeletal muscle biology, and techniques for free radical detection.

(ESR) for detection of reactive oxygen radicals (resulting from oxidative activity in living systems). Though our laboratories have been intimately involved with in vivo electron spin resonance (ESR) since its infancy [1], and we believe it continues to have an important role in studying free radical biology, we have written this review from a particularly provocative stance. In doing so, it is our intention to challenge the community to consider new approaches that utilize the strengths of spin trapping techniques, coupled with methodologies that show some promise in overcoming some of the shortcomings encountered with current spin trapping methodology.

In deciding on any given probe or technique to study free radical oxidative activity in living organisms there are a few general specifications or criteria that can be applied for comparative evaluation. These criteria, which are listed below, are applicable to a wide variety of methods, not just spin traps. To our knowledge, no ideal probe or method exists today that entirely meets these criteria.

Reprinted from: *Free Radical Biology & Medicine*, Vol. 30, No. 5, pp. 489–499, 2001

Sensitivity of the measuring device

Since many free radicals occur at low concentration in biology with short half-lives due to their high reactivity with other reactive species and metabolites, one requires extremely sensitive measuring techniques. This relates to instrument specifications and the intrinsic sensitivity of the spectroscopic method.

Stability of the reaction products

In order to overcome instrument sensitivity limits, the measurement generally necessitates that signals *accumulate* in the tissue until a detection threshold level is reached. In order for the products of reaction (i.e., between spin traps and oxy- or other radicals) to accumulate to high enough concentrations to be detected, these reaction products must be relatively stable in living systems, (i.e., tissue).

Specificity of free radical or oxidant reactions

The spin trap/probe must exhibit highly specific reactions with oxidants or radicals. Furthermore, the products must not be affected by normal enzymatic, metabolic, or other reactions, in or out of the cell (see also previous item).

Localization

Ideally, the probe will target specific cells or body compartments and, after reaction with radicals and/or oxidants, will remain and accumulate in the same compartment, at least until reaching sufficient concentrations to be detected. Also important is the hydrophobic vs. hydrophilic partitioning of the spin trap/probe and/or products and their extracellular, intracellular, and membrane compartmentalization.

Toxicity and invasiveness

The spin trap/probe must not interfere substantially with normal cell or organism function, which affects free radical or oxidant production. In addition, the instrumentation should ideally operate noninvasively or, at the worst, in a minimally invasive mode.

EVALUATION OF IN VIVO ESR DETECTION USING SPIN PROBES

In vivo ESR provides a window to the quantification of free radical production in real time and within living organisms [2]. However, the technical application of in vivo ESR to practical problems involving oxidative stress has been challenging. Using the yardsticks described above, the limitations of in vivo ESR, to date, can be evaluated.

Sensitivity and specificity of ESR spin trapping

No other technique has proven to be as sensitive or as specific for free radical detection from a purely analytical point of view. Therefore, in this category in vivo ESR (coupled with the use of spin traps) is the technique of choice. However, other factors detract from the practical sensitivity, which are discussed in the following paragraphs.

Stability of free radical adducts

In general, products from spin trap/free radical reactions have half-lives on the order of several minutes to hours, depending on the specific nitrone and the adduct formed. Most of the published half-lives were measured in buffer solutions, not in a whole animal or living organism. Spin traps have been very useful in studying the O_2 burst of neutrophils and other inflammatory cells bathed in buffer solutions [3–5]. However, measurements of oxygen radical adduct formation under similar conditions in whole blood, with nitrone spin probes is rarely successful because the signal rapidly disappears [6]. For example, the estimated half-life of DEPMPO/ $^•OH$ is approximate 1–2 min in vivo [7]. The low level of radical adducts in vivo reflects both competing reactions for the primary radical and the capacity of red blood cells and other blood cells to reduce nitrone adducts and other radicals to ESR silent diamagnetic products [8]. This is most prevalent intracellularly where bioreduction is facile [8]. There have, however, been some exceptionally impressive results in vivo with direct detection of nitrone adducts [7,9] as well as with a spin trapped hemoglobin thiyl radical in rats [10,11], and with quite persistent lipid radical adducts with nitrosobenzene and nifedipine [12]. However, under less severe conditions than reported above [7,9] the most success in obtaining ESR-detectable oxyradical spin adducts from living systems are usually in buffer-perfused organs or cells. In whole animals, the maximum detectable adduct levels are generally in the range of 10–100 μM.

Frequently, this quantitation is complicated by the kinetics of competing pathways of adduct formation and degradation. The advantage of using spin traps is that the radical is "stabilized" as a unique, stable paramagnetic molecule whose characteristic ESR resonances can be assigned to a particular radical species. In practice, the concentration of adduct at any point in time is the net result of the rates of reaction of the nitrone with the radical (which also competes with other antioxidants and radical scavengers) and the decay rate of the nitrone radical adduct. Since the decay rate depends on adduct concentration, at high concentrations these adducts actually degrade more rapidly by various bimolecular decay processes (disproportionation, reduction) than at low concentrations where only reduction will be dominant. Therefore, the net concentration of a nitrone radical adduct will be a complex function of these processes [13]. Consequently, although some in vivo ESR measurements of free radical reactions have been successful, quantitative interpretation remains more difficult.

Localization

Another critical issue is the cellular distribution of specific spin traps/probes and their reaction products, an area where further research is critical. Since a spin trap or a radical product varies in its relative water/lipid solubility, both accessibility to the source of free radical production and the distribution of adduct formed are additional variables to consider. Common nitrone spin probes, such as 5,5-dimethylpyrroline-1-oxide (DMPO) and DEPMPO are very water-soluble, with partition coefficients in octanol/water of 0.15 and 0.13, respectively (V. Khramtsov, unpublished observations), whereas other probes, such as PBN and its analogs, are very lipid soluble. Considering these distribution coefficients, it is most likely that DMPO or DEPMPO are more efficient at trapping extracellular radicals that might arise from, for example, inflammatory cell activation, or radicals produced at the cell membrane surface. Although intracellular trapping of radicals is also believed to occur with DMPO and DEPMPO [14,15], this is more difficult to characterize in vivo. On the other hand, spin traps/probes that are highly lipid soluble will tend to concentrate in cell membranes. For example, PBN reacts with intracellular free radicals buried in membranes and will most likely react with lipophilic radicals. Since the lipid phase is less accessible to most (water-soluble) bioreductants we find that carbon-based radicals in particular remain stable in the membrane for extended periods and can even be detected in excised samples. For example, ex vivo detection of PBN radical adducts was demonstrated by a number of investigators [16–18]. In our laboratory we evaluated aspects of oxidative stress induced by respiratory failure by using PBN post-trapping protocols [19] modeled after the work of Mergner et al. [18]. However, this approach has yet to be successful for direct in vivo measurement since radical adduct levels in the membrane were too low to be easily measured by ESR unless the adducts were extracted into organic media (e.g., chloroform/methanol), evaporated, concentrated 50-fold or more and measured at X-band. In ad-

dition, there are issues related to RF penetration. Most in vivo ESR studies at L-band have used small animals such as the mouse or weanling rat [2]. Although larger instruments are now available down to 250 MHz in both continuous wave and FT-pulsed operation [20], the theoretically promised sensitivity is similar to L-band [21]. The tradeoff in going to L-band is a general loss of sensitivity compared to X-band. L-band has the advantage of a much larger number of spins, since a larger sample can be accommodated. X-band takes advantage of the very high frequency sensitivity and cavity Q for small chemical or biological samples. The use of noninvasive in vivo ESR for measurements of radical production in any human application would likely be restricted to the application of surface coils where practical use is limited to 1 cm or so below the skin [22]. Alternatively, somewhat more invasive procedures involving the use of needle probes or catheters can be targeted to specific tissues. This holds particular promise for future free radical applications as well as the measurement of oxygen tension [22]. Certainly, the most sensitive approach has been the monitoring of spin traps by X-band ESR on excised (ex vivo) samples [16,17,19].

Toxicity and practical applications of spin trap dosing for in vivo applications

Most spin traps are relatively nontoxic and some, such as PBN, are even being tested effectively as therapeutic agents for septic shock [23] and ischemia, etc. Nonetheless, there is a finite limit to the practical concentrations that can be employed in vivo. For example, at concentrations above 5 mM, PBN strongly inhibits skeletal muscle function [24]. While this inhibition is completely reversible and similar effects have been noted in both heart [25] and smooth muscle [26], dosing at these levels could result in respiratory or cardiac failure. Liu and colleagues [9] have studied the in vivo pharmacodynamics of DEPMPO at distribution concentrations <10 mM in the whole mouse without undue toxicity; however, 20 mM DEPMPO resulted in death in <1 h. Up to 13 mM appears to be safe for DMPO [27]. The tradeoff with using lower concentrations of spin traps, in vivo, is that this is a second order reaction. In addition, the spin trap must compete with other targets for the radical, including an enormous reservoir of antioxidants and radical scavenging agents. Nevertheless, Liu et al. [9] were able to detect significant radical levels in vivo for the $SO_3^{\cdot-}$ radical adduct at concentrations of 2.5 mM DEPMPO. Fortunately, the DEPMPO/$SO_3^{\cdot-}$ adduct is reasonably stable, making accurate quantification more possible than with many other biologically relevant radicals. On the other hand, concentrations of 25 mM or more are frequently required with DEPMPO or DMPO for satisfactory trapping efficiency and quantification of oxygen radicals from human neutrophils in buffer solutions, but this is obviously impractical for in vivo experiments because of toxicity concerns.

In summary, in vivo ESR techniques have much to offer [7,9]. They are certainly the gold standard, especially since they remain the only method specific for free radical characterization. The problem is that sometimes we cannot easily quantify radical production in vivo if the radicals or their spin adducts undergo rapid decomposition reactions in vivo. We describe below alternative methods which offer some potential advantages for quantification, but also with some tradeoffs in specificity and/or sensitivity. While these complementary techniques are in the early stages of development, they hold promise in overcoming some of the obstacles which can occur with ESR spin trapping in vivo.

NEW APPROACHES WITH IN VIVO SPIN TRAPPING

NMR spin trapping

Why NMR for free radicals? Normally, if one performs NMR on a radical, the spectrum is paramagnetically broadened to almost the baseline, preventing any detailed characterization or even detection of the resonance. As described above, many of the degradation/decomposition reactions of spin trap adducts can lead to diamagnetic products. Proton NMR is frequently too rich in resonance lines to easily resolve anything useful. However, two alternative stable isotopes with comparable sensitivity are ^{31}P and ^{19}F which, if incorporated into spin traps, should give simple, resolvable spectra of any diamagnetic reaction products. Hence, we could potentially salvage some valuable information from these degradation reactions.

Recently, we have demonstrated this approach, termed *NMR spin trapping* (ST-NMR), by following the ^{31}P-NMR spectra of a phosphorus containing nitrone spin trap, DEPMPO, after reaction with a variety of free radical species [13]. A similar approach was developed in the late 1980s with fluorine-based probes for radical reactions in organic solvent [28]. ST-NMR retains the

icals results in uniquely shifted ^{31}P-NMR lines that reflect radical adducts that were subsequently converted into secondary diamagnetic "NMR visible" degradation products. An example of the NMR spectra from the reaction of DEPMPO with $^{\bullet}CH_3$ are shown in Fig. 1. Here the starting "parent" DEPMPO (23.67 ppm), converted to a pair of related lines at 32.31 ppm and 30.83 ppm, to a product that reflected reaction with $^{\bullet}CH_3$ radical. These lines are easily resolved from other in vivo phosphorus metabolites (e.g., ATP, ADP, inorganic phosphate, creatine phosphate, etc.) and represent only *diamagnetic* forms of the DEPMPO/adduct degradation/ reduction and/or disproportionation products. The experiment in Fig. 1 was performed in the presence of ascorbate, which results in a "recycling" of Fenton chemistry, yielding a steady accumulation of product over time (insert), which potentially simulates in vivo conditions. Another advantage here is that ascorbate rapidly reduces the radical adduct(s) to their respective hydroxylamines, limiting pathways to other diamagnetic products.

For carbon-based radicals these diamagnetic products have extremely long lifetimes, whereas the corresponding paramagnetic radical adducts decay relatively rapidly (i.e., in a few minutes). Sensitive detection of these comparatively stable diamagnetic products from degradation of paramagnetic spin adducts is accomplished through accumulation of these products at levels amenable to NMR. (e.g., $>100~\mu M$ for ^{31}P-NMR) [13]. Thus, the possibility of accumulating stable, diamagnetic spin adduct degradation products (Fig. 1) decreases the gap in sensitivity between NMR and ESR remarkably. Moreover, this may be particularly important for in vivo experiments, since cellular systems have a high capacity for bioreduction [6,8,29].

In principle, the ability to distinguish and quantify multiple degradation products by NMR also provides insight into the specific degradation pathways, which can be extremely useful and may be unique to particular in vivo environments. In the case of DEPMPO spin trapping, radical adduct(s) RA_1 and RA_2 are formed according to Scheme 1:

$$\text{(1)}$$

most important feature of ESR spin trapping, namely sensitivity to the structure of the trapped radical.

The reaction of DEPMPO with a variety of free rad-

RA_1 and RA_2 are two stereoisomers that can be distinguished by ESR for all species such as DEPMPO/$O_2^{\bullet-}$ or DEPMPO/$CH_3OO^{\bullet}$ [13]. In high concentrations, RA_1

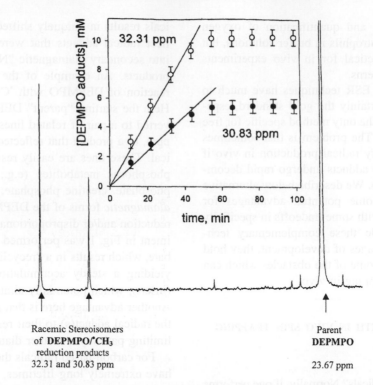

Fig. 1. ^{31}P-NMR spectra of reaction products from exposure of DEPMPO (100 mM), with a methyl radical producing mixture (i.e., 2mM DTPA, 0.2 mM FeSO$_4$, 20 mM H$_2$O$_2$, and 10% DMSO). Experiments were performed in the presence of 20 mM ascorbate, which causes recycling of Fe^{+3} to Fe^{+2} and an accumulation of product. The spectra shows the "parent" DEPMPO nitrone (right) and the racemic hydroxylamine reduction products formed from methyl radical production. Inset: Time course of product accumulation. (Adapted from [13] with permission.)

and RA$_2$ generally decay by bimolecular disproportion-ation and/or reduction to the corresponding hydroxyl-amines, HA (i.e., HA$_1$ and HA$_2$) and a new nitrone, NA, according to Scheme 2 [13]:

27.05 ppm) resulting from reaction of DEPMPO with either hydroxyl or superoxide radicals, which suggests some additional conversion of one or both of the prod-ucts in Scheme 2. At low oxygen radical concentrations,

$$\text{(2)}$$

In Fig. 1, only the hydroxylamine products were ob-served because the radical adducts formed were quickly reduced by ascorbate to their corresponding hydroxyl-amines, bypassing bimolecular disproportionation to the nitrone. As shown in Scheme 2, all of the products have been characterized by NMR for a variety of free radical species [13].

Not surprisingly, we have encountered some practical problems to date with these systems, requiring caution, particularly for detection of oxygen-centered radicals. For example, we have observed the same ^{31}P-line (at

this ultimately results in elimination of the NMR signals due to hydroxylamine formation, which regenerates the parent nitrone, DEPMPO [13]. Interestingly, this can provide some insight into the antioxidant properties of these spin probes, which may have medical significance [14]. The major obstacle is still the comparatively low sensitivity of ^{31}P-NMR spectroscopy, even taking into account product accumulation. Nevertheless a recent pa-per reported a characteristic NMR peak, shifted 3.3 ppm down from the parent DEPMPO, taken 1 h after warm ischemia and reperfusion of isolated rat liver [30]. The

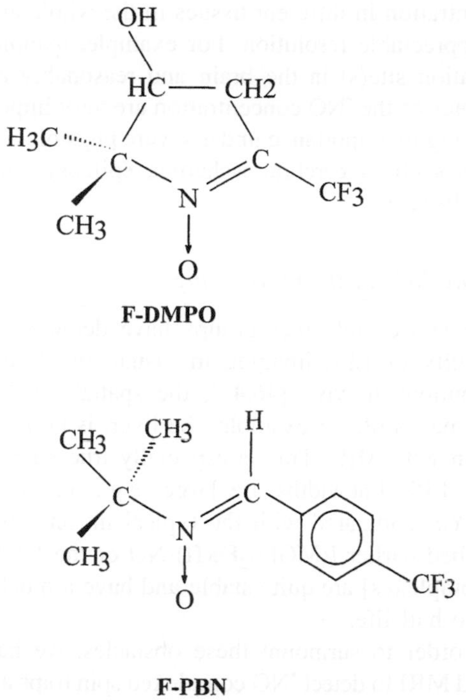

Fig. 2. Chemical structures of fluorine spin traps for NMR spin trapping. (Adapted from [31] with permission.)

identity of this peak, relative to that observed for hydroxyl-generating systems [13], suggests assignment to the diamagnetic products of DEPMPO-$\cdot$OH reaction products, most probably N-hydroxy-pyrrolidone. Consequently, we can trace free radical reactions from their unique, diamagnetic degradation products. This "observation" of free radical generation in this ischemic injury model supports the potential use of NMR spin trapping in biomedicine. Recently, we have begun exploring ^{19}F-based spin traps (Fig. 2), which have the advantage of greatly improved signal to noise and essentially no background fluorine-containing compounds in biological systems [31,32]. In summary, by using ^{31}P- or ^{19}F-NMR we eliminate the multitude of overlapping NMR signals that would occur with the more common ^{1}H or ^{13}C nuclei. A potential disadvantage, the almost 1000-fold reduced sensitivity of NMR versus ESR, can be overcome by accumulating these stable diamagnetic products from the spin adduct breakdown. Importantly, the inertness of these diamagnetic products to the reducing environment gives ST-NMR its most important advantage in biological systems.

Nitric oxide: Ex vivo and in vivo detection by ESR

In this section, we review progress, problems, and developments in the monitoring of perhaps the most important free radical in biology, $\cdot$NO. We maintain the

same provocative theme of finding alternative, yet complementary spectroscopic methods that offer advantages over present techniques. Nitric oxide has been implicated in many diverse physiological processes, including smooth muscle relaxation, inhibition of platelet adhesion, neurotransmission, and many others [33]. Its importance in oxidative stress has recently become of major interest with respect to mitochondrial survival and apoptosis [34]. It is critical to be able to observe real time $\cdot$NO generation at the site of production. Several methods of detecting $\cdot$NO have been developed including chemiluminescence, oxyhemoglobin, GC-MS, chemical measurements of the end products NO_2^-/NO_3^-, fluorescent dye, and nitrosyl-hemoglobin formation by EPR [35–39]. As far as we know, however, none of these methods can be easily applied in vivo to isolated tissue or to experimental animals in order to determine real time $\cdot$NO production.

Hence it has been of intense interest to find sensitive, ESR in vivo measurements for $\cdot$NO even though we have known for years that the $\cdot$NO adduct of hemoglobin (or myoglobin) gives characteristic ESR spectra. These complexes are best observed in liquid nitrogen or helium, where a mouse would be highly stressed! Furthermore, isolation of the myoglobin or hemoglobin is preferable since the spectrum can be observed without contamination of other heme-iron compounds. The real breakthrough in ESR detection of $\cdot$NO occurred when Lai and Komarov [40] and Komarov and Lai [41] recognized the value of dithiocarbamate:Fe (II), particularly N-methyl-D-glucamine) (MGD) complexes as good spin traps for $\cdot$NO. Actually, these are chelated *complexes* that stabilize the paramagnetic $\cdot$NO as a strong, three line (^{14}N isotope) spectrum at room temperature. Consequently, it is technically a spin *complex*, not a spin trap in the pure sense.

Fujii, Koscielniak, and Berliner [42] applied this to whole-body, live animal studies at L-band by injecting $(MGD)_2$-Fe(II) complex into mice undergoing septic shock. Figure 3 shows comparative EPR control spectrum (a) after injection of exogenous $(MGD)_2$Fe(II)-NO complex to mice, followed by a spectrum (b) reflecting physiological $\cdot$NO generation induced 6 h earlier with lipopolysaccharide (LPS), then injected with $(MGD)_2$Fe(II). The $\cdot$NO levels accumulated to a maximum of about 100 μM in the liver, while the levels in the brain were very low; in fact, the small level of $\cdot$NO-complex detected was from the tissue area surrounding the head region since $(MGD)_2$Fe(II)-NO cannot pass the blood-brain barrier.

Vanin and coworkers first employed the water insoluble N,N-diethyldithiocarbamate (DETC)-Fe(II) complex to detect $\cdot$NO in septic-shock mice by injecting DETC and Fe(II) separately in the animal and examining

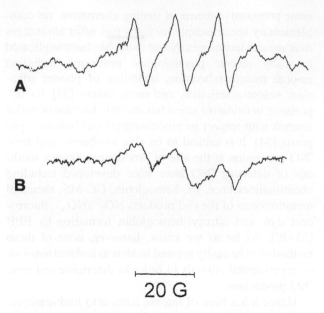

20 G

Fig. 3. (A) L-band EPR spectra of preformed $(MGD)_2$-Fe(II)-NO in vivo in the mouse. $(MGD)_2$-Fe(II)-NO complex measured at the abdomen 0.4 ml of $(MGD)_2$-Fe(II)-NO solution was injected i.v. in the lateral tail vein of the mouse. The mouse (ca. 20 g) was anesthetized with sodium pentobarbital (50 mg/kg) 2 h after injection and the L-band EPR spectra were measured as shown. The $(MGD)_2$-Fe(II)-NO complex was prepared by mixing 2 mM NO (saturated saline) solution with $(MGD)_2$-Fe(II) in saline [MGD:Fe(II), 100 mM:20 mM]. (B) In vivo L-band EPR spectra of $(MGD)_2$-Fe(II)-NO in an LPS-treated mouse. The animal (ca. 20 g) was injected i.p. with LPS in saline (1 mg/0.3 ml), and after 6 h a subcutaneous injection with 0.4 ml of $(MGD)_2$-Fe(II) complex in saline [MGD:Fe(II), 100 mM:20 mM]. After 2 h following the last injection the EPR spectrum of the mouse was measured. Spectrometer conditions were: frequency, 1.256 GHz; applied magnetic field, 425 gauss; microwave power, 10 mW; modulation, 0.5 gauss; sweep rate, 50 gauss/min. (Adapted from [42] with permission.)

several organs including brain tissue [43,44]. More recently $(DETC)_2$-Fe(II)-NO detection in the brain of septic-shock mice was also confirmed in vivo and ex vivo using ESR at L-band [45]. The overall results suggested that lipophilic DETC alone is able to cross the blood brain barrier and then complex (consecutively) with Fe(II) from the tissue, followed by complexing with ˙NO, also generated within the lipophilic brain tissue.

Although ESR has been a powerful method for detecting this free radical over long periods of time with reasonable sensitivity under severe pathological conditions, ˙NO is actually difficult to quantitate because it is produced in small amounts under basal physiological conditions by the constitutive nitric oxide synthase (NOS) and has a short lifetime in oxygenated aqueous media. The ESR experiments require quite high boluses of dithiocarbamate: Fe(II) (i.e., 100 mM:20 mM) and the possibility of oxidation to the Fe(III) complex and resulting toxicity cannot be totally ruled out. In addition, it has not been easy to make a detailed examination of ˙NO

concentration in different tissues in the whole animal at any appreciable resolution. For example, mapping ˙NO generation site(s) in the brain and reasonably accurate estimates of the ˙NO concentration are very important in clarifying its importance under severe physiological conditions such as cerebral ischemia, epilepsy, and other brain diseases.

Imaging NO: MRI spin trapping

While we and other groups have demonstrated the feasibility of EPR imaging in visualizing free radical distributions in vivo [46,47], the spatial resolution of EPR images of, for example, the liver, is poor by comparison with MRI. This is especially true since the intrinsic EPR linewidths are large (e.g., 3.5–4.0 gauss) [48]. Yet compared with the typical nitrone spin traps described earlier [$(MGD)_2$-Fe(II)-NO or the DETC analog complexes] are quite stable and have a much longer in vivo half-life.

In order to surmount these obstacles, we have employed MRI to detect ˙NO complexed spin traps and have evaluated the feasibility of mapping ˙NO distributions in septic-shock rats [49]. For some years, stable nitroxyl radicals have been evaluated as potential MRI contrast agents, despite their susceptibility to bioreduction [50, 51]. $(MGD)_2$-Fe(II)-NO complex shows good proton relaxation enhancement since the unpaired electron promotes both spin-lattice (T_1) and spin-spin (T_2) relaxation of the surrounding water (and other protons), resulting in a decrease in their nuclear spin relaxation times, which can be exploited as enhanced signal in T_1 or T_2 weighted MR images. T_1 and T_2 relaxation times of $(MGD)_2$-Fe(II)-NO in aqueous media at 20 MHz and 85 MHz yields T_1 relaxivity of 0.31 and 0.27 (1/mM · s) at 20 MHz and 85 MHz, respectively. T_2 relaxivity corresponded to 0.31 and 0.35 (1/mM · s), respectively, at the same two frequencies. The T_1 relaxivity of the uncomplexed spin trap $(MGD)_2$-Fe(II) alone was 0.044 (1/mM · s) at both frequencies. The distinct increase in relaxivity occurring after *complexation* of ˙NO with $(MGD)_2$-Fe(II) suggests a feasible contrast agent for visualizing regions in vivo where the ˙NO was produced (trapped) [49]. In septic rats (induced with LPS 6 h earlier) the resultant $(MGD)_2$-Fe(II)-NO complex, which concentrated in the liver, displayed significant contrast in the hepatic vein and inferior vena cava vascular structure (Fig. 4). Additionally, the source of ˙NO was verified as NOS in these rats by administering the competitive inhibitor, N-monomethyl L-arginine, where the image showed significantly reduced enhancement. In addition, the ˙NO complex was more stable in vivo and proved a more effective MRI contrast agent than other stable nitrogen containing radicals, such as nitroxides.

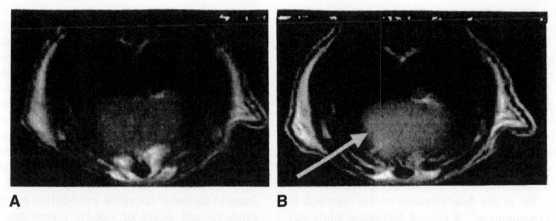

A   **B**

Fig. 4. Transverse T_1-weighted MR images focused on a selected region of the liver in LPS-treated rats. Six hours after LPS injection, the NO spin trap (3 ml of (MGD)$_2$-Fe(II), MGD: 100 mM, Fe: 20 mM) was administered i.p. The MR images were measured at (A) 35 min and (B) 90 min after injection of (MGD)$_2$-Fe(II). This observation was consistent with our previous EPR observations where the signal intensity reached a maximum at 90–120 min after administration of spin trap to LPS-injected mice. The arrow in the right image designates the most enhanced region of the liver. MR images of the anesthetized rats ($n = 3$) were obtained with a Signa Horizon 1.5T scanner (version 5.6), GE Medical System, Milwaukee, WI, USA. Parameters for T_1-weighted spin echo images were TR 500 ms, TE 10 ms, 2NEX, 4 mm slice thickness, 1 mm slice gap, field-of-view 12 × 12 cm, and matrix, 256 × 256. A 6 cm in diameter birdcage coil was used. (Adapted from [49] with permission.)

This approach, called *MRI spin trapping*, encompasses the class of paramagnetic contrast agents described above, but could potentially involve the use of other probes as well. Hence, it is similar, but not identical to NMR spin trapping. In summary, we have outlined two alternative magnetic resonance approaches, suitable not only for mapping ·NO, but also for detection of other important kinds of free radical reactions in vivo when combined with specific spin trapping techniques. The use of other NMR compounds that are not spin traps but are highly redox-sensitive also have potential applications for these methods.

Detection of thiol status using ESR and NMR

The key role of thiols and thiol-disulfide status of the cell in response to oxidative stress has been amply documented [52]. Among other low-molecular-weight compounds, glutathione (GSH) is particularly important because it is present in all animal cells and blood in concentrations up to a few millimolar. The [GSH]/[GSSG] ratio is generally believed to be an important marker of oxidative stress status [52]. Optical and chromatographic methods have been used for quantitative determination of sulfhydryl groups [53–55]. Their limitations are the requirement of optical transparency of the samples and labor-consuming procedures in case of chromatographic methods, in particular HPLC. ESR methods for quantitative determination of sulfhydryl groups [56–58] potentially overcome some of these limitations. The method is based on the application of disulfide biradicals (another spin trap-like compound), which participate in thiol-disulfide exchange reactions

with thiols (as shown in Fig. 5), resulting in distinct alterations in the ESR spectrum, since a three-line monoradical spectrum ("thiol spin label adduct") results from the five-line biradical starting material. For example, the biradical R$_1$S-SR$_1$ has been used to measure GSH content in erythrocytes [56], Chinese hamster ovary cells [59], murine neuroblastoma, malignant melanoma cells [60], and tumor HeLa cells [61] during the treatment by anticancer antibiotics. Noninvasive measurement of intracellular GSH by ESR is based on the dominant contribution of GSH reacting with the biradical, fast diffusion of the biradical across the cell membrane and its tendency for comparatively low reduction susceptibility. The sensitivity of the method is sufficiently high to perform the measurements in a few (~100) cells [61] or in whole tissues [62]. ESR studies of thiols in human and rat blood showed increased levels of oxidized GSH in the plasma under oxidative stress, including some human

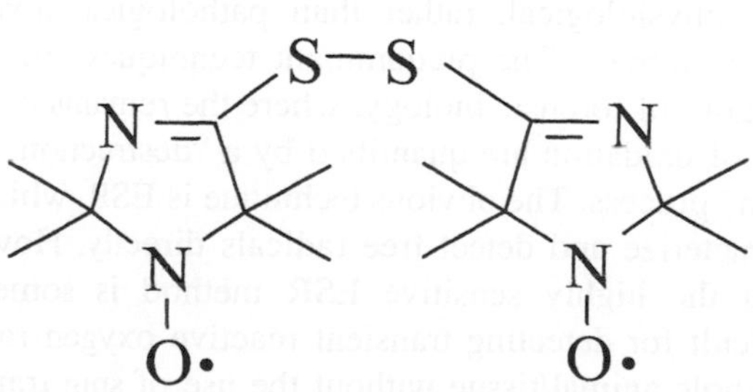

R$_1$S-SR$_1$

GlutathioneSH + R$_1$S-SR$_1$ → GlutathioneS-SR$_1$ + R$_1$S·

Fig. 5. Chemical structure of disulfide biradical R$_1$S-SR$_1$ and its reaction with glutathione.

pathologies, such as kyphoscoliosis [63]. The ESR assay for thiol measurements in the blood meets the requirements of a sensitive, convenient method that does not need complicated sample preparation.

Nohl et al. [62] described the application of this biradical for the determination of thiol levels in isolated perfused hearts undergoing oxidative stress induced by ischemia/reperfusion. They measured the monoradicals released into the perfusate that resulted from reaction of the biradicals with tissue GSH during constant infusion of the biradical. Following ischemia/reperfusion, clear differences in thiol levels were seen compared with controls [62]. One of the disadvantages of this approach is that the "consumption" of critical biological thiols can irreversibly damage the system under study and therefore toxicity becomes a significant concern. However, recent studies with a new imidazolidine disulfide biradical allows the application of much lower concentration of the label without significant consumption of thiols [57].

NMR approaches for noninvasive estimation of glutathione status in humans and other living organisms, which take advantage of the high cellular concentration of this species, are also under investigation. Current techniques do not use spin trapping agents but they could eventually be applied in the future. Recent studies have measured total brain glutathione using a double quantum coherence filtering technique of ^{1}H -NMR [64]. These methods, again in their infancy, show promise toward development of noninvasive monitoring of redox status for diagnostic testing.

CONCLUSIONS AND FUTURE DIRECTIONS

The importance of monitoring reactive oxygen species or related radicals in medicine cannot be understated because of their strong participation in pathology and disease. The future is in new methodology. Despite the progress over the last 20 years, our current noninvasive techniques do not yet have the sensitivity to adequately detect reactive oxygen and nitrogen species at steady-state physiological, rather than pathological levels in living animals. The predominant techniques fall in the category of forensic biology, where the remains of products of oxidation are quantified by a "destruction evaluation" process. The obvious technique is ESR, which can characterize and detect free radicals directly. However, even the highly sensitive ESR method is sometimes difficult for detecting transient reactive oxygen radicals in whole animal/tissue without the use of spin traps. As discussed earlier, while spin trap adducts should accumulate to detectable levels of paramagnetic product with time, this is frequently complicated by bioreduction, adduct breakdown, and diffusion out of the tissue. Nonetheless, the experiments described above have demon-

strated success where very high levels of radical production were induced. Alternatively, the somewhat more invasive procedures involving the use of needle probes or catheters combined with stable inert paramagnetic materials targeted to specific tissues hold much promise for the measurement of oxygen tension in tissue, which may reflect oxidative stress in certain situations [22].

The applications of NMR and MRI to studying radical reactions and oxidative status have been demonstrated as proof of concept. However, again, concentration and sensitivity issues are a major limitation, especially given the lower sensitivity for nuclear spins. The major challenge to the use of both ESR and NMR in this field is the synthesis and design of suitable probes that meet the criteria described in the introduction. Given that the total number of useful spin traps has yet to reach a baker's dozen, the community direly needs a major multidisciplinary effort, directed at the synthesis of new classes of compounds that exploit the advantages listed above. By using NMR, these do not actually have to be spin probes at all, but rather compounds that react with high fidelity and redox specificity in a biological environment. Hence, we have pointed out both the advantages and disadvantages of these two magnetic resonance techniques (ESR and NMR) and have shown how the "marriage" of the two may yield the most profitable approach. Although considerable progress has been made recently in developing new probes for fluorescence microscopy, again we are left to a handful of candidates, each with its own significant limitations. Hence, many of the same challenges exist for fluorescence techniques as apply to magnetic resonance spectroscopy.

Acknowledgements — Work was supported by USPHS (HL53333, GM-58772) Ministry of Education, Science, Sports and Culture, Japan (11670923) and the Russian Federation of Basic Research: NU-99-04049921.

REFERENCES

[1] Berliner, L. J.; Fujii, H. Magnetic resonance imaging of biological specimens by electron paramagnetic resonance of nitroxide spin labels. *Science* **227:**517–519; 1985.

[2] Berliner, L. J. In vivo EPR(ESR): Theory and applications. In: Berliner, L. J., ed. *Biological magnetic resonance, vol. 18.* New York: Kluwer Academic/Plenum Publishing Corp.; 2000:in press.

[3] Roubaud, V.; Sankarapandi, S.; Kuppusamy, P.; Tordo, P.; Zweier, J. Quantitative measurement of superoxide generation using the spin trap 5-(diethoxyphosphoryl)-5-methyl-1-pyrroline-N-oxide. *Anal. Biochem.* **247:**404–411; 1997.

[4] Janzen, E. G.; Jandrisits, L. T.; Barber, D. L. Studies on the origin of hydroxyl spin adduct of DMPO produced from the stimulation of neutrophils by phorbol-12-myristate-13-acetate. *Free Radic. Res. Commun.* **4:**115–123; 1987.

[5] Tanigawa, T.; Kotake, Y.; Reinke, L. A. Spin trapping of superoxide radicals following stimulation of neutrophils with fMLP is temperature dependent. *Free Radic. Biol. Med.* **15:**425–433; 1993.

[6] Samuni, A.; Swartz, H. M. The cellular-induced decay of DMPO

spin adducts of hydroxyl and superoxide. *Free Radic. Biol. Med.* **6:**179–183; 1989.

[7] Timmins, G. S.; Liu, K. J.; Bechara, E. J. G.; Kotake, Y.; Swartz, H. M. Trapping of free radicals with direct in vivo EPR detection: a comparison of 5,5-dimethyl-1-pyrroline-N-oxide and 5-diethoxyphosphoryl-5-methyl-1-pyrroline-N-oxide as spin traps for HO and SO_4^-. *Free Radic. Biol. Med.* **27:**329–333; 1999.

[8] Samuni, A.; Carmichael, A. J.; Russo, A.; Mitchell, J. B.; Riesz, P. On the spin trapping and ESR detection of oxygen-derived radicals. *Proc. Natl. Acad. Sci. USA* **83:**7593–7597; 1986.

[9] Liu, K. J.; Miyake, M.; Panz, T.; Swartz, H. M. Evaluation of DEPMPO as a spin trapping agent in biological systems. *Free Radic. Biol. Med.* **26:**714–721; 1999.

[10] Berliner, L. J. The development and future of ESR imaging and related techniques. *Physica Acta* **5:**63–75; 1989.

[11] Jiang, J. J.; Liu, K. J.; Jordan, S. J.; Swartz, H. M.; Mason, R. P. Detection of free radical metabolite formation using in vivo EPR spectroscopy: evidence of rat hemoglobin thiyl radical formation following administration of phenylhydrazine. *Arch. Biochem. Biophys.* **330:**266–270; 1996.

[12] Fujii, H.; Berliner, L. J. In vivo EPR evidence for free radical adducts of nifedipine. *Magn. Reson. Med.* **42:**691–694; 1999.

[13] Khramtsov, V.; Berliner, L. J.; Clanton, T. L. NMR spin trapping: detection of free radical reactions using a phosphorus-containing nitrone spin trap. *Magn. Reson. Med.* **42:**228–234; 1999.

[14] Pietri, S.; Liebgott, T.; Frejaville, C.; Tordo, P.; Culcasi, M. Nitrone spin traps and their pyrrolidine analogs in myocardial reperfusion injury: hemodynamic and ESR implications. *Eur. J. Biochem.* **254:**256–265; 1998.

[15] Cova, D.; De Angelis, L.; Monti, E.; Piccinini, F. Subcellular distribution of two spin trapping agents in the rat heart: possible explanation for their different protective effects against doxorubicin-induced cardiotoxicity. *Free Radic. Res. Commun.* **193:**878–885; 1992.

[16] Mason, R. P.; Kadiska, M. B. Ex vivo detection of free radical metabolites of toxic chemicals and drugs by spin trapping. In: Berliner, L. J., ed. *Biological magnetic resonance, vol. 18.* New York: Kluwer Academic/Plenum Publishing Corp.; 2000:in press.

[17] Burkitt, M. J.; Mason, R. P. Direct evidence for in vivo hydroxyl-radical generation in experimental iron overload: an ESR spin-trapping investigation. *Proc. Natl. Acad. Sci. USA* **88:**8440–8444; 1991.

[18] Mergner, G. W.; Weglicki, W. B.; Kramer, J. H. Postischemic free radical production in the venous blood of the regionally ischemic swine heart: effect of desferoxamine. *Circulation* **84:**2079–2090; 1991.

[19] Hartell, M. G.; Borzone, G.; Clanton, T. L.; Berliner, L. J. Detection of free radicals in blood by electron spin resonance (ESR) in a model of respiratory failure in the rat. *Free Radic. Biol. Med.* **17:**467–472; 1994.

[20] Subramanian, S.; Mitchell, J. B.; Krishna, M. C. Time domain radio frequency EPR imaging. In: Berliner, L. J., ed. *Biological magnetic resonance, Vol. 18.* New York: Kluwer Academic/ Plenum Publishing; 2000:in press.

[21] Rinard, G. A.; Quine, R. W.; Harbridge, J. R.; Song, R.; Eaton, G. R.; Eaton, S. S. Frequency dependence of EPR signal-to-noise. *J. Magn. Reson.* **140:**218–227; 1999.

[22] Swartz, H. M.; Walczak, T. Developing in vivo EPR oximetry for clinical use. In: Hudetz, A. G.; Bruley, D. F., eds. *Oxygen transport to tissue XX.* New York: Plenum Publishing; 1998:243–252.

[23] French, J. F.; Thomas, C. E.; Downs, T. R.; Ohlweiler, D. F.; Carr, A. A.; Dage, R. C. Protective effects of a cyclic nitrone antioxidant in animal models of endotoxic shock and chronic bacteremia. *Circ. Shock* **43:**130–136; 1994.

[24] Andersen, K. A.; Diaz, P. T.; Wright, V. P.; Clanton, T. L. N-*tert*-butyl-alpha-phenylnitrone: a free radical trap with unanticipated effects on diaphragm function. *J. Appl. Physiol.* **80:**862–868; 1996.

[25] Li, X. Y.; Sun, J.; Bradamante, S.; Piccinini, F.; Bolli, R. Effects of the spin trap alpha-phenyl-N-tert-butyl nitrone on myocardial

function and flow: a dose-response study in the open-chest dog in the isolated rat heart. *Free Radic. Biol. Med.* **14:**277–285; 1993.

[26] Anderson, D. E.; Yuan, X.; Tseng, C.; Rubin, L. J.; Rosen, G. M.; Tod, M. L. Nitrone spin traps block calcium channels and induce pulmonary artery relaxation independent of free radicals. *Biochem. Biophys. Res. Commun.* **193:**878–885; 1993.

[27] Janzen, E. G.; Poyer, J. L.; Schaefer, C. F.; Downs, P. E.; DuBose, C. M. Biological spin trapping II. Toxicity of nitrone spin traps: dose-ranging in the rat. *J. Biochem. Biophys. Methods* **30:**239–247; 1995.

[28] Selinsky, B. S.; Levy, L. A.; Motten, A. G.; London, R. E. Development of fluorinated, NMR-active spin traps for studies of free radical chemistry. *J. Magn. Reson.* **81:**57–67; 1989.

[29] Swartz, H. M.; Sentjurc, M.; Kocherginsky, N. Metabolism and distribution of nitroxides in vivo. In: Kocherginsky, N.; Swartz, H. M., eds. *Nitroxide spin labels: reactions in biology and chemistry.* Boca Raton, FL: CRC Press; 1995:153–173.

[30] Delmas-Beauvieux, M. C.; Pietri, S.; Culcasi, M.; Leducq, N.; Valeins, H.; Liebgott, T.; Diolez, P.; Canioni, P.; Gallis, J. L. Use of spin-traps during warm ischemia-reperfusion in rat liver: comparative effect on energetic metabolism studied using ^{31}P nuclear magnetic resonance. *MAGMA* **5:**45–52; 1997.

[31] Khramtsov, V.; Clanton, T. L.; Berliner, L.; Reznikov, V. A.; Sergeeva, S. V.; Litkin, A. K. ^{19}F-NMR spin trapping: detection of free radical reactions with new fluorinated spin traps. *Free Radic. Biol. Med.* **27:**132 abstr.; 1999.

[32] Mooney, E. F. *An introduction to 19F NMR spectroscopy.* London: Heyden; 1970.

[33] Ignarro, L. J. Physiology and pathophysiology of nitric oxide. *Kidney Int. Suppl.* **55:**2–5; 1996.

[34] Liu, L.; Stamler, J. S. NO: an inhibitor of cell death. *Cell Death Differ.* **6:**937–942; 1999.

[35] Westenberger, U.; Thanner, S.; Ruf, H. H.; Gersonde, K.; Sugger, G.; Trentz, O. Formation of free radicals and nitric oxide derivative of hemoglobin in rats during shock syndrome. *Free Radic. Biol. Med.* **11:**167–178; 1990.

[36] Wang, J. F.; Komarov, P.; Sies, H.; de Groot, H. Contribution of nitric oxide synthase to luminol-dependent chemiluminescence generated by phorbol-ester-activated Kupffer cells. *Biochem. J.* **279:**311–314; 1991.

[37] Kosaka, H.; Watanabe, M.; Yoshihara, H.; Harada, N.; Shiga, T. Detection of nitric oxide production in lipopolysaccharide-treated rats by ESR using carbon monoxide hemoglobin. *Biochem. Biophys. Res. Commun.* **184:**1119–1124; 1992.

[38] Lancaster, J. R. Jr.; Hoffman, R. A.; Simmons, R. L. EPR detection of heme and nonheme iron-containing protein nitrosylation by nitric oxide during rejection of rat heart allograft. *J. Biol. Chem.* **267:**10994–10998; 1992.

[39] Kojima, H.; Nakatsubo, N.; Kikuchi, K.; Urano, Y.; Higuchi, T.; Tanaka, J.; Kudo, Y.; Nagano, T. Direct evidence of NO production in rat hippocampus and cortex using a new fluorescent indicator: DAF-2 DA. *Neuroreport* **9:**3345–3348; 1998.

[40] Lai, C. S.; Komarov, A. M. Spin trapping of nitric oxide produced in vivo in septic-shock mice. *FEBS Lett.* **345:**120–124; 1994.

[41] Komarov, A. M.; Lai, C. S. Detection of nitric oxide production in mice by spin trapping electron paramagnetic resonance spectroscopy. *Biochim. Biophys. Acta* **1272:**29–36; 1995.

[42] Fujii, H.; Koscielniak, J.; Berliner, L. J. Determination and characterization of nitric oxide generation in mice by in vivo L-band EPR spectroscopy. *Magn. Reson. Med.* **38:**565–568; 1997.

[43] Vanin, A. F.; Liu, X.; Samouilov, A.; Stukan, R. A.; Zweier, J. L. Redox properties of iron-dithiocarbamates and their nitrosyl derivatives: implications for their use as traps of nitric oxide in biological systems. *Biochim. Biophys. Acta* **1474:**365–377; 2000.

[44] Mikoyan, V. D.; Kubrina, L. N.; Serezhenkov, V. A.; Stukan, R. A. Complexes of Fe^{2+} with diethyldithiocarbamate or N-methyl-D-glucamine as traps of nitric oxide in animal tissues: comparative investigations. *Biochim. Biophys. Acta* **1336:**225–234; 1997.

[45] Fujii, H.; Berliner, L. J. Ex vivo detection of nitric oxide in brain tissue. *Magn. Reson. Med.* **42:**599–602; 1999.

[46] Fujii, H.; Berliner, L. J. One- and two-dimensional EPR imaging studies on phantoms and plant specimens. *Magn. Reson. Med.* **2**:275–282; 1985.

[47] Kuppusamy, P.; Chzhan, M.; Vij, K.; Lefer, D.; Giannelaa, E.; Zweier, J. L. Three-dimensional spectral-spatial EPR imaging of free radicals in the heart: a technique for imaging tissue metabolism and oxygenation. *Proc. Natl. Acad. Sci. USA* **91**:3388–3392; 1994.

[48] Yoshimura, T.; Yokoyama, H.; Fujii, S.; Takeyama, F.; Oikaw, K.; Kamada, H. In vivo EPR detection and imaging of endogenous nitric oxide in lipopolysaccharide-treated mice. *Nat. Biotechnol.* **14**:992–994; 1996.

[49] Fujii, H.; Wan, X.; Zhong, J.; Berliner, L. J.; Yoshikawa, K. In vivo imaging of spin-trapped nitric oxide in rats with septic shock: MRI spin trapping. *Magn. Reson. Med.* **42**:235–239; 1999.

[50] Bennett, H. F.; Brown, R. D. I.; Keana, J. F.; Koenig, S. H.; Swartz, H. M. Interactions of nitroxides with plasma and blood: effect on $1/T_1$ of water protons. *Magn. Reson. Med.* **14**:40–55; 1990.

[51] Bennett, H. F.; Swartz, H. M.; Brown, R. D. I.; Koenig, S. H. Modification of relaxation of lipid protons by molecular oxygen and nitroxides. *Invest. Radiol.* **22**:502–507; 1987.

[52] Sies, H. Glutathione and its role in cellular function. *Free Radic. Biol. Med.* **27**:916–921; 1999.

[53] Boyne, A. F.; Ellman, G. L. A methodology for analysis of tissue sulfhydryl components. *Anal. Biochem.* **46**:639–653; 1972.

[54] Kosower, N. S.; Kosower, E. M. Diamide: an oxidant probe for thiols. *Methods Enzymol.* **251**:123–133; 1995.

[55] Newton, G. L.; Fahey, R. C. Determination of biothiols by bromobiamane labeling and high-performance liquid chromatography. *Methods Enzymol.* **251**:148–166; 1995.

[56] Khramtsov, V. V.; Yelinova, V. I.; Weiner, L. M.; Berezina, T. A.; Martin, V. V.; Volodarsky, L. B. Quantitative determination of SH groups in low- and high-molecular-weight compounds by an electron spin resonance method. *Anal. Biochem.* **182**:58–63; 1989.

[57] Khramtsov, V. V.; Yelinova, V. I.; Glazachev, Y.; Reznikov, V. A.; Zimmer, G. Quantitative determination and reversible modification of thiols using imidazolidine biradical disulfide label. *J. Biochem. Biophys. Methods* **35**:115–128; 1997.

[58] Khramtsov, V. V.; Volodarsky, L. B. Imidazoline nitroxides in studies of chemical reaction: ESR measurements of the concentration and reactivity of protons, thiols and nitric oxide. In: Berliner, L. J., ed. *Biological magnetic resonance, vol. 14.* New York and London: Plenum Press; 1998:109–180.

[59] Weiner, L. M.; Hu, H.; Swartz, H. M. EPR method for the measurement of cellular sulfhydryl groups. *FEBS Lett.* **290**:243–246; 1991.

[60] Busse, E.; Zimmer, G.; Schopohl, B.; Kornhuber, B. Influence of alpha-lipoic acid on intracellular glutathione in vitro and in vivo. *Arzneimittelforschung* **42**:829–831; 1992.

[61] Busse, E.; Zimmer, G.; Kornhuber, B. Plasma-membrane fluidity studies of murine neuroblastoma and malignant melanoma cells under irradiation. *Strahlenther. Onkol.* **168**:419–422; 1992.

[62] Nohl, H.; Stolze, K.; Weiner, L. M. Noninvasive measurement of thiol status levels in cells and isolated organs. *Methods Enzymol.* **251**:191–203; 1995.

[63] Yelinova, V. I.; Glazachev, Y.; Khramtsov, V. V.; Kudryashova, L.; Rykova, V.; Salganik, R. Studies of human and rat blood under oxidative stress: changes in plasma thiol level, antioxidant enzyme activity, protein carbonyl content and fluidity of erythrocyte membrane. *Biochem. Biophys. Res. Commun.* **221**:300–303; 1996.

[64] Trabesinger, A. H.; Weber, O. M.; Duc, C. O.; Boesiger, P. Detection of glutathione in the human brain in vivo by means of double quantum coherence filtering. *Magn. Reson. Med.* **42**:283–289; 1999.

FORUM ON OXIDATIVE STRESS STATUS (OSS)—THE SEVENTH SEGMENT

WILLIAM A. PRYOR

Biodynamics Institute, Louisiana State University, Baton Rouge, LA, USA

This is the seventh segment of the ongoing Forum on Oxidative Stress Status (OSS). All of our previous Forum segments have included a minimum of three articles. However, *FRBM* will now publish all articles of this type as soon as they are accepted, rather than holding up the publication of one manuscript in order to wait for others to be ready and accepted to "complete the set." We hope this will result in an improved publication time for each article.

Previous segments of the ongoing Forum on OSS can be found as follows:

- The first set of seven articles in *FRBM* 27:11/12 (December, 1999), pp. 1135–1196.

- The second set of four articles in *FRBM* 28:4 (February, 2000), pp. 503–536.
- The third set of six articles in *FRBM* 28:6 (March 2000), pp. 837–886.
- The fourth set of four articles in *FRBM* 29:5 (September, 2000), pp. 387–415.
- The fifth set of six articles in *FRBM* 29:11 (December, 2000), pp. 455–499.
- The sixth set of three articles in *FRBM* 30:5 (March, 2001), pp. 455–499.

The seventh segment consists of one article by Luca Valgimigli, Gian Franco Pedulli and Moreno Paolini, entitled "Measurement of oxidative stress by EPR radical-probe technique."

Address correspondence to: Dr. William A. Pryor, Biodynamics Institute, 711 Choppin Hall, Louisiana State University, Baton Rouge, LA 70803, USA.

Reprinted from: *Free Radical Biology & Medicine, Vol. 31, No. 6, pp. 707, 2001*

Bioassays for Oxidative Stress Status (BOSS). Edited by W.A. Pryor

MEASUREMENT OF OXIDATIVE STRESS BY EPR RADICAL-PROBE TECHNIQUE

LUCA VALGIMIGLI,* GIAN FRANCO PEDULLI,* and MORENO PAOLINI[†]

*Department of Organic Chemistry "A. Mangini" and [†]Department of Pharmacology, Biochemical Toxicology Unit, University of Bologna, Bologna, Italy

(Received 19 October 2000; Accepted 2 February 2001)

Abstract—An EPR method for the measurement of the oxidative stress status in biological systems is described. The method is based on the X-band EPR detection of a persistent nitroxide generated under physiological or pseudo-physiological conditions by oxidation of a highly lipophylic hydroxylamine probe. The probe employed is bis(1-hydroxy-2,2,6,6-tetramethyl-4-piperidinyl)-decandioate which is administrated as hydrochloride salt. This probe is able to give a fast reaction with the majority of radical species involved in the oxidative stress. Furthermore, it crosses cell membranes and distributes in a biological environment without the need to alter or destroy compartmentation. The method is therefore suitable for quantitative measurements of ROS and can be applied to human tissues in real clinical settings. It has been successfully employed in systems of growing complexity and interest, ranging from subcellular fractions to whole animals and human liver. © 2001 Elsevier Science Inc.

Keywords—Free radical, Oxidative stress, EPR, Human liver, β-carotene, Cytochrome P450

INTRODUCTION

Free radicals are involved in several pathological conditions [1–3] and, at subtoxic concentration, they have recently been shown to play several crucial physiological roles in biological systems [4,5]. Unfortunately, even at toxic levels, such as those involved in the oxidative stress process, their concentration is so low that the detection or

Luca Valgimigli is research associate at the Department of Organic Chemistry, University of Bologna. He received his Ph.D. from the University of Bologna in 1998. His research interests include kinetics, thermochemistry, and mechanisms of natural and synthetic radical scavengers, EPR, and the measurement of oxidative stress in biological systems.

Gian Franco Pedulli is full professor of "Physical methods in organic chemistry" at the University of Bologna, Faculty of Pharmacy. His research involved EPR spectroscopy, EPR-imaging, and thermochemistry, kinetics, and mechanisms of free radical reactions. His current interests are: study of antioxidants, supramolecular chemistry of free radicals, and applications of EPR in biology.

Moreno Paolini is associate professor of "Pharmacology" at the University of Bologna, Faculty of Science. His research interests include inducible enzymes and their role in carcinogenesis, free radical toxicology, bioactivation of xenobiotics and their role in oxidative stress and cancer promotion, and measurement of oxidative stress in biological systems.

Address correspondence to: Luca Valgimigli, Department of Organic Chemistry "A. Mangini", University of Bologna, Via S. Donato 15, I-40127, Bologna, Italy; Tel: +39 (051) 243218; Fax: +39 (051) 244064; E-Mail: valgimig@alma.unibo.it.

quantification of these very short-lived species is extremely challenging. Even the most specific techniques available for the study of odd-electron species like electron paramagnetic resonance (EPR) fail to reveal free radicals in very complex environments such as biological systems.

The "spin trapping" technique aims to circumvent this problem by adding a suitable diamagnetic compound, the "trap," to the system under investigation in order to capture any transient radical that may be present and form a much longer-lived radical species [6,7]. The most commonly employed traps are nitrones (like PBN), cyclic nitrones (like DMPO), or nitroso compounds (like MNP): upon reaction with a radical species, these compounds generate relatively persistent nitroxide radicals that then accumulate in the systems, reaching a sufficiently high concentration for detection and characterization by EPR. The spin probe technique can usually prove the involvement of radical species in a process or condition, and it provides a relatively safe [8] way of identifying species actually present in the system [9,10]. Furthermore, individual traps usually display a marked selectivity for particular species, thereby permitting modulation of the sensitivity toward a given radical. The major drawback of the method is that spin adducts are only relatively persistent, and their life time, especially in biological environments where they are subject to degradation by several enzymatic systems, is usually too short to allow reliable quantitative measurements.

Newly developed spin traps yielding spin adducts with longer life times [11,12] have made quantitative measurements of free radicals in biological systems possible, albeit troublesome [13]. Another way of making oxidative stress status detectable by EPR involves the use of exogenous nitroxides as probes of the red-ox balance in a given environment. Cyclic nitroxides such as hydroxyl-TEMPO and carboxyl-PROXYL are very persistent in water or organic solution, but when used in vivo or in a biological sample they are reduced to the parent hydroxylamine by several enzymatic processes mainly involving ascorbate or glutathione. The rate at which the nitroxide is reduced to the diamagnetic hydroxylamine, which can be evaluated by EPR, is related to the reducing capacity of the organism and hence to its oxidative status [14].

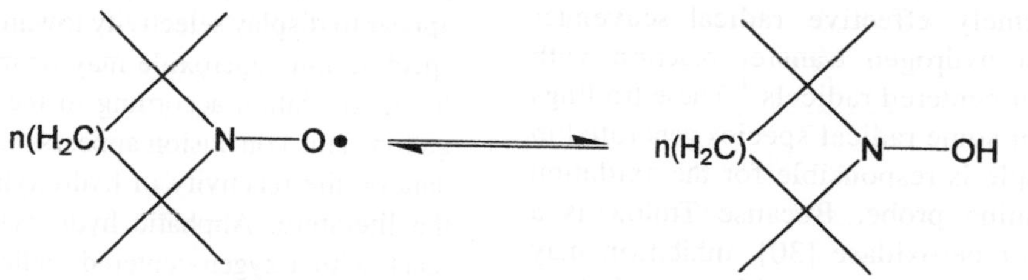

innovative probe with interesting physical-chemical properties [23] and we have investigated its use in a number of biological environments of growing complexity, from animal sub-cellular fractions to human liver. Herein we will briefly review our knowledge of the mechanism, potentiality and limitations of the use of this new probe together with a selection of interesting applications.

Mechanism, advantages and limitations

In all the approaches mentioned above, the probe represents indeed the only link between the multitude of enzymatic and nonenzymatic processes responsible for the OSS and the actual EPR measurement. As a consequence, the choice of the probe is of crucial importance for the significance and reliability of the measurements. The probe we have developed is the bis(1-hydroxy-2,2,6,6-tetramethyl-4-piperidinyl)decandioate (I): al-

Scheme 1

Furthermore, it has been shown that processes that affect the oxidative stress status of a biological system, like X-irradiation, also affect the life-time of the nitroxide in vivo [15]. This approach, which has been recently reviewed in this journal [15,16], has already allowed some very interesting investigations, including in vivo monitoring of the OSS and ESR-computed tomography of the oxidative stress in the experiment animal.

The first time this principle was successfully used to assess the oxidative stress status (OSS) was actually back in 1979 [17] when Rosen's group proposed a way of observing the same phenomenon from the other side. The method was based on EPR monitoring of a persistent nitroxide radical generated by oxidation of a suitable hydroxylamine probe in a biological system. Over the years the basic principle has been further developed [18–24], and various probes have been employed (see Scheme 2).

During the last 5 years we have been developing an

though not commercially available it can easily be prepared from the parent piperidine, according to the literature (Scheme 3) [25]. Briefly, one equivalent of bis(2,2,6,6-tetramethyl-4-piperidinyl)decandioate (Tinuvin 770, CIBA Specialty Chemicals, Bologna, Italy) is

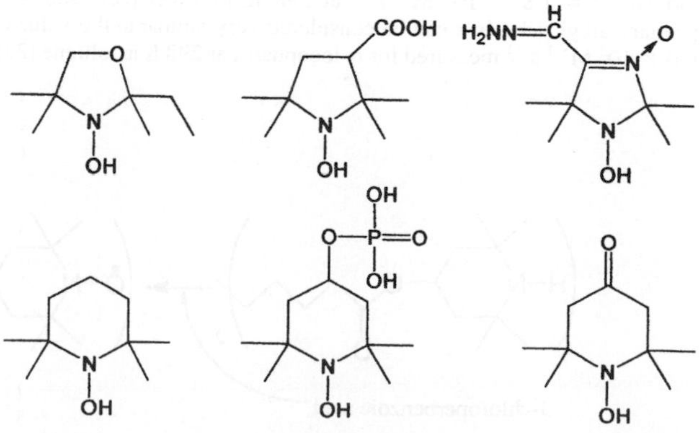

Scheme 2

reacted at room temperature with five equivalents of 3-chloroperbenzoic acid (Aldrich, Milan, Italy) in dichloromethane for 5 h. The resulting bis-nitroxide (yield 80%) is reduced to the hydroxylamine I by catalytic hydrogenation in a Parr reactor and is recovered and crystallized as hydrochloride salt (yield 90%).

The addition of a physiological solution containing the hydroxylamine I to a biological sample (e.g., hepatic or lung microsomes, hepatic cell cultures or liver tissue samples) in the presence of a heavy-metals chelating agent (like EDTA or deferoxamine), inside the cavity of a X-band EPR spectrometer results in the formation of a 3-line EPR signal. On the basis of its spectral parameters ($a_N = 15.52$ G, $g = 2.0062$) this signal can be assigned to the nitroxide II.

In all cases, signal intensities are markedly higher than those obtained from blank samples prepared by leaving the same solution of hydroxylamine exposed to atmospheric oxygen in the absence of the biological sample under the same experimental conditions. The signal formation can be completely abolished by adding 0.1 M Trolox to the medium [24]. This water-soluble mimic of α-tocopherol [26] is known to behave as an extremely effective radical scavenger giving a very fast hydrogen transfer reaction with oxygen- and carbon-centered radicals.[1] These findings clearly indicate that some radical species generated in the biological sample is responsible for the oxidation of the hydroxylamine probe. Because Trolox is a known substrate for peroxidase [30], inhibition may arise also from Trolox competing with the hydroxylamine I for peroxidases present in the system. However, addition of Catalase (1000 U/ml) did not abolish the EPR signal, which was reduced only by less than 10%, indicating that oxidation of I by peroxidase is not a major process leading to the nitroxide II under our experimental conditions.

Although in early investigations experiments per-

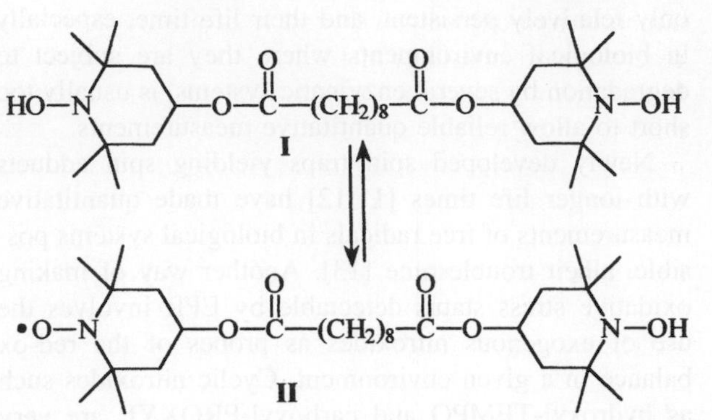

As an example, the rate constant for the reaction with peroxyl radicals is 1.1×10^6 M^{-1}s^{-1} at 303 K in styrene [27], with alkoxyl radicals $k = 1.8 \times 10^9$ M^{-1}s^{-1} at 298 K in water [28], and with primary alkyl radicals it can be considered very similar to the value of 6.0×10^5 M^{-1} s^{-1} measured for α-tocopherol at 298 K in toluene [29].

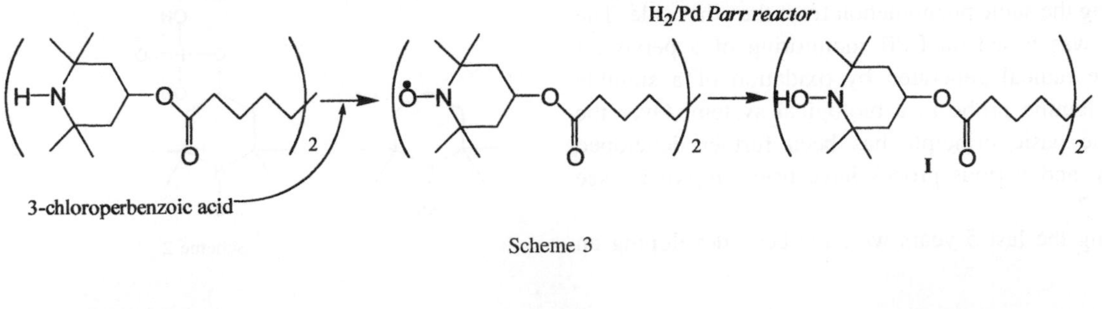

Scheme 4

formed upon addition of SOD to the system suggested that our probe would selectively detect superoxide in hepatic microsome preparations from mice [23], subsequent studies revealed that this is not actually the case [24]. Indeed, the addition of thermally denatured SOD (and SOD + Catalase) produced about the same signal inhibition as the active enzymes. The hydroxylamine probe I (and conceivably any hydroxylamine) does not appear to display selectivity toward any particular radical species, and superoxide may or may not be responsible for its oxidation according to the system under investigation. This conclusion appears reasonable in light of the data on the reactivity of hydroxylamines available from the literature. Aliphatic hydroxylamines are known to react with oxygen-centered radicals to generate stabilized nitroxides. Thus the rate constant for hydrogen abstraction by peroxyl radicals has been reported as $k = 3 \times 10^5$ M^{-1}s^{-1} at 323 K from diethyl hydroxylamine [31] and as $k = 5 \times 10^5$ M^{-1}s^{-1} at 338 K from 1-hydroxyl-4-oxo-2,2,6,6-tetramethyl-piperidine [32]. TEMPOH is oxidized to the corresponding nitroxide by superoxide at pH 7.8 and room temperature with a rate constant of 1.7×10^3 M^{-1}s^{-1} [18], and this value is expected to increase to 10^4 M^{-1}s^{-1} under more acidic conditions where the superoxide exists predominantly in the more reactive HOO• form. Alkoxyl radicals are known to give a very fast hydrogen abstraction from

Scheme 3

aliphatic hydroxylamines],[2] and hydroxyl radicals will react with rate constants of $10^9 \, M^{-1}s^{-1}$ or higher. Cyclic hydroxylamines can be oxidized to the corresponding nitroxides by peroxidase/H_2O_2 systems [34] and they may also undergo autoxidation in biological systems due to the presence of metal ions such as Cu^{2+} and Fe^{3+} [18, 35]. Furthermore, the nitroxide can be reduced back by ascorbate, GSH, and by some radical reactions such as trapping of alkyl radicals. Apart from autoxidation, which can be neglected under our experimental conditions due to the presence of metal chelators (EDTA and deferoxamine) in all samples, any of the other processes might potentially contribute to determine the total amount of nitroxide detected by EPR in our measurements. Should this method be chosen to assess the impact of vitamin C on OSS, the capacity of ascorbate to reduce persistent nitroxides must be taken into account. On the other hand, 100-fold excess Trolox, and consequently vitamin E, does not reduce the nitroxide II and the rate constant for hydrogen abstraction from Trolox by II is < $1 \, M^{-1}s^{-1}$ at 298 K [24]. Therefore there is no particular problem for the assessment of the effect of vitamin E on OSS.

Because the nitroxide II produced by oxidation of I bears no structural information regarding the radical species that has actually been detected by the probe, unlike spin trapping, the present radical-probe technique does not provide a means to identify a particular radical species, but should rather be regarded as a general way to measure the oxidative status of the system under investigation. However, because hydroxylamines display a very high reactivity toward oxygen-centered radicals, the technique may also provide a method to quantify ROS in a biological environment.

It should also be noted that cyclic hydroxylamines give very fast hydrogen abstraction with peroxynitrite. The rate constant for reaction of TEMPONE-H with peroxynitrite to give the stable nitroxide TEMPONE was measured as $6 \times 10^9 \, M^{-1}s^{-1}$ at room temperature [19]. Because peroxynitrite is known to be generated in vivo by the very fast reaction of NO with superoxide ($k = 6.7 \times 10^9 \, M^{-1}s^{-1}$ [36]), the technique also constitutes a means to detect reactive nitrogen intermediates.

The lack of selectivity is certainly the major disadvantage and limitation of the method. However, if what one wants to obtain is a comprehensive quantitative evaluation of OSS, this factor may actually be an advantage. Another advantage of our probe as compared with more traditional systems is given by its peculiar chemi-

cal-physical properties. The majority of biological systems from simple microsome fractions to living animals are characterized by a high degree of compartmentation, which creates a very heterogeneous environment. Enzymatic and nonenzymatic processes capable of producing ROS under physiological or pseudo-physiological conditions are potentially located in any of these compartments and, due to their reactivity, radical species would, in general, react close to their site of formation. As a consequence, the hydrophylicity/lipophylicity balance of a given radical probe or spin trap physically confines its location to just one compartment and enables it to specifically detect only the radicals produced therein. When this constitutes a limitation, free distribution of the probe can be obtained by homogenation. Unfortunately, however, this cannot always be accomplished without significantly altering the system under investigation (e.g., in vivo) and creating artifactual results.

With our hydroxylamine probe, the protonated form administrated to the biological system is completely water-soluble and will distribute in the extracellular compartments, but it is also in equilibrium with the free form, which is highly lipophilic (the calculated [37] logP is 4.01) and will readily cross the cell membrane to allow distribution in any compartment where the production of free radicals can take place. It can therefore provide a global view of the oxidative stress, which is desirable for quantitative measurements, without destroying compartmentation.

The radical probe technique in the detection of cytochrome P450 overexpression in vitro and in vivo

The cytochrome P450 (CYP) multienzymatic system is a family of enzymes involved in the phase 1 metabolism of xenobiotics, its purpose being that of protecting living organisms by facilitating the deactivation and removal of various toxins. However, CYP isoenzymes can also endanger the organism by activating oxygen to the superoxide ($O_2^{-\bullet}$) and hydroxyl ($^\bullet OH$) radicals [38]. CYP overexpression can be caused both by genetic polymorphism and by induction by xenobiotics such as alcohol, drugs, pollutants, etc., and is a recognized risk factor for cancer [39]. Thus the induction of certain CYP isoforms increases the biotransformation of ubiquitous pretoxic or precarcinogenic substances to DNA-damaging products [40–42].

In order to evaluate a possible correlation between CYP induction and OSS, the production of reactive oxygen species was measured by the radical probe method at three different levels of biological complexity, namely subcellular fractions, cell cultures, and whole animals, with or without previous treatment with a series of known inducers of CYP isoforms. Subcellular hepatic

---

For example, the ratio of the rate constant for the reaction $(CH_3)_3CO^\bullet + Et_2N\text{-}OH$ over the rate of β-cleavage for butoxyl radical at 388 K has been reported as $k_H/k_\beta = 1.4 \times 10^3 \, M^{-1}$ [31] and k_β can be calculated as $5.5 \times 10^4 \, s^{-1}$ at the same temperature [33], hence as $k_H = 7.7 \times 10^7 \, M^{-1}s^{-1}$.

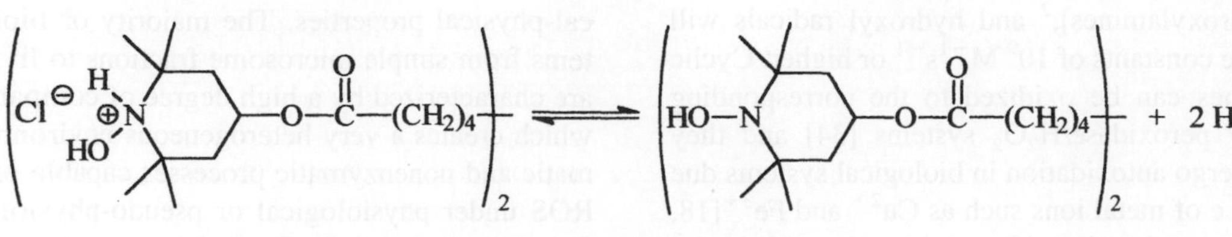

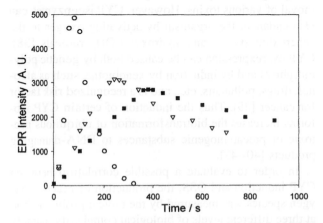

Scheme 5

fractions (microsomes) prepared from Swiss Albino CD1 mice maintained on a standard diet were incubated at 37°C in a capillary glass EPR tube with the hydroxyl-amine probe I (1 mM), NADP$^+$ (0.06 mM), G6P (3.33 mM), MgCl$_2$ (4 mM) and G6PDH (0.93 U/ml), and spectra were recorded at regular intervals immediately after mixing. The time evolution of the EPR signal intensity due to nitrocide II was compared to that obtained with microsomes from animals that had previously been treated intraperitoneally daily for 3 consecutive days with five specific inducers of different CYPs, namely: sodium phenobarbital (PB, CYP 2B1), β-naphthoflavone (βNF, CYP 1A1), isosafrole (IS, CYP 1A2), pregnenolone-16α-carbonitrile (PCN, CYP 3A), and clofibrate (CL, CYP 4A), or allowed to assume ad libitum ethanol (EtOH, CYP 2E1).

As can be seen in Fig. 1, treatment with CYPs inducers strongly affects the time evolution of the EPR signal. Thus, both the rate of formation of the nitrocide and its maximum concentration are increased in induced samples. For all the six inducers the normalized slope (P$_0$N) of the first segment of the EPR signal time course was significantly ($p < .01$) higher than the value obtained for uninduced animals: this clearly indicates that a higher oxidative stress level is associated with the induction of specific CYP isoforms. Indeed, the P$_0$N ratio between

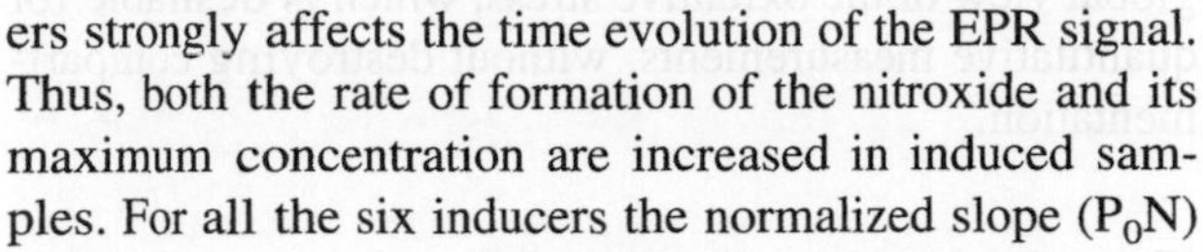

Fig. 1. Time evolution of the EPR signal observed in uninduced (■) and pregnenolone 16α-carbonitrile- (PCN, ○), or isosafrole- (IS, ▽) induced microsomes from Swiss Albino CD1 mice during incubation at 37°C with 1 mM hydroxylamine probe.

treated and untreated mice ranged from ~2 for βNF, CL, and IS to ~4 for EtOH, ~5 for PB, and as much as ~6 for PCN, the order of the relative nitrocide formation being PCN > PB > EtOH > βNF > CL > IS. Actual overexpression of CYPs after treatment with the above-mentioned inducers was also independently assessed, and it is noteworthy that a good linearity ($r = 0.98$) could be established between the amount of CYP and the rate of nitrocide production.

Similar results were obtained with whole animals using the whole-body-harvest approach. This allowed measurements to be performed using an X-band EPR spectrometer. Ten CD1 mice for each group treated with CYP inducers as previously described and 10 noninduced animals received a single intraperitoneal injection of the hydroxylamine I (200 mg/kg). Urine was collected every 2 h up to 10 h and spectra of the nitrocide product were recorded. In all cases the maximum concentration of nitrocide II was reached after 10 h, but the actual value, obtained after calibration of the spectrometer response, was significantly affected by the inducer. Thus, concentration of II in urine from treated mice ranged from $(6.1 \pm 0.3) \times 10^{-9}$ mol/ml to $(1.41 \pm 0.2) \times 10^{-8}$ mol/ml, as compared with the average value of $(1.29 \pm 0.2) \times 10^{-9}$ mol/ml for nontreated mice (see Fig. 2). The relative effect on OSS recorded for the various CYP inducers administred in vivo nearly paralleled that previously observed in microsomes (with the exception of PCN, which was more effective in vivo), suggesting an interesting parallelism between OSS in vivo and cancer promoting potentiality of CYP inducers [43]. Experiments performed on a nontransformed murine liver epithelial line C2.8 incubated in a nutrient medium with the addition of the hydroxylamine probe (0.5 mM) provided experimental evidence that the probe possesses exactly the right hydrophylicity/lipophylicity balance to cross the cell membrane and reach intracellular compartments.

Evaluation of the role of diet supplementation with β-carotene on the OSS

The antioxidant activity of β-Carotene (pro-vitamin A) has been thoroughly investigated. Epidemiological as well as animal studies on vitamin A and its analogs suggest that suitable forms of dietary supplementation

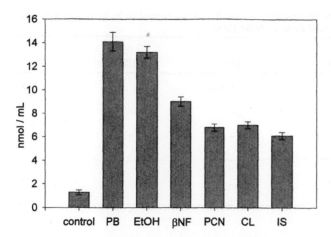

Fig. 2. Levels of nitroxide (nmol/ml) detected in urine from variously (PCN, PB, EtOH, βNF, CL, and IS) induced mice 10 h after intraperitoneal injection of hydroxylamine probe (200mg/kg in corn oil). Values are means of 5 determinations each on two mice. Vertical error bars represent ± SD. All differences were statistically significant by Wilcoxon's analysis ($p < .01$ as compared with controls).

could help prevent cancer in humans [44]. However, it has also been shown that under certain experimental conditions (i.e., high partial pressure of oxygen) β-carotene might instead display pro-oxidant activity [45,46]. Indeed clinical trials regarding the efficacy of α-tocopherol plus β-carotene (ATBC) and β-carotene plus retinol (CARET) suggested that high-dose β-carotene supplementation actually increases the risk of lung cancer in heavy smokers and asbestos workers [47,48]. Since β-carotene is also known to be an anti-genotoxic agent [49], its potential co-carcinogenic activity could be due to some epigenetic mechanisms such as the induction of specific cytochrome P450 isoenzymes and a pro-oxidant activity at high doses in vivo [43].

As an experimental model, 10 Sprague Dawley male rats maintained on a standard diet received by mouth 500 mg/kg of β-carotene in corn oil for 5 consecutive days. Ten controls, housed under the same conditions, received only corn oil. Lung microsomes were analyzed by the radical probe technique under the usual conditions (vide supra) after incubation at 37°C with 1 mM hydroxylamine I [50]. The animals treated with β-carotene displayed significantly ($p < .01$) higher levels of oxidative stress, which was assessed by measuring both the rate of nitroxide II production and its maximum concentration. The expression of several CYP isoenzymes of particular toxicological relevance was also evaluated by means of specific assays. Noteworthy highly significant ($p < .01$) increases were observed with all the carcinogen-metabolizing enzymes tested. To evaluate the CYP isoenzyme activity the following probes were employed: ethoxyresorurufin O-deethylase activity (CYP1A1), testosterone 7α-hydroxylation (CYP1A1/2 and CYP3A1),

testosterone 2β-hydroxylation (CYP1A1 and CYP3A1), and 17-testosterone hydroxylase activity (CYP2B1 and CYP3A1). All these probes displayed increases in the activity ranging from ~300% to ~1500%. In particular, the activity of CYP1A1 (activating aromatic amines, polychlorinated biphenyls, and dioxins) was increased by ~800%. This is particularly intriguing because previous studies have associated an increased risk in lung cancer with polymorphisms in CYP1A1 [50]. The pro-oxidant and CYPs induction activity of high-dose β-carotene supplementation appears therefore to be a *possible* explanation for its co-carcinogenic effect in humans. Clearly, our experimental conditions were not intended to mimic a trial situation, but merely to shed some light on the co-carcinogenic potential of β-carotene. It is worth stressing that, regardless of the experimental conditions, carcinogenesis is a highly species-specific process and transferring results obtained with rats to humans would be, at best, tentative.

Measurement of ROS in healthy and diseased human liver

The rationale for the involvement of oxygen-centered free radicals in the inflammatory process stems from their involvement in the cellular signaling system [4,5], with the possibility that they may represent a chemiotactic factor for immune system cells, along with their role in phagocytic cells where they serve the purpose of damaging the pathogenic agent [51]. Nevertheless, the actual role of ROS in human inflammation, especially in connection with chronic and acute disease, has never been clearly defined due to the difficulty of quantifying ROS or oxidative stress in humans [52].

We set out to find a proper way to apply the radical-probe technique to human tissues in real clinical settings [24]. Liver disease was chosen as the prototype of a pathology in which the involvement of inflammatory processes has a major role in the evolution of the disease [53]. Thirty-two subjects, including 10 healthy controls, were enrolled after giving informed consent. Ten of the 22 patients had hepatitis C, 3 had hepatitis B, while the remainder had a variety of diseases characterized by an autoimmune nature which, for statistical purposes, were clustered in a group called nonviral liver diseases (NVLD). The method we developed was quite simple and only moderately invasive: 2–3 mg of liver biopsy (obtained by the fine needle technique) were weighted and incubated for 5 min at 37°C with a physiological solution of the hydroxylamine I (1 mM) containing a metal chelating agent. After incubation, the sample was quickly frozen in liquid nitrogen to denature enzymes and stop any reaction, and subsequently warmed at room temperature prior to the EPR measurement. For practical

Table 1. ROS Median Value (Range is Given in Parentheses) Measured in the Liver Tissue. The *p* Values Reported Refer to the Mann-Whitney Test for the Difference in ROS Found Between Different Groups

Groups	ROS Median Mol/mg	Controls	LD	CHB	CHC
Controls	1.8×10^{-11} $(3 \times 10^{-13} - 4.4 \times 10^{-10})$	—	$p < .00001$	$p < .02$	$p < .0003$
LD	3.2×10^{-9} $(1.9 \times 10^{-10} - 2.6 \times 10^{-8})$	$p < .00001$	—	–	–
CHB	5.8×10^{-10} $(4.7 \times 10^{-10} - 1.8 \times 10^{-9})$	$p < .02$	–	–	n.s.
CHC	2.7×10^{-9} $(1.9 \times 10^{-10} - 7.7 \times 10^{-9})$	$p < .0003$	–	n.s.	–
NVLD	9.8×10^{-9} $(2.6 \times 10^{-9} - 2.6 \times 10^{-8})$	$p < .005$	–	$p < .04$	$p < .04$

LD = All enrolled liver diseases considered as a single group.

reasons, we monitored the maximum concentration of nitroxide instead of the full time evolution. As exemplified in Fig. 3, diseased tissue provided a more oxidizing environment than healthy liver. Furthermore, the nature of the disease affected the oxidative status.

The effect of the various experimental conditions on the final result, including length of incubation, time from tissue extraction to addition of the probe, and time from incubation to the EPR measurement, were systematically investigated in order to set the optimal standardized experimental conditions [24]. Interestingly, our results revealed that homogenization of the tissue is unnecessary since the signal measured immediately after homogenization in the presence of the probe was very close to that obtained after 5 min incubation with the whole biopsy. After calibration of the spectrometer response it was possible to obtain quantitative values for the oxidative stress. Our results indicated that the oxidative stress level in diseased liver is several orders of magnitude higher than in healthy controls, and the differences were highly significant (Fig. 4).

Last but not least, we have found marked differences among the different pathologic conditions (Table 1), and in those cases where the number of samples analyzed was sufficiently high, the differences were statistically significant. This is particularly intriguing because it gives diagnostic value to our method and also provides some clinical insight into the pathophysiology of human liver disease.

CONCLUSIONS

The radical probe technique described herein provides a versatile way of gaining rapid and reproducible quantitative measurements of OSS in biological systems of variable complexity. Due to its simplicity and limited invasiveness it could be applied to human liver in real clinical settings. While its major limitation is certainly the lack of selectivity in detecting one particular radical species, it has the advantage that, thanks to the peculiar physical-chemical properties of the hydroxylamine probe employed, it can provide a global picture of the oxidative stress status.

Besides the few examples of applications reported in this account, the technique has also been successfully applied in animal models to evaluate the role of high-dose vitamin C dietary supplementation [54] and the risk for human health associated with exposure to a widely used fungicide [55]. Further work is in progress to extend applications of the technique to the evaluation of other fashionable diet supplements and molecules of toxicological interest. Our major current challenge, however, is to extend its application to human tissues other than liver biopsy, and in particular to human blood. As well as simplifying experimental protocols, this would further

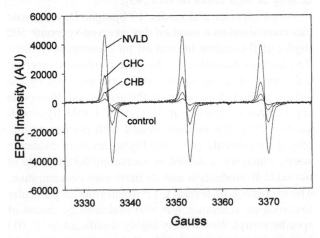

Fig. 3. EPR spectra of the nitroxide generated by incubation of the hydroxylamine spin-probe (1 mM) in physiologic solution with 1 mg of healthy liver (control) or with the same amount of liver biopsy from patients affected by chronic hepatitis B, chronic hepatitis C, or nonviral liver disease. Signals are superimposed in the same scale for comparison. (Reproduced from ref. [24].)

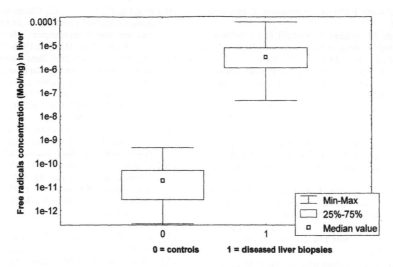

Fig. 4. Levels of nitroxide (nmol/ml) detected in urine from variously (PCN, PB, EtOH, βNF, CL, and IS) induced mice 10 h after intraperitoneal injection of hydroxylamine probe (200mg/kg in corn oil). Values are means of 5 determinations each on two mice. Vertical error bars represent ± SD. All differences were statistically significant by Wilcoxon's analysis ($p < .01$ as compared with controls). (Reproduced from ref. [24].)

reduce the invasiveness of the technique and expand its clinical relevance as diagnostic tool.

Acknowledgements — Financial support from the University of Bologna, CNR (Rome) and MURST (Rome) is gratefully acknowledged. The authors thank *Free Radical Research* for reproduction of Figs. 3 and 4. We are grateful to Robin M. T. Cooke for editing.

REFERENCES

[1] Diaz, M. D.; Frei, B.; Vita, J. A.; Keaney, V. Antioxidants and atherosclerotic heart diseases. *N. Engl. J. Med.* **337:**408–416; 1997.

[2] Cerutti, P. A. Oxy-radicals and cancer. *Lancet* **344:**862–863; 1994.

[3] Giacosa, A.; Filiberti, R. Free radicals, oxidative damage and degenerative diseases. *Eur. J. Cancer Prev.* **5:**307–312; 1996.

[4] Suzuki, Y. J.; Forman, H. J.; Sevanian, A. Oxidants as stimulators of signal transduction. *Free Radic. Biol. Med.* **22:**269–285;1997.

[5] Clément, M.-V.; Pervaiz, S. Reactive oxygen intermediates regulate cellular response to apoptotic stimuli: an hypothesis. *Free Radic. Res.* **30:**247–252; 1999.

[6] Britigan, B. E.; Cohen, M. S.; Rosen, G. M. Detection of the production of oxygen-centered free radicals by human neutrophils using spin trapping techniques: a critical perspective. *J. Leukoc. Biol.* **41:**349–362; 1987.

[7] Halliwell, B.; Gutteridge, J. M. C. *Free radicals in biology and medicine* (3rd ed.). New York: Oxford University Press Inc.; 1999.

[8] Dikalov, S. I.; Mason, R. P. Reassignment of organic peroxyl radical adducts. *Free Radic. Biol. Med.* **27:**864–872; 1999.

[9] Bolli, R.; Patel, B. S.; Jeroudi, M. O.; Lai, E. K.; McCay, P. B. Demonstration of free radical generation in "stunned" myocardium of intact dogs with the use of the spin trap alpha-phenyl-N-*tert*-butyl nitrone. *J. Clin. Invest.* **82:**476–485; 1988.

[10] Kadiiska, M. B.; Burkitt, M. J.; Xiang, Q. H.; Mason, R. P. Iron supplementation generates hydroxyl radical in vivo. An ESR spin-trapping investigation. *J. Clin. Invest.* **96:**1653–1657; 1995.

[11] Tuccio, B.; Zeghdaoui, A. H.; Finet, J.-P.; Cerri, V.; Tordo, P. Use of new beta-phosphorylated nitrones for the spin trapping of free radicals. *Res. Chem. Intermed.* **22:**393–404; 1996.

[12] Roubaud, V.; Lauricella, R.; Tuccio, B.; Bouteiller, J.-C.; Tordo,

P. Decay of superoxide spin adducts of new PBN-type phosphorylated nitrones. *Res. Chem. Intermed.* **22:**405–416; 1996.

[13] Roubaud, V.; Sankarapandi, S.; Kuppusami, P.; Tordo, P.; Zweier, J. L. Quantitative measurement of superoxide generation using the spin trap 5-(diethoxyphosphoryl)-5-methyl-1-pyrroline-N-oxide. *Anal. Biochem.* **247:**404–411; 1997.

[14] Miura, Y.; Utsumi, H.; Hamada, A. Effects of inspired oxygen concentration on in vivo redox reaction of nitroxide radicals in whole mice. *Biochem. Biophys. Res. Commun.* **182:**1108–1114; 1992.

[15] Miura, Y.; Ozawa, T. Noninvasive study of radiation-induced oxidative damage using in vivo electron spin resonance. *Free Radic. Biol. Med.* **28:**854–859; 2000.

[16] Togashi, H.; Shinzawa, H.; Matsuo, T.; Takeda, Y.; Takahashi, T.; Aoyama, M.; Oikawa, K.; Kamada, H. Analysis of hepatic oxidative stress by electron spin resonance spectroscopy and imaging. *Free Radic. Biol. Med.* **28:**846–853; 2000.

[17] Rauckman, E. J.; Rosen, G. M.; Kitchell, B. B. Superoxide radical as an intermediate in the oxidation of hydroxylamines by mixed function amine oxidase. *Mol. Pharmacol.* **15:**131–137; 1979.

[18] Rosen, G. M.; Finkelstein, E.; Rauckman, E. J. A method for the detection of superoxide in biological systems. *Arch. Biochem. Biophys.* **215:**367–378; 1982.

[19] Dikalov, S.; Skatchkov, M.; Bassenge, E. Quantification of peroxynitrite, superoxide, and peroxyl radicals by a new spin trap hydroxylamine 1-hydroxy-2,2,6,6-tetramethyl-4-oxo-piperidine. *Biochem. Biophys. Res. Commun.* **230:**54–57; 1997.

[20] Dikalov, S.; Skatchkov, M.; Stalleicken D.; Bassenge, E. Formation of reactive oxygen species by pentaerithrityltetranitrate and glyceryl trinitrate in vitro and development of nitrate tolerance. *J. Pharmacol. Exp. Ther.* **286:**938–944; 1998.

[21] Matsuo, T.; Shinzawa, H.; Togashi, H.; Aoki, M.; Sugahara, K.; Saito, K.; Saito, T.; Takahashi, T.; Yamaguchi, I.; Aoyama, M.; Kamada, H. Highly sensitive hepatitis B surface antigen detection by measuring stable nitroxide radical formation with ESR spectroscopy. *Free Radic. Biol. Med.* **25:**929–935; 1998.

[22] Fink, B.; Dikalov, S.; Bassenge, E. A new approach for extracellular spin trapping of nitroglicerin-induced superoxide radicals both in vitro and in vivo. *Free Radic. Biol. Med.* **28:**121–128; 2000.

[23] Paolini, M.; Pozzetti, L.; Pedulli, G. F.; Cipollone, M.; Mesirca, R.; Cantelli-Forti, G. Paramagnetic resonance in detecting carci-

nogenic risk from cytochrome P450 overexpression. *J. Investig. Med.* **44:**470–473; 1996.

[24] Valgimigli, L.; Valgimigli, M.; Gaiani, S.; Pedulli, G. F.; Bolondi, L. Measurement of oxidative stress in human liver by EPR spin-probe technique. *Free Radic. Res.* **33:**167–178; 2000.

[25] Behrens, R. A.; Seltzer, A. Inventors, CIBA Geigy Corporation, USA, assignee. Hydroxylamines derived from hindered amines. US Pat. No. 4,691,015. 1987 Aug 1.

[26] Scott, J. W.; Cort, W. M.; Harley, H.; Parrish, D. R.; Saucy, G. J. 6-Hydroxy-chroman-2-carboxylic acids: novel antioxidants. *J. Am. Oil Chem. Soc.* **51:**200–203; 1974.

[27] Burton, G. W.; Doba, T.; Gabe, E. J.; Hughes, L.; Lee, F. L.; Prasad, L.; Ingold, K. U. Autoxidation of biological molecules. 4. Maximizing the antioxidant activity of phenols. *J. Am. Chem. Soc.* **107:**7053–7065; 1985.

[28] Valgimigli, L.; Ingold, K. U.; Lusztyk, J. Antioxidant activities of vitamin E analogues in water and a Kamlet-Taft β-value for water. *J. Am. Chem. Soc.* **118:**3545–3549; 1996.

[29] Franchi, P.; Lucarini, M.; Pedulli, G. F.; Valgimigli, L.; Lunelli, B. Reactivity of substituted phenols toward primary alkyl radicals. *J. Am. Chem. Soc.* **121:**507–514; 1999.

[30] Nakamura, M.; Hayashi, T. Oxidation mechanism of vitamin E analogue (Trlox C, 6-hydroxy-2,2,5,7,8-pentamethylchroman) and vitamin E by horseradish peroxidase and myoglobin. *Arch. Biochem. Biophys.* **299:**313–319; 1992.

[31] Abuin, E.; Encina, M. V.; Diaz, S.; Lissi, E. A. On the reactivity of diethyl hydroxyl amine toward free radicals. *Int. J. Chem. Kin.* **10:**677–686; 1978.

[32] Bownlie, I. T.; Ingold, K. U. The Inhibited autoxidation of styrene. Part VII. Inhibition by nitroxides and hydroxylamines. *Can. J. Chem.* **45:**2427–2432; 1967.

[33] Carlsson, D. J.; Ingold, K. U. Reactions of alcoxy radicals. IV. The kinetics and absolute rate constants for some *t*-butyl hypochlorite chlorinations. *J. Am. Chem. Soc.* **89:**4891–4894; 1967.

[34] Moore, K. L.; Moronne, M. M.; Mehlhorn, R. J. Electron spin resonance study of peroxidase activity and kinetics. *Arch. Biochem. Biophys.* **299:**47–56; 1992.

[35] Dikalov, S. I.; Vitek, M. P.; Maples, K. R.; Mason, R. P. Amyloid β peptides do not form peptide-derived free radicals spontaneously, but can enhance metal catalyzed oxidation of hydroxylamines to nitroxides. *J. Biol. Chem.* **274:**9392–9399; 1999.

[36] Huie, R. E.; Padmaja, S. The reaction rate of nitric oxide with superoxide. *Free. Radic. Res. Commun.* **18:**195–199; 1993.

[37] Crippen G. M. Why energy embedding works. *J. Phys. Chem.* **91:**6341–6343; 1987.

[38] Ekstrom, G.; Ingelman-Sunderberg, M. Rat liver microsomal NADPH-supported oxidase activity and lipid peroxidation dependent on ethanol inducible cytochrome P450 (P450IIE1). *Biochem. Pharmacol.* **38:**1313–1319; 1989.

[39] Paolini, M.; Barone, E.; Corsi, C.; Paganin, C.; Revoltella, R. P. Expression and inducibility of drug metabolizing enzymes in novel murine liver epithelial cell lines and their ability to activate procarcinogens. *Cancer Res.* **51:**301–309; 1991.

[40] Paolini, M.; Biagi, G. L.; Bauer, C.; Cantelli-Forti, G. On the nature of non-genotoxic carcinogens. A unified theory including NGCs, co-carcinogens, and promoters. *Mutat. Res.* **281:**245–246; 1992.

[41] Paolini, M.; Legator, M. S. Healthy broccoli? *Nature* **357:**448; 1992.

[42] Gonder, J. C.; Proctor, R. A.; Will, J. A. Genetic differences in oxygen toxicity are correlated with cytochrome P450 inducibility. *Proc. Natl. Acad. Sci. USA* **82:**6315–6319; 1985.

[43] Paolini, M.; Biagi, G. L.; Cantelli-Forti, G.; Bauer, C. Further mechanisms of non-genotoxic carcinogenesis. *Trends Pharmacol. Sci.* **15:**322–323; 1994.

[44] Peto, R.; Doll, R.; Buckey, Y. D.; Sporn, M. D. Can dietary beta carotene materially reduce human cancer rate? *Nature* **290:**201–208; 1989.

[45] Burton, G. W.; Ingold, K. U. Beta-carotene: an unusual type of lipid antioxidant. *Science* **224:**569–573; 1984.

[46] Palozza, P.; Luberto, C.; Calviello, G.; Ricci, P.; Bartoli, G. M. Antioxidant and prooxidant role of beta-carotene in murine normal and tumor thymocytes: effects of oxygen partial pressure. *Free Radic. Biol. Med.* **22:**1065–1073; 1997.

[47] The Alpha-Tocopherol, Beta-Carotene Cancer Prevention Study Group. The effect of vitamin E and beta carotene on the incidence of lung cancer and other cancers in male smokers. *N. Engl. J. Med.* **330:**1029–1035; 1994.

[48] Rowe, P. M. CARET and ATBC refine conclusions about β-carotene. *Lancet* **348:**1369; 1996.

[49] Azuine, M. A.; Goswami, U. C.; Kalial, J. J.; Bhide, S. V. Antimutagenic and anticarcinogenic effects of carotenoids and dietary polin oil. *Nutr. Cancer* **17:**287–295; 1992.

[50] Paolini, M.; Cantelli-Forti, G.; Perocco, P.; Pedulli, G. F.; Abdel-Rahman, S. Z.; Legator, M. S. Co-carcinogenic effect of β-carotene. *Nature* **398:**760–761; 1999.

[51] Badway, J. A.; Karnovsky, M. L. Active oxigen species and the function of phagocytic leukocytes. *Annu. Rev. Biochem.* **49:**695–726; 1980.

[52] Pryor, W. A.; Godber, S. S. Non-invasive measure of oxidative stress status in humans. *Free Radic. Biol. Med.* **10:**177–184; 1991.

[53] Akaike, T.; Suga, M.; Maeda, H. Free radicals in viral pathogenesis: molecular mechanisms involving superoxide and NO. *Proc. Soc. Exp. Biol. Med.* **198:**721–727; 1991.

[54] Paolini, M.; Pozzetti, L.; Pedulli, G. F.; Marchesi, E.; Cantelli-Forti, G. The nature of the prooxidant activity of vitamin C. *Life Sci.* **64:**273–278; 1999.

[55] Paolini, M.; Barillari, J.; Trepidi, S.; Valgimigli, L.; Pedulli, G. F.; Cantelli-Forti, G. Captan impairs CYP-catalyzed drug metabolism in the mouse. *Chem. Biol. Interact.* **123:**149–170; 1999.

ABBREVIATIONS

ROS—Reactive Oxygen Species

NVLD—Nonviral Liver Disease

HBV—Hepatitis B Virus

HCV—Hepatitis C Virus

HDV—Hepatitis D Virus

EDTA—Ethylenediaminetetraacetic acid

TEMPO—2,2,6,6-Tetramethyl-1-piperidinyloxyl

TEMPOH—1-Hydroxyl-2,2,6,6-tetramethylpiperidine

TEMPO—2,2,6,6-Tetramethyl-4-oxo-1-piperidinyloxyl

TEMPONE-H—1-Hydroxyl-2,2,6,6-tetramethyl,4-oxopiperidine

carboxyl-PROXYL—3-Carbamoyl-2,2,5,5-tetramethylpyrrolidine-1-oxyl

Trolox—6-Hydroxy-2,5,7,8-tetramethylchroman-2-carboxylic acid

SOD—Superoxide Dismutase

EPR—Electron Paramagnetic Resonance

PBN—α-Phenyl-N-*tert*-butylnitrone

DMPO—5,5-Dimethyl-1-pyrroline *n*-oxide

MNP—2-Methyl-2-nitrosopropane

CYP—Cytochrome P450

PB—Sodium phenobarbital

βNF—β-naphtoflavone

IS—Isosafrole

PCN—Pregnenolone-16α-carbonitrile

CL—Clofibrate

logP—logarithm of the octanol/water partition coefficient

AUTHOR INDEX

KEYWORD INDEX